2D Nanomaterials

2D Nanomaterials
Chemistry and Properties

Edited by
Ram K. Gupta

CRC Press is an imprint of the
Taylor & Francis Group, an **informa** business

First edition published 2022
by CRC Press
6000 Broken Sound Parkway NW, Suite 300, Boca Raton, FL 33487-2742

and by CRC Press
2 Park Square, Milton Park, Abingdon, Oxon, OX14 4RN

CRC Press is an imprint of Taylor & Francis Group, LLC

Library of Congress Cataloging-in-Publication Data
Names: Gupta, Ram K., editor.
Title: 2D nanomaterials : chemistry and properties / edited by Ram K. Gupta.
Description: First edition. | Boca Raton : CRC Press, 2022. |
Includes bibliographical references and index.
Identifiers: LCCN 2021058577 (print) | LCCN 2021058578 (ebook) |
ISBN 9781032013947 (hardback) | ISBN 9781032013992 (paperback) |
ISBN 9781003178453 (ebook)
Subjects: LCSH: Two-dimensional materials.
Classification: LCC TA418.9.T96 A123 2022 (print) | LCC TA418.9.T96
(ebook) | DDC 620.1/12—dc23/eng/20220120
LC record available at https://lccn.loc.gov/2021058577
LC ebook record available at https://lccn.loc.gov/2021058578

ISBN: 978-1-032-01394-7 (hbk)
ISBN: 978-1-032-01399-2 (pbk)
ISBN: 978-1-003-17845-3 (ebk)

DOI: 10.1201/9781003178453

Typeset in Times
by codeMantra

I would like to dedicate this book to my parents who taught me the importance of electricity (energy).

Contents

Preface

Energy serves as a focal point for every activity that is carried out in our daily lives. The human brain, for instance, is considered the most efficient processing system, faster than that of any supercomputer ever developed to date and, consumes energy as low as 20 W. "Aurora" the supercomputer being developed by Intel is claimed to hit a computing speed of 1 exaflop – equivalent to one quintillion floating-point computations per second. The United States Department of Energy estimates electricity target to power "Aurora" will be 40 mW. Although we are trying to hit the extravagant efficiencies closer to the brain, the devices that are as efficient to the low energy consumption of the brain remain a challenge. The answer to all these problems is hidden in the material science of electronic devices that are being fabricated. The development of materials has allowed us to fit billions of transistors in small-sized chips and allow progress per Moore's law. The major contribution to these materials includes two-dimensional (2D) nanomaterials, which are widely used materials in energy applications due to their unique structure at low cost.

2D nanomaterials have emerged as promising candidates for energy devices owing to their superior electrochemical properties, surface area, nano-device integration, multifunctionality, printability, and mechanical flexibility. In this book, we summarize research aspects for different materials in the 2D category such as graphene and its derivatives, transition metal chalcogenides, and transition metal oxides/hydroxides for energy applications such as in batteries, supercapacitors, solar cells, and fuel cells. Apart from conventional techniques, this book explores new aspects of synthesizing 2D nanomaterials beyond traditionally layered structures such as 2D metal oxides and polymers, broadening the vision for readers to explore novel material systems for enhanced energy applications.

The book is divided into two volumes to cover synthesis, properties, and applications of 2D nanomaterials for energy. The first volume (this volume) covers basic concepts, chemistries, the importance of 2D nanomaterials for energy along theoretical consideration in designing new 2D nanomaterials. The effect of doping, structural variation, phase, and exfoliation on structural and electrochemical properties of 2D nanomaterials are discussed for their applications in the energy sector. Types of energy devices and their working principles are covered in detail. Some of the advanced applications such as flexible photodetectors, batteries, fuel cells, and photovoltaics are discussed. The second volume covers a wide range of applications of 2D nanomaterials for energy.

Ram K. Gupta, Associate Professor
Department of Chemistry
Kansas Polymer Research Center
Pittsburg State University
Pittsburg, Kansas, United States

Author

Dr. Ram K. Gupta is an Associate Professor at Pittsburg State University. Dr. Gupta's research focuses on conducting polymers and composites, green energy production and storage using biowastes and nanomaterials, optoelectronics and photovoltaics devices, organic-inorganic heterojunctions for sensors, bio-based polymers, flame-retardant polymers, bio-compatible nanofibers for tissue regeneration, scaffold and antibacterial applications, corrosion inhibiting coatings, and bio-degradable metallic implants. Dr. Gupta has published over 235 peer-reviewed articles, made over 300 national, international, and regional presentations, chaired many sessions at national/international meetings, edited many books, and written several book chapters. He has received over two and a half million dollars for research and educational activities from many funding agencies. He is serving as Editor-in-Chief, Associate Editor, and editorial board member for numerous journals.

Contributors

Arpana Agrawal
Department of Physics
Shri Neelkantheshwar Government Post-Graduate College
Khandwa, India

Junaid Ahmad
Department of Physics
Division of Science and Technology
University of Education
Lahore, Pakistan

Khuram Shahzad Ahmad
Department of Environmental Sciences
Fatima Jinnah Women University
Rawalpindi, Pakistan

Saif Ali
Department of Physics
Division of Science and Technology
University of Education
Lahore, Pakistan

Tabitha A. Amollo
Department of Physics
Njoro Campus
Egerton University
Egerton Nakuru, Kenya

Daniel Nframah Ampong
Department of Materials Engineering
College of Engineering, Kwame Nkrumah
University of Science and Technology
Kumasi, Ghana

Sourabh Barua
Department of Physics
Birla Institute of Technology Mesra
Ranchi, India

K. Brijesh
Department of Physics
National Institute of Technology Karnataka (NITK)
Surathkal, India

Faheem K. Butt
Department of Physics
Division of Science and Technology
University of Education
Lahore, Pakistan

Yu-Hsu Chang
Department of Materials and Mineral Resources Engineering
Institute of Mineral Resources Engineering
National Taipei University of Technology
Taipei, Taiwan

Jyotsna Chaturvedi
Department of Inorganic and Physical Chemistry
Indian Institute of Science
Bangalore, India

Felipe de Souza
Kansas Polymer Research Center
Pittsburg State University
Pittsburg, Kansas

Nitika Devi
School of Physics & Material Science
Shoolini University
Solan, India

P. Dhanasekaran
Fuel Cells Division
CSIR-Central Electrochemical Research Institute-Madras Unit
Chennai, India

Mangesh Diware
Center for Novel State of Complex Materials Research
Department of Physics and Astronomy
Seoul National University
Seoul, Republic of Korea

Pawan Kumar Dubey
Electroptics Laboratory
Department of Electrical and Electronics Engineering
Ariel University
Ariel, Israel

Sumit Dutta
Department of Physics
Birla Institute of Technology
Mesra
Ranchi, India

Mashal Firdous
Department of Physics
Division of Science and Technology
University of Education
Lahore, Pakistan

Shraddha Ganorkar
Advanced Mechanics and Material Design Lab
Department of Mechanical Engineering
Sungkyunkwan University
Suwon, Republic of Korea

Gibin George
Department of Mechanical Engineering
SCMS School of Engineering and Technology
Ernakulam, India

Charu Goyal
Department of Chemistry
GLA University
Mathura, India

Yue Hao
The State Key Discipline Laboratory of Wide Band Gap Semiconductor Technology
Xidian University
Xi'an, China
and
Shaanxi Joint Key Laboratory of Graphene
Xidian University
Xi'an, China

Nagaraja Hosakoppa
Department of Physics
National Institute of Technology Karnataka (NITK)
Surathkal, India

Linrui Hou
School of Materials Science & Engineering
University of Jinan
Jinan, PR China

Shaan Bibi Jaffri
Department of Environmental Sciences
Fatima Jinnah Women University
Rawalpindi, Pakistan

Ashok Kumar Kakarla
Department of Electronics and Information Convergence Engineering
Institute for Wearable Convergence Electronics
Kyung Hee University
Yongin-si, Republic of Korea

Anuj Kumar
Department of Chemistry
GLA University
Mathura, India

Chia-Jyi Liu
Department of Physics
National Changhua University of Education
Changhua, Taiwan

Yang Liu
School of Materials Science & Engineering
University of Jinan
Jinan, PR China

Joseph Chennemkeril Mathew
Department of Physics
Dayananda Sagar College of Engineering
Bangalore, India

Kwadwo Mensah-Darkwa
Department of Materials Engineering
College of Engineering, Kwame Nkrumah
University of Science and Technology
Kumasi, Ghana

S. Mohanapriya
Bionanomaterials Lab
Department of Chemistry
Periyar University
Salem, India

Sreejesh Moolayadukkam
Centre for Nano and Soft Matter Sciences (CeNS)
Bangalore, India

Arun Prasad Murthy
Department of Chemistry
School of Advanced Sciences
Vellore Institute of Technology
Vellore, India

Jing Ning
The State Key Discipline Laboratory of Wide Band Gap Semiconductor Technology
Xidian University
Xi' an, China
and
Shaanxi Joint Key Laboratory of Graphene
Xidian University
Xi' an, China

Vincent O. Nyamori
School of Chemistry and Physics
Westville Campus
University of KwaZulu-Natal
Durban, South Africa

S. Olutunde Oyadiji
Department of Mechanical, Aerospace and Civil Engineering
School of Engineering, Faculty of Science and Engineering
The University of Manchester
Manchester, United Kingdom

Deepthi Panoth
School of Chemical Sciences
Kannur University
Payyanur, India

Anjali Paravannoor
School of Chemical Sciences
Kannur University
Payyanur, India

Aruna Pattipati
Department of Physics
Dayananda Sagar College of Engineering
Bangalore, India

Immanuel Paulraj
Department of Physics
National Changhua University of Education
Changhua, Taiwan

Nihila Rahamathulla
Department of Chemistry
School of Advanced Sciences
Vellore Institute of Technology
Vellore, India

Alok Kumar Rai
Department of Chemistry
University of Delhi (North Campus)
New Delhi, India

Harsha Rajagopalan
Department of Physics
School of Advanced Science
Vellore Institute of Technology
Vellore, India

Sajid Ur Rehman
School of Science, Optoelectronic Research Center
Minzu University of China
Beijing, China

Zia Ur Rehman
School of Physics
College of physical Science and Technology and School of Environmental Science and Engineering
Yangzhou University
Yangzhou, PR China

Chen Shen
Department of Mechanical
Aerospace and Civil Engineering, School of Engineering, Faculty of Science and Engineering
The University of Manchester
Manchester, United Kingdom

Rajesh Kumar Singh
School of Physical and material Sciences
Central University of Himachal Pradesh
Dharamshala, India

Zeeshan Tariq
School of Science
Optoelectronic Research Center
Minzu University of China
Beijing, China

Laxmi Narayan Tripathi
Department of Physics
School of Advanced Science
Vellore Institute of Technology
Vellore, India

Emmanuel Acheampong Tsiwah
Department of Materials Science and Engineering
School of Chemistry and Materials Science
University of Science and Technology of China
Hefei, China

Sami Ullah
Department of Physics
Division of Science and Technology
University of Education
Lahore, Pakistan

S. Vinod Selvaganesh
Indian Institute of Technology-Madras
Chennai, India

Dong Wang
The State Key Discipline Laboratory of Wide Band Gap Semiconductor Technology
Xidian University
Xi'an, China
and
Shaanxi Joint Key Laboratory of Graphene
Xidian University
Xi'an, China

Tian Wang
Department of Electronics and Information Convergence Engineering
Institute for Wearable Convergence Electronics
Kyung Hee University
Yongin-si, Republic of Korea

Yuyan Wang
School of Materials Science & Engineering
University of Jinan
Jinan, PR China

Maoyang Xia
The State Key Discipline Laboratory of Wide Band Gap Semiconductor Technology
Xidian University
Xi' an, China
and
Shaanxi Joint Key Laboratory of Graphene
Xidian University
Xi' an, China

Daniel Yeboah
Department of Materials and Manufacturing
Institute of Industrial Research - Council for Scientific and Industrial Research
Accra, Ghana

Jae Su Yu
Department of Electronics and Information Convergence Engineering
Institute for Wearable Convergence Electronics
Kyung Hee University
Yongin-si, Republic of Korea

Changzhou Yuan
School of Materials Science & Engineering
University of Jinan
Jinan, PR China

Jincheng Zhang
The State Key Discipline Laboratory of Wide Band Gap Semiconductor Technology
Xidian University
Xi'an, China
and
Shaanxi Joint Key Laboratory of Graphene
Xidian University
Xi'an, China

1 Chemistry of 2D Materials for Energy Applications

Charu Goyal and Anuj Kumar
GLA University

Ram K. Gupta
Pittsburg State University

CONTENTS

1.1 INTRODUCTION

In modern society, significant energy needs and environmental crises have accelerated the development of energy storage technology. Due to their high efficiency and environmental acceptability, different electrochemical energy storage technologies have attracted recent interest. Supercapacitors, especially, can collect and generate electricity at excessive power densities, joining the power and energy gap between batteries and capacitors. Alkali metal (Li, Na, K) -ion batteries have high energy densities and wider operating voltages, making them suitable for use in mobiles, laptops, and e-mobility. Metal-air batteries, which have a potential energy density equivalent to gasoline, can power long-distance electric motors [1]. Without a question, these gadgets offer a lot of promise; nevertheless, meeting the growing demand for practical

DOI: 10.1201/9781003178453-1

applications is still a challenge. One of the most significant barriers is the electrode materials, which are either too costly or inactive. As a result, innovative materials for electrodes having excessive efficiency and being cheap are urgently required.

Two-dimensional (2D) materials, which are a type of independent planar material, have a large size to thickness ratio. They show many unique physicochemical characteristics such as (a) high specific surfaces to enhance ion adsorption and improve capacitance, (b) high conductivity to speed up electron transfer, (c) high ion intercalation to provide high energy density, and (d) tunable active sites to provide rich electrocatalytic properties for enhanced energy storage. Graphene is a characteristic 2D substance that is a carbon allotrope in which the atoms are arranged in a honeycomb lattice. It got wide attention in 2004 when micromechanical cleavage using Scotch tape was used to prepare graphene from graphite [2]. Since then, graphene has been widely investigated, and numerous intriguing characteristics, such as good mechanical qualities, large thermal and electrical conductivity, amazing surface area, and exceptional optical characteristics have been discovered. Graphene has been used in a variety of disciplines, including energy, catalysis, optoelectronic devices, and healthcare, due to its unique characteristics.

Several types of 2D materials have been noted in the wake of graphene's discovery, containing mono-elemental analogues (MEAs) of graphene, transition metal dichalcogenides (TMDs), carbides, nitrides, and carbonitrides (MXene) of transition metals. The typical architecture of popular 2D materials is shown in Figure 1.1. MEAs are 2D materials that are made up of only one kind of element with 2D structures similar to graphene. MX_2, in which M is a transition metal atom (including Mo, Nb, V, or W) and X is represented as a chalcogen atom, is the chemical formula for TMDs (including S, Se, or Te). 2D transition metal dichalcogenides have a wide range of applications in high-end electronics, spin-manipulated electronics technology, optoelectronic devices, and energy storage and conversion due to their sturdy spin–orbit coupling and favorable electrical as well as mechanical characteristics [3]. MXenes are typically made by carefully removing the 'A' components from MAX phases, which have the chemical stoichiometry of $M_{n+1}AX_n$ (where M is an earlier transition metal, A is a group 12–16 element, and X is carbon or nitrogen). Large electrical conductivity, appropriate hydrophilic behavior, good thermal stability,

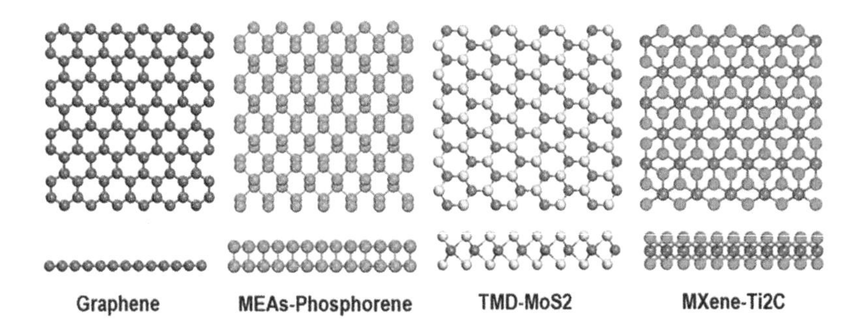

FIGURE 1.1 Structural representation of some common 2D materials. Adapted with permission from [5]. Copyright (2021), Wiley-VCH.

broad interlayer spacing, with readily adjustable structure have all been discovered in MXenes [4], which makes it a potential material for electrochemical energy storage.

The newest advancements in 2D materials for electrochemical power storage are the subject of this study. We want to improve our intellect of the link between 2D materials' arrangements, characteristics, and applications, as well as encourage the use of high-efficiency 2D materials. The applications of energy to supercapacitors, alkali metal-ion/metal-air batteries are then discussed. Lastly, some broad ideas for future research into new 2D materials for electrochemical power storage are suggested.

1.2 CHEMISTRY, STRUCTURES, AND PROPERTIES OF 2D MATERIALS

1.2.1 Transition Metal Di-Chalcogenides-based 2D Materials

Since molybdenum disulfide (MoS_2) has a direct bandgap, high mobility, and quantum confinement, it is a major figure among 2D stacked materials in the field of nano-electronics and optoelectronics. This will be very useful for low-power devices in post-Si technologies. For example, MoS_2 is useful in photocatalysts and solar cells [6]. MoS_2 and other TMDC isomorphs have been intensively investigated in this context for accessible applications in water and air purification, hydrogen evolution reaction (HER), as well as power storage [7]. Usually, MoS_2 has a non-centrosymmetric crystal structure having long-range periodic order and is a direct bandgap material. It is an sp^3 linked solid made up of two covalent bonds of Mo-S in a layered form with each layer joined by van der Waals interaction in between the layers [8]. The fundamental building block of MoS_2 is a hexagonal (honeycomb-like) system of Mo and S atoms that looks like graphene. Bi-/trilayer MoS_2 is formed when two or three single layers of S-Mo-S are stacked on top of each other and held together by weak van der Waals forces. It's possible to create three-dimensional (3D) bulk MoS_2 by layering over 100 single layers of MoS_2.

A revised MoS_2 structure was computed using density-based theory and local density approximations. The MoS_2 cell has been found to consist of a honeycomb-like lattice structure having a lattice factor of 3.12 Å and vertical spacing between layers of sulfur of 3.11 Å [9]. MoS_2 is classed as per a d-orbital material since Mo d-orbits both constitute the highest bands of valence and the lowest bands of conduction [10,11]. In addition, the external d-electron interactions lead to a gap in the band (approx. 1 eV) between the occupied and unoccupied d-states of 2D MoS_2, making it distinguishable with other nanostructures that are sp-hybridized semiconductors. Remarkably, when the 3D bulk structures transform into 2D constructions, it converts from indirect to direct bandgap [12]. The indirect bandgap in 3D MoS_2 sits underneath the direct bandgap. Nevertheless, when the number of layers is reduced, the direct bandgap changes higher, leading to a transversal to a direct gap mate in the individual MoS_2 layer [12]. This shift to the bandgap happens to the 3D structure because of the substantial quantum containment of the 2D structure. This is due to transitions of direct gap between the split-valence band maxima and the conduction band minima around the K point of the Brillouin zone that contributes to the photoluminescence characteristics [12]. This is the principal cause why MoS_2 shows photoluminescence among its 2D forms but lacks photoluminescence in its 3D crystal structures [8,13]. Similarly, atomic layer thickness TMDC materials are distinguishable from other sp^3 bonded semiconductor materials by the quantum captivity they display [8]. Dimension reliant on bandgap changes besides the visible to near-IR light system, on the other hand, are expressions of carrier production, implying that these materials might be used in sustainable solar-energy conversion systems [14].

1.2.2 MXenes-Based 2D Materials

MXene's crystal shape resembles that of its MAX-phase predecessor, with a hexagonal close-packed arrangement. Here, M atoms are organized in a close-packed arrangement, while X atoms occupy octahedral positions. Similar to other 2D materials, neighboring stacked units are linked by van der Waals forces. MXenes, particularly acidic fluorides, are generally made in aqueous solutions. As a result, MXenes have a combination of OH, F, and O terminations on their surfaces. These molecules are abbreviated as $M_{n+1}X_nT_x$, where T denotes the surface termination. There are MXenes with the structures M_2XT_x, $M_3X_2T_x$, and $M_4X_3T_x$ (Figure 1.2) [15]. The latest computer simulations show that the surface termination of MXenes has a big influence on their characteristics. Hu et al. [16], for instance, have systematically investigated Bader charging analysis and thermodynamic calculations on the chemical origin of termination-specific MXenes; the material has demonstrated stability order such as $Ti_3C_2O_2 > Ti_3C_2F_2 > Ti_3C_2(OH)_2 > Ti_3C_2H_2 > Ti_3C_2$, attributable to the scatter of the 3d strongly degenerated surface orbits of Ti. In a separate study, Fu et al. [17] methodically examined the impact of many functional groups (such as Cl, F, H, O, and OH) on the stabilization, electronic structures, and mechanical properties of an MXene (Ti_3C_2); the researchers noted that oxygen-functionalized Ti_3C_2 has stronger thermodynamic stabilization and toughness than its other counterparts

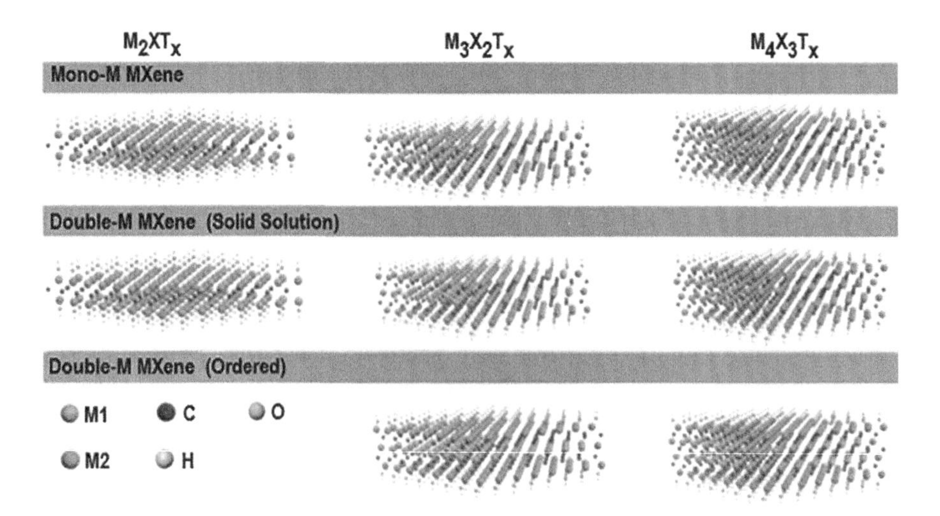

FIGURE 1.2 Three different formulas (M_2XT_x, $M_3X_2T_x$, and $M_4X_3T_x$) and compositions (mono-MMXenes and double-M MXenes) of MXenes. Reproduced with permission from [15]. Copyright (2021), Wiley-VCH.

because of cation exchange between the inner bond and outer surface. Surface termination of MXenes such as $Ti_3C_2T_x$ and V_2CT_x has also been studied utilizing experimental approaches in addition to theoretical investigations. Wang et al. [18], for example, used aberration-corrected scanning transmission electron microscopy (STEM) to reveal the surface atomic scale of $Ti_3C_2T_x$; the group discovered that surface functional groups (e.g., OH, F, and O) are randomly distributed on MXene surfaces and prefer to occupy the top sites of the central Ti atom. Karlsson's group [19] used aberration-corrected STEM coupled with electron energy loss spectroscopy (EELS) to study individual and double sheets of Ti_3C_2, revealing sheet coverage, inherent defects, and TiO_x adatom complexes. Surface termination of MXenes such as $Ti_3C_2T_x$ and V_2CT_x is also being studied utilizing experimental approaches in addition to theoretical investigations. Wang and co-workers [17], for example, used aberration-corrected STEM to reveal the surface atomic scale of $Ti_3C_2T_x$; the group discovered that surface functional groups (e.g., OH, F, and O) are randomly spread on MXene surfaces and favor to enter the top sites of the center atom of Ti. Karlsson's group [19] used aberration-corrected STEM-EELS to study individual and double sheets of Ti_3C_2, revealing sheet coverage, inherent imperfections, as well as TiO_x adsorption complexes. Sang and colleagues used STEM to discover various point defects in monolayer Ti_3C_2 nanosheets using the minimum intense layer delamination technique [20]. Hope and colleagues [21] used 1H and 19F nuclear magnetic resonance (NMR) studies to quantify the surface functional groups of $Ti_3C_2T_x$ and discovered that the proportions of distinct surface terminations are significantly dependent on the material's production process. Harris and co-workers used stable NMR to directly quantify the surface termination groups of V_2CT_x MXenes [22].

The inherent good electrical characteristics of MXene materials are primarily responsible for their use in energy storage and electrocatalysis. The influence of various M, X, and surface functional groups on the electrical characteristics of most MXenes has been investigated in recent theoretical computational research. Because MXenes contain a variety of transition metals, their electrical characteristics might range from metallic to semiconducting [23]. Since they may also contain heavy transition metals like Mo, W, and Cr, some MXenes could be topological insulators [15,24]. Surface termination may also affect the electrical characteristics of uncoated MXenes. For instance, Fredrickson and co-workers used density-functional theory (DFT) calculations to study the structural and electrical characteristics of layered bulk Ti_2C and Mo_2C with various functional groups in aqueous environments [25]. Surface functional groups and water intercalation have a significant impact on the out-of-plane lattice parameter of bulk MXenes. Bulk MXenes (Ti_2C and Mo_2C) were functionalized by one monolayer of O at no applied voltage. However, regardless of the applied potential, bare MXenes were unstable. Furthermore, with an applied potential, changes in the surface functional groups of Ti_2C from O-covered to H-covered might induce metal-insulator transition. Tang and co-workers found that uncoated MXenes (such as Ti_3C_2) have metallic characteristics, but functionalizing Ti_3C_2 with various groups (such as –OH, –F, and –I) results in semiconductor qualities with small bandgaps [26]. It was also shown that the M layer has a significant impact on the electrical characteristics of the final material. Remarkably, MXenes containing Mo have semiconductor characteristics, while $Ti_3C_2T_x$ is metallic. DFT calculations revealed that Ni_2N MXenes had inherent half-metallicity [27]. The nanostructures of MXene have been linked to their electrical characteristics. For example, Enyashin and Ivanovskii [28] anticipated that hydroxylated Ti_3C_2 nanotubes will have metallic-like properties. Ti_3C_2 nanoribbons, according to Zhao et al. [29], show unique electrical characteristics from nanosheets of MXenes. The electronic characteristics of MXenes have been studied in many experiments. Just the electrical characteristics of a few MXenes, such as Ti_2CT_x, $Ti_3C_2T_x$, and Mo_2CT_x, have been examined experimentally thus far [30]. Halim and colleagues [31], for instance, studied the electrical conductivity of $Ti_3C_2T_x$ and Mo_2CT_x films. The electrical characteristics of monolayer $Ti_3C_2T_x$ flakes were measured by Lipatov and colleagues [32]. The outstanding electrical characteristics of 2D Ti_2CT_x were discovered by Lai et al. [33]. MXenes offer outstanding electronic characteristics, as demonstrated by computational and experimental studies, and are attractive candidate materials for electrochemistry, energy storage, and electrocatalysis.

1.2.3 Graphene-Based 2D Materials

Graphene is a carbon allotrope with 2D characteristics that is crystalline. Carbon atoms are tightly packed in graphene in an atomic-scale sp^2-bonded hexagonal arrangement. A one-atom-thick sheet of graphite is referred to as graphene. Other allotropes, such as graphite, charcoal, carbon nanotubes, and fullerenes, have it as a structural constituent. Graphene is a promising active material in electrocatalysis due to its unique features of high electrical conductivity and huge surface area. Because of the original graphene's low catalytic activity, [34] graphene with a large planer conjugated structure is generally coupled with components capable of electrocatalytic activity through intermolecular interactions at the start. Many attempts have lately been made to improve graphene's intrinsic performance by developing active sites and electronic control, therefore increasing the catalytic process of graphene hybrids via strong electronic coupling between graphene and other materials [35–37]. Regulation methods for graphene-based electrocatalysts may be categorized into four groups (Figure 1.3): (a) graphene's inherent defect control, (b) non-metallic element doping, (c) graphene coupling with active components (graphene-based composites), and (d) graphene-based single-atom catalyst. The use of various techniques in electrocatalytic areas is thoroughly summarized in this section.

The variety of heteroatoms and matrix materials provides an ever-increasing number of design possibilities for active materials. Graphene, with its 2D monolayer structure, has a huge specific surface area (SSA), many defect sites, and oxygen-containing functional groups on both plane and edge, making it an excellent matrix material for heteroatom doping. Heteroatom doping can significantly enhance the electronic structure of graphene, resulting in improved catalytic performance. Doping heteroatoms, such as N, P, S, B, and others, can offer numerous active sites and modify the local electronic structure, therefore increasing the system's reaction activity [38]. Non-metal heteroatoms doping is extensively utilized in a variety of energy-related sectors, including lithium-ion batteries, sodium-ion batteries, lithium-sulfur batteries [39], supercapacitors [40], electrocatalysis [38,41,42], photocatalysis [43], and many more.

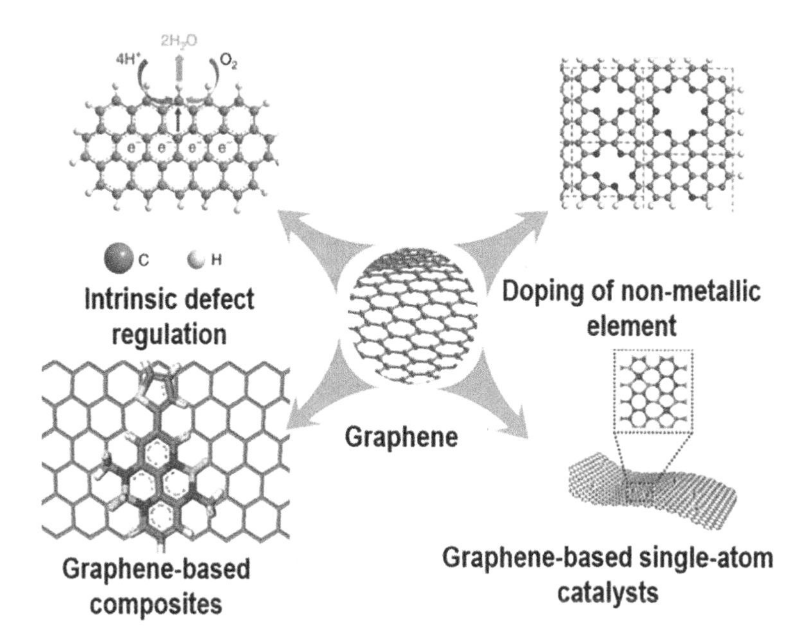

FIGURE 1.3 Graphene-based catalysts are regulated by introducing defects, nonmetallic elements, single atoms, and other active components into the graphene system. Adapted from Ref. [44]. Copyright (2021) Nature. Adapted from Ref. [45]. Copyright 2021 American Chemical Society. Adapted from Ref. [46]. Copyright (2021) Wiley-VCH. Adapted from Ref. [47]. Copyright (2021) Tsinghua University Press and Springer-Verlag GmbH Germany.

1.3 ENERGY APPLICATION OF 2D MATERIALS

In addition to the larger SSA, functionalization of surface, reversible ion and transport of mass, and dependable mechanical resilience, controlled construction of 2D materials at the nanoscale into desired designs have released valuable approaches. A thorough grasp of current structural engineering methodologies is also useful for the best design of suitable structures, which may be employed for practical energy purposes [48–50]. Particles or molecules can be intercalated in 2D materials by manipulating their layer spacing. When it comes to secondary batteries and supercapacitors, the capacity of the 2D host to trap charged particles/molecules in a regulated reversible manner is crucial for successful power storage. For example, chemical composition, surface wetness, effective surface area, and the charge transfer characteristics of an electrode structure may all influence the amount of energy that can be stored by 2D materials (i.e. the extreme energy density). The surface chemistry/geometry and charge and/or mass transport characteristics of a catalytic structure, on the other hand, will influence the primary performance in energy conversion systems after being paired with diverse energy sources, such as current density and onset voltage [51–54]. The major challenge of energy conversion technologies nowadays is to discover new electrocatalysts. 2D framework materials are very capable electrocatalysts with high conversion efficiency at low energy input due to their precise composition and/or structure tunability, large surface area, dense exposed catalytic sites, also enhanced conductivity compared to bulk counterparts. HER, oxygen evolution reaction (OER), oxygen reduction reaction (ORR), and CO_2 reduction reaction (CO_2RR) are only a few of the key energy conversion processes that use 2D framework electrocatalysts.

1.3.1 Electrochemical Hydrogen Evolution Reaction

It has been universally accepted that electrocatalytic hydrogen generation via water splitting produces high-grade H_2 in a clean and sustainable method. Integration of transition metal (Fe, Co, Ni, etc.) into 2D framework catalysts has been studied in the past to increase the HER catalytic activity [55–60]. Using a liquid–liquid interfacial method, the Marinescu group successfully synthesized a cobalt dithiolene-fused 2D metal-organic frameworks (MOF) [57]. With a HER overpotential of 0.3 V at 10 mA cm^{-2} in an acidic medium, the target MOS electrocatalyst showed outstanding performance due to its high catalytic site loading and remarkable stability. For electrocatalytic HER, different metal dithiolene-based 2D framework electrocatalysts were created. Dong et al. rationally integrated metal dithiolene–diamine (MS_2N_2, M=Co/Ni) into carbon-rich 2D MOFs as a model carbon electrocatalyst for HER in addition to metal dithiolene sites [56]. In the sequence, $MS_2N_2 > MN_4 > MS_4$, the electrocatalytic HER activity of these 2D MOFs changed for the various metal coordination modes. Nanosheet protonation favors M–N sites situated in the MS_2N_2 coordination spheres. They helped us get a better knowledge of various metal complexes (MN_X and MS_X) as well as shed insight on the development of high-performance catalysts. For electrocatalytic HER, 2D covalent-organic frameworks (COFs) without metal having numerous functional groups were also produced. The Pradhan group was the first to publish a metal-free HER electrocatalyst based on imine-based conjugated pyrene and porphyrin-based 2D COF (such as SB-PORPy COF) [61]. As a catalytic center, the imine nitrogen sites in the SBPORPy COF showed exceptional activity of catalysts with a low onset potential of 50 mV and good stability in acid electrolyte. Due to their poorer stability and overall conductivity, contemporary 2D framework-based electrocatalysts have lower catalytic activity than the industry-recognized inorganic catalysts, despite substantial advances.

1.3.2 Electrochemical Oxygen Evolution Reaction

2D frameworks for oxygen evolution have demonstrated amazing performance equivalent to the best OER electrocatalysts available today. According to Du's team, a 2D nickel phthalocyanine MOF (NiPc-MOF) with a low onset potential of 1.48 V with the high mass activity of 883 A g^{-1} in alkaline medium has been produced from the bottom up [61]. The 2D structure and high conductivity of 2D MOFs had a major role in the excellent OER activity. As compared to the benchmark noble metal-based catalyst, the ultrathin NiCo bimetallic framework nanosheets produced by Tang and colleagues needed an exceptionally low overpotential of 250 mV at 10 mA cm^{-2} in OER under alkaline circumstances (IrO_2 and RuO_2). Because of their high electrocatalytic activity, the unsaturated metal atoms and the coupling effect between Ni and Co metals are essential. MOFs have also been proposed to increase OER catalytic activity by fabricating ultrathin or lattice-strained nanosheet arrays of 2D MOFs on current collectors, as well as hybrid 2D MOFs.

1.3.3 Electrochemical Oxygen Reduction Reaction

Electrocatalytic ORR in 2D frameworks has shown promising development, even if most relevant instances only address their non-exfoliated variants [62–66]. Thin films of the conductive MOF $Ni_3(HITP)_2$ (thickness of around 120 nm, coated on glassy carbon electrodes) with well-defined NiN units as an ORR electrocatalyst in alkaline electrolyte were reported in 2016 by the Dinc group [63]. The electrocatalyst showed an onset potential of 0.82 V with good stability over protracted polarization. This highlights the importance of layer-conductive MOFs in the development of tuneable and rational electrocatalysts. When Ni/Co metal sources with 2,3,6,7,10,11-hexaiminotriphenylene (HITP) are used in various ratios, 2D MOFs are produced [65]. To compensate for the unpaired $3d_Z{}^2$ electron in Co, the Co_3HITP_2's quadrilateral structure deformed and its conductivity dropped. In alkaline electrolyte, however, it showed remarkable ORR performance (onset potential of 0.9 V and electron transfer number of 3.97) because the unpaired electron on Co $3d_Z{}^2$ in the catalytically active CoN_4 center is beneficial for promoting the binding of oxygen intermediates and thus accelerates the ORR energetics despite the reduced electric conductivity. As opposed to weakly bound Ni_3HITP_2, which travels through a two-electron ORR route with reduced catalytic activity. A highly active electrocatalyst for ORR may be developed by optimizing the architectural and electrical structure of layered MOF materials.

1.3.4 Electrochemical Carbon Dioxide Reduction Reaction

To regulate the activity of the catalyst, efficiency, and selectivity in one system with excellent performance toward the target products, tailored 2D frameworks electrocatalysts would be an ideal choice in CO_2RR electrocatalysis. As active centers, cobalt porphyrins were used to fabricate ultrathin 2D MOF nanosheets (TCPP(Co)/Zr-BTB), which had a turn of frequency (TOF) of 4768/h at −0.919 V vs. RHE. In addition, a post-modification technique was used to enhance the catalytic performance of the catalyst. Modified samples of p-sulfamidobenzoic acid showed an enhanced TOF of 5315/h at −0.769 V with a faradaic efficiency of 85.7% for CO and long-term durability, compared to unmodified samples [67]. As a result, increased use of active sites and optimal post-modification are viable ways to improve CO_2RR performance. An electrocatalyst that is extremely active and selective for CO_2RR was created by the Feng group, for example, using a bimetallic conjugated MOF (2D c-MOFs). It had a CO selectivity of 88% and a TOF of 0.39/s in the 2D c-MOF (PcCu-O-Zn). This has been demonstrated by in situ X-ray absorption spectroscopy, surface-enhanced infrared absorption petro-electrochemistry, and theoretical calculation investigations. This innovative method for creating MOF-based electrocatalyst uses ZnO_4 complexes as CO_2RR catalytic sites, while CuN_4 centers enhance the protonation of adsorbed CO_2 during CO_2RR [68]. In addition, the Lan group proposed directed efficient electron transmission by metalloporphyrins, to accomplish extremely active electrocatalytic CO_2 to CO conversion by the application of tetrathiafulvalene as an electron donator [69]. When Co-TTCOF was exfoliated, it showed an impressive faradic efficiency for CO (above 90% across a larger potential range), with the highest value reaching up to 99.7%, which is better than the existing state-of-the-art catalysts.

1.3.5 2D Materials for Advanced Batteries

The majority of secondary battery research has focused on improving power density and charging and/or discharging kinetics because of their critical share as a power source for electronic devices/vehicles. This has been driven by the current appearance of novel 2D materials with large surface area and easy charge transportation. These advantages can be attributed in part to the electrode structure's optimum design. Although there have been some recent advances, there are still significant obstacles to the rapid use of 2D materials without structural modification. Limited stability of electrochemical window (connected to redox potential) and small initial coulombic efficiency are some of the problems (correlated to Faraday effect) [70,71]. Designing 3D buildings using flexible networks made of 2D materials is a viable method for addressing the problems outlined above. However, these structures also handle the volume shift that occurs after repeated charging/discharging cycles, eliminating capacity fading. Li^+ or Na^+ ions may be able to travel short distances through these materials' porous frameworks, which might affect the voltage window and columbic efficiency.

Using 2D graphene as an encapsulant for active materials for the electrode (such as Si nanoparticles), Lee et al. published a groundbreaking study in this field (Figure 1.4a) [72]. Accordingly, Zhang and co-workers examined the potential of graphene interweaved as anode material in the same way [73]. Such nanoarchitectures, as can be shown in Figure 1.4b, might effectively limit the accumulation of active materials and buffer the mechanical strain which comes from significant volume shifts. Furthermore, nanoparticles enclosed in 2D materials can contribute to structural integrity. As a result of the capacity to enclose 0D materials, new cathode architectures have been developed as well. Of the use of silica spheres as templates, Zhu and co-workers produced 3D macro-porous composites with Li_2FeSiO_4 and 2D materials.

The constant cycle windows for Na^+ ions in some TMDCs, such as TiS_2 and NbS_2, have been observed [74,75]. Another promising Li–S battery material is MXene, a group of transition metals, nitrides, carbides, and carbonitrides. Among these improvements is the use of dipole-type polaritonic excitation for effective trapping of polysulfide with rich surface chemistry. $Ti_3C_2T_x$ nanosheets with a negative charge tend to be superior at forming certain microstructures when combined with positively charged melamine, as illustrated in Figure 1.4c [76]. Nitrogen insertion into the MXene framework allows polysulfide to easily adsorb on the MXene surface, which is extremely desired for Li–S battery applications, as well. Few studies have been carried out in 1D nanoscale assembly for secondary batteries because surface contact in 1D geometry is challenging [77]. It's clear from recent developments, however, that new battery topologies, such as those based on wires and fabrics, as well as new approaches for manipulating 2D materials structurally, are not far [78].

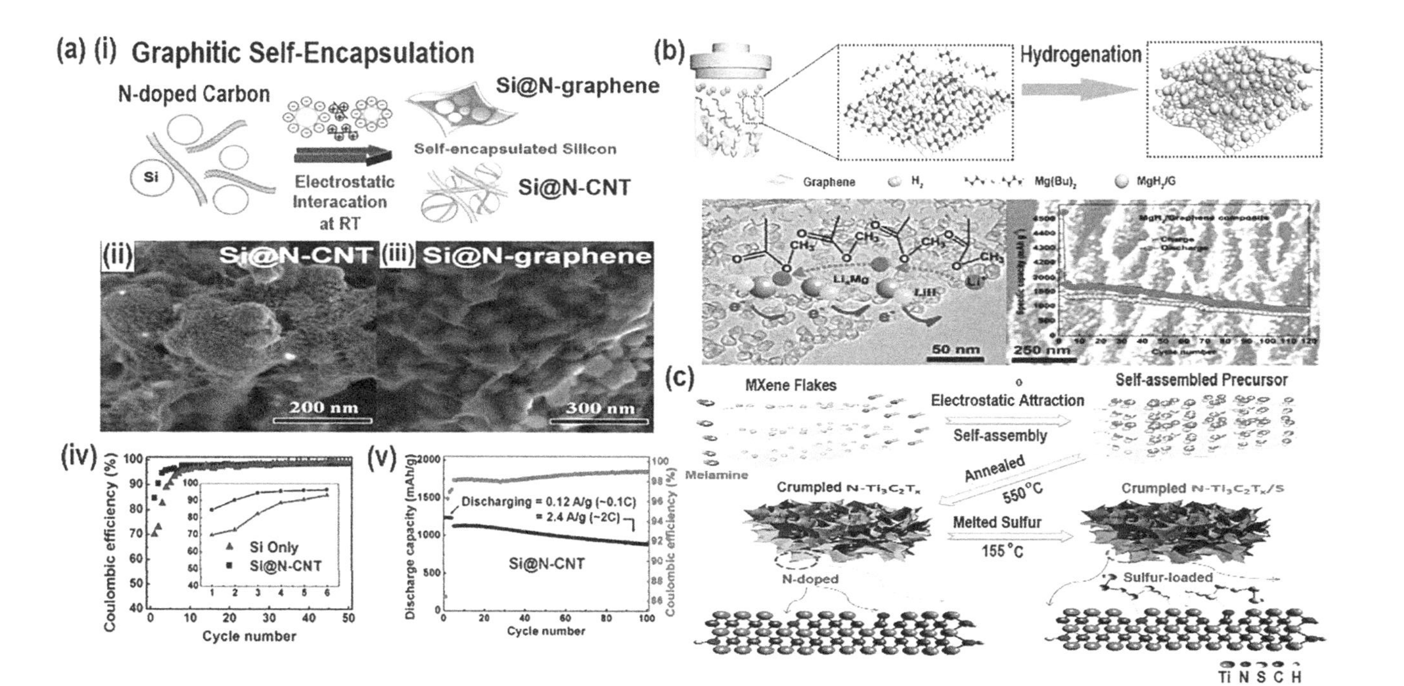

FIGURE 1.4 Secondary batteries made from 2D materials. (a) Graphite's self-encapsulation (i) scheme of graphite's self-encapsulation is shown, where the pH was adjusted for electrostatic interaction between SiO_2 surface and N-doped sites at graphitic carbons. (ii) HR-SEM. (iii) HR-TEM pictures of Si particles with captured N-doped CNTs and graphene. (iv) Coulombic effecienecy vs cycle number, and (v) Discharge current vs cycle number for samples. Adapted with permission 2021, [72] Royal Society of Chemistry copyright (2021). (b) Production of MgH_2 nanoparticles (NPs) that are evenly attached to graphene as anode materials for Li-ion batteries via bottom-up self-assembly (LIBs). (c) Preparation of N-$Ti_3C_2T_x$/S composites. Adapted with permission [76], (Copyright) 2021, Wiley-VCH.

1.3.6 2D Materials for Supercapacitor Devices

As opposed to secondary batteries, supercapacitors are the usual energy systems that have excellent power performance. Due to their very basic charge storage process of keeping a charged particle under an electrical field, they exhibit high power density. Most recently, a new device mechanism based on the nanoscale assemblage of 2D materials has been developed, which uses a 3D network structure to store charge in a faradic (such as pseudocapacitors) or nonfaradic method (such as electric double layer capacitor, EDLC). Studies in the literature provide more details on how e^- and ions can be kept effectively on the surface of 2D materials [79,80]. High-power supercapacitors require interpenetrating electron and ion transport routes, which provide effective interfacial ion storage or intercalated pseudocapacitance, to function properly. Due to strong interactions, 2D layers tend to recollect when manufactured into a structure of electrode without consistent shape engineering, reducing the available surface area and limiting the transfer of charged species [81]. Idealized supercapacitor electrode architectures may be fabricated using nanoscale-constructed 2D materials that exhibit desirable properties like out-of-plane flexibility and dipole-type polaritonic activity. The excellent gravimetric capabilities of these supercapacitor electrode topologies have already benefitted several early research initiatives concentrating on the creation of hierarchically porous structures. Supercapacitors are hampered by their low packing density, which results in limited volumetric capacity and weak networks in practice. It follows from this that, in order to optimize the design of supercapacitors based on 2D materials, the permeability and volume density of electrode structure must be carefully balanced.

Core architectures are extremely advantageous in the case of fiber-type supercapacitors, where two distinct constituents at the core and shell may play different essential roles. For example, conductive polymer core and 2D material shells in a fiber shape provide interfacial charge storage and intercalated pseudocapacitance, as well as mechanical robustness. It was possible to produce the same synergistic effect in a hydrogel structure by adding metal cations to 2D material dispersants (Figure 1.5) [82]. Cooperation-based crosslinking of multivalent metal cations in 2D materials can lead to stable hydrogel formation with strong EDLC performance, as well as increased pseudocapacitance with existing cations. Because graphene is chemically inert, the EDLC mechanism dominates in TMDC, MXene, and black phosphorous electrochemical energy storage. MXene-like V_2CT_x, produced using the Cation Exchange Method, has good rate capability and great gravimetric capacity (Figure 1.6a) [83]. Cation exchange enhances electrical conductivity and structural integration in the various MXene families. By inserting new low-dimensional materials, the dominating organization of 2D materials may be changed. Unfortunately, block phosphorous' weak electron conduction and restacking tendencies preclude it from being utilized in superconducting capacitors, which have a wider interlayer spacing. Structured BP/CNT (CNT in the form of buckypaper) has shown a great potential for high energy density performance (96.5 mWh cm^{-3}) in comparison to previously reported 2D material-based structures (MoS_2-based structures, 1.6 mWh cm^{-3} graphene-based structures, 6.3 mWh cm^{-3} MnO_2-based structures, 11.1 mWh cm^{-3} and MXene-based, 32.6 mWh cm^{-3}) (Figure 1.6b) [84].

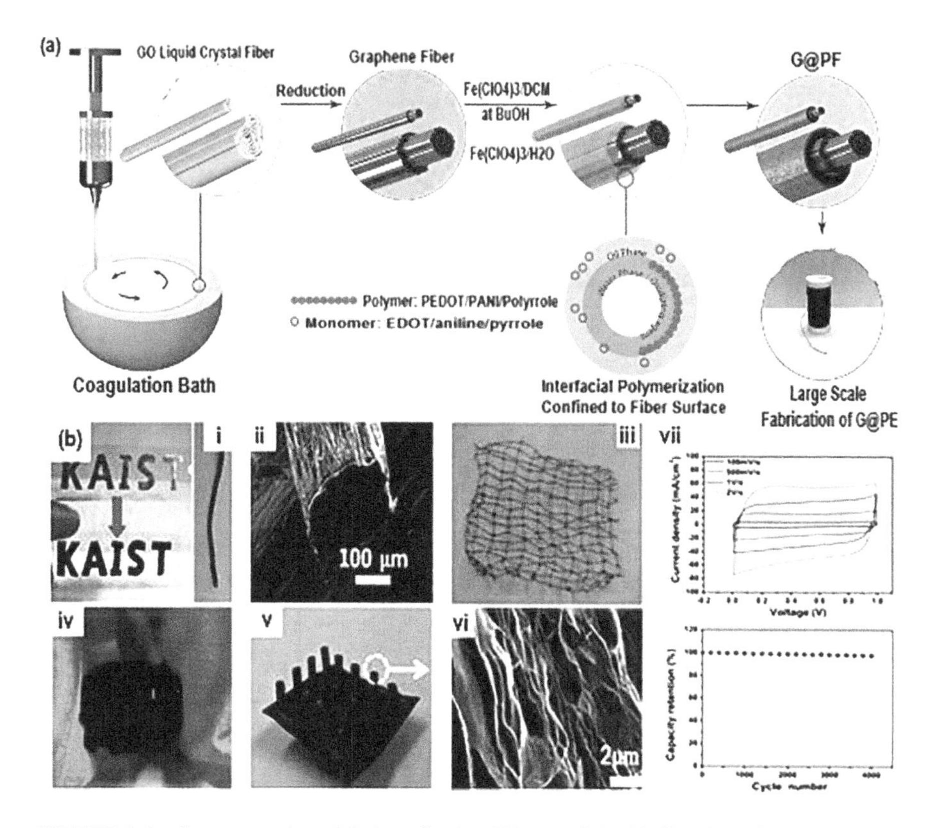

FIGURE 1.5 Supercapacitors fabricated using 2D materials. (a) Graphene@polymer core–shell fibers (G@PFs) manufacturing process. Photograph of three G@PEDOT fiber supercapacitors that light up a green LED when serially connected. Adapted with permission from [85] Copyright (2021), American Chemical Society. (b) (i) Optical picture of local graphene gelation on a prepatterned Zn substrate. (ii) Image of graphene gel tube produced with Z wire in SEM, (iii) Monolithic graphene gel fiber net optical picture, (iv, v) 3D graphene gel optical pictures, (vi) SEM image, (vii) Image of vertical gel tube taken with a SEM microscope graphene gel and capacity retention curves after 4,000 cycles. Adapted with permission from [82], Copyright (2021), American Chemical Society.

1.4 CONCLUSION

With fast advancements in novel exfoliation processes, the development of 2D framework materials is seeing a highly dynamic expansion, opening the door for diverse technical applications in the energy industry. Many studies that examine the effects on materials characteristics of nano-sizing along one axis describe just one ultra-thin material, which is frequently not fully described using acceptable methodologies (i.e. microscopy). So that researchers may better grasp the underlying impacts, they should compare a wider range of materials, rather than just a single undefined exfoliated material. In addition, most of the study is focused on the behavior of a small number of well-understood materials systems, which opens up an exciting opportunity to examine the applicability of principles established thus far to more

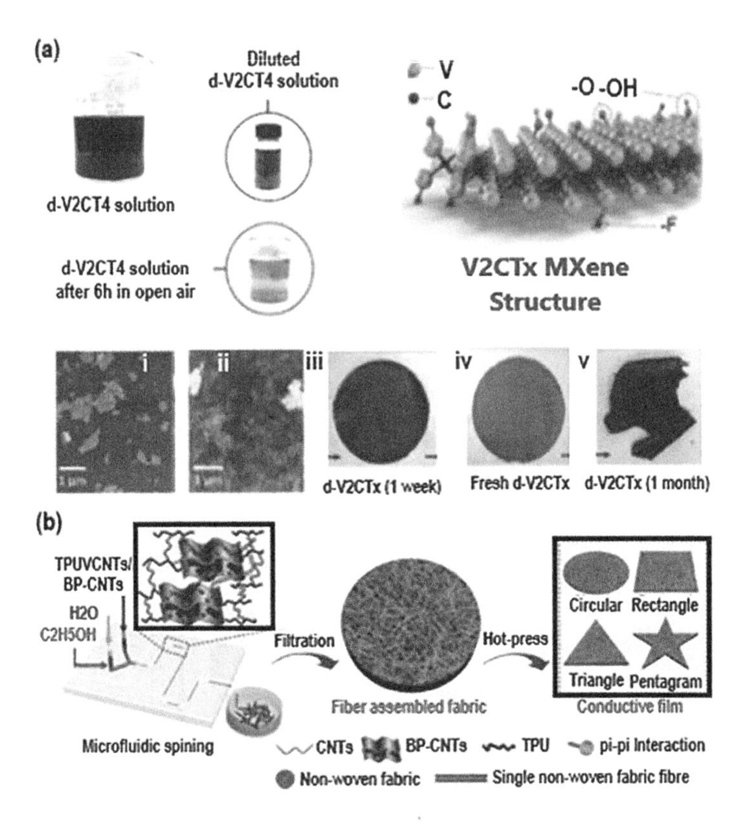

FIGURE 1.6 (a) Assembling pseudocapacitive electrodes with improved electrochemical characteristics from normally unstable 2D V2CTx MXene flakes via a cation-driven assembly method. (i–v) microstructure and optical images of the sample. Adapted with permission from [83] Copyright (2021) Wiley-VCH. (b) Microfluidic-spinning-technique (MST) production of BP/CNT macrofibers. Published under the Creative Commons Attribution 4.0 International License (CC-BY). Adapted with permission from [84], Copyright (2021), Nature.

complicated systems customized for specific applications. The compatibility of flat 2D sheets with a wide range of device configurations opens up a wide range of interesting applications. Material-specific compounds with low packing density should not be overvalued. There is already a lot of promising research being conducted in this area, and we expect that with a better understanding of and control over the formation of few-layer structures, even better results will be possible.

REFERENCES

1. Zhang, X., X.-G. Wang, Z. Xie, and Z. Zhou, Recent progress in rechargeable alkali metal–air batteries. *Green Energy & Environment*, 2016. **1**(1): pp. 4–17.
2. Novoselov, K.S., A.K. Geim, S.V. Morozov, D. Jiang, Y. Zhang, S.V. Dubonos, I.V. Grigorieva, and A.A. Firsov, Electric field effect in atomically thin carbon films. *Science*, 2004. **306**(5696): pp. 666–669.

3. Majidi, L., P. Yasaei, R.E. Warburton, S. Fuladi, J. Cavin, X. Hu, Z. Hemmat, S.B. Cho, P. Abbasi, M. Vörös, and L. Cheng, New class of electrocatalysts based on 2D transition metal dichalcogenides in ionic liquid. *Advanced Materials*, 2019. **31**(4): p. 1804453.
4. Hart, J.L., K. Hantanasirisakul, A.C. Lang, B. Anasori, D. Pinto, Y. Pivak, J. Tijn van Omme, S.J. May, Y. Gogotsi, and M.L. Taheri, Control of MXenes' electronic properties through termination and intercalation. *Nature Communications*, 2019. **10**(1): pp. 1–10.
5. Cui, H., Y. Guo, W. Ma, and Z. Zhou, 2 D materials for electrochemical energy storage: Design, preparation, and application. *ChemSusChem*, 2020. **13**(6): pp. 1155–1171.
6. Du, G., Z. Guo, S. Wang, R. Zeng, Z. Chen, and H. Liu, Superior stability and high capacity of restacked molybdenum disulfide as anode material for lithium ion batteries. *Chemical Communications*, 2010. **46**: 1106–1108.
7. Bhandavat, R., L. David, and G. Singh, Synthesis of surface-functionalized WS_2 nanosheets and performance as Li-ion battery anodes. *The Journal of Physical Chemistry Letters*, 2012. **3**(11): pp. 1523–1530.
8. Splendiani, A., L. Sun, Y. Zhang, T. Li, J. Kim, C.-Y. Chim, G. Galli, and F. Wang, Emerging photoluminescence in monolayer MoS_2. *Nano Letters*, 2010. **10**(4): pp. 1271–1275.
9. Cao, T., G. Wang, W. Han, H. Ye, C. Zhu, J. Shi, Q. Niu, P. Tan, E. Wang, B. Liu, and J. Feng, Valley-selective circular dichroism of monolayer molybdenum disulphide. *Nature Communications*, 2012. **3**(1): pp. 1–5.
10. Mattheiss, L., Band structures of transition-metal-dichalcogenide layer compounds. *Physical Review B*, 1973. **8**(8): pp. 3719.
11. Li, T. and G. Galli, Electronic properties of MoS_2 nanoparticles. *The Journal of Physical Chemistry C*, 2007. **111**(44): pp. 16192–16196.
12. Mak, K.F., C. Lee, J. Hone, J. Shan, and T.F. Heinz, Atomically thin MoS_2: A new direct-gap semiconductor. *Physical Review Letters*, 2010. **105**(13): p. 136805.
13. Jiang, H., Electronic band structures of molybdenum and tungsten dichalcogenides by the GW approach. *The Journal of Physical Chemistry C*, 2012. **116**(14): pp. 7664–7671.
14. Abrams, B. and J. Wilcoxon, Nanosize semiconductors for photooxidation. *Critical Reviews in Solid State and Materials Sciences*, 2005. **30**(3): pp. 153–182.
15. Li, Z. and Y. Wu, 2D early transition metal carbides (MXenes) for catalysis. *Small*, 2019. **15**(29): p. 1804736.
16. Hu, T., Z. Li, M. Hu, J. Wang, Q. Hu, Q. Li, and X. Wang, Chemical origin of termination-functionalized MXenes: Ti_3C_2 T_2 as a case study. *The Journal of Physical Chemistry C*, 2017. **121**(35): pp. 19254–19261.
17. Fu, Z.H., Q.F. Zhang, D. Legut, C. Si, T.C. Germann, T. Lookman, S.Y. Du, J.S. Francisco, and R.F. Zhang, Stabilization and strengthening effects of functional groups in two-dimensional titanium carbide. *Physical Review B*, 2016. **94**(10): p. 104103.
18. Wang, X., X. Shen, Y. Gao, Z. Wang, R. Yu, and L. Chen, Atomic-scale recognition of surface structure and intercalation mechanism of Ti_3C_2X. *Journal of the American Chemical Society*, 2015. **137**(7): pp. 2715–2721.
19. Karlsson, L.H., J. Birch, J. Halim, M.W. Barsoum, and P.O. Persson, Atomically resolved structural and chemical investigation of single MXene sheets. *Nano Letters*, 2015. **15**(8): pp. 4955–4960.
20. Sang, X., Y. Xie, M.-W. Lin, M.. Alhabeb, K.L. Van Aken, Y. Gogotsi, P.R.C. Kent, K. Xiao, and R.R. Unocic, Atomic defects in monolayer titanium carbide ($Ti_3C_2T_x$) MXene. *ACS Nano*, 2016. **10**(10): pp. 9193–9200.
21. Hope, M., A. Forse, K. Griffith, M.R. Lukatskaya, M. Ghidiu, Y. Gogotsi, and C. Grey, NMR reveals the surface functionalisation of Ti_3C_2 MXene. *Physical Chemistry Chemical Physics*, 2016. **18**: 5099–5102.
22. Harris, K.J., M. Bugnet, M. Naguib, M.W. Barsoum, and G.R. Goward, Direct measurement of surface termination groups and their connectivity in the 2D MXene V_2CT_x using NMR spectroscopy. *The Journal of Physical Chemistry C*, 2015. **119**(24): pp. 13713–13720.

23. Zheng, S., C.J. Zhang, F. Zhou, Y. Dong, X. Shi, V. Nicolosi, Z.-S. Wu, and X. Bao, Ionic liquid pre-intercalated MXene films for ionogel-based flexible micro-supercapacitors with high volumetric energy density. *Journal of Materials Chemistry A*, 2019. **7**(16): pp. 9478–9485.
24. Anasori, B., M.R. Lukatskaya, and Y. Gogotsi, 2D metal carbides and nitrides (MXenes) for energy storage. *Nature Reviews Materials*, 2017. **2**(2): pp. 1–17.
25. Fredrickson, K.D., B. Anasori, Z.W. Seh, Y. Gogotsi, and A. Vojvodic, Effects of applied potential and water intercalation on the surface chemistry of Ti_2C and Mo_2C MXenes. *The Journal of Physical Chemistry C*, 2016. **120**(50): pp. 28432–28440.
26. Tang, Q., Z. Zhou, and P. Shen, Are MXenes promising anode materials for Li ion batteries? Computational studies on electronic properties and Li storage capability of Ti_3C_2 and $Ti_3C_2X_2$ (X= F, OH) monolayer. *Journal of the American Chemical Society*, 2012. **134**(40): pp. 16909–16916.
27. Wang, G. and Y. Liao, Theoretical prediction of robust and intrinsic half-metallicity in Ni_2N MXene with different types of surface terminations. *Applied Surface Science*, 2017. **426**: pp. 804–811.
28. Enyashin, A. and A. Ivanovskii, Atomic structure, comparative stability and electronic properties of hydroxylated Ti_2C and Ti_3C_2 nanotubes. *Computational and Theoretical Chemistry*, 2012. **989**: pp. 27–32.
29. Zhao, S., W. Kang, and J. Xue, MXene nanoribbons. *Journal of Materials Chemistry C*, 2015. **3**(4): pp. 879–888.
30. Bi, S., C. Lu, W. Zhang, F. Qiu, and F. Zhang, Two-dimensional polymer-based nanosheets for electrochemical energy storage and conversion. *Journal of Energy Chemistry*, 2018. **27**(1): pp. 99–116.
31. Halim, J., M.R. Lukatskaya, K.M. Cook, J. Lu, C.R. Smith, L.Å. Näslund, S.J. May, L. Hultman, Y. Gogotsi, P. Eklund, and M.W. Barsoum, Transparent conductive two-dimensional titanium carbide epitaxial thin films. *Chemistry of Materials*, 2014. **26**(7): pp. 2374–2381.
32. Lipatov, A., M. Alhabeb, M.R. Lukatskaya, A. Boson, Y. Gogotsi, and A. Sinitskii, Effect of synthesis on quality, electronic properties and environmental stability of individual monolayer Ti_3C_2 MXene flakes. *Advanced Electronic Materials*, 2016. **2**(12): p. 1600255.
33. Lai, S., J. Jeon, S.K. Jang, J. Xu, Y.J. Choi, J.-H. Park, E. Hwang, and S. Lee, Correction: Surface group modification and carrier transport properties of layered transition metal carbides (Ti_2CT_x, T:–OH, –F and–O). *Nanoscale*, 2015. **8**(2): pp. 1216–1216.
34. Kruusenberg, I., N. Alexeyeva, K. Tammeveski, J. Kozlova, L. Matisen, V. Sammelselg, J. Solla-Gullón, and J.M. Feliu, Effect of purification of carbon nanotubes on their electrocatalytic properties for oxygen reduction in acid solution. *Carbon*, 2011. **49**(12): pp. 4031–4039.
35. Zhang, J. and L. Dai, Nitrogen, phosphorus, and fluorine tri-doped graphene as a multifunctional catalyst for self-powered electrochemical water splitting. *Angewandte Chemie International Edition*, 2016. **55**(42): pp. 13296–13300.
36. Chen, P., K. Xu, T. Zhou, Y. Tong, J. Wu, H. Cheng, X. Lu, H. Ding, C. Wu, and Y. Xie, Strong-coupled cobalt borate nanosheets/graphene hybrid as electrocatalyst for water oxidation under both alkaline and neutral conditions. *Angewandte Chemie International Edition*, 2016. **55**(7): pp. 2488–2492.
37. Han, Q., Z. Cheng, J. Gao, Y. Zhao, Z. Zhang, L. Dai, and L. Qu, Mesh-on-mesh graphitic-C_3N_4@ graphene for highly efficient hydrogen evolution. *Advanced Functional Materials*, 2017. **27**(15): p. 1606352.
38. Chen, P., T. Zhou, M. Zhang, Y. Tong, C. Zhong, N. Zhang, L. Zhang, C. Wu, and Y. Xie, 3D nitrogen-anion-decorated nickel sulfides for highly efficient overall water splitting. *Advanced Materials*, 2017. **29**(30): pp. 1701584.

39. Li, Q., Y. Song, R. Xu, L. Zhang, J. Gao, Z. Xia, Tian, Z., Wei, N., Rümmeli, M.H., Zou, X. and Sun, J., Biotemplating growth of nepenthes-like N-doped graphene as a bifunctional polysulfide scavenger for Li–S batteries. *ACS Nano*, 2018. **12**(10): pp. 10240–10250.
40. Zhou, F., H. Huang, , Xiao, C., Zheng, S., Shi, X., Qin, J., Fu, Q., Bao, X., Feng, X., Müllen, K. and Wu, Z.S., Electrochemically scalable production of fluorine-modified graphene for flexible and high-energy ionogel-based microsupercapacitors. *Journal of the American Chemical Society*, 2018. **140**(26): pp. 8198–8205.
41. Zang, Y., S. Niu, Y. Wu, X. Zheng, J. Cai, J. Ye, Y. Xie, Y. Liu, J. Zhou, J. Zhu, and X. Liu, Tuning orbital orientation endows molybdenum disulfide with exceptional alkaline hydrogen evolution capability. *Nature Communications*, 2019. **10**(1): pp. 1–8.
42. Song, J., X. Guo, J. Zhang, Y. Chen, C. Zhang, L. Luo, F. Wang, and G. Wang, Rational design of free-standing 3D porous MXene/rGO hybrid aerogels as polysulfide reservoirs for high-energy lithium–sulfur batteries. *Journal of Materials Chemistry A*, 2019. **7**(11): pp. 6507–6513.
43. Meng, L., D. Rao, W. Tian, F. Cao, X. Yan, and L. Li, Simultaneous manipulation of O-Doping and metal vacancy in atomically thin $Zn_{10}In_{16}S_{34}$ Nanosheet arrays toward improved photoelectrochemical performance. *Angewandte Chemie International Edition*, 2018. **57**(51): pp. 16882–16887.
44. Chu, S. and A. Majumdar, Opportunities and challenges for a sustainable energy future. *Nature*, 2012. **488**(7411): pp. 294–303.
45. Roger, I., M.A. Shipman, and M.D. Symes, Earth-abundant catalysts for electrochemical and photoelectrochemical water splitting. *Nature Reviews Chemistry*, 2017. **1**(1): pp. 1–13.
46. Guo, J., X. Zhang, Y. Sun, L. Tang, and X. Zhang, Self-template synthesis of hierarchical $CoMoS_3$ nanotubes constructed of ultrathin nanosheets for robust water electrolysis. *Journal of Materials Chemistry A*, 2017. **5**(22): pp. 11309–11315.
47. Meyer, J.C., A.K. Geim, M.I. Katsnelson, K.S. Novoselov, T.J. Booth, and S. Roth, The structure of suspended graphene sheets. *Nature*, 2007. **446**(7131): pp. 60–63.
48. Shi, L. and T. Zhao, Recent advances in inorganic 2D materials and their applications in lithium and sodium batteries. *Journal of Materials Chemistry A*, 2017. **5**(8): pp. 3735–3758.
49. Armand, M. and J.-M. Tarascon, Building better batteries. *Nature*, 2008. **451**(7179): pp. 652–657.
50. Goodenough, J.B. and K.-S. Park, The Li-ion rechargeable battery: A perspective. *Journal of the American Chemical Society*, 2013. **135**(4): pp. 1167–1176.
51. Kumar, K.S., N. Choudhary, Y. Jung, and J. Thomas, Recent advances in two-dimensional nanomaterials for supercapacitor electrode applications. *ACS Energy Letters*, 2018. **3**(2): pp. 482–495.
52. Peng, L., Y. Zhu, H. Li, and G. Yu, Chemically integrated inorganic-graphene two-dimensional hybrid materials for flexible energy storage devices. *Small*, 2016. **12**(45): pp. 6183–6199.
53. Martínez-Periñán, E., M.P. Down, C. Gibaja, E. Lorenzo, F. Zamora, and C.E. Banks, Antimonene: A novel 2D nanomaterial for supercapacitor applications. *Advanced Energy Materials*, 2018. **8**(11): p. 1702606.
54. Palaniselvam, T. and J.-B. Baek, Graphene based 2D-materials for supercapacitors. *2D Materials*, 2015. **2**(3): p. 032002.
55. Dong, R., M. Pfeffermann, H. Liang, Z. Zheng, X. Zhu, J. Zhang, and X. Feng, Large-area, free-standing, two-dimensional supramolecular polymer single-layer sheets for highly efficient electrocatalytic hydrogen evolution. *Angewandte Chemie International Edition*, 2015. **54**(41): pp. 12058–12063.
56. Dong, R., Z. Zheng, D.C. Tranca, J. Zhang, N. Chandrasekhar, S. Liu, X. Zhuang, G. Seifert, and X. Feng, Immobilizing molecular metal dithiolene–diamine complexes on 2D metal–organic frameworks for electrocatalytic H_2 production. *Chemistry–A European Journal*, 2017. **23**(10): pp. 2255–2260.

57. Clough, A.J., J.W. Yoo, M.H. Mecklenburg, and S.C. Marinescu, Two-dimensional metal–organic surfaces for efficient hydrogen evolution from water. *Journal of the American Chemical Society*, 2015. **137**(1): pp. 118–121.
58. Downes, C.A. and S.C. Marinescu, Understanding variability in the hydrogen evolution activity of a cobalt anthracenetetrathiolate coordination polymer. *ACS Catalysis*, 2017. **7**(12): pp. 8605–8612.
59. Zhou, Y.-C., W.-W. Dong, M.-Y. Jiang, Y.-P. Wu, D.-S. Li, Z.-F. Tian, and J. Zhao, A new 3D 8-fold interpenetrating 66-dia topological Co-MOF: Syntheses, crystal structure, magnetic properties and electrocatalytic hydrogen evolution reaction. *Journal of Solid State Chemistry*, 2019. **279**: p. 120929.
60. Downes, C.A., A.J. Clough, K. Chen, J.W. Yoo, and S.C. Marinescu, Evaluation of the H_2 evolving activity of benzenehexathiolate coordination frameworks and the effect of film thickness on H_2 production. *ACS Applied Materials & Interfaces*, 2018. **10**(2): pp. 1719–1727.
61. Jia, H., Y. Yao, J. Zhao, Y. Gao, Z. Luo, and P. Du, A novel two-dimensional nickel phthalocyanine-based metal–organic framework for highly efficient water oxidation catalysis. *Journal of Materials Chemistry A*, 2018. **6**(3): pp. 1188–1195.
62. Zhong, H., K.H. Ly, M. Wang, Y. Krupskaya, X. Han, J. Zhang, J. Zhang, V. Kataev, B. Büchner, I.M. Weidinger, and S. Kaskel, A phthalocyanine-based layered two-dimensional conjugated metal–organic framework as a highly efficient electrocatalyst for the oxygen reduction reaction. *Angewandte Chemie International Edition*, 2019. **58**(31): pp. 10677–10682.
63. Miner, E.M., T. Fukushima, D. Sheberla, L. Sun, Y. Surendranath, and M. Dincă, Electrochemical oxygen reduction catalysed by Ni_3 (hexaiminotriphenylene) 2. *Nature Communications*, 2016. **7**(1): pp. 1–7.
64. Guo, J., C.-Y. Lin, Z. Xia, and Z. Xiang, A pyrolysis-free covalent organic polymer for oxygen reduction. *Angewandte Chemie International Edition*, 2018. **57**(38): pp. 12567–12572.
65. Lian, Y., W. Yang, C. Zhang, H. Sun, Z. Deng, W. Xu, L. Song, Z. Ouyang, Z. Wang, J. Guo, and Y. Peng, Unpaired 3d electrons on atomically dispersed cobalt centres in coordination polymers regulate both oxygen reduction reaction (ORR) activity and selectivity for use in zinc–air batteries. *Angewandte Chemie International Edition*, 2020. **59**(1): pp. 286–294.
66. Miner, E.M., L. Wang, and M. Dincă, Modular O_2 electroreduction activity in triphenylene-based metal–organic frameworks. *Chemical Science*, 2018. **9**(29): pp. 6286–6291.
67. Zhang, X.-D., S.-Z. Hou, J.-X. Wu, and Z.-Y. Gu, Two-dimensional metal–organic framework nanosheets with cobalt-porphyrins for high-performance CO_2 electroreduction. *Chemistry–A European Journal*, 2020. **26**(7): pp. 1604–1611.
68. Zhong, H., M. Ghorbani-Asl, K.H. Ly, J. Zhang, J. Ge, M. Wang, Z. Liao, D. Makarov, E. Zschech, E. Brunner, and I.M. Weidinger, Synergistic electroreduction of carbon dioxide to carbon monoxide on bimetallic layered conjugated metal-organic frameworks. *Nature Communications*, 2020. **11**(1): pp. 1–10.
69. Zhu, H.-J., M. Lu, Y.-R. Wang, S.-J. Yao, M. Zhang, Y.-H. Kan, J. Liu, Y. Chen, S.-L. Li, and Ya-Qian Lan, Efficient electron transmission in covalent organic framework nanosheets for highly active electrocatalytic carbon dioxide reduction. *Nature Communications*, 2020. **11**(1): pp. 1–10.
70. Saito, T., Y. Tatematsu, Y. Yamaguchi, S. Ikeuchi, S. Ogasawara, N. Yamada, R. Ikeda, I. Ogawa, and T. Idehara, Observation of dynamic interactions between fundamental and second-harmonic modes in a high-power sub-terahertz gyrotron operating in regimes of soft and hard self-excitation. *Physical Review Letters*, 2012. **109**(15): p. 155001.
71. Liao, M., Y. Zang, Z. Guan, H. Li, Y. Gong, K. Zhu, X.P. Hu, D. Zhang, Y. Xu, Y.Y. Wang, and K. He, Epitaxial growth of two-dimensional stanene. *Nature Materials*, 2015. **14**(10): pp. 1020–1025.

72. Lee, W.J., T.H. Hwang, J.O. Hwang, H.W. Kim, J. Lim, H.Y. Jeong, J. Shim, T.H. Han, J.Y. Kim, J.W. Choi, and S.O. Kim, N-doped graphitic self-encapsulation for high performance silicon anodes in lithium-ion batteries. *Energy & Environmental Science*, 2014. **7**(2): pp. 621–626.
73. Zhang, B., G. Xia, D. Sun, F. Fang, and X. Yu, Magnesium hydride nanoparticles self-assembled on graphene as anode material for high-performance lithium-ion batteries. *ACS Nano*, 2018. **12**(4): pp. 3816–3824.
74. Yang, E., H. Ji, and Y. Jung, Two-dimensional transition metal dichalcogenide monolayers as promising sodium ion battery anodes. *The Journal of Physical Chemistry C*, 2015. **119**(47): pp. 26374–26380.
75. Park, J., J.-S. Kim, J.-W. Park, T.-H. Nam, K.-W. Kim, J.-H. Ahn, G. Wang, and H.-J. Ahn, Discharge mechanism of MoS_2 for sodium ion battery: Electrochemical measurements and characterization. *Electrochimica Acta*, 2013. **92**: pp. 427–432.
76. Bao, W., L. Liu, C. Wang, S. Choi, D. Wang, and G. Wang, Facile synthesis of crumpled nitrogen-doped mxene nanosheets as a new sulfur host for lithium–sulfur batteries. *Advanced Energy Materials*, 2018. **8**(13): p. 1702485.
77. Hoshide, T., Y. Zheng, J. Hou, Z. Wang, Q. Li, Z. Zhao, R. Ma, T. Sasaki, and F. Geng, Flexible lithium-ion fiber battery by the regular stacking of two-dimensional titanium oxide nanosheets hybridized with reduced graphene oxide. *Nano Letters*, 2017. **17**(6): pp. 3543–3549.
78. Pomerantseva, E. and Y. Gogotsi, Two-dimensional heterostructures for energy storage. *Nature Energy*, 2017. **2**(7): pp. 1–6.
79. Romanitan, C., P. Varasteanu, I. Mihalache, D. Culita, S. Somacescu, R. Pascu, E. Tanasa, S.A. Eremia, A. Boldeiu, M. Simion, and A. Radoi, High-performance solid state supercapacitors assembling graphene interconnected networks in porous silicon electrode by electrochemical methods using 2, 6-dihydroxynaphthalen. *Scientific Reports*, 2018. **8**(1): pp. 1–14.
80. Wang, X., G. Sun, P. Routh, D.-H. Kim, W. Huang, and P. Chen, Heteroatom-doped graphene materials: Syntheses, properties and applications. *Chemical Society Reviews*, 2014. **43**(20): pp. 7067–7098.
81. Gao, S., L. Yang, Z. Liu, J. Shao, Q. Qu, M. Hossain, Y. Wu, P. Adelhelm, and R. Holze, Carbon-coated SnS nanosheets supported on porous microspheres as negative electrode material for sodium-ion batteries. *Energy Technology*, 2020. **8**(7): p. 2000258.
82. Lee, K.E., J.E. Kim, U.N. Maiti, J. Lim, J.O. Hwang, J. Shim, J.J. Oh, T. Yun, and S.O. Kim, Liquid crystal size selection of large-size graphene oxide for size-dependent N-doping and oxygen reduction catalysis. *ACS Nano*, 2014. **8**(9): pp. 9073–9080.
83. Guan, Y., S. Jiang, Y. Cong, J. Wang, Z. Dong, Q. Zhang, G. Yuan, Y. Li, and X. Li, A hydrofluoric acid-free synthesis of 2D vanadium carbide (V2C) MXene for supercapacitor electrodes. *2D Materials*, 2020. **7**(2): p. 025010.
84. Wu, X., Y. Xu, Y. Hu, G. Wu, H. Cheng, Q. Yu, K. Zhang, W. Chen, and S. Chen, Microfluidic-spinning construction of black-phosphorus-hybrid microfibres for non-woven fabrics toward a high energy density flexible supercapacitor. *Nature Communications*, 2018. **9**(1): pp. 1–11.
85. Padmajan Sasikala, S., K.E. Lee, J. Lim, H.J. Lee, S.H. Koo, I.H. Kim, H.J. Jung, and S.O. Kim, Interface-confined high crystalline growth of semiconducting polymers at graphene fibers for high-performance wearable supercapacitors. *ACS Nano*, 2017. **11**(9): pp. 9424–9434.

2 Advanced 2D Materials for Energy Applications

Immanuel Paulraj and Chia-Jyi Liu
National Changhua University of Education

CONTENTS

2.1 INTRODUCTION

Over the past decade, two-dimensional (2D) materials have been the hot spot for extensive research efforts, particularly MXenes, transition metal dichalcogenides (TMDs), and graphene due to their unique chemical, structural, morphological, optical, electronic, thermal, and mechanical properties. Compared to 0D and 1D materials, 2D materials have fascinating properties such as low electrical resistivity, large surface area, and robustness in mechanical strength. These properties make 2D materials to be developed as multifunctional nanomaterials in applications of electronic devices, energy storage, tissue engineering, and biomedicine. As fossil reserves are rapidly declining, it is imperative to develop and integrate materials and devices for energy harvesting and energy storage. Due to the rapid growth of research interest in energy applications of 2D materials, it would be advantageous to outline the research progress of layered materials.

Thermal energy can be stored using a class of materials called phase transition materials or phase change materials (PCMs). PCMs could absorb or release a significant amount of latent heat during phase transition. PCMs exhibit large storage densities with 5–10 times more than sensible heat storage capacity. Delcase and Raymond studied

DOI: 10.1201/9781003178453-2

the PCMs in 1940 and pioneered the research on PCMs as energy storage materials for heat management during the 1973–1974 energy crisis. The selection of the proper phase-changing materials is very important in the design of latent heat thermal energy storage system. PCMs can be categorized into three types: (a) inorganic salt hydrates ($A_xB_y{\cdot}nH_2O$), (b) organic paraffins (C_nH_{2n+2}), lipids, and sugar alcohols, and (c) eutectic system (inorganic–inorganic, organic–organic, and inorganic–organic). Rashid et al. [1] used the PCMs of polyvinylpyrrolidone, polyethylene glycol, and carboxymethyl cellulose sodium salt to study the efficiency of thermal energy storage. Their results reveal that PCMs can store thermal energy 25%–40% more efficiently than those without using PCMs. Zendehboudi et al. [2] studied the thermal conductivity of nano-PCM-based graphene by response surface methodology. By adding graphene flakes as a filler in a PCM of Paraffin Wax, the thermal conductivity of the nanocomposite increases. Hence the energy storage capacity is significantly enhanced. To improve the overall performance of PCMs, however, there are still a few issues to address concerning thermal stability, energy storage capacity, and preparation methods.

Since the isolation of one atom thick, 2D crystal graphene from graphite in 2004, numerous 2D materials have been discovered including TMD monolayers, 2D hexagonal boron nitrides (2D-hBN), and MXenes. The 2D materials MXenes consist of few-atoms-thick layers of metal carbides (MCs), nitrides, or carbonitrides. MXenes are composed of $M_{n+1}X_n$ stacked sheets, which are held together by van der Waals interactions and/or hydrogen bonds. MXenes are used in a wide range of applications because of their unique and rich properties such as metallic conductivity, tunable surfaces, excellent mechanical strength, transparency, high transmittance, and acting as a host for intercalation. MXenes have three structures and are inherited from the parent MAX phases: M_2X, M_3X_2, and M_4X_3. In 2011, Naguib et al. [3] developed a method to produce 2D nanosheets composed of a few Ti_3C_2 layers and conical scrolls by selectively etching out "A" elements (Al) from a MAX phase (Ti_3AlC_2) at room temperature in hydrofluoric acid. The MAX phases have a hexagonal layered structure with the general formula of $M_{n+1}AX_n$, (MAX) where A is an element from group 13 or 14 of the periodic table. Since the surfaces of MXene sheets are often terminated by a functional group such as O, F, OH, or Cl, they are represented by a general formula $M_{n+1}X_nT_x$, where $n = 1$ to 4, M is an early transition metal, X either carbon and/or nitrogen and T a functional group. MXene-based PCMs have recently attracted attention due to their good thermal performance. MXene-based PCM composite films are easy to make and have high electrical conductivity; they are also stable in water, strong and stiff.

Different strategies are adopted to enhance the properties and applications of MCs. Besides, the chemical functionalization of F, OH, and O groups on the surface of MXenes could be used to create new electronic and magnetic properties of 2D transition metal carbides (TMCs) and nitrides (TMNs). Dinh et al. [4] gave a comprehensive evaluation on the preparation of nanostructured TMCs, transition metal borides, transition metal phosphides, TMNs. Li et al. [5] summarized the latest development of 2D materials of TMCs and their nanocomposites for tuning the mechanical, structural, electrical, and chemical properties of TMCs. Shahzad et al. [6] summarized recent developments of using 2D TMCs for electrochemical sensing of diverse types of valuable analytes such as small molecules, heavy metals, dyes, and biomolecules. Regarding 2D materials for energy applications, we specifically cover the topics on the preparation methods of advanced 2D materials MXene and briefly discuss conjugated microporous polymer (CMP)/MXene composites, supercapacitor (SC), and thermoelectric (TE) materials [7]. The synthesis of 2D MXenes includes chemical vapor deposition (CVD), hydrofluoric acid (HFA) etching process, molten salt methods, and electrochemical etching methods.

2.2 PREPARATION METHODS OF MXenes

In general, there are two approaches to preparing MXenes.

1. Bottom-up approach
2. Top-down approach

   CVD is one of the bottom-up methods of preparing uniform and high-quality thin 2D materials. The ball milling method is the top-down approach alternative to preparing MXenes. However, some accounts are given to the following methods of preparing MXenes.

   - CVD
   - HFA
   - Molten salts
   - Electrochemical etching

2.2.1 Chemical Vapor Deposition (CVD)

CVD is a chemical deposition technique used to produce solid materials with high purity and good performance. The semiconductor industry uses this technology to grow thin films. This process uses a gaseous form of reagents to carry out thermally induced chemical reactions at the surface of a heated substrate. The well-developed CVD technique is adopted to fabricate a huge variety of 2D ultrathin crystals such as tungsten carbide (WC) and tantalum carbide (TaC). Ultrathin tantalum carbide (TaC), nitride (TaN), and boride (TaB) were grown on Cu surface by CVD using acetylene, boron, and ammonia reacting with a heated Ta-Cu bilayer, respectively [8]. Compared to other methods, however, the CVD process is not a cost-effective and facile method to synthesize MXenes.

2.2.2 Hydrofluoric Acid

MCs have recently been produced using the HFA etching process. Due to relatively weakly bonded "A" layers as compared to the M-X bonding, MCs are synthesized by selectively etching the "A" layers from the parent-layered MAX phase with the control of the reaction time and acid concentration at low temperatures. Naguib et al. [9] immersed MAX powders ($M_{n+1}AlX_n$) in concentrated HFA for 2 hours at room temperature to form MXenes ($M_{n+1}X_n$). As a result, the immersed powders exfoliate to form 2D TMCs and transition metal carbonitrides. Since then the acid etching approach has been routinely followed. However, HFA is highly corrosive and can penetrate our body such as bones, skin, and muscle tissue, causing handling and disposal hazards. Recent research tends to avoid using concentrated HFA in the synthesis of MCs. For instance, Ghidiu et al. [10] developed a safer and faster route to synthesize MXenes with high yield using a combination of inexpensive hydrochloric acid and fluoride salts such as lithium fluoride LiF. In this process, Ti_3AlC_2 powders are slowly added to HCl containing LiF. The resulting solution is then heated up to 40°C for 45 hours. Consequently, a greater lateral size enlargement is observed except nanosized defects that are regularly observed samples corroded by HFA. With this method, a huge number of single-layer and large-sized MC can be easily attained. Guan et al. [11] obtained high-quality and uniform multilayer 2D V_2C MXene using LiF/HCl as etchants to etch V_2AlC powders at 90°C for different lengths of duration. The above two synthetic processes are the most mature etching routes in making multilayer MXene with different shapes and surface chemistry. Unfortunately, both methods cause environmental pollution because of the presence of fluorine. Green strategies for synthesizing MXenes are currently explored to prevent polluting our environments.

2.2.3 Molten Salts

Both Ti_3SiC_2 and Cr_2AlC had been synthesized by using molten salt methods; however, their micron-sized grain size is too big to become 2D materials for applications. Nanosized Ti_3AlC_2 powders can now be synthesized using molten salt methods. Li et al. [12] synthesized a series of novel nanolaminated MAX phases (Ti_3ZnC_2, Ti_2ZnC, Ti_2ZnN, and V_2ZnC) and Cl-terminated MXenes ($Ti_3C_2Cl_2$ and Ti_2CCl_2) based on element exchange reaction between traditional MAX phase such as Ti_3AlC_2, Ti_2AlC, Ti_2AlN, V_2AlC, and molten $ZnCl_2$ and subsequent exfoliation. Both the A-site element exchange reaction and A-site element etching process are top-down routes to synthesize novel MAX phase and Cl-terminated MXenes, respectively. The Cl-functionalized MXenes have shown enhanced electrochemical performance and could have potential applications in energy storage.

2.2.4 Electrochemical Etching

Electrochemical etching is a very convenient and effective method to synthesize MXenes because it is fluorine-free and environmentally friendly. Molten fluoride salt methods have difficulty to completely remove the fluoride or other residues in etching the MAX phase, which can be easily dealt with using electrochemical etching methods. Fluorine-free Ti_2CT_x (MXene), T = Cl, O, and OH were synthesized using electrochemical etching in diluted HCl aqueous electrolyte [13]. However, over-etching of the parent MAX phases to carbide-derived carbon on the surface should be avoided by careful balances in etching parameters. In another report, it was advised that the choice of electrolytes determines whether one can selectively remove the Al layers while preserving the 2D structure of Ti_3C_2 [14].

2.3 CONJUGATED MICROPOROUS POLYMER/ MXene COMPOSITES

Even though 2D MXenes have received great attention due to their remarkable physical and chemical properties, however, they are readily deteriorated by restacking of MXene flakes and by the oxidation between MXenes and oxygen in the ambient environment, thereby losing their functional features and limiting their applications.

Therefore, it would be beneficial to develop strategies of preventing restacking and agglomeration of nanosheets and keeping away from oxidation. CMPs are a subclass of porous materials and are composed of inflexible conjugated molecules via covalent bonds. Compared to organic polymers, CMPs possess a large surface area and porous structure. The size of both the surface area and porosity of CMPs can be tailored for various applications. The properties of CMPs can be optimized by tuning the diversified π units in their building blocks. Besides, the synthesis of CMPs is cost-effective. Therefore, CMPs have extensive attention in energy storage. Recent studies have mainly focused on fabricating CMP composites. Yang et al. [15] reported an effective strategy to synthesize MXene-based CMPs by covalently sandwiching MXene between CMPs using *p*-iodophenyl functionalized MXene as templates. The MXene-based CMPs exhibit 2D architecture and high electrical conductivity inherited from MXene and hierarchical porous structure and large specific surface area inherited from CMPs. Both the hierarchical porous structure and large specific surface area play a key role in electrochemical energy storage. In general, polymers exhibit amorphous structure with agglomerated microstructure. Nevertheless, after MXene compositing with CMPs, the MXene-based CMPs composite exhibits a sheet-like structure based on morphology observation using scanning electron microscope. Furthermore, no pristine CMP or MXene nanosheets are observed, which indicates most of the polymer units are grafted on the surface of MXene. Furthermore, sandwiching MXene between CMPs brings about the effects of effectively increasing the interlayer distance of MXene, increasing available active sites of CMPs, and shortening the ion diffusion path. Accordingly, the electrochemical performance is enhanced. Since the drawbacks of restacking and poor oxidation stability of MXene nanosheets are overcome by sandwiching MXene between CMPs, the MXene-based CMPs composite exhibits potential as SC electrode for applications in energy storage and conversion.

2.4 SUPERCAPACITORS

Supercapacitors (SCs), also known as electrochemical capacitors, have high specific capacitance, high power density, fast charging and discharging characteristics, compact design, and low maintenance cost [16,17]. They can charge and discharge faster than batteries, store much more energy per unit mass than electrolytic capacitors, and tolerate much more life cycles than rechargeable batteries. SCs are used in diversified applications requiring many rapid charge/discharge cycles rather than long-term compact energy storage. SCs could be used to store the electricity for portable electronic devices, to store energy temporarily in energy-harvesting systems, to stabilize the power supply, to power uninterruptible power supplies, to deliver 500 J in defibrillators, to supply short-term high currents for motor startup and buffer batteries, to deliver power for photographic flashes, to provide bursts of power in vehicles, to recover breaking energy, and to deliver starting energy in transportation.

Figure 2.1 shows different categories of the SCs. There are three main mechanisms for energy storage in SCs, depending on the accumulation of charge or reversible redox reactions: (a) electric double layer capacitors (EDLCs); (b) pseudocapacitor; (c) hybrid supercapacitor (HSC) – the combination of EDLC and pseudocapacitor. EDLCs refer to their charges being adsorbed electrostatically on the interface of

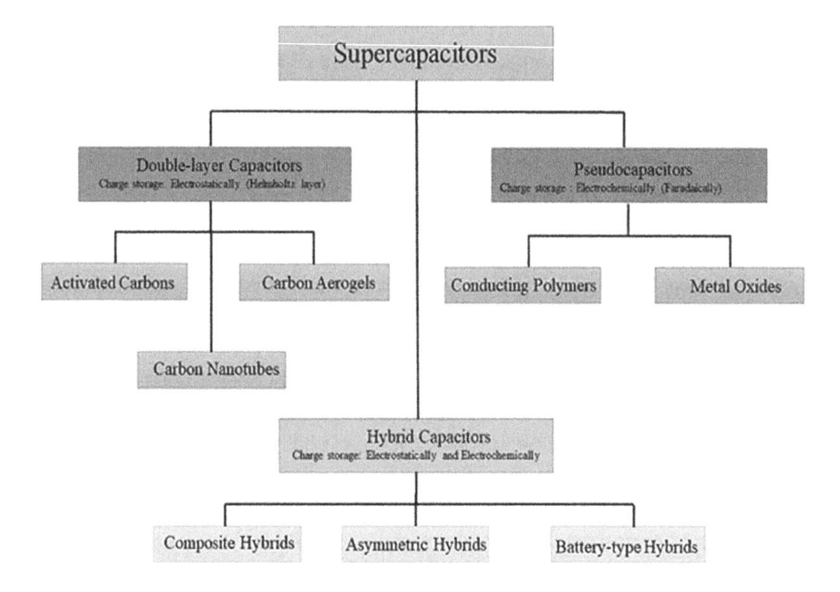

FIGURE 2.1 Classification of supercapacitors.

electrode/electrolyte; pseudocapacitors refer to their energy being stored through redox reaction; HSCs refer to the charge storage via a combination of the above two mechanisms. The active materials of EDLC include activated carbon, graphene, multi-walled carbon nanotubes (MWCNTs), TMDs, etc. There are two types of redox reactions for pseudocapacitors: (a) surface redox reactions and (b) ion intercalation.

For the energy storage applications, the 2D TMDs such as VS_2, MoS_2, ReS, WS_2, and VSe_2 are the candidates to be used as active electrode materials in SCs due to their large surface area and variable oxidation states. Since the electrical conductivity of 2D TMDs is low and hence low power density. To enhance the electrical conductivity of TMDs, which is one of the major factors influencing the electrochemical performance of SCs, the following strategies are approached: (a) hybridizing with highly conducting materials, e.g., carbon nanotube (CNT), graphene, and organic conductive polymers (OCPs); (b) tuning the electronic structure; and (c) defect engineering.

MoS_2 is one of the attractive SC electrode materials in 2D TMDs due to its graphene-like morphology and large surface area. The charge storage mechanisms of MoS_2 include interlayer EDLC, intralayer EDLC between individual MoS_2 atoms, and faradaic charge transfer on the Mo centers arising from the various oxidation states of Mo. It has been proven effective to overcome the drawbacks of intrinsic low conductivity of MoS_2 by adding conductive carbon material and hence enhance its capacitance and energy density. To enhance the electrochemical performance, MoS_2 are wrapped onto conductive network such as CNT, reduced graphene oxide (RGO) and conductive polymers (CP) to construct a hierarchical porous network composite, to increase electrode conductivity, and to constrain the volume charge. Kumuthini et al. [18] synthesized MoS_2@C composite nanofibers via electrospinning technique followed by stabilization using iodine as a stabilizer and carbonization at 900°C for 2 hours. The resulting MoS_2@C nanofibers exhibited a high specific capacitance of

355.6 F g^{-1} at a scan rate of 5 mV s^{-1} in 6 M KOH electrolyte as compared to 36.1 F g^{-1} for the bulk MoS_2, and long cycling stability with specific capacitance retention of 93% up to 2,000 cycles. Sari et al. [19] adopted a microwave-assisted hydrothermal method under an acidic condition to fabricate a composite electrode by directly growing MoS_2 nanowalls onto highly conductive vapor grown carbon nanofibers and obtained high electrical conductivity and good specific capacitance of 284 F g^{-1} at 5 mVs^{-1}. The resulting composite electrode exhibited excellent electrochemical stability with a retention of 96% after 1,000 cycles at a high charge rate of 200 mV s^{-1}. Sun et al. [20] synthesized a hierarchical graphene-wrapped CNT@MoS_2 composite electrode in two stages. At the first stage, CNT@MoS_2 composites were first synthesized via solvothermal reaction, followed by surface modification via dispersing CNT@MoS_2 into 1 wt% poly(diallyldimethylammoniuurm chloride) (PDDA) solution. At the second stage, the graphene-wrapped CNT@MoS_2 composite was synthesized via electrostatic interaction between the PDDA-modified CNT@MoS_2 dispersion and graphene oxide (GO) suspension, followed by chemical reduction using $NaBH_4$ maintained at 40°C for 10 hours. The composite electrode exhibited a high specific capacitance of 498 F g^{-1} and demonstrated long-term cycle-life stability with only 5.7% loss of its initial capacitance after 10,000 cycles at a high current density of 5 A g^{-1}. Tiwari et al. [21] used magnetron sputtering to directly grow atomically thin hierarchal MoS_2 nanoflakes on CVD grown CNT as electrodes for all solid-state symmetric SCs. The resulting MoS_2@CNT electrodes possessed high areal capacitance of 131 mF cm^{-2}, volumetric capacitance of 2.9 F cm^{-3}, and high cyclic stability of 97.6% after 2,500 cycles. Sun et al. [22] fabricated MoS_2-rGO/MWCNT fibers as the anode and rGO/MWCNT fibers as the cathode for solid-state flexible asymmetric supercapacitors (ASCs), which can operate in a wide potential window of 1.4 V with high coulombic efficiency, high energy density, good rate, and cycling stability. ASCs refer to hybrid capacitors composed of two dissimilar electrodes of a capacitor kind and a battery kind electrode. Sun et al. [23] adopted pulsed laser deposition to fabricate defect-rich MoS_2@CNTs/Ni core/shell-structured electrode. The MoS_{2-x} nanolayer was deposited on CNTs surface as the substrate, which was in situ grown on Ni mesh. The resulting electrode has a specific capacitance of 512 F g^{-1} at 1 A g^{-1} and long cycle life without any decay after 2,000 cycles in 1 M Na_2SO_4 electrolyte. An ASC with MoS_{2-x}@CNTs/Ni as the positive electrode and CNT networks as the negative electrode exhibited a large energy density of 63 Wh kg^{-1} at 850 W kg^{-1} and a power density of 25.5 kW kg^{-1} at 44.2 Wh kg^{-1}.

2.5 THERMOELECTRIC MATERIALS

TE materials can directly convert waste heat energy into electric voltage and therefore can harvest waste heat or exhausted heat, which are often dissipated to the environment. The efficiency of a TE material depends on the TE dimensionless figure of merit $zT = \frac{S^2\sigma}{k_e + k_l}T$, where S is the thermopower or Seebeck coefficient, σ the electrical conductivity, $k = k_e + k_l$ the total thermal conductivity, k_e the electronic thermal conductivity contributed from charge carriers, k_l lattice thermal conductivity contributed from atomic vibration, and T the absolute temperature [24,25]. To enhance the efficiency of a given TE material requires high electrical conductivity, large thermopower, and low total thermal conductivity. However, the three transport parameters of electrical conductivity, thermopower, and thermal conductivity depend mutually based on the Boltzmann transport equation, the high electrical conductivity would be accompanied by a small thermopower for materials with one type of charge carrier. Moreover, the Wiedemann–Franz law governs the relationship between electrical conductivity and electronic thermal conductivity at a given temperature, provided that a common relaxation time exists for electrical and thermal processes. It is thereby intrinsically challenging to develop TE materials with high zT values.

Due to the unique density of states of confined electrons and holes, low dimensionality (1D and 2D) provides novel ways to high TE power factors. Low dimensional materials have larger $\frac{d \ln N(E)}{dE}$ near Fermi level than the bulk, their thermopower (S) could be significantly enhanced according to $S \propto \left[\frac{d \ln N(E)}{dE}\right]_{E=E_F}$. Besides, thermal transport of 2D materials behaves differently from bulk materials due to interfaces, sample size, strain, and defects. Thermal transport of 2D materials can be tuned by their thickness and the feature size in the basal planes. Chemical doping via surface functionalization or intercalation can be also used to tune thermal transport of 2D materials. Calculations were carried out to study the zT of Bi_2Te_3 layers in a quantum-well superlattice structure under the assumptions of one-band material with parabolic shape and a constant relaxation time [26]. It was theoretically found that a 10 Å thick Bi_2Te_3 film prepared with layers in the *a-c* planes has zT enhancement by a factor of 13 over the bulk value; if the multilayers are prepared in the *a-b* direction, the zT is enhanced by a factor of 3.

The wearable devices in the recent rapid development only require low power energy consumption. Large-area flexible TE devices can be fabricated to serve for sensing and Internet of Things (IoT) applications by using the temperature difference between body heat and the ambient environment. Therefore, TE energy conversion based on large area flexible materials has attracted much attention. Atomically thin (<1 nm) graphene and akin 2D materials are rigid against the bending and intrinsically flexible, they are potential candidates for practical flexible TE devices as long as their TE properties are optimized. Therefore, 2D materials graphene, black phosphorus, and TMDs have attracted great attention due to their unique geometric structure and TE properties [27]. The recent progress on the TE performance of 2D TE materials of graphene, black phosphorus, and TMDs are presented as follows. Figure 2.2 shows the classification of 2D TE materials.

2.5.1 Graphene

As shown in Figure 2.3, graphene is a single layer of carbon atoms bonded together in a 2D hexagonal honeycomb lattice with features of zero bandgap semiconductor and massless Dirac fermion characteristics. Charge transport of graphene is essentially governed by Dirac's (relativistic) equation and exhibits ambipolar characteristics with exceptional carrier mobility [28,29]. External electric and magnetic fields can be used to control the Dirac electrons.

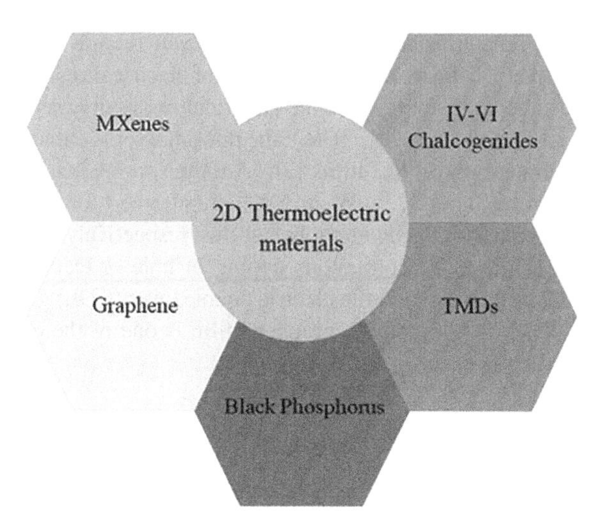

FIGURE 2.2 Classification of 2D thermoelectric materials.

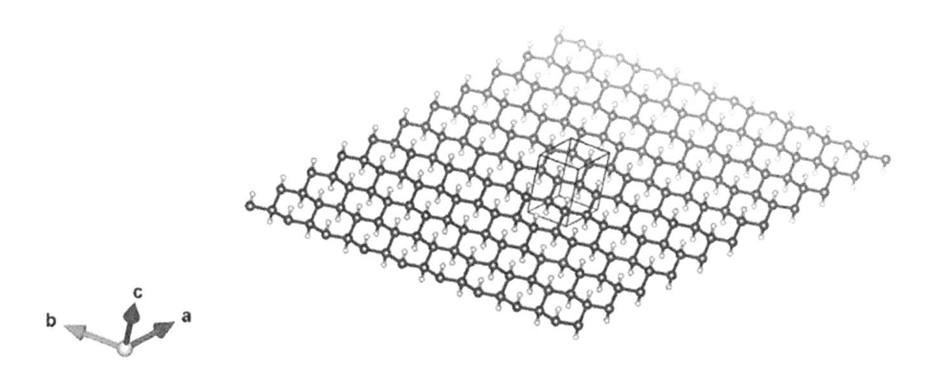

FIGURE 2.3 Crystal structure of graphene.

Graphene attracts much attention for TE applications due to its remarkable transport properties and large TE power factor. Since the CVD process has been developed to grow large-area single-crystalline and poly-crystalline graphene films with ultra-high uniformity, graphene is one of the candidates for TE generators. The maximum power factor of ~45 μW cm^{-1} K^{-2} is attained at 575 K for few layers graphene (FLG) films after the CVD-grown graphene is treated with oxygen plasma for a few seconds [30]. It is about a 15-fold enhanced power factor of the pristine FLG. This enhancement is attributed to generation of local disordered carbon that is opening the bandgap. However, the oxygen plasma treatment is not suitable for single-layer graphene films and reduced graphene oxide (rGO) films since it results in no enhancement of the power factor. Moreover, the pristine FLG films grown by the CVD process have a thermopower of ~40 μV K^{-1}. The thermopower is increased to ~180 μV K^{-1} after the FLG films treated with 1,10-azobis(cyanocyclohexane)

(ACN) or 1,3,6,8-pyrenetetrasulfonic acid (TPA). Even though the attachment of ACN or TPA onto the FLG films led to an increase of electrical resistivity, the resulting power factor is about 4.5-fold, while the electrical resistivity increase is depending on the treating concentration. As a result, the power factor is enhanced by 4.5-fold by molecular attachment onto FLG films [31]. Another promising report is that the peak power factor of 69.3 and 32.9 μW cm^{-1} K^{-2} are attained for *p*- and *n*-type carrier doping of large-area CVD-grown graphene films, respectively. The enhancement of power factor is achieved by continuous doping of hole or electron carriers and modulation of the Fermi energy via the electric double layer (EDL) gating technique. Therefore, the CVD-grown large-area graphene film is one of the candidates to be applied for the power generator of IoT devices [32].

2.5.2 Black Phosphorus

Although having superior power factors, graphene possesses high thermal conductivity and hence its zT value is not good enough. Another interesting 2D material is the black phosphorus. Like graphene, the black phosphorus atoms are strongly bonded within a single layer while adjacent layers are weakly stacked together via van der Waals forces in the black phosphorus lattice. Figure 2.4 shows the crystal structure of black phosphorous. Unlike in graphite, black phosphorus forms a puckered honeycomb lattice due to the phosphorus atom bonding together via sp^3 hybridized orbitals [33]. The black phosphorus exhibits zigzag and armchair structural anisotropy and hence anisotropic electrical and thermal properties in the zigzag and armchair directions. Moreover, graphene is a zero-gap semiconductor, while the black phosphorus is a semiconductor with a sizeable gap depending on its thickness of the layers. Monolayer black phosphorus (phosphorene) has a predicted direct bandgap of ca. 2 eV at the Γ point of the first Brillouin zone. The size of the bandgap decreases with the number of layers due to interlayer interaction for few-layer phosphorene and would be 0.3 eV for bulk black phosphorus [34]. Bulk black phosphorous is experimentally determined to have thermopower of $S = 335 \pm 10$ μV K^{-1} at room temperature and increases with increasing

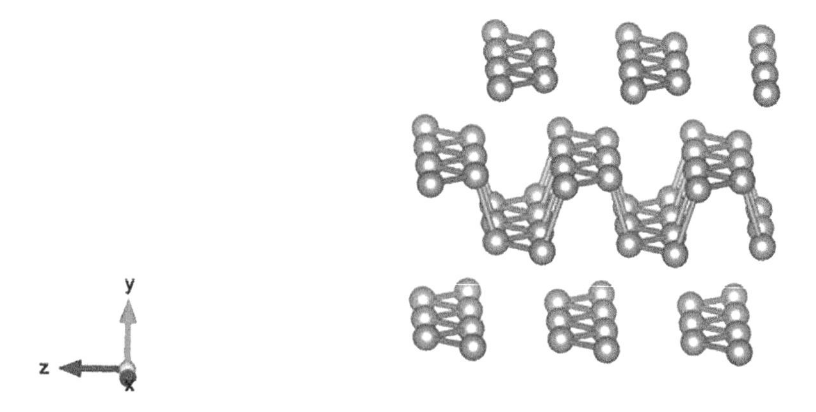

FIGURE 2.4 Crystal structure of black phosphorous.

temperature reaching the maximum value of $S_{max} = 415$ μV K^{-1} at 380 K [35]. The energy gap of bulk black phosphorous is estimated to be 0.32 eV using $S_{max} = E_g/2eT$, where e is the electron charge. Besides, the electrical resistance drops 40% as the temperature increases from room temperature to 385 K. The thermopower of 10–30 nm thick black phosphorous is ca. 400 μV K^{-1} near room temperature [36]. It is also found that 2D Mott's variable range hopping dominates the transport at low temperatures.

Since the lone pair electrons of phosphorous react easily with oxygen in the air, black phosphorus is not stable in ambient conditions. This instability causes changes in composition and physical properties and consequently prevents black phosphorus from TE applications. Mechanically exfoliated black phosphorus is very tiny. Considering the TE applications, liquid-exfoliated polycrystalline black phosphorus materials seem to be more suitable due to their availability of surface modification and chemical doping. Au nanoparticles are used to functionalize the surface of exfoliated black phosphorus via a redox reaction of Au precursors. As a result, a significant enhancement of TE performance and long-term stability in the air is achieved. The electrical conductivity of the Au-decorated black phosphorus increases from 0.001 to 63.3 S cm^{-1} by enhancing the carrier concentration [37]. Consequently, its power factor increases significantly to 0.685 μV cm^{-1} K^{-2}, which is quite small and smaller than graphene; however, is much greater than that of the pristine black phosphorus. A power output of 79.3 nW with $\Delta T = 2$°C is attained using Au-decorated BP as p-legs in a vertical TE generator.

2.5.3 TMDs

Another alternative for potential TE applications is the 2D TMDs with gapped and tunable band structure. The TMDs with the general formula of MX_2 such as M = Mo, W, Ti and X = S, Se, Te have recently attracted extensive interest due to their rich and interesting characteristics in chemistry and physics such as large effective mass, a wide-range bandgap, valley degeneracy, high mobility, chemical stability, excellent mechanical properties, and confinement effects. TMDs are hexagonally packed structure composed of three atomic planes with metals sandwiched between two layers of chalcogens. The metal coordination environment is either octahedral or trigonal prismatic. The nature of interaction between layers is weak van der Waals force, while it has more covalent bonding character between metals and chalcogenides. Among TMDs, TE properties have been investigated for MoS_2, SnS_2, TiS_2, WSe_2, $PtSe_2$, and 1T′-$MoTe_2$.

MoS_2 has a large intrinsic bandgap of 1.8 eV and a large in-plane mobility of 200–500 cm^{2}V^{-1}s^{-1}. TE properties of mechanical-exfoliated and CVD-grown single crystalline MoS_2 monolayers have been investigated. Due to the quantum-well structure, a large thermopower of −5160 μVK^{-1} at a resistivity of 490 Ω m is observed in single-layer MoS_2 by electric field modulation, [38] which is consistent with the Mott's formula for thermopower, i.e., $S \propto \left[\frac{dlnN(E)}{dE}\right]_{E=E_F}$ [39,40]. High power factor of 8,500 μW m^{-1} K^{-2} of bilayer MoS_2 flakes is attained at 300 K in the metallic regime with high carrier concentration. The enhanced power factor comes from the high electrical conductivity and a large thermopower due to its high valley degeneracy of $g_v = 6$ and large effective mass of 0.68 m_0 at conduction band minimum [41].

However, low thermal conductivity is still required to obtain high zT for practical TE applications. The measured thermal conductivity values for the single-layer and multilayer MoS_2 are usually larger than 30 W mK^{-1} [42]. Nevertheless, high power factor itself might find its application in in-plane Peltier cooling [43].

Tin diselenide $SnSe_2$ has a CdI_2-type structure and a bandgap of ~1.6 eV at room temperature. The $SnSe_2$ nanosheets pellets are fabricated by spark plasma sintering and obtain preferential orientation of (00l) facets. Its in-plane power factor at room temperature is enhanced to 8 μW cm^{-1} K^{-2} through simultaneously introducing Se deficiency and chlorine doping. $zT = 0.63$ is attained at 673 K [44]. Intercalation of TiS_2 can be readily achieved to modify the transport properties of the pristine TiS_2. For Cu-intercalated $Cu_{0.02}TiS_2$, $zT \approx 0.45$ at 800 K [45]. A random stacking of WSe_2 film is synthesized by modulated elemental reactants method [46]. An ultra-low cross-plane thermal conductivity of 0.05 W m^{-1} K^{-1} is attained at 300 K. This magnitude is 30-fold smaller than the *c*-axis thermal conductivity of single-crystal WSe_2 and 6-fold smaller than the calculated minimum thermal conductivity for WSe_2. The ultra-low thermal conductivity is attributed to the localization of lattice vibrations induced by the random stacking of 2D crystalline WSe_2 sheets.

Moreover, the field-effect-modulated TE properties of CVD-grown large-area WSe_2 monolayers are investigated by using EDL gating [47]. The charge carrier type of WSe_2 is tuned via electrostatic doping to control the energy band filling. Due to the atomically thin structure, the maximum absolute thermopowers of 380 μVK^{-1} and 250 μVK^{-1} are attained for the *p*-type the *n*-type WSe_2, respectively. The maximum power factor is 3 μWcm^{-1} K^{-2} for the *p* type and 1 μWcm^{-1} K^{-2} for the *n* type. As compared to the bulk sample, the enhancement of TE properties is attributed to the low-dimensional effects.

2.6 SUMMARY

We briefly describe the recent research progress on 2D materials of MXenes, graphene, black phosphorus, TMDs, and their applications in supercapacitors and TEs. With progressive efforts, the reported data have shown excellent properties with the advanced fabrication methods. These 2D materials are promising for the next generation energy applications.

ACKNOWLEDGMENT

This work was supported by the ministry of science and technology of Taiwan (MOST) under grant 107–2112-M-018-006-MY3.

REFERENCES

1. Rashid F.L., Shareef A.S., Alwan H.F. Enhancement of fresh water production in solar still using new phase change materials. *J Adv Res Fluid Mech Therm Sci.* 2019, 61(1), 63–72.
2. Zendehboudi A., Aslfattahi N., Rahman S., Rahman S., Mohd Sabri M.F., Said S.M., Arifutzzaman A., Che Sidik N.A. Optimization of thermal conductivity of NanoPCM-based graphene by response surface methodology. *J Adv Res Fluid Mech Therm Sci.* 2020, 75(3), 108–125.

3. Naguib M., Kurtoglu M., Presser V., Lu J., Niu J., Heon M., Hultman L., Gogotsi Y., Barsoum M.W. Two-dimensional nanocrystals produced by exfoliation of Ti_3AlC_2. *Adv Mater.* 2011, 23(37), 4248–4253.
4. Dinh K.N., Liang Q., Du C.-F., Zhao J., Tok A.I.Y., Mao H., Yan Q. Nanostructured metallic transition metal carbides, nitrides, phosphides, and borides for energy storage and conversion. *Nano Today.* 2019, 25, 99–121.
5. Li Z., Wu Y. 2D early transition metal carbides (MXenes) for catalysis. *Small.* 2019, 15(29), 1804736.
6. Shahzad F., Zaidi S.A., Naqvi R.A. 2D transition metal carbides (MXene) for electrochemical sensing: A review. *Critical Rev Anal Chem.* 2020, 1–17. https://doi.org/10.1080/10408347.2020.1836470
7. Geng D., Cheng Y., Zhang G., Sun X.A. *Layered Materials for Energy Storage and Conversion*, 2019. Royal Society of Chemistry, London.
8. Wang Z., Kochat V., Pandey P., Kashyap S., Chattopadhyay S., Samanta A., Sarkar S., Manimunda P., Zhang X., Asif S. Metal immiscibility route to synthesis of ultrathin carbides, borides, and nitrides. *Adv Mater.* 2017, 29(29), 1700364.
9. Naguib M., Mashtalir O., Carle J., Presser V., Lu J., Hultman L., Gogotsi Y., Barsoum M.W. Two-dimensional transition metal carbides. *ACS Nano.* 2012, 6(2), 1322–1331.
10. Ghidiu M., Lukatskaya M.R., Zhao M.-Q., Gogotsi Y., Barsoum M.W. Conductive two-dimensional titanium carbide 'clay'with high volumetric capacitance. *Nature.* 2014, 516(7529), 78–81.
11. Guan Y., Jiang S., Cong Y., Wang J., Dong Z., Zhang Q., Yuan G., Li Y., Li X. A hydrofluoric acid-free synthesis of 2D vanadium carbide (V2C) MXene for supercapacitor electrodes. *2D Materials.* 2020, 7(2), 025010.
12. Li M., Lu J., Luo K., Li Y., Chang K., Chen K., Zhou J., Rosen J., Hultman L., Eklund P. Element replacement approach by reaction with Lewis acidic molten salts to synthesize nanolaminated MAX phases and MXenes. *J Am Chem Soc.* 2019, 141(11), 4730–4737.
13. Sun W., Shah S., Chen Y., Tan Z., Gao H., Habib T., Radovic M., Green M. Electrochemical etching of Ti_2AlC to Ti_2CT_x (MXene) in low-concentration hydrochloric acid solution. *J Mater Chem A.* 2017, 5(41), 21663–21668.
14. Yang S., Zhang P., Wang F., Ricciardulli A.G., Lohe M.R, Blom P.W, Feng X. Fluoride-free synthesis of two-dimensional titanium carbide (MXene) using a binary aqueous system. *Angew Chem.* 2018, 130(47), 15717–15721.
15. Yang W., Huang B., Li L., Zhang K., Li Y., Huang J., Tang X., Hu T., Yuan K., Chen Y. Covalently sandwiching MXene by conjugated microporous polymers with excellent stability for supercapacitors. *Small Methods.* 2020, 4(10), 2000434.
16. Gupta H., Mothkuri S., McGlynn R., Carolan D., Maguire P., Mariotti D., Jain P., Rao T.N., Padmanabham G., Chakrabarti S. Activated functionalized carbon nanotubes and 2D nanostructured MoS_2 hybrid electrode material for high-performance supercapacitor applications. *Phys Status Solidi A* 2020, 217(10), 1900855.
17. Immanuel P., Senguttuvan G., Chang J., Mohanraj K., Kumar N.S. Effect of Cr doping on Mn_3O_4 thin films for high-performance Supercapacitors. *J Mater Sci: Mater Electron.* 2021, 32(3), 3732–3742.
18. Kumuthini R., Ramachandran R., Therese H., Wang F. Electrochemical properties of electrospun MoS_2@ C nanofiber as electrode material for high-performance supercapacitor application. *J Alloys Compd.* 2017, 705, 624–630.
19. Sari F.N.I., Ting J.-M. Direct growth of MoS_2 nanowalls on carbon nanofibers for use in supercapacitor. *Sci Rep* 2017, 7(1), 5999.
20. Sun T., Liu X., Li Z., Ma L., Wang J., Yang S. Graphene-wrapped CNT@ MoS_2 hierarchical structure: Synthesis, characterization and electrochemical application in supercapacitors. *New J Chem.* 2017, 41(15), 7142–7150.

21. Tiwari P., Jaiswal J., Chandra R. Hierarchal growth of MoS_2@ CNT heterostructure for all solid state symmetric supercapacitor: Insights into the surface science and storage mechanism. *Electrochim Acta.* 2019, 324, 134767.
22. Sun G., Zhang X., Lin R., Yang J., Zhang H., Chen P. Hybrid fibers made of molybdenum disulfide, reduced graphene oxide, and multi-walled carbon nanotubes for solid-state, flexible, asymmetric supercapacitors. *Angew Chem.* 2015, 127(15), 4734–4739.
23. Sun P., Wang R., Wang Q., Wang H., Wang X. Uniform MoS_2 nanolayer with sulfur vacancy on carbon nanotube networks as binder-free electrodes for asymmetrical supercapacitor. *Appl Surf Sci.* 2019, 475, 793–802.
24. Immanuel P., Liang T.-F., Yang T.-S., Wang C.-H., Chen J.-L., Wang Y.-W., Liu C.-J. Enhanced power factor of PEDOT: PSS films post-treated using a combination of ethylene glycol and metal chlorides and temperature dependence of electronic transport (325–450 K). *ACS Appl Energy Mater.* 2020, 3(12), 12447–12459.
25. Sidharth D., Alagar Nedunchezhian A., Akilan R., Srivastava A., Srinivasan B., Immanuel P., Rajkumar R., Yalini Devi N., Arivanandhan M., Liu C.-J., Anbalagan G., Shankar R., Jayavel R. Enhanced thermoelectric performance of band structure engineered $GeSe_{1-x}Te_x$ alloys. *Sustain. Energ Fuels.* 2021, 5(6), 1734–1746.
26. Hicks L.D, Dresselhaus M.S. Effect of quantum-well structures on the thermoelectric figure of merit. *Phys Rev B.* 1993, 47(19), 12727.
27. Juntunen T., Jussila H., Ruoho M., Liu S., Hu G., Albrow-Owen T., Ng L.W., Howe R.C., Hasan T., Sun Z. Inkjet printed large-area flexible few-layer graphene thermoelectrics. *Adv Funct Mater.* 2018, 28(22), 1800480.
28. Novoselov K.S., Geim A.K., Morozov S.V., Jiang D., Zhang Y., Dubonos S.V., Grigorieva I.V., Firsov A.A. Electric field effect in atomically thin carbon films. *Science.* 2004, 306(5696), 666–669.
29. Neto A.C., Guinea F., Peres N.M., Novoselov, K.S., Geim A.K. The electronic properties of graphene. *Rev Mod Phys.* 2009, 81(1), 109.
30. Xiao N., Dong X., Song L, Liu D, Tay Y, Wu S, Li L-J, Zhao Y, Yu T, Zhang H. Enhanced thermopower of graphene films with oxygen plasma treatment. *ACS Nano.* 2011, 5(4), 2749–2755.
31. Sim D., Liu D., Dong X., Xiao N., Li S., Zhao Y., Li L.-J., Yan Q., Hng H.H. Power factor enhancement for few-layered graphene films by molecular attachments. *J Phys Chem C.* 2011, 115(5), 1780–1785.
32. Kanahashi K., Ishihara M., Hasegawa M., Ohta H., Takenobu T. Giant power factors in p- and n-type large-area graphene films on a flexible plastic substrate. *2D Mater Appl.* 2019, 3(1), 44.
33. Castellanos-Gomez A. Black phosphorus: Narrow gap, wide applications. *J Phys Chem Lett.* 2015, 6(21), 4280–4291.
34. Li L., Yu Y., Ye G.J., Ge Q., Ou X., Wu H., Feng D., Chen X. H., Zhang Y. Black phosphorus field-effect transistors. *Nat. Nanotechnol.* 2014, 9(5), 372–377.
35. Flores E., Ares J.R., Castellanos-Gomez A., Barawi M., Ferrer I.J., Sánchez C. Thermoelectric power of bulk black-phosphorus. *Appl Phys Lett.* 2015, 106(2), 022102.
36. Choi S.J., Kim B.-K., Lee T.-H., Kim Y.H., Li Z., Pop E., Kim J.-J., Song J.H., Bae M.-H. Electrical and thermoelectric transport by variable range hopping in thin black phosphorus devices. *Nano Lett.* 2016, 16(7), 3969–3975.
37. An C.J., Kang Y.H., Lee C., Cho S.Y. Preparation of highly stable black phosphorus by gold decoration for high-performance thermoelectric generators. *Adv Funct Mater.* 2018, 28(28), 1800532.
38. Dobusch L., Furchi M.M., Pospischil A., Mueller T., Bertagnolli E., Lugstein A. Electric field modulation of thermovoltage in single-layer MoS_2. *Appl Phys Lett.* 2014, 105(25), 253103.

39. Mott N.F., Davis E.A., *Electronic Processes in Non-crystalline Materials*, 1979. Clarendon Press, Bristol.
40. Wu C.-A., Chang K.-C., Lin F.-H., Yang, Z.-R., Gharleghi A., Wei T.-Z., Liu C.-J. Low thermal conductivity and enhanced zT values of porous and nanostructured $Cu_{1-x}Ni_x$ alloys. *Chem Eng Sci*. 2019, 368, 409–416.
41. Hippalgaonkar K., Wang Y., Ye Y., Qiu D.Y., Zhu H., Wang Y., Moore J., Louie S.G., Zhang X. High thermoelectric power factor in two-dimensional crystals of MoS_2. *Phys Rev B*. 2017, 95(11), 115407.
42. Gu X., Yang R. Phonon transport in single-layer transition metal dichalcogenides: A first-principles study. *Appl Phys Lett*. 2014, 105(13), 131903.
43. Sinha S., Goodson K.E. Multiscale thermal modeling in nanoelectronics. *Int J Multiscale Comput Eng*. 2005, 3(1), 107–133.
44. Luo Y., Zheng Y., Luo Z., Hao S., Du C., Liang Q., Li Z., Khor K.A., Hippalgaonkar K., Xu J. n-type $SnSe_2$ oriented-nanoplate-based pellets for high thermoelectric performance. *Adv Energy Mater*. 2018, 8(8), 1702167.
45. Guilmeau E., Bréard Y., Maignan A. Transport and thermoelectric properties in Copper intercalated TiS_2 chalcogenide. *Appl Phys Lett*. 2011, 99(5), 052107.
46. Chiritescu C., Cahill D.G., Nguyen N., Johnson D., Bodapati A., Keblinski P., Zschack P. Ultralow thermal conductivity in disordered, layered WSe_2 crystals. *Science*. 2007, 315(5810), 351–353.
47. Pu J., Kanahashi K., Cuong N.T., Chen C.-H., Li L.-J., Okada S., Ohta H., Takenobu T. Enhanced thermoelectric power in two-dimensional transition metal dichalcogenide monolayers. *Phys Rev B*. 2016, 94(1), 014312.

3 Top-Down Synthesis of 2D Nanomaterials

Arpana Agrawal
Shri Neelkantheshwar Government Post-Graduate College

CONTENTS

3.1 INTRODUCTION

Two-dimensional (2D) nanomaterials including graphene, hexagonal boron nitride (h-BN), transition metal dichalcogenides (TMDs) such as MoS_2, $MoSe_2$, WS_2, WSe_2, TaS_2, and $TaSe_2$ are one of the emerging classes of nanomaterials which exhibits ultra-thin sheet-like characteristics whose size may vary from 1 nanometer (nm) to few micrometers. In bulk form, these materials demonstrate insulating behavior and if it is possible to reduce the dimensionality somehow either by fluorination, hydrogenation, oxygen functionalization, and switch those from bulk form to 2D nanosheets (NSs), nanotubes (NTs), or nanoribbons (NRs), then the functionalities of these materials get significantly enhanced. They generally show intriguing edge magnetism, encouraging electrically conductive behavior, excellent optical properties, and many more.

Remarkable progress in the field of 2D nanomaterials was marked after 2004 when 2D counterpart of carbon, i.e., graphene was mechanically exfoliated from graphite using a scotch tape, for which the 2009 Nobel prize in Physics has also been awarded [1]. Graphene is a 2D allotrope of carbon which is one atomic thick layer having a thickness of 0.34 nm with sp^2 hybridized carbon atoms that are held together in a hexagonal closed pack. One atomic thick layer of graphene (zero bandgap material) can transmit 97.7% of light and hence possess outstanding transparency. Hence, knowing the transmittance of the graphene layer, one can estimate the number of graphene layers. However, a more accurate determination of the number of layers can be inferred from the Raman spectra of graphene. Apart from graphene, other graphene-like 2D nanomaterials including h-BN, TMDs, etc. are also engendered

DOI: 10.1201/9781003178453-3

great scientific research interest from a fundamental to a technological viewpoint. BN is isostructural and isoelectronic to carbon having an equivalent composition of nitrogen and boron atoms which are bonded together with a covalent bond within each layer and a single h-BN layer is analogous to graphene and known as white graphene [2]. Also, TMDs are chemical compounds with formula MX_2 (M = transition metal ion; X = Chalcogen), where each TMD monolayer is made up of one transition metal layer sandwiched between two chalcogen layers (X-M-X) [3]. However, the stacked layers of all these 2D nanomaterials are held together with weak van der Waal force. Such materials are potentially important candidates for several attractive device applications, e.g., flexible electronics, carrier (hole/electron) transport layer in photovoltaics/solar cells, transparent conducting electrodes, sensor applications, and optoelectronic applications [4–10].

It is worth mentioning here that the synthesis methods of these nanomaterials are one of the decisive factors for their utility in various specific device applications. The technological importance of these nanomaterials significantly encourages researchers to work for the development of efficient synthesis approaches for the mass production of high-quality ultrathin sheets of 2D nanomaterials. Several well-established synthesis approaches for the growth of these nanomaterials include mechanical exfoliation/cleavage, ball milling, electrochemical exfoliation (ECE), sonication-assisted liquid-phase exfoliation (LPE), and unzipping of the corresponding NTs. However, each of these techniques has its synthesis conditions and requirement and hence the benefits and limitations. Accordingly, the present chapter presents an inclusive overview of various synthesis approaches particularly top-down methods for the growth of high-quality 2D nanomaterials along with the recent studies being carried out on 2D nanomaterials.

3.2 TOP-DOWN SYNTHESIS OF 2D NANOMATERIALS

It is worth stressing that the properties of a material can be effectively altered by reducing the dimensions of material. Among the reduced dimensional materials, 2D nanomaterials have attracted enormous research interest in today's technological era. Additionally, the properties of such 2D nanomaterials can be further manipulated physically and/or chemically. The basic idea behind the top-down approach for the synthesis of 2D nanomaterials is the breakup of bulk graphite materials to produce nanosized graphene layers. The layered nature of the 2D nanomaterials facilitates them to be easily exfoliated into thin 2D NSs via several top-down-based exfoliation methods including scotch tape-based mechanical exfoliation/cleavage [1,3,11], ball milling [12–14], sonication-assisted LPE [15,16], ECE [17–19], unzipping of corresponding NTs via various routes such as chemical etching or ion intercalation [2,20], etc.

3.2.1 Scotch-Tape-Based Micromechanical Cleavage

Scotch tape-based mechanical exfoliation/cleavage is the very first approach that was developed by Geim and Novoselov in 2004 to isolate the 2D graphene layer from bulk graphite [1]. They applied an adhesive tape onto the graphite and then peeled it off from its surface so that an atomically thin 2D graphene layer has been

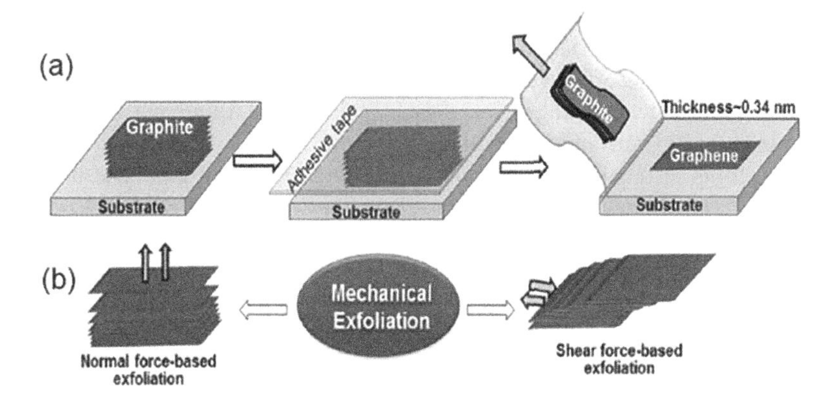

FIGURE 3.1 (a) Scotch tape-based mechanical exfoliation of graphene layer from bulk graphite. (b) Normal-force and shear-force-based mechanical exfoliation for 2D nanomaterials. Adapted with permission from [21]. Copyright (2020) Elsevier.

obtained. Figure 3.1a schematically illustrates the exfoliation of ultrathin atomic-layered graphene having a thickness of 0.34 nm from graphite employing an adhesive tape. Since then, it has become the most commonly used and simplest approach that has been adopted for other 2D materials as well. Li et al. [3] have reported the growth and characterization of various TMDs including WSe_2, TaS_2, and $TaSe_2$ on Si/SiO_2 substrate using scotch tape-based mechanical exfoliation/cleavage approach. They have isolated the TMD NSs from the respective bulk single crystals and deposited them onto 90 nm/300 nm Si substrate coated with SiO_2 oxide layer. The same mechanical cleavage method has also been extended to synthesize atomically thin h-BN NSs by Alem et al. [11]. However, the isolation of monolayer h-BN is quite difficult due to the strong interplanar bonding in h-BN. Accordingly, Alem et al. [11] have adopted a mechanical exfoliation followed by a reactive ion etching process for the growth of any desired number of h-BN NSs. Depending upon the direction of the force applied on the bulk material for their exfoliation to synthesize 2D nanomaterials, the mechanical exfoliation/cleavage process can be either normal-force- or shear-force-based exfoliation and is schematically shown in Figure 3.1b.

3.2.2 Ball Milling Exfoliation

The ball milling exfoliation method is based on the shear force that was applied by the metallic balls on the bulk crystal/material, loaded together in a rotating steel chamber, and milled under suitable conditions. As a consequence of the impact of the metallic balls, the particle size reduces continuously leading to the formation of exfoliated 2D-layered materials. For the synthesis of graphene, h-BN and TMDs, bulk graphite, bulk h-BN crystals, and bulk TMDs, respectively, are employed as precursors. This shear force deteriorates the interlayer van der Waal's attraction leading to rupturing of bulk materials and hence results in 2D nanomaterials. Several milling agents and solvents/surfactants are reported to be utilized as the milling agent including salts (NaCl) [22] or melamine (Na_2SO_4) [23], kerosene [24],

N-methyl-2-pyrrolidone (NMP) [14], and di-methyl-formamide (DMF) [25] for the synthesis of 2D nanomaterials. Apart from this, one has to take care of the several milling parameters for effective ball milling exfoliation which includes the milling revolution per minute, grinding material such as size, static filling rate, etc., graphite powder to ball ratio, milling time, ball diameter, etc.

Ball milling exfoliation of graphite to prepared exfoliated graphene has been reported by several groups [22–25]. A correlation between the defect density in ball-milled graphite and the total oxygen content of graphene oxide produced by oxidizing ball-milled graphite has been reported by Mohanta et al. [12]. Xing et al. [26] have demonstrated the growth of 2D nanomaterial mainly, graphene, h-BN, and MoS_2 via ball milling exfoliation of their bulk crystals in the presence of two different gases (argon and ammonia) individually for various milling times. The bulk crystals were milled together with four steel balls having a diameter of 2.5 cm in the presence of argon (Ar) and ammonia (NH_3) gases in a milling jar rotating at a speed of 150 rpm for different milling times. Figures 3.2a–f represent the x-ray diffraction (XRD) profiles of graphite, h-BN, and MoS_2 ball milled for different times in Ar and NH_3 gases, respectively. The XRD peaks in Figures 3.2a and b corresponding to (+), (O), and (*) signs show the peaks of graphite, stainless steel, and hardened steel, respectively, and evidence shows the different crystal structures of the grown 2D nanomaterials. As compared to the ball milling results in NH_3 gas, (002) diffraction peak shows broadening in the case of Ar gas.

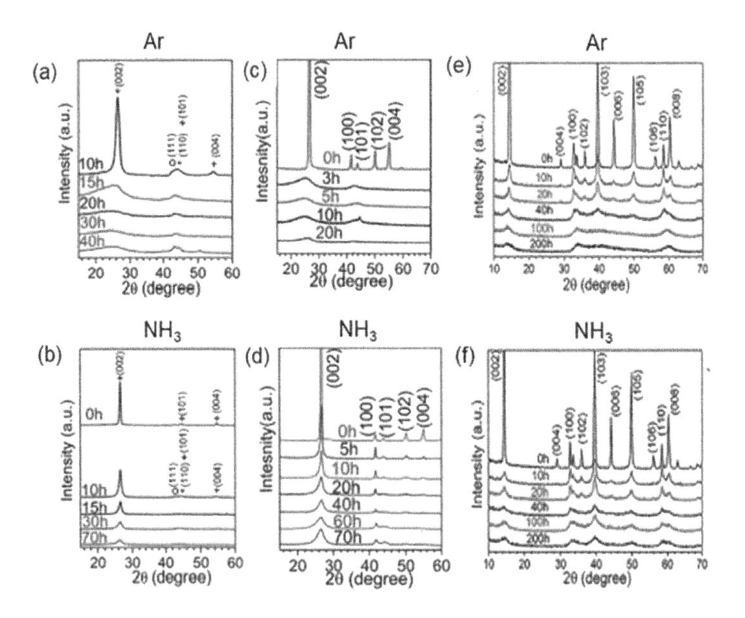

FIGURE 3.2 XRD profiles graphite (a, b), h-BN (c, d), and MoS_2 (e, f) in the presence of Ar and NH_3 gases for various milling times, respectively. Adapted with permission from [26]. Copyright (2016) The Authors, some rights reserved; exclusive licensee [Springer Nature]. Distributed under a Creative Commons Attribution License 4.0 (CC BY) https://creativecommons.org/licenses/by/4.0/.

Recently, xylitol-assisted planetary ball milling exfoliation of h-BN NSs was reported by Wang et al. [27], where BN powder and xylitol crystals were loaded in a 500 ml agate tank rotating at a speed of 500 rpm for 12 hours. The obtained product contains residual xylitol, which can be washed away using simple washing by distilled water followed by sonication of collect BN paste dispersed in 280 mL DI water for 30 minutes. Finally, the residual h-BN was collected by centrifugation method carried out at 200 rpm for 30 minutes giving a high production yield (80.4 wt.%) of h-BN. Large-scale mechanical peeling via low energy ball milling process for the production of BN NSs has been reported by Li et al. [28]. They have manipulated the ball milling conditions to obtain the high yield of quality NSs of BN with enhanced efficiency. Herein, a planetary mill containing 0.5 g h-BN, 6 mL benzyl benzoate (effective lubricant), and 50 Steel balls (12.7 mm diameter) were milled together in a steel chamber, which was rotated at speed of 150 rpm due to which a shear force induces, resulting in exfoliated h-BN NSs. Finally, the milled sample was sonicated for 0.5 hours and centrifuge to remove unexfoliated BN particles. The same group has also studied the photoluminescence of BN NSs produced using the same approach except the milling time was increased to 15 hours with 50 steel balls of the same diameter. A vortex fluidic device was also discussed by Chen et al. [29], for the exfoliation of h-BN suspension in NMP solvent within an inclined glass tube rotating at a speed of 7,000 rpm for 30 minutes. A thin layer of NMP forms at the inner wall of the tube and rotating shaft to provide sufficient shear force to weaken the interlayer van der Waal force leading to the formation of h-BN NSs. Small lateral-sized, few-layered BN NSs have also been reported to be prepared using the wet ball milling process [13].

The ball milling process has been further extended to synthesize 2D TMDs. Dai et al. [30] have reported the synthesis of exfoliated WS_2 by ball milling of WS_2 powder with agate balls (90 g) using sodium dodecyl sulfate solution as a surfactant. The mixture was milled for a constant time of 12 hours with varying rotation speeds of 100, 200, and 250 rpm. All the obtained suspensions were then sonicated and centrifuged at 8,000 rpm. The residual surfactant was then washed away by using ethyl alcohol and deionized water followed by drying at 60°C. Coupled ultrasonication milling process was proposed by Dong et al. [14] to produce high yield MoS_2 NSs via direct exfoliation of naturally occurring molybdenite powders in NMP with polyvinylpyrrolidone molecules.

3.2.3 Sonication-Driven Liquid Phase Exfoliation

The LPE method involves the ultrasonication of bulk material (either in the form of flakes or powders) in a suitable solvent under suitable conditions. Surface tension (ST) matching between the electrolyte and the 2D nanomaterial that is to be obtained is the foremost requirement of the solvent utilized for such exfoliations. In general, the ST of 2D materials including graphene, h-BN, and TMDs lies in between 40 and 65 mJ m^{-2} and hence the solvents that are commonly used for exfoliation of such nanomaterials include NMP (boiling point (BP) = 203°C; ST = 40 mJ m^{-2}), *N, N* dimethylacetamide (BP = 165°C; ST = 36.7 mJ m^{-2}), ortho-dichlorobenzene (ODCB) (BP = 181°C; ST = 37 mJ m^{-2}), *N, N* dimethylformamide (BP = 154°C;

ST = 37.1 mJ m^{-2}), Y-butyroloctane (BP = 204°C; ST = 46.5 mJ m^{-2}), dimethylsulfoxide (BP = 189°C; ST = 42.9 mJ m^{-2}), and acetonitrile. Such ST matched solvents relax the exfoliation process because of the minimum mixing enthalpy. However, the high BP of such solvents needs an additional purification process to remove the solvent after 2D nanomaterial's production as low BP solvents facilitate the easy purification process. Apart from the proper choice of solvent, several sonication parameters such as temperature, sonication time, and intensity are also very crucial for effective exfoliation because they offer suitable cavitational effects for producing defect-free 2D nanomaterials.

Sonication-assisted LPE was first employed by Coleman et al. [31] for the production of 2D graphene layers via the exfoliation of graphite where they have dispersed graphite powder/flakes in NMP solvent, which undergoes sonication and is finally purified using the centrifugation process. Efficient exfoliation of graphite in a mixture of *n*-octylbenzene (NOTBZ) and organic solvents, e.g., NMP or ODCB has also been demonstrated by Haar et al. [15], where NOTBZ acts as a graphene-dispersion stabilizing agent. The graphite flakes were separately sonicated in the solvent mixtures (NMP and NOTBZ; ODCB and NOTBZ). Initially, the dispersion of graphite flakes in NMP and ODCB was prepared via sonication at 600 W and 40°C in an ultrasonic bath for 6 hours. After this, various amounts of NOTBZ were added in the prepared dispersions and again sonicated to form a homogenous gray liquid, followed by centrifugation at 10,000 rpm for 30 minutes to eliminate unexfoliated graphite flakes. Figures 3.3a and b show the photographs of graphene dispersion obtained by graphite exfoliation in NMP and ODCB with different volume percentage of NOTBZ, respectively. The variation of percentage of graphene concentration dispersed in NMP and ODCB corresponding to various volume percentages of NOTBZ has been shown in Figures 3.3c and d. Figure 3.3e illustrates the Raman spectra of prepared dispersion after adding NOTBZ along with reference graphite, which shows the growth of single- and few-layer graphene formation. Figure 3.3f statistically represents the thickness distribution of graphite flakes in NMP and ODCB solvents before and after adding NOTBZ. High-resolution transmission electron microscope (HR-TEM) and TEM images for the monolayer graphene have also been shown in Figures 3.3g and h, respectively.

Growth of poly (3-hexalthiophene) (P3HT)/graphene composite in toluene has also been reported [32]. Graphene is dispersed in P3HT via ultrasonication process in toluene solvent at a temperature of 80°C for 30 minutes. To remove the unexfoliated or thick graphite flakes, the prepared dispersion then undergoes centrifugation process and is finally collected as graphene dispersion. This dispersion was then filtered using a filter paper (pore size ~ 4 μm) followed by drying at 100°C for 5 minutes to prepare the P3HT/graphene composite film paper. Efficient sonication-assisted LPE of h-BN in surface energy matched *n* poly(vinyl alcohol) and epoxy matrices are reported by Song et al. [33]. A mixed solvent approach has also been demonstrated for the LPE of h-BN by Cao et al. [16] who have illustrated a sonication-assisted LPE for the growth of 2D h-BN NSs in a solution of isopropylalcohol (IPA) (44–65 mJ m^{-2}) and NH_3 (NH_3:IPA = 4:1; 2:3; 3:2; 1:4; 1:1). Commercial bulk h-BN undergoes sonication for different sonication times (1, 5, 10, 30, 35, 40, 50, 65 hours) where the high yield was obtained for the optimized time of 35 hours and then the dispersion was configured.

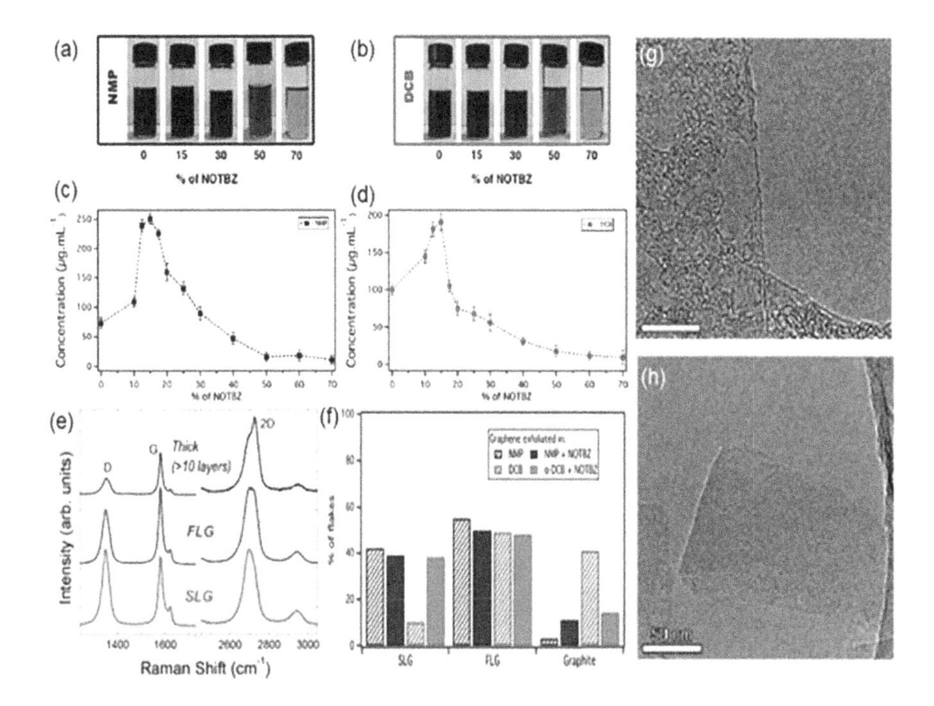

FIGURE 3.3 Photographs of graphene dispersions obtained from graphite exfoliation in NMP (a) and ODCB (b) with various volume percentages of NOTBZ. Average concentration of graphene dispersed in NMP (c) and ODCB (d) with various volume percentage of NOTBZ. (e) Raman spectra of prepared dispersion after adding NOTBZ. (f) Statistical variation of thickness distribution of prepared dispersion before and after adding NOTBZ. (g) and (h) HRTEM and TEM images of the prepared monolayer, respectively. Adapted with permission from [15]. Copyright (2015) The Authors, some rights reserved; exclusive licensee [Springer Nature]. Distributed under a Creative Commons Attribution License 4.0 (CC BY) https://creativecommons.org/licenses/by/4.0/.

Efficient exfoliation of bulk h-BN powder in thionyl chloride ($SoCl_2$) to synthesize ultrathin h-BN NSs with a very high production yield (20 wt.%) has also been reported by Sun et al. [34], without utilizing any dispersion agent. A mixture of h-BN powder and $SoCl_2$ was sonicated for an optimized sonication time of 20 hours and then the collected product undergoes centrifugation process for 5 minutes at a speed of 2,000–25,000 rpm. The obtained product was then vacuum dried at 70°C for 24 hours.

LPE synthesis of WS_2 NSs has also been reported by Han et al. [35], via direct dispersion of WS_2 powder in DMF (exfoliation solvent) followed by ultrasonication process. In this technique, bulk WS_2 powder was dispersed in a 50 mL beaker containing 30 mL DMF solution, which was then undergoing sonication for 5 hours. The resultant dispersion was left undisturbed for 72 hours. To remove the unexfoliated WS_2 (if any), the obtained product was then finally centrifuged for 45 minutes at a speed of 1,500 rpm. Adilbekova et al. [5] have demonstrated the LPE of bulk MoS_2 and WS_2 powder in aqueous NH_3 (NH_3(aq.)), which can produce a high concentration (0.5–1.0 mg mL^{-1}) of MoS_2 and WS_2 NSs. It should be noted that the ST of NH_3(aq.)

matches well with that of TMDs and hence allows efficient exfoliation of TMDs in this solvent; however, it is less environmentally friendly. The powder was sonicated in 10 mL NH_3(aq.) solution for 3 hours in a water bath cooling system to maintain a constant temperature of 5°C. The resultant dispersions were then centrifuged at different rotation speeds varying from 3,000 to 8,000 rpm.

Epigallocatechin-assisted LPE of MoS_2 and WS_2 to produce 2D MoS_2 and WS_2 NSs has been illustrated by Zhao et al. [8]. Sonication (950 W) was done for 10–720 minutes in a water-cooled bath followed by high-speed centrifugation at 10,000 rpm for 30 minutes. NSs of MoS_2 (E-MoS_2) and WS_2 (E-WS_2) were prepared in 2 mg mL^{-1} EGCG solution for a constant sonication time of 12 hours. SEM images for bulk MoS_2, E-MoS_2 NSs, and E-WS_2 NS under different magnifications are shown in Figures 3.4(a, a'), (b, b'), (c, c'), respectively. Figures 3.4(d, d') and (e, e') represent the TEM micrographs (different magnification) for MoS_2 NSs and E-WS_2 NS, respectively, while the atomic force microscope (AFM) images of MoS_2 NSs and E-WS_2 NS are illustrated in Figures 3.4f and g, respectively.

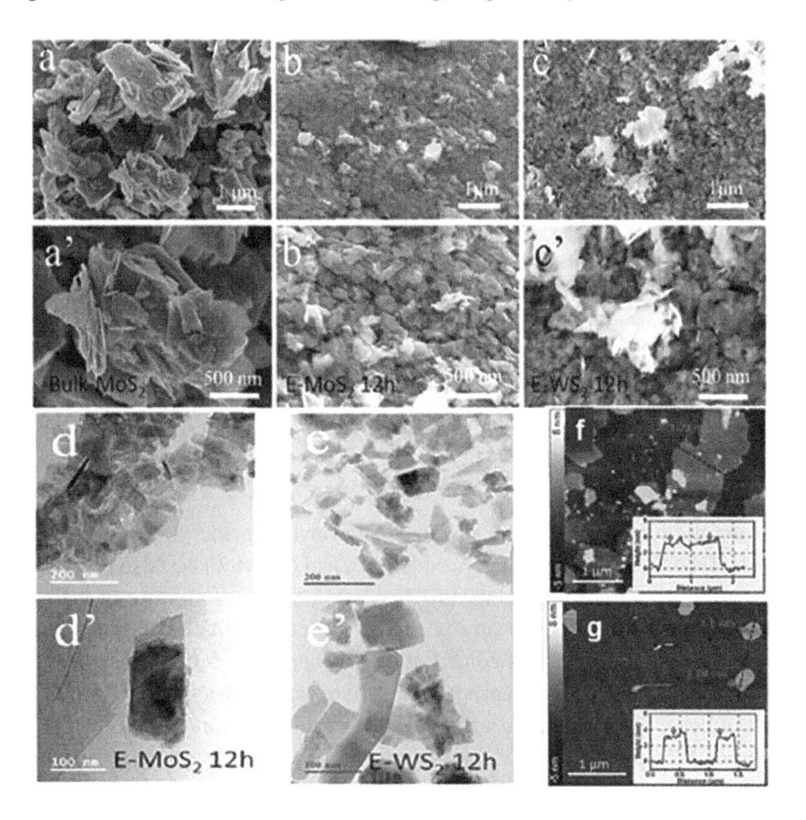

FIGURE 3.4 (a, a'), (b, b'), (c, c') SEM images with different magnification for bulk MoS_2, E-MoS_2 NSs, and E-WS_2 NSs, respectively. (d, d') and (e, e') TEM images different magnification for MoS_2 NSs and E-WS_2 NSs, respectively. (f) and (g) AFM images of MoS_2 NSs and E-WS_2 NSs, respectively. E-MoS_2 and E-WS_2 NSs are prepared in 2 mg mL^{-1} EGCG solution for a constant sonication time of 12 hours. Adapted with permission from [8]. Copyright (2019) Elsevier.

An atomically thin layer of $MoSe_2$ was also reported to be synthesized by LPE carried out in deionized water [36]. Sonication of bulk $MoSe_2$ dispersed in deionized water was performed at 50°C for 24 hours. The collected dark green suspension was centrifuged at a speed of 8,000 rpm for 20 minutes. Pan et al. [37] have also utilized water as an exfoliation solvent for the synthesis of exfoliated 2D WS_2 NSs. Grayfer et al. [38] have presented a compilation of several reports for the exfoliation of MoS_2, $MoSe_2$, and WS_2 powders in colloidal solution to produce the colloidal dispersion solutions containing their respective 2D NSs. Growth of 2D WSe_2 NSs was also reported via LPE of bulk WSe_2 in a mixture of water and *n*-propanol (1:1 v/v) [39].

3.2.4 Electrochemical Exfoliation Approach

ECE is an important technique that ensures the high-quality production of 2D nanomaterials with a high yield. Generally, in this technique, the 2D material that is to be obtained is in the form of an electrode named a working electrode and immersed in an electrolyte containing a counter electrode, which is made up of platinum (Pt) wire/plate. A suitable potential is applied to the working electrode, which facilitates its exfoliation leading to the formation of exfoliated 2D material. The basic idea behind this exfoliation is the intercalation of ionic species present in the electrolyte in between the layers of bulk material (working electrode), which weakens the van der Waals force between the layers resulting in its expansion and hence gets exfoliated. So far, several electrolytes are being utilized for this purpose. Parvez et al. [40] have presented a compilation of several electrolytes such as $(NH_4)_2SO_4$, K_2SO_4, and Na_2SO_4 for the exfoliation of graphite to produce 2D graphene layers. Synthesis of graphene layers has also been reported by Rao et al. [17]. They have employed an anodic ECE approach to synthesize graphene layers where graphite and Pt electrodes were dipped in sodium hydroxide/hydrogen peroxide/water ($NaOH/H_2O_2/H_2O$) electrolyte. The same group has also utilized organic compounds in electrolyte namely glycine (Gly). HSO_4 ionic complex significantly improves the exfoliation efficiency by facilitating the intercalation of SO_4 and HSO ions into the graphite electrode [41]. The ECE method of compressed graphite has also been employed for the production of high-quality graphene NSs by Achee et al. [42].

You et al. [43] have discussed a facile ECE route for large-scale production of high-quality MoS_2 NSs. They have employed a Pt wire and MoS_2 bulk crystal acting as counter and working electrodes, respectively, dipped in an aqueous solution of H_2SO_4 serving as an electrolyte. Initially, a static voltage of +1 V is applied to the working electrode, which was later increased to +10 V and kept constant for 30 minutes till the exfoliation process gets completed. Figure 3.5a schematically shows the experimental setup for the ECE of MoS_2. The basic idea behind this exfoliation is the intercalation of the anionic $SO_4{}^{2-}$ and OH^- species into the bulk MoS_2 crystal to produce SO_2 or O_2 gas bubbles, which generate a force for MoS_2 to overcome the weak van der Waal force acting between the MoS_2 layers leading to its expansion and finally the exfoliation to MoS_2 NSs. The mechanism of the overall ECE of MoS_2 via ion insertion has been depicted in Figure 3.5b. The exfoliated MoS_2 NSs were then coated on Si substrate covered with 300 nm thick SiO_2 by simply dipping the substrate into prepared MoS_2 dispersion. Figure 3.5c illustrates the AFM

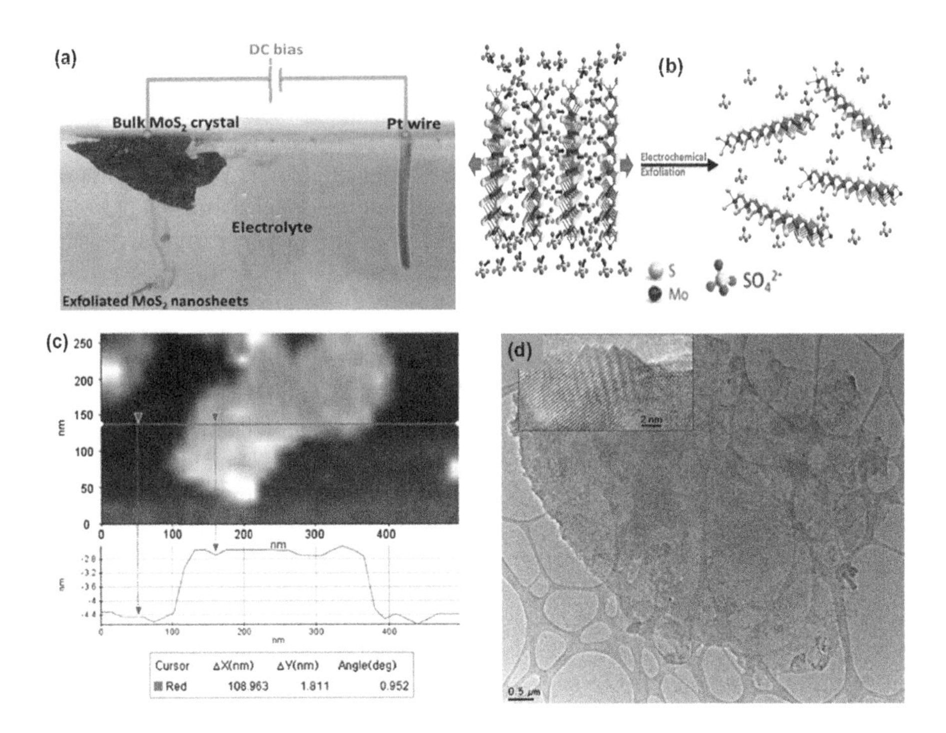

FIGURE 3.5 (a) Experiment setup for the ECE of MoS_2; (b) Schematic diagram illustrating the mechanism of ECE of MoS_2 using ions insertion. (c) Atomic force microscope image of MoS_2 NSs/Si/SiO_2 (300 nm). (d) TEM image of exfoliated MoS_2 NSs on a carbon grid. Inset shows the higher resolution TEM image. Adapted with permission from [43]. Copyright (2014) Elsevier.

image of exfoliated MoS_2 NSs on Si/SiO_2 substrate, which reveals the average topographic height of 1.8 nm for the exfoliated MoS_2 NSs deposited on Si/SiO_2 substrate. A high degree of crystallinity was confirmed from the TEM image of the exfoliated MoS_2 NSs as shown in Figure 3.5d; inset shows the corresponding higher resolution TEM image.

A very similar approach for the growth of MoS_2 NSs via ECE has also been demonstrated by Liu et al. [18]. They have also reported the growth of large lateral size (50 μm) atomically thin MoS_2 NSs by exfoliation of bulk MoS_2 crystal using Pt wire (counter electrode) and bulk MoS_2 as working electrode loaded in 0.5 M Na_2SO_4 electrolytic solution. At the initial stages, a constant potential of +2 V is applied for 10 minutes to the bulk MoS_2 crystal, which was then increased to +10 V for a period of 0.5–2 hours for the exfoliation of bulk MoS_2 crystal to synthesize 2D MoS_2 NSs. The vacuum-filtered exfoliated MoS_2 was then dispersed in NMP solvent so that a uniform MoS_2 NS solution can be obtained with a yield of 5%–9%. The overall mechanism responsible for such an exfoliation lies in the production of OH and O radicals and/or SO_4^{2-} anions, which insert into the bulk MoS_2 crystal leading to a lowering in the van der Waal interaction of MoS_2 interlayers. As a result, expansion of MoS_2 interlayer takes place causing the detachment of MoS_2 flakes from bulk MoS_2 crystal.

This method has also been utilized for the 2D growth of h-BN NSs. Wang et al. [19] have reported ECE of bulk h-BN by applying a constant +10 V DC potential to two platinum driving electrodes dipped in bulk h-BN dispersed in sodium sulfate, which acts as h-BN source, supports electrolyte, respectively.

3.2.5 Unzipping of the NTs

Unzipping of NTs (either single-walled or multi-walled) is one of the fascinating techniques to synthesize 2D nanomaterials. Graphene, h-BN, or TMDs layers can be prepared by the unwrapping of their corresponding NTs, e.g., unzipping of carbon NTs (CNTs) gives graphene and unrolling of boron nitride NTs produces boron nitride NS. Unrolling of CNTs has been reported by several authors [44–46] and warrants the dissociation of C-C bonds either in the axial direction or in the longitudinal direction and hence requires a high level of strain. Hence several routes are employed for this purpose, e.g., unzipping can be initiated via metal intercalation, chemical, and plasma etching treatments. In plasma etching treatment, a polymer/CNT matrix is formed by embedding the CNTs in a suitable polymer. Generally, polymethyl-methacrylate (PMMA) is employed for this purpose. Once the polymer is formed, they are deposited on some substrate and then peeled off using KOH solution and finally undergoes argon treatment. On the other hand, chemical unrolling involves the treatment of CNTs with H_2SO_4 followed by oxidation using $KMnO_4$ to split the C-C bonds in axial/longitudinal directions. Metal intercalation is one of the best approaches to rupture the NT walls where the foreign atoms/ions/molecules are forced to intercalate within the walls of the NTs to produce enough stress leading to the formation of 2D nanomaterials.

Li et al. [44] have reported an effective intertube and intratube intercalation-assisted longitudinal unzipping of multi-walled carbon NTs (MWCNTs) for producing graphene NRs. They have treated the MWCNTs with 1 g KNO_3 and varying concentrations of H_2SO_4 with continuous magnetic stirring at 300 rpm for 2 hours, which allows the intertube and intratube intercalation of K^+, NO^{3-}, and $SO_4^{\;2-}$ ions within the bundled CNTs. After this, 0.5 g $KMnO_4$ (oxidant) is added slowly with further stirring for the next 2 hours. This initiates longitudinal unzipping by weakening the interlayer van der Waals attraction giving rise to graphene layers.

Unzipping process has been further extended for the growth of 2D TMDs (MX_2) layered nanomaterials. Nethravati et al. [20] have reported the chemical unzipping of WS_2 NT to synthesize 2D WS_2 NRs. The WS_2 NTs employed were 10 μm long and 20–80 nm wide. The overall process involves two steps: (a) Lithiation of WS_2 NTs to form $LixWS_2$ –NTs and (b) unzipping of $LixWS_2$ –NTs to form WS_2 NRs. Once the $LixWS_2$ –NTs were prepared, they are then dispersed in solvents such as water, methanol, ethanol, octanethiol, and dodecanethiol and undergo sonication for 1 minute to form a stable colloidal dispersion. The chemical reaction process of lithiated $LixWS_2$ –NTs in water and ethanol leads to the formation of WS_2 NRs along with the liberation of H_2 gas and can be expressed by Eqs. (3.1) and (3.2) [20].

$$Li_xWS_2 + xH_2O \rightarrow WS_2 + xLiOH + x/2H_2 \tag{3.1}$$

$$Li_xWS_2 + xC_2H_5OH \rightarrow WS_2 + xC_2H_5OLi + x/2H_2 \tag{3.2}$$

The dispersion was then coagulated by adding acetone followed by filtration and vacuum drying. To remove the residual Li salts, the obtained mixture was repeatedly washed using a mixture of ethanol and water to obtain high-quality WS_2 NRs. Kvashnin et al. [47] have theoretically analyzed the experimental results obtained by Nethravati et al. [20] on the unzipping of multiwalled WS_2 NTs via Li intercalation using ab-initio calculations.

Recently, Sasikala et al. [48] have also demonstrated the longitudinal unzipping of 2D TMDs via two-step process. The first step involves the intercalation of Li^+ ions into bulk MX_2 to produce Li_xMX_2, followed by the second process of obtaining a stable suspension via ultrasonication in oxygenated water due to which Li_xMX_2 experiences instantaneous longitudinal unzipping.

Figures 3.6a and b depict the SEM images with (scale bars: 0.5 μm (a) and 0.1 μm (b)) displaying the initiation of unzipping after 5 minutes of ultrasonication from

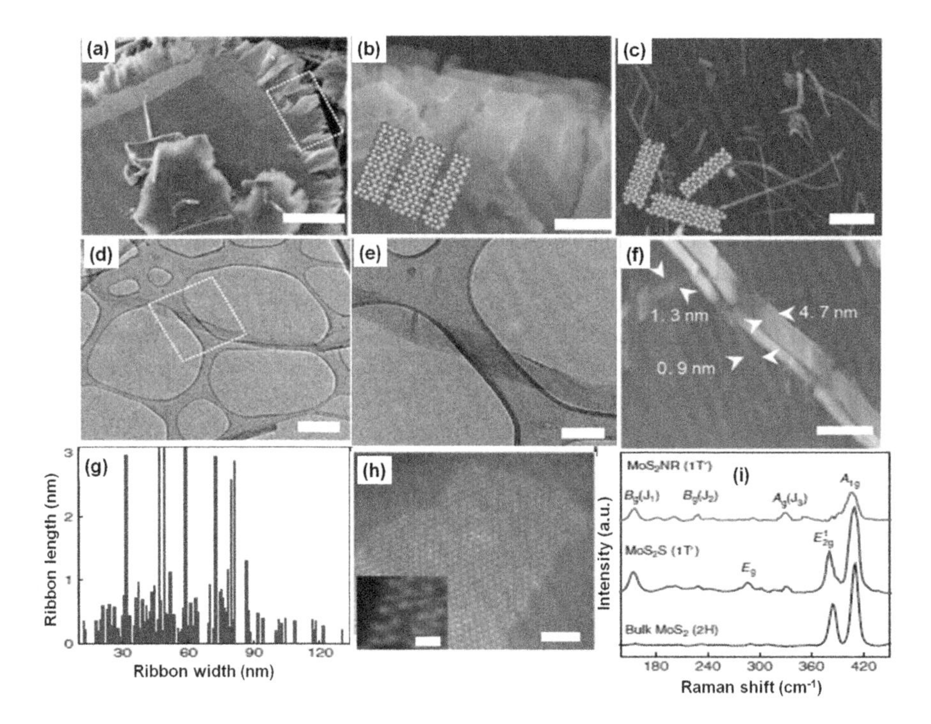

FIGURE 3.6 SEM images of unzipped 2D MoS_2 after 5 minutes of ultrasonication (Scale bars: 0.5 μm (a) and 0.1 μm (b)). (c) SEM image of MoS_2 NRs after complete unzipping of MoS_2 crystal. (d, e) TEM images of isolated MoS2 NRs (Scale bars: 0.2 μm (d) and 0.1 μm (e)). (f) AFM image of MoS_2 NRs (Scale bar: 0.2 μm). (g) Length of NRs versus width obtained from TEM. (h) STEM image of MoS_2 NR with lattice image shown in the inset. (i) Raman spectra of bulk MoS_2, MoS_2 sheet with 1Ti crystal structure, and MoS_2 NRs. Adapted with permission from [48].

the edges of bulk MoS_2 crystal. The SEM image after complete unwrapping of MoS_2 crystal to produce MoS_2 NRs is shown in Figure 3.6c. Figures 3.6d and 3.6e show the TEM images (scale bars: 0.2 μm (d) and 0.1 μm (e)) of the unzipped MoS_2 NRs. The AFM image of the grown NRs is shown in Figure 3.6f. The variation of length of the NRs as a function of width obtained from TEM results has been illustrated in Figure 3.6g. Figure 3.6h displays the STEM image of MoS_2 NRs with a lattice image, which shows Mo-terminated zigzag edge structure. Raman spectra of bulk MoS_2, MoS_2 sheet with 1Ti crystal structure, and MoS_2 NRs are depicted in Figure 3.6i.

Similar to the unwrapping of CNTs to obtain graphene, one of the effective ways to obtain h-BN NSs is via the unwrapping of BN NTs. However, it is worth stressing here that h-BN is less susceptible to intercalants as compared to graphite and hence unzipping of BN NTs is quite difficult. The very first study of unzipping of BN nanotube to form BN NS was reported by Zeng et al. [2] by using the polymer etching method where BNNT is dispersed in PMMA polymer. They have prepared a dispersion of BN NT in PMMA polymer and spin-coated it on Si substrate to form a polymer matrix on a substrate surface. It is then peeled off from the substrate followed by etching using plasma treatment, which ensures the cutting of BN NT leading to the formation of BN NS. Erickson et al. [49] have demonstrated the unzipping of BN NTs by alkali metal intercalation where vaporized potassium metal ions get inserted within the NT walls to induce lengthwise splitting of the BN NTs. This whole process was carried out at 300°C in vacuum ambient for 72 hours and 1 μm along with 5 nm wide h-BN NSs were obtained.

3.3 CONCLUSIONS

In the last two decades, there has been immense progress in the research area of 2D nanomaterials including graphene, h-BN, and TMDs from fundamental to technological advancements. Each of these 2D nanomaterials can be synthesized via a wide range of well-established growth techniques. The properties of these 2D nanomaterials can be effectively altered by varying the growth technique as well as the growth parameters, which may induce several defects, strain, or can vary the surface properties of these materials. It should be noted that there has been an explosive evolution in the field of graphene synthesis; however, there is still immatureness in the development of top-down approaches for the synthesis of other 2D nanomaterials beyond graphene. Therefore, most of the 2D nanomaterials apart from the graphene are being prepared using the well-developed synthesis methods adopted for graphene-based nanomaterials. Moreover, for the commercialization of such 2D nanomaterials, several challenges arise from the various synthesis techniques such as production cost, production yield, quality of the growth material, mass production, and production rate among these challenges, large scale production without compromising the quality is one of the major factors to fulfill the requirements from the viewpoint of device applications. Such device applications generally warrant the fabrication of hybrid structures and/or homo-/heteronanostructures employing such promising 2D nanomaterials. Hence more novel approaches are required to be developed, which will be one of the most promising futures research directions.

REFERENCES

1. A.K. Geim, K.S. Novoselov, The rise of graphene, *Nat. Mater.* 6 (2007) 183–191.
2. H. Zeng, C. Zhi, Z. Zhang, X. Wei, X. Wang, W. Guo, Y. Bando, D. Golberg, "White Graphenes": Boron nitride nanoribbons via boron nitride nanotube unwrapping, *Nano Lett.* 10 (2010) 5049–5055.
3. H. Li, G. Lu, Y. Wang, Z. Yin, C. Cong, Q. He, L. Wang, F. Ding, T. Yu, H. Zhang, Mechanical exfoliation and characterization of single and few-layer nanosheets of WSe_2, TaS_2, and $TaSe_2$, *Small* 9 (2013) 1974–1981.
4. H. Oh, J. Park, W. Choi, H. Kim, Y. Tchoe, A. Agrawal, G.-C. Yi, Vertical ZnO nanotube transistor on graphene film for flexible inorganic electronics, *Small* 17 (2018) 1800240.
5. B. Adilbekova, Y. Lin, E. Yengel, H. Faber, G. Harrison, Y. Firdaus, A. El-Labban, D.H. Anjum, V. Tung, T.D. Anthopoulos, Liquid phase exfoliation of MoS_2 and WS_2 in aqueous ammonia and their application in highly efficient organic solar cells, *J. Mater. Chem. C* 8 (2020) 5259.
6. H. Hu, X. Yang, X. Guo, K. Khaliji, S.R. Biswas, F.J.G. de Abajo, T. Low, Z. Sun, Q. Dai, Gas identification with graphene plasmons, *Nat. Commun.* 10 (2019) 1–7.
7. K. Patel, P.K. Tyagi, Multilayer graphene as a transparent conducting electrode in silicon heterojunction solar cells, *AIP Adv.* 5 (2015) 077165.
8. H. Zhao, H.W.J. Wua, J. Li, Y. Wang, Y. Zhang, H. Liu, Preparation of MoS_2/WS_2 nanosheets by liquid phase exfoliation with assistance of epigallocatechin gallate and study as an additive for high-performance lithium-sulfur batteries, *J. Colloid Interface Sci.* 552 (2019) 554–562.
9. A. Agrawal, G.-C. Yi, Database on the nonlinear optical properties of graphene based materials, *Data-in-Brief* 28 (2020) 105049.
10. A. Agrawal, Y. Tchoe, J.Y. Park, P. Sen, G.-C. Yi, Unravelling absorptive and refractive optical nonlinearities in CVD grown graphene layers transferred onto foreign quartz substrate, *Appl. Surf. Sci.* 505 (2020) 144392.
11. N. Alem, R. Erni, C. Kisielowski, M. Rossell, W. Gannett, A. Zettl, Atomically thin hexagonal boron nitride probed by ultrahigh-resolution transmission electron microscopy, *Phys. Rev. B* 80 (2009) 155425.
12. Z. Mohanta, H.S. Atreya, C. Srivastava, Correlation between defect density in mechanically milled graphite and total oxygen content of graphene oxide produced from oxidizing the milled graphite, *Sci. Rep.* 8 (2018) 15773.
13. L.H. Li, Y. Chen, B.-M. Cheng, M.-Y. Lin, S.-L. Chou, Y.-C. Peng, Photoluminescence of boron nitride nanosheets exfoliated by ball milling, *Appl. Phys. Lett.* 100 (2012) 261108.
14. H. Dong, D. Chen, K. Wang, R. Zhang, High-yield preparation and electrochemical properties of few-layer MoS_2 nanosheets by exfoliating natural molybdenite powders directly via a coupled ultrasonication-milling process, *Nanoscale Res. Lett.* 11 (2016) 409.
15. S. Haar, M. El Gemayel, Y. Shin, G. Melinte, M.A. Squillaci, O. Ersen, C. Casiraghi, A. Ciesielski, P. Samorì, Enhancing the liquid-phase exfoliation of graphene in organic solvents upon addition of n-octylbenzene, *Sci. Rep.* 5 (2015) 16684.
16. L. Cao, S. Emami, K. Lafdi, Large-scale exfoliation of hexagonal boron nitride nanosheets in liquid phase, *Mater. Express* 4 (2014) 165–171.
17. K.S. Rao, J. Senthilnathan, Y.F. Liu, M. Yoshimura, Role of peroxide ions in formation of graphene nanosheets by electrochemical exfoliation of graphite, *Sci. Rep.* 4 (2014) 4237.
18. N. Liu, P. Kim, J.H. Kim, J.H. Ye, S. Kim, C.J. Lee, Large-area atomically thin MoS_2 nanosheets prepared using electrochemical exfoliation, *ACS Nano* 8 (2014) 6902–6910.

19. Y. Wang, C.C. Mayorga-Martinez, X. Chia, Z. Sofer, M. Pumera, Nonconductive layered hexagonal boron nitride exfoliation by bipolar electrochemistry, *Nanoscale* 10 (2018) 7298.
20. C. Nethravathi, A.A. Jeffery, M. Rajamathi, N. Kawamoto, R. Tenne, D. Golberg, Y. Bando, Chemical unzipping of WS_2 nanotubes, *ACS Nano*, 7 (2013) 7311–7317.
21. A. Agrawal, G.-C. Yi, Sample pretreatment with graphene materials, Editor: C.M. Hussain, *Analytical Applications of Graphene for Comprehensive Analytical Chemistry*, Volume 91, 2020, Elsevier, Cambridge.
22. B. Alinejad, K. Mahmoodi, Synthesis of graphene nanoflakes by grinding natural graphite together with NaCl in a planetary ball mill, *Funct. Mater. Lett.* 10 (2017) 1750047.
23. V. Leon, M. Quintana, M.A. Herrero, J.L.G. Fierro, Few layer graphenes from ball milling of graphite with melamine, *Chem. Commun.* 47 (2011) 10936–10938.
24. A.S.A. Sherbini, M. Bakr, I. Ghoneim, M. Saad, Exfoliation of graphene sheets via high energy wet milling of graphite in 2-ethylhexanol and kerosene, *J. Adv. Res.* 8 (2017) 209–215.
25. W. Zhao, F. Wu, H. Wu, G. Chen, Preparation of colloidal dispersions of graphene sheets in organic solvents by using ball milling, *J. Nanomater.* 2010 (2010) 528235.
26. T. Xing, S. Mateti, L.H. Li, F. Ma, A. Du, Y. Gogotsi, Y. Chen, Gas protection of two-dimensional nanomaterials from high-energy impacts, *Sci. Rep.* 6 (2016) 35532.
27. Z. Wang, Y. Zhu, H. Yu, Z. Li, Simultaneously environmental-friendly exfoliation of boron nitride nanosheets and graphene and the preparation of high thermal conductivity nano-mixture composite membranes, *Mater. Charact.* 168 (2020) 110508.
28. L.H. Li, Y. Chen, G. Behan, H. Zhang, M. Petravic, A.M. Glushenkov, Large-scale mechanical peeling of boron nitride nanosheets by low-energy ball milling, *J. Mater. Chem.* 21 (2011) 11862–11866.
29. X. Chen, J.F. Dobson, C.L. Raston, Vortex fluidic exfoliation of graphite and boron nitride, *Chem. Commun.* 48 (2012) 3703–3705.
30. Y. Dai, L.-L. Wei, M. Chen, J.-J. Wang, J. Ren, Q. Wang, Y.-Z. Wu, Y.-P. Wang, X.-N. Cheng, X.-H. Yan, Liquid exfoliation and electrochemical properties of WS_2 nanosheets, *J. Nanosci. Nanotechnol.* 18 (2018) 3165–3170.
31. J.N. Coleman, M. Lotya, A. O'Neill, S.D. Bergin, P.J. King, U. Khan, K. Young, A. Gaucher, S. De, R.J. Smith, I.V. Shvets, S.K. Arora, G. Stanton, H.-Y. Kim, K. Lee, G.T. Kim, G.S. Duesberg, T. Hallam, J.J. Boland, J.J. Wang, J.F. Donegan, J.C. Grunlan, G. Moriarty, A. Shmeliov, R.J. Nicholls, J.M. Perkins, E.M. Grieveson, K. Theuwissen, D.W. McComb, P.D. Nellist, V. Nicolosi, Two dimensional nanosheets produced by liquid exfoliation of layered materials, *Science* 331 (2011) 568–571.
32. H. Iguchi, C. Higashi, Y. Funasaki, K. Fujita, A. Mori, A. Nakasuga, T. Maruyama, Rational and practical exfoliation of graphite using well-defined poly(3-hexylthiophene) for the preparation of conductive polymer/graphene composite, *Sci. Rep.* 7 (2017) 39937.
33. W.-L. Song, P. Wang, L. Cao, A. Anderson, M.J. Meziani, A.J. Farr, Y.-P. Sun, Polymer/boron nitride nanocomposite materials for superior thermal transport performance, *Angew. Chem. Int. Ed. Engl.* 51 (2012) 6498–6501.
34. W. Sun, Y. Meng, Q. Fu, F. Wang, G. Wang, W. Gao, X.-C. Huang, F. Lu, High-yield production of boron nitride nanosheet and its uses as a catalyst support for hydrogenation of nitroaromatics, *ACS Appl. Mater. Interfaces* 8 (2016) 9881.
35. G.-Q. Han, Y.-R. Liu, W.-H. Hu, B. Dong, X. Li, Y.-M. Chai, WS_2 nanosheets based on liquid exfoliation as effective electrocatalysts for hydrogen evolution reaction, *Mater. Chem. Phys.* 167 (2015) 271–277.
36. Y.-T. Liu, X.-D. Zhu, X.-M. Xie, Direct exfoliation of high-quality, atomically thin $MoSe_2$ layers in water, *Adv. Sustain. Syst.* 2 (2017) 1700107.

37. L. Pan, Y.-T. Liu, X.-M. Xie, X.-Y. Ye, Facile and green production of impurity-free aqueous solutions of WS_2 nanosheets by direct exfoliation in water, *Small* 12 (2016) 6703–6713.
38. E.D. Grayfer, M.N. Kozlova, V.E. Fedorov, Colloidal 2D nanosheets of MoS_2 and other transition metal dichalcogenides through liquid-phase exfoliation, *Adv. Colloid Interface Sci.* 245 (2017) 40–61.
39. P. Iamprasertkun, W. Hirunpinyopas, V. Deerattrakul, M. Sawangphruk, C. Nualchimplee, Controlling the flake size of bifunctional 2D WSe_2 nanosheets as flexible binders and supercapacitor materials, *Nanoscale Adv.* 3 (2021) 653–660.
40. K. Parvez, Z.-S. Wu, R. Li, X. Liu, R. Graf, X. Feng, K. Müllen, Exfoliation of graphite into graphene in aqueous solutions of inorganic salts, *J. Am. Chem. Soc.* 136 (2014) 6083–6091.
41. K.S. Rao, J. Sentilnathan, H.-W. Cho, J.-J. Wu, M. Yoshimura, Soft processing of graphene nanosheets by glycine-bisulfate ionic-complex-assisted electrochemical exfoliation of graphite for reduction catalysis, *Adv. Funct. Mater.* 25 (2014) 298–305.
42. T.C. Achee, W. Sun, J.T. Hope, S.G. Quitzau, C.B. Sweeney, S.A. Shah, T. Habib, M.J. Green, High-yield scalable graphene nanosheet production from compressed graphite using electrochemical exfoliation, *Sci. Rep.* 8 (2018) 14525–14533.
43. X. You, N. Liu, C.J. Lee, J.J. Pak, An electrochemicalroutetoMoS_2 nanosheets for device applications, *Mater. Lett.* 121 (2014) 31–35.
44. Y.-S. Li, J.-L. Liao, S.-Y. Wang, W.-H. Chiang, Intercalation-assisted longitudinal unzipping of carbon nanotubes for green and scalable synthesis of graphene nanoribbons, *Sci. Rep.* 6 (2016) 22755.
45. N.L. Rangel, J.C. Sotelo, J.M. Seminario, Mechanism of carbon nanotubes unzipping into graphene ribbons, *J. Chem. Phys.* 131 (2009) 031105.
46. C.S. Tiwary, B. Javvaji, C. Kumar, D.R. Mahapatra, S. Ozden, P.M. Ajayan, K. Chattopadhyay, Chemical-free graphene by unzipping carbon nanotubes using cryo-milling, *Carbon* 89 (2015) 217–224.
47. D.G. Kvashnin, L. Yu. Antipina, P.B. Sorokin, R. Tenne, D. Golberg, Theoretical aspects of WS_2 nanotube chemical unzipping, *Nanoscale* 6 (2014) 8400.
48. S.P. Sasikala, Y. Singh, L. Bing, T. Yun, S.H. Koo, Y. Jung, S.O. Kim, Longitudinal unzipping of 2D transition metal dichalcogenides, *Nat. Commun.* 11 (2020) 1–8.
49. K.J. Erickson, A.L. Gibb, A. Sinitskii, M. Rousseas, N. Alem, J.M. Tour, A.K. Zettl, Longitudinal splitting of boron nitride nanotubes for the facile synthesis of high quality boron nitride nanoribbons, *Nano Lett.* 11 (2011) 3221–3226.

4 Bottom-Up Synthesis of 2D Nanomaterials for Energy Applications

Felipe de Souza and Ram K. Gupta
Pittsburg State University

CONTENTS

4.1 INTRODUCTION

There is an increasing demand for energy storage technologies as the dependence on electronic devices is becoming more intrinsic to modern society. Such conditions lead to a higher level of dependence on the materials used to fabricate the components and push academia as well as industry to develop new approaches that can deliver better performance accompanied by lower cost. In this manner, stationary storage grids, transport systems, and electronic devices are powered by some type of battery or supercapacitor. Their working principle is based on rechargeable electrochemical batteries. In that sense, lithium-ion batteries (LIB) are the dominant technology, which consists of two electrodes (anode and cathode) that are separated by an electrolyte. The redox process is carried out based on the electrochemical potential difference between the anode and the cathode. Because of that, two-dimensional (2D) materials were brought into the picture to further improve the performance of current energy storage devices. The 2D structure improves the surface area, which facilitates the diffusion of ions from the electrolyte as well as the redox process due to more exposure of active sites and it can facilitate the ionic or electronic transport over its surface. Along with that, the 2D structure can also present enhanced mechanical properties compared with their bulky counterparts in terms of Young's modulus and flexibility. Through that, these types of materials can be used for more

DOI: 10.1201/9781003178453-4

robust applications that require a material to be exposed to strain or to be bendable, which is the current trend of the latest wearable devices. In another sense, 2D materials can be applied as a coating over an electrode's surface to prevent the formation of Li metal dendrites or potentially prevent the shuttling effect of polysulfides, which are a recurrent challenge on LIB that can cause them to short-circuit. On top of that, they can play other roles such as improving the surface area for better capacitance, conductibility to facilitate charge transfer, dissipate heat from the battery, and function as a composite to reinforce specific properties. Through that, they can be employed as components for either cathode or anode to increase the cycle stability along with gravimetric and volumetric capacitances. Based on that, 2D materials are valuable components to further improve the much-desired performance of the current energy storage devices.

Currently, commercially available LIB have their cathode active materials composed of $LiCoO_2$, $LiNi_xMn_yCo_zO_2$, and $LiFePO_4$ and anode active materials such as $Li_4Ti_5O_{12}$, which provide satisfactory performance, yet still suffer from high polarization, low cycling stability, and safety concerns. A viable option to improve their overall performance has been widely studied through graphene, which has remarkable properties such as a high conductibility of 10^6 S cm^{-1}, a charge carrier mobility of 200,000 cm^2/V.s, chemical stability, and flexibility [1]. Another aspect is that a lower resistance in the electron transport in the cathode is important to improve the diffusion of Li^+. Based on that, several carbon-based materials alongside graphene such as carbon nanotubes (CNT), carbon black, and acetylene black can be employed as additives in electrodes. When graphene was added into $LiFePO_4$ it surpassed other carbon-based materials such as CNT and acetylene black by a large marge in terms of the diffusion coefficient and lower interface resistance reaching 5.9×10^{-9} cm^2 s^{-1} and 130 Ω cm^{-2}, respectively [2]. In addition, lower amounts of graphene (between 0.2 and 5 wt.%) were required to reach optimal properties. This effect is related to the uniform dispersion of graphene and its long-range pathway for electron transport, which decreases the overall resistance of the system. In other cases, 2D $Ni(OH)_2$ and V_2O_5 have delivered high specific capacitance, fast charge/discharge process, and low volume expansion due to Li (de)intercalation, which was attributed to shorter paths for ion diffusion along with a larger number of sites for ion intercalation [3,4]. The architecture of two-dimensional can vary to some degree being completely planar, puckered, or buckled, which are important to either facilitate electron flow or to improve the surface area for better adsorption of ions. The representation of some of the structures of regularly employed 2D nanomaterials is provided in Figure 4.1. With these concepts in place, the following sessions cover the main synthetic approaches to obtain 2D materials by briefly showing top-down approaches followed by a more in-depth discussion of bottom-up synthetic methods.

4.2 APPROACHES FOR THE SYNTHESIS OF 2D MATERIALS

The top-down approach is described as a process in which a bulky material is converted into a single layer while maintaining most of its integrity in the process. It can be performed in several ways, which roughly include mechanical, chemical, or liquid exfoliation approaches. It is convenient in the sense of being generally cheaper and

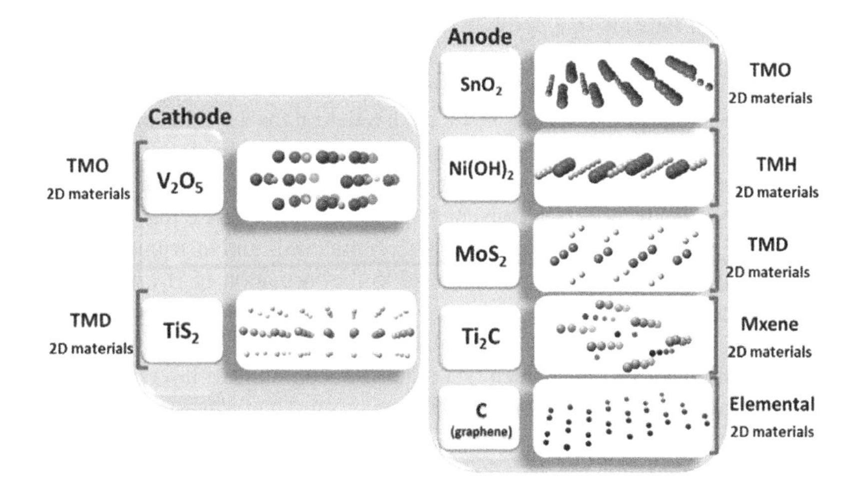

FIGURE 4.1 Widely used 2D materials to compose electrodes for LIB as either anodes or cathodes depending on the working potential against Li/Li^+. Adapted with permission from reference [5], Copyright (2020), American Chemical Society.

reproducible with the trade-off that it could be some crystallinity disruption or property change. The mechanical exfoliation approach consists of breaking weak intermolecular interactions such as van der Waals' through physical force. Micromechanical cleavage is one method based on this concept at which a 2D material is fabricated by using a scotch tape that is adhered to the surface of the bulky material and peeled off to obtain a single or fewer layer stacked. Since this method functions by weakening the intermolecular interactions, it maintains the chemical structure of the parent material, its crystallinity along a clean surface. Graphene, for instance, can be synthesized through this method by using graphite as the bulky parent material. Ball milling exfoliation is another convenient process that breaks the chemical interactions through shear and compressive force, which can promote chemical reactions in the absence of solvent. 2D nanolayers of less than 200 nm of materials such as boron nitride, MoS_2, and graphene can be obtained through this approach.

Liquid exfoliation is a method that makes use of organic solvents, surfactants, and/or ionic liquids. The process can be performed by thinning the bulky materials through sonication and ball milling. The follow-up process consists of a chemical etching at which fluoride acid (HF) is largely employed. This approach is widely used to synthesize MXenes, which consists of a bulky $M_{n+1}AX_n$ material where M is a transition metal, A is the interlayered atom that can be etched such as Al, and X is usually C or N. One example for this case is Ta_4C_3 for instance, which can be obtained from the chemical etching of bulky Ta_4AlC_3. Through that, several 2D materials can be obtained with high efficiency with the drawback of the relatively expensive cost of HF.

The mechanical force-assisted technique is another top-down procedure that can rely on both sonication and shear force. Through this process, ultrasonic waves are created, which induce liquid cavitation on the bulky material leading to the formation of bubbles. Once they burst, localized heat is released, which promotes the formation of either hydroxyl radicals or a pyrolytic process, exfoliating the bulky material into 2D nanolayers. This approach has been used to synthesize MoS_2 nanolayers from 5 to 15 nm of thickness and around 40 to 70 nm of length and width [6,7]. The Ion-intercalation technique is a milder approach that relies on the diffusion of small metallic cations such as Li^+, Na^+, and K^+ within the structure of bulky material and increases the interlayer spacing between the nanosheets. Then, gentle sonication can be applied to separate the nanolayers. It is a convenient approach in the sense that water can be used as a solvent, and requires relatively low-cost materials and instrumentation. The presence of cations under sonication can lead to the formation of H_2, which further improves the exfoliation [8]. A follow-up ion exchange process can also be adopted, at which a cation with a larger radius can be used to replace the smaller cation and thus increase the interlayer spacing, which facilitates the separation of layers.

Oxidation-assisted method or modified Hummer's method consists of using strong oxidizing agents that can introduce oxygenated groups causing the bulky structure to become more polarized and therefore more soluble in water. After that, sonication becomes more effective to separate the nanolayers. This approach is commonly used to synthesize graphene oxide (GO) through the strong oxidative process over graphite with $KMnO_4$ and concentrated H_2SO_4. It has a relatively high yield; however, it promotes a drastic change in the chemical structure of the chemically exfoliated nanolayered material when compared to the bulky parent. Hence, a thermal or chemical reduction process can be performed to obtain reduced graphene oxide (rGO). Lastly, cryo exfoliation of 2D quantum dots is a method that relies on low temperature under the presence of solvents to promote physical exfoliation, with the advantage of preserving the structure of the bulky starting materials since no chemical reactions take place [9]. Some of the top-down approaches are displayed in Figure 4.2.

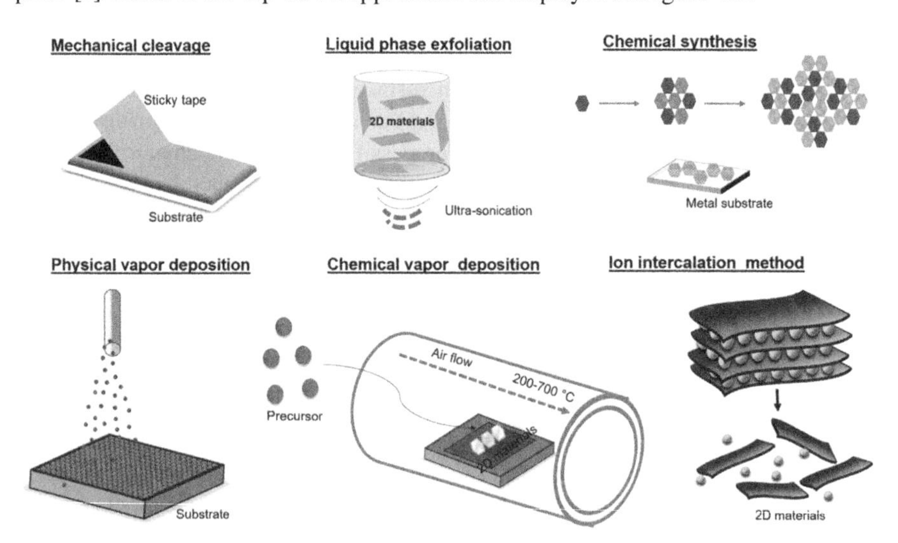

FIGURE 4.2 Widely used synthetical approaches for top-down such as mechanical cleavage, liquid-phase exfoliation, and ion intercalation. Bottom-up approaches displayed such as chemical synthesis, physical vapor deposition, and chemical vapor deposition. Adapted with permission from reference [9], Copyright (2019), Elsevier.

4.3 BOTTOM-UP SYNTHESIS OF 2D NANOMATERIALS

The bottom-up approach consists of the chemical build-up of a nanolayered structure taking an atom or a smaller molecule as starting reagent, named precursors. It can be performed through several techniques such as chemical or physical vapor deposition (PVD), ultra-high vacuum (UHV)-assisted synthesis, hydro or solvothermal, among other methods. These technologies are valuable in the sense that they provide 2D nanomaterials through mostly facile, inexpensive, and efficient methods.

4.3.1 Chemical Vapor Deposition

Chemical vapor deposition (CVD) is a convenient method used to grow thin films of organic or inorganic compounds through deposition over a substrate. It allows experimental control as scientists can tune properties such as morphology, phases, number of layers, doping, insertion of defects, and particle sizes. Such possibility yields 2D nanomaterials with high surface area and exposed active sites. These properties can be defined through precise control of parameters such as chamber pressure, temperature, the flow of carrier gas, concentration of reagents precursor, and distance between the starting materials and substrate. Such control of parameters was studied by Wu and colleagues, which demonstrated a method that targeted the supply of gas to a focal point, which created one nucleation site with a carbon precursor that allowed it to grow up to a few centimeters forming a single graphene crystal [10].

The concentration of precursor, which functions as the reagent for a CVD, defines the deposited number of atoms over the substrate, which is usually measured in standard 1 cm^3 $minute^{-1}$, which can be converted to 7.4×10^{-7} mol s^{-1}. Also, several carrier gases such as H_2, Ar, SiH_4, or SiH_2Cl_2 can be used to control the environment in different ways, for example, H_2 can be used to reduce undesired reactive groups or dandling double bonds to obtain samples with high quality. A common example is CVD used to synthesize graphene through CH_4 and H_2 as precursors. Hence, through a controlled ratio of precursor and carrier gas, proper control of size, morphology, and structure can be achieved. Also, doping with P and N can be performed by adding gases such as PH_3 or NH_3. MoS_2, $MoSe_2$, WS_2, WSe_2, among others can also be synthesized through CVD by using precursors such as MoO_3, $MoCl_5$, or WO_3 under the presence of S or Se powders. In another case, precursors such as H_2S, $CS(CH_3)_2$, or $Mo(CO)_6$ can lead to a more uniform structure of transition metal dichalcogenides (TMDC) [11].

Temperature influences the flow of carrier gas, the kinetics of chemical reactions, and deposition rate as when higher temperatures are employed a faster diffusion is promoted. This effect can result in a gradient of concentration, which varies the mass transport. Through that, the uniformity of the deposited 2D material can change. For the case of TMDC, the temperature has a considerable impact if compared with graphene, as a small variation in this parameter can cause a change in saturation pressure, varying the material's growth. Based on previous experiments it has been noted that temperatures around 800°C contribute to a higher deposition rate as it can lead to multilayered materials.

Substrates are the surface at which the precursor materials grow. It plays an important role in the catalysis of the process since active Ni or Cu, for instance, possess active sites that enhance the reaction rate at their surface, which are widely used for the growth of graphene. For the case of TMDCs substrates such as polyimide, mica, Si/SiO_2, Au, or W can be used. Nucleation sites to start the material's growth can also be employed, which include the addition of rGO or compounds such as perylene-3,4,9,10-tetracarboxylic dianhydride (PTCDA) or perylene-3,4,9,10-tetracarboxylic acid that have the goal to promote catalytic activity over inert substrates like Si/SiO_2. The substrate's orientation is also a factor that can change the morphology as a previous study demonstrated the difference between a CVD performed with the substrate at vertical rather than horizontal orientation, which promoted a film with uniform formation without further need of volatile solvents of higher vacuum [12].

Pressure greatly influences the reaction progression, and it provides more control over the system when high vacuum and low precursor concentration are used, enabling more accurate deposition over the substrate. This effect has been observed by varying the partial pressure of $Mo(CO)_6$ as a precursor for MoS_2 at which under low pressure the deposition occurs more accurately at the grain boundaries. However, when high pressure is employed the deposition occurs randomly, which leads to a non-uniform surface of mono and multilayer nanomaterials [13]. Other variations of CVD can also influence the quality of the 2D films such as inductively coupled plasma CVD, which can be performed at lower temperatures. It aids to decrease cost and is more efficient toward the deposition of polymeric film's growth over glass substrates [13]. There are other relevant factors for the optimization for the 2D film, which can be summarized in grain size, layers, growth direction, phase, morphology, degree, and effect of doping, and defects on the structure. During a film formation, it is expected that it should be close to a single crystal instead of a structure with several grain boundaries, which has been performed for several known 2D materials that include $MoTe_2$, WSe_2, WS_2, and MoS_2. The latter has been studied by employing an Au foil as substrate and WO_3 and S_2 as a precursor for the formation of the MoS_2 film [14,15]. In this case, Au played a catalytic role as it dissociates S_2 into single S atoms over its surface, which facilitates the follow-up sulfurization of MoO_3. This condition led to the formation of regular films over the substrate with an edge length of around 80 μm.

The higher number of layers causes a decrease in the bandgap of most TMDCs. Based on that, when WS_2, MoS_2, and $MoSe_2$ reach one layer of dimension there is a transition from indirect to direct bandgaps. This characteristic allows these materials to be applicable as in photovoltaic cells, laser diodes, and light-emitting diodes. Hence, properly controlling the stacking of nanolayers is an important step to control the properties. Based on that, some approaches such as oxygen-plasma treatment and increase of layer deposition by increasing temperature in the sense that one-, two-, and three-layered structures of MoS_2 were obtained for the temperatures of 750°C, 825°C, and 900°C, respectively, have been employed [16,17].

The orientation of the precursor over an aligned substrate is another parameter that correlates with the properties as it describes the nucleation model during the growing process. However, if the precursor is not oriented with substrate there is a higher tendency of more grain boundaries, uneven distribution of the film, and more defects. That is a relevant matter for the film growth as graphene films on the scale of meters have been reported when the deposition process occurred with domains under 60° of rotation [18]. Yet, an effort to reach such dimension is still being made for TMDCs.

To attain more control of orientation growth it is necessary to use a substrate that can properly interact with the precursor, usually through van der Waals interaction, match of lattice between the components, and symmetrical lattices. For that, facet sapphire, c-plane, has been often employed to obtain crystals of high quality.

The influence of morphology in the properties of 2D nanomaterials used on sensors, batteries, supercapacitors, and electrocatalysts is a considerable matter for performance improvement. In that perspective, the catalytic properties of electrodes for water splitting are greatly improved by the density of edge sites [19]. The symmetric threefold lattice commonly observed in the TMDC's structure that takes the shape of truncated triangles, dendritic, hexagons, three- or six-pointed stars are factors that contribute to the edge effect. The morphology can vary based on several factors such as the ratio of precursors, flow rate, and temperature. The effect of change in these parameters in the morphology was studied by Wang et al. [20]. In the study the authors showed that the distance of that MoO_3 precursor is from the substrate for the synthesis of MoS_2 provides a relation between the ratio of Mo and S in the film. This situation led to a change in growth rate as it followed the order of medium-sized truncated triangles, to larger sized, again medium-sized, diminishment size of hexagons, and small triangles as displayed in Figure 4.3a. This variation occurred due to the change in growth rate in the MoS_2's different facets. Also, a higher concentration of Mo leads to a suggestive Mo edge denominated as zigzag Mo edge (Mo-zz). A similar phenomenon occurs when S was in excess leading to zigzag S edge (S-zz) as shown in Figure 4.3b.

Phase control is an important matter for the definition of properties. For the case of MX_2, there is an arrangement of X–M–X on the X and Y axis, along with those chalcogen elements that are aligned vertically on the Z-axis forming a trigonal prismatic phase, referred to as the 2H phase (semiconductor). In another way, the adjacent chalcogen atom can align toward each other, forming an octahedral phase, referred to as the 1T phase (metallic). Hence, controlling the phase change allows controlling the material's properties without introducing new atoms. Based on that, recent studies showed that to obtain metastable TMDC phases some post-synthesis techniques such as electron beam irradiation, intercalation with alkaline metals, mechanical force, or plasma treatment as viable options [20,21]. In that sense, Li salts intercalated with MoS_2 1T phase have been obtained and its stability in the air has been improved by hydrogenation of the intercalated Li leading to the formation of LiH [22]. Despite being an important factor for property enhancement, phase change has not been widely performed by CVD, which still requires further studies.

In another perspective, doping is a process at which a heteroatom is introduced into the structure to tune the material's properties. Since the metals are less exposed and stacked in between chalcogen atoms it deems simpler to dope the chalcogens, which can be performed through CVD with the current challenge of defining proper concentration and dopant position during the process [13]. Yet, synthesizing ternary TMDC by doping with metals has been performed, as $Mo_{1-x}W_xS_2$ was obtained by laser-plasma CVD where the precursors were WCl_6 and MoO_3 for W and Mo sources, respectively. The reaction between Mo and S was performed at 700°C [23]. This type of approach promotes the decrease of the bandgap, facilitating electron transitions through a relatively simple technique. Based on this discussion several factors influence the overall properties of 2D materials as CVD is a highly versatile technique in

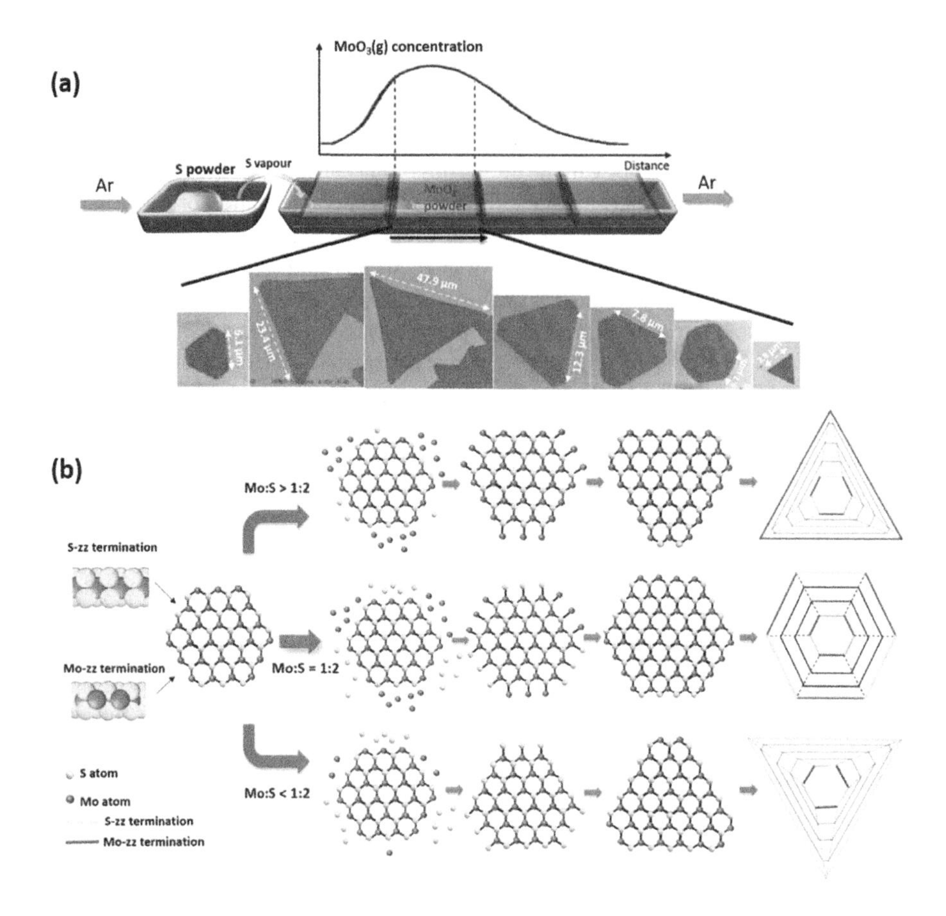

FIGURE 4.3 (a) Effect of the distance of precursor distance between precursors as a factor for change in morphology according to the rate of growth. (b) Change in morphology according to the cases of higher concentration of Mo or S to form MoS_2 film. Adapted with permission from reference [20], Copyright (2014), American Chemical Society.

that regard. For that, the key factors for the proper control of CVD to allow optimal growth of 2D nanomaterials are expressed in Figure 4.4a. Also, Figure 4.4b provides an overall summary of the effect of the main parameters for CVD synthesis.

4.3.2 Physical Vapor Deposition

PVD is a process that has some advantages over CVD such as more reproducibility, lower temperature, and tendency to cover larger areas of substrate with more uniformity. It functions under the same principles with the difference that it relies on the disruption of intermolecular interactions, mostly van der Waals, instead of the formation of new chemical bonds as it occurs in CVD. The PVD process can be mainly divided into methods that are sputtering and laser deposition, which is known to be the most suitable to obtain 2D materials with satisfactory performance [24].

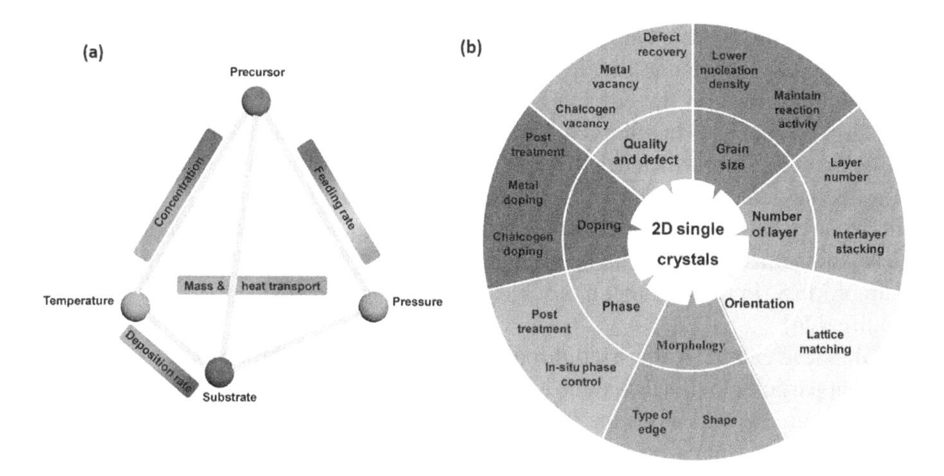

FIGURE 4.4 (a) Correlation between the main parameters for CVD approach such as precursor, pressure, substrate, and temperature. (b) Key factors for the proper control of CVD technique for the growth of 2D nanomaterials. Adapted with permission from reference [13], Copyright (2018), American Chemical Society.

The sputtering PVD consists of exposing the bulky material that already has the desired composition to vacuum and a large negative potential that can range from −1,000 to −100 V. Then, after highly pressurized inert or reactive gas is introduced in the system within the range of 70–7,000 Pa. Ar is commonly used for this purpose as it is an inert gas that when exposed to both negatively high voltage and pressure it is ionized releasing electrons and forming cations. The positively charged gas species are accelerated into a negatively charged substrate in which the bulky material (target) is placed. Once the cationic species hit the substrate atoms are released, which vaporize the bulky material. Controlling the voltage influences the amount of sputtered yield, which is the number of atoms on the target that leaves the bulky structure when bombarded by one cationic gas atom. Yet it is important to note that voltages above 1,000 V cause the incident cations to be buried too deep into the target and be unable to sputter other atoms. Hence, usually, voltages of 500 V under 1.3 kPa are applied, which allow proper control of the kinetic energy of the incident gaseous cations [25]. The technique is meant to be performed on the bulky equivalent of the 2D desired materials as the final composition of the layered product tends to present a proportional number of atoms of those of the bulky one. It is an effect since observed on TMDCs, as the chalcogens present higher vapor pressure than metals as well as similar size compared to the cationic incident species, which maximizes the momentum transfer for the ejection of chalcogen atoms. Thus, chalcogens leave the bulky structure faster than metals, which may cause a stoichiometric imbalance in the 2D structure. To address that, the deposition pressure can be increased along with the distance between target and substrate.

Another parameter of study is the mechanism of TMDCs growth, which provides information on the exposure of active sites. Taking the case of MoS_2 for instance scientists observed that growth occurred through a layer-by-layer mechanism at the basal planes (002), which are parallel to the surface. The low energy surface toward the center of the film and the higher energy surface on the edges, due to the driving force of atoms in the edge to bond, is what directs this type of film to grow. Misalignment of growing films can occur on the basal plane due to impurities such as moisture or oxygen, which can disrupt the deposition of other layers. Yet, purposely introducing defects into an inert basal plane is a convenient strategy to improve the surface area, which can be performed through incident ions to create anionic vacancies. Another aspects that can be used to control the 2D material's structure are the ratio of Mo:S. In the case that more S needs to be introduced due to its deficiency in the film [26].

Graphene can be synthesized through pulsed laser PVD; however, the energy of incident particles in this technique is often strong enough to cause damage to the lattice structure, which causes the structure to deteriorate [27,28]. For that, low-energy electron beam evaporation or thermal treatment is more favorable to induce an advantageous number of defects, and therefore preserves the electron transport properties of graphene [29]. Another strategy consists of using reactive gases, such as a mixture of N_2 as the reactive species along with Ar, over a metal target that can function as a cathode. This approach can be employed to synthesize many metal nitrides such as AlN, ScN, TiN, among others. The other type of PVD can be performed through pulsed laser deposition, which consists of the incidence of a laser with a wavelength of 248 nm (ultraviolet) as the source of energy, instead of ion bombardment as it occurs in the case of sputtering. Based on that, this type of approach is suitable to obtain 2D electrically insulator materials from their bulky counterparts. The reason for that is because conductive materials present higher thermal conductivity and reflectivity leads to a decreased energy adsorption that weakens the effect of the incident laser as the heat diffuses faster through the structure. The main parameters for this approach are based on the laser's energy, length, and absorption on the target material. The synthesis of boron nitride (BN) through this method has yielded ultrathin film that ranged from 5 to 10 nm, which maintained its high dielectric constant, whereas it can be performed over several substrates such as metals, polymers, or ceramics. BN has promising applications due to its chemical and thermal stability, which allows it to be used in devices to prevent the deterioration of other active electrode materials.

TMDC can also be synthesized through this technique, which has been known to yield high-quality films. Yet, the same phenomenon of a faster release rate of chalcogens is observed. To counter that Serrao and colleagues used tetrathiomolybdate (MoS_4) to compensate for the higher release rate of S as a means to prevent excessive anionic vacancies that could damage the structure [30]. Through that, thin MoS_2 films were obtained, which could be deposited over either semiconducting or insulating films as the process was performed at 700°C. Based on these concepts several research areas make proper use of PVD, among them, there is the fabrication of organic photovoltaics that are promising materials suitable for solar cells. With recent progress in this area there has been the combination of graphene as a substrate along with metal phthalocyanines (MP), which present tunable properties, chemical stability, and a high coefficient of adsorption in the visible spectrum, hence highly attractive for this field [31]. Through that, growing MPs over C-plane sapphire, graphene substrate can provide a good combination of MP's properties along with optical transparency, flexibility, and high conductivity of graphene, for instance. This idea was tested by Mirabito et al. [32] by performing a PVD of ZnPc over different substrates such as graphene, Cu, and hexagonal boron nitride (h-BN). The authors observed that high lengthwise nanowires were formed over graphene, Cu, or h-BN. Yet, the morphology of ZnPc was highly influenced by the structure of graphene as it tended to form continuous films when single-layered graphene was used as substrate whereas it formed nanowires when stacked graphene nanolayers were used. This variation was likely due to the more exposed surface area of single-layer graphene, which allowed more deposition of ZnPc. For photovoltaic activity, it is desired that materials deposited over the substrate present a smaller grain size to facilitate the formation of excitons and electron transfer steps. This process is tunable and dependent on the substrate in which the MC is grown as it can define its morphology, orientation, and structure. The change in morphology when ZnPc is grown in the different substrates can be seen in Figure 4.5.

4.3.3 Ultra-High Vacuum-Assisted Synthesis

Both CVD and PVD are widely explored methods due to the 2D material's properties that can be obtained and tuned through them. Yet, UHV-assisted synthesis differs in the sense that it can avoid contaminants as it allows techniques for on-surface synthesis and molecular beam epitaxy. On top of that, the UHV technique can be used to synthesize specific 2D materials. Ogikubo et al. took advantage of this technique to synthesize a hetero 2D structure based on germanene and stanine [33]. These compounds present a similar structure compared to graphene. Based on that, finding ways to properly synthesize nanolayered compounds from the Group 14 can provide a new array of opportunities for materials that can potentially surpass graphene's properties and perhaps serve as a viable option within several research areas including energy applications. For this case, the synthesis was performed at UHV adopting both molecular beam epitaxy (MBE) and atomic segregation epitaxy, which led to a precise multijunction of Ge and Sn atoms over an Ag (111) film. The authors observed that

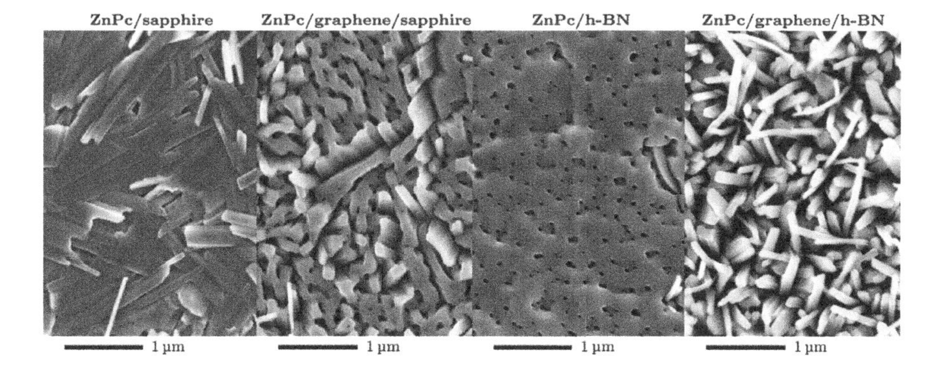

FIGURE 4.5 Scanning electron microscopy (SEM) images for the PVD-grown ZnPc over different substrates and the changes in ZnPc's morphology. Adapted with permission from reference [32], Copyright (2021), American Chemical Society.

both germanene and stanene could be fabricated without intermixing the nanosheets through scanning tunneling microscopy (STM). The precise control for this type of approach may provide a satisfactory performance on electronic devices [33].

Another approach that has been first developed more than 100 years ago is the Ullmann reaction that consists of C-C coupling performed through the reaction between an aryl halide along with high surface area Cu at temperatures higher than 200°C to yield a biaryl and a Cu halide byproduct. Despite its complex mechanism, this approach is still employed today under UHV conditions to grow a thin layer of carbon-based materials over a Cu substrate, even though this approach does not yield large molecules. Yet, large polymerizations with highly aligned structures were obtained through Ullmann's coupling by using 2,5-diiodo-3,4-ethylene dioxythiophene with the advantage that when it was performed over Cu (110) substrate there was a Cu-S interaction that promoted the alignment of the polymer leading to all-cis geometry, which was also attributed to the anisotropic nature of Cu (110) that created a driving force for one-directional growth [34]. This approach deems a convenient process as it can form accurately structured 2D films over a metallic substrate as it has been used to synthesize porous graphene. The synthesis was performed by 575 K annealing of hexaiodo-cyclohexa-*m*-phenylene (CHP) for 5 minutes. The precise nanostructured layered material had a pore spacing of 7.4 Å. On top of that, iodine atoms were released from the structure by thermal treatment at 875 K without damaging the 2D structure. The reaction that describes this process can be seen in Figure 4.6.

The formation of a regular and atomically precise film is influenced by the condition imposed by UHV but also the substrate. With that in mind, Bieri and colleagues proposed the synthesis of a 2D polyphenylene over Ag (111), Au (111), and Cu (111) to form a honeycomb-like structure. Through this study, the authors observed that substrate plays a major role in the amount of C–C coupling once the optimal activation temperature is reached. This driving force for the film growth is based on the transportation that the metal substrate allows along with its exposure of reactive sites to promote the reaction. In that sense, Cu presented the most favorable surface for reactions to carry over, meaning that radicals formed were quickly bonding with

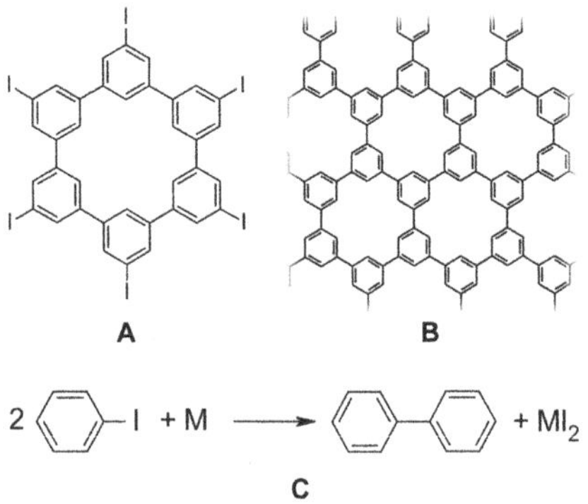

A B

2 ⟨Ph⟩–I + M ⟶ ⟨Ph⟩–⟨Ph⟩ + MI_2

C

FIGURE 4.6 (a) Represents the CHP's chemical structure. (b) Chemical structure of the polyphenylene network. (c) The chemical reaction between aryl-halides forms an aryl–aryl bond. Adapted with permission from reference [35], Copyright (2010), American Chemical Society.

each other, that effect also caused a decrease in the mobility as it presented a more "opened" branched structure. For the case of Ag, the opposite behavior was observed as this metal allowed faster transport of reactive species, which led to the formation of denser polymeric networks over Ag's surface, yet with relatively lower exposure of active sites. Au on the other hand presented a behavior that was in between that of Cu and Ag by yielding a mix of compact clustered networks and branched domains. It is worth mentioning that the annealing temperatures for each of the metals were 475, 525, and 575 K for Cu, Au, and Ag, respectively. The structure growth patterns for the polyphenylene over the different metals are presented through STM images displayed in Figure 4.7. These approaches show one of the major advantages of UHV systems, which allows the formation of C–C bonds under atomic precision to grow graphene under different structures and porosities as it can be controlled by selecting the desired starting halide-aryl. On top of that, the synthesis of this 2D layered structure is relatively shorter if compared with other methods. The chemistry of C–C coupling under UHV can be performed in other ways such as cyclodehydrogenation, which consists of the bond between two C–H bonds through the loss of H_2 followed by cyclization. It can be catalyzed with $FeCl_3$ or $AlCl_3$ [36]. Yet, other approaches that make use of annealing at 400°C of polyanthrylenes followed up by thermal treatment at 440°C to perform the cyclization step. The inherent drawback of this process is that due to the large distance a lower amount of starting material that is sublimated can reach the substrate, which may account for around 2%.

The use of UHV systems for the synthesis of fullerene (C_{60}) and triazafullerene ($C_{57}N_3$) was performed over a Pt (111) substrate at 750 K to yield the desired 0D structures. This type of structure has potential applications for storage energy due to its high surface area, which may improve the catalytic activity or energy storage properties.

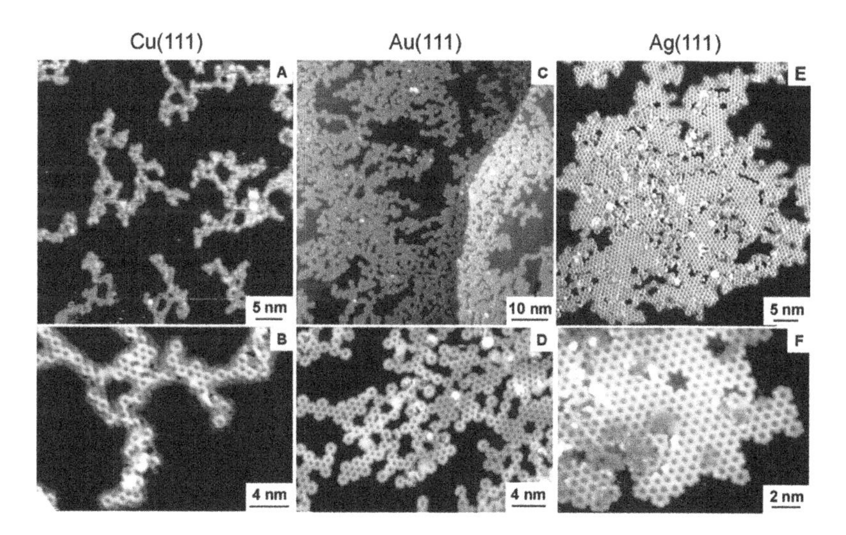

FIGURE 4.7 (a, c, e) The figures display the overview for STM images and (b, d, f) the figures display high-resolution STM for the networks of polyphenylene over Cu (111), Au (111), and Ag (111), respectively. Adapted with permission from reference [35], Copyright (2010), American Chemical Society.

It can be also envisioned that, if this method is performed under the presence of guest species, its properties can be functionalized and perhaps obtain enhanced properties [37]. Based on these examples it is notable that synthesis at UHV condition allows several types of reactions besides Ullman and dehydrogenation, which also include Sonogashira, radical (CH_2•), imination, amidation, urea linkage, dehydration, esterification under the presence of boron-based compounds, and amino-fullerenes [36].

4.3.4 Hydrothermal/Solvothermal Synthesis

Hydrothermal and solvothermal synthesis are procedures performed under a reactor that is hermetically sealed or an autoclave. Due to the controlled pressure within the reaction system, higher temperatures can be achieved without causing the solvents or components to evaporate. This condition prompts the precursor or monomers to deposit over the substrate surfaces forming 2D nanosheets. It is a widely used method due to its low-cost, high yield, and scalable method for the synthesis of several types of 2D materials, although it tends to form structures that present more than a single layer of 2D material, it is effective to properly bind the synthesized material over the substrate's surface. The process allows the dissolution of the precursor in a solvent that would be poorly soluble under normal conditions. Hence, when the reaction system is submitted to this high temperature and pressurized environment, the system becomes homogeneous, and the precursor recrystallizes over the substrate usually presenting a higher degree of crystallinity. Another convenient variation for this method is the microwave-assisted type, which greatly reduces the reaction time due to exposure of high frequency. Yet, it applies to polarizable materials that can absorb microwaves [38]. Such convenience of this method allows it to be adopted for the synthesis of several types of materials besides thin films such as simple oxides, Perovskites, bioceramics, vanadates, garnets, among others.

Based on that, the synthesis of 2D nanomaterials suitable for energy storage applications can be widely explored through the hydrothermal method. With that insight, 2D nanolayers of MoS_2 were synthesized by Huang and colleagues [39]. MoS_2 is a vastly researched material because it presents satisfactory electrochemical properties whereas relatively lower cost if compared with other materials. Some of its properties arise from its planar structure that similarly to graphene's allows a facile electron transport over its surface, which gives it appreciable optical, physical, and electrical properties. The synthesis was carried out by dissolving the precursors that were $Na_2MoO_4{\cdot}2H_2O$ as Mo source and L-cysteine as S source at $pH=6.5$. The high heat and pressurized environment provided by the hydrothermal reaction yielded the MoS_2 nanosheets that were deposited over Ni foam. The strong interaction that was formed between the MoS_2 nanolayers and the Ni foam improved the overall electrochemical performance of the electrode, despite the relatively high resistance of MoS_2, which showed a stable capacitance up to 500 cycles with retention of 85.1%. This effect was likely due to the hydrothermal method approach that provided the dissolution of both MoO_4^{2-} and H_2S, formed during the decomposition of L-cystine, under the hydrothermal reactor environment. Through that, the MoS_2 was grown over Ni foam providing a strong interfacial contact between them, which improved its stability. Because of that, the electrical double-layer mechanism for charge storage was also optimized as there was a minor detachment of active material from the electrode's surface. The MoS_2 nanolayers reached 129.2 F g^{-1} of specific capacitance, which were satisfactory values for straightforward adhesion of a TMDC without any other additive.

The hydrothermal method is constantly used to synthesize composites for energy storage applications. In the previous case, neat nanolayers of MoS_2 were obtained, which deliver satisfactory properties when considered as a standalone electrode material. Yet, its properties can be further optimized by introducing other nanomaterials such as N-doped graphene (NG) as it was performed by Xie et al. [40]. For this case, the one-step hydrothermal method was used to obtain nanoflowers of MoS_2 over NG's surface. The authors compared the performance of a composite electrode material based on MoS_2 over graphene nanosheets (MoS_2/GN) as well as NG (MoS_2/NG). The procedure was based on a typical synthetical approach for the hydrothermal method. It was notable that the doped nitrogen in graphene's structure promoted a considerable change in the composite's morphology even though the amount of precursor, Na_2MoO_4 and GO, were the same for both cases. On top of that, the load of MoS_2 deposited over graphene's surface remained nearly unaltered when comparing the neat graphene and N-doped. Yet, the ratio of precursors is an important matter since the high concentration of MoS_2 over graphene's surface decreases the interfacial contact between graphene and the electrolyte, which partially hinders the ionic transport for the charge/discharge process. On the other hand, a lower amount of Mo would lead to a 3D structure as it would facilitate the restacking of graphene nanolayers. Based on these events, it is also deemed important to understand the growth of MoS_2/Graphene as it takes place through a reductive process where the chemical driving force is provided by thiourea (NH_2CSNH_2), which simultaneously reduced both $MoO_4{}^{2-}$ and GO into a layered structure of MoS_2 deposited over rGO. The proper electrostatic interaction between both components promotes a more even distribution of MoS_2. The description of the synthetical process along with the change in morphology of the GO compared to the MoS_2/GN and MoS_2/NG is presented in Figure 4.8. This change in morphology and electrochemical behavior was observed during the CV analysis as a synergistic effect between rGO providing an electric double-layer capacitance whereas MoS_2 provided a redox process based on the Mo^{4+}/Mo^{5+} pair. This process was enhanced due to the presence of N, which facilitates the electron transfer step. This improvement in electrochemical properties was suggested by the appearance of a covalent bond of C–O–Mo, which provides better performance for the electrode. That effect was observed by the electrochemical analysis results at which MoS_2/NG containing 1.5 ratios of Mo source provided the highest capacitance when compared to NG and MoS_2, with the respective values of 227, 155, and 70 F g^{-1}. Through that, it was notable the influence of an N-doped conductive substrate and the proper arrangement of MoS_2 over graphene through the hydrothermal method, which reinforces the convenience of this method. Based on these concepts, it is notable that the hydrothermal approach is a convenient method to perform the growth of precursors into larger materials through a convenient and facile process that yields optimized materials usually with proper adhesion between the active materials and the substrate. On top of that, it's a versatile process that can be widely applied in several cases.

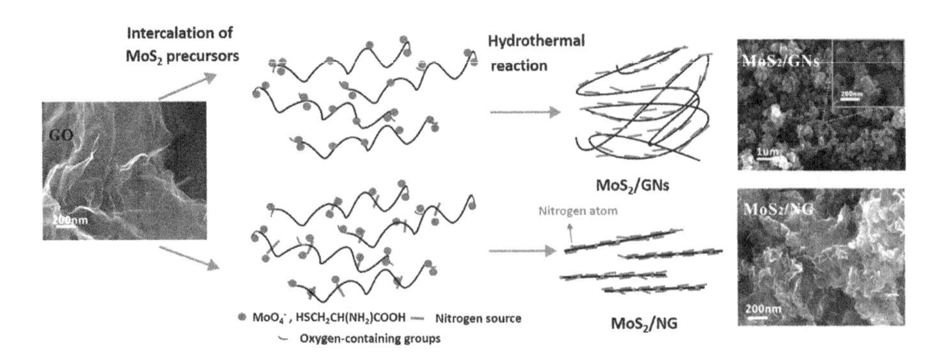

FIGURE 4.8 Schematics for the synthesis of nanosheets of graphene functionalized with MoS_2 (MoS_2/NG) and N-doped graphene functionalized with MoS_2 (MoS_2/GN) under the presence of thiourea for the chemical reduction of the precursors performed through the hydrothermal method. Adapted with permission from reference [40], Copyright (2016), Elsevier.

4.4 CONCLUSION

In conclusion to this chapter, it is notable the bottom-up approaches promote a viable option to obtain high-performance materials, which were vastly used in the energy storage field. That is due to the control over the system that bottom-up methods offer in comparison to other procedures. For instance, either CVD or PVD allows accurate control over the amount of material that is deposited over a substrate based on several parameters such as precursor concentration, carrier gas, temperature, and distance from the substrate. Even though lower quantities are produced the variation in these parameters can considerably change properties in terms of morphology and surface area, which are core aspects for the optimization of electrochemical properties. Also, UHV-assisted system promoted an atom accurate process as graphene can be grown over diverse surfaces with its pore size pre-defined. Such prediction is possible by selecting the desired precursor for the process as well as growing the 2D nanomaterial over a specific substrate, which are usually metals such as Cu, Ag, Au, or Pt. The latter plays an important role as it defines the nanostructure of the growth film, which effectively changes the film's morphology. Another widely used method for the synthesis of nanofilms is the hydrothermal or solvothermal method, which among the other methods presents a relatively lower cost and usually yields materials with satisfactory electrochemical performances. Based on that, bottom-up approaches are valuable techniques that allow more control over the reaction system for the growth of nanofilms, which provides high surface area and exposure of active sites, which are inherently important aspects for the enhancement energy storage properties.

REFERENCES

1. S. Mukherjee, Z. Ren, G. Singh, Beyond graphene anode materials for emerging metal ion batteries and supercapacitors, *Nano-Micro Lett.* 10 (2018) 70.
2. D. Zhao, Y. Feng, Y. Wang, Y. Xia, Electrochemical performance comparison of $LiFePO_4$ supported by various carbon materials, *Electrochim. Acta* 88 (2013) 632–638.

3. H.T. Tan, W. Sun, L. Wang, Q. Yan, 2D transition metal oxides/hydroxides for energy-storage applications, *ChemNanoMat* 2 (2016) 562–577.
4. C.C. Leong, H. Pan, S.K. Ho, Two-dimensional transition-metal oxide monolayers as cathode materials for Li and Na ion batteries, *Phys. Chem. Chem. Phys.* 18 (2016) 7527–7534.
5. R. Rojaee, R. Shahbazian-Yassar, Two-dimensional materials to address the lithium battery challenges, *ACS Nano* 14 (2020) 2628–2658.
6. Y. Li, X. Yin, W. Wu, Preparation of few-layer MoS_2 nanosheets via an efficient shearing exfoliation method, *Ind. Eng. Chem. Res.* 57 (2018) 2838–2846.
7. Y. Huang, J. Guo, Y. Kang, Y. Ai, C.M. Li, Two dimensional atomically thin MoS_2 nanosheets and their sensing applications, *Nanoscale* 7 (2015) 19358–19376.
8. M.A. Tsiamtsouri, P.K. Allan, A.J. Pell, J.M. Stratford, G. Kim, R.N. Kerber, P.C.M.M. Magusin, D.A. Jefferson, C.P. Grey, Exfoliation of layered na-ion anode material $Na_2Ti_3O_7$ for enhanced capacity and cyclability, *Chem. Mater.* 30 (2018) 1505–1516.
9. C. Murugan, V. Sharma, R.K. Murugan, G. Malaimegu, A. Sundaramurthy, Two-dimensional cancer theranostic nanomaterials: Synthesis, surface functionalization and applications in photothermal therapy, *J. Control. Release* 299 (2019) 1–20.
10. T. Wu, X. Zhang, Q. Yuan, J. Xue, G. Lu, Z. Liu, H. Wang, H. Wang, F. Ding, Q. Yu, X. Xie, M. Jiang, Fast growth of inch-sized single-crystalline graphene from a controlled single nucleus on Cu–Ni alloys, *Nat. Mater.* 15 (2016) 43–47.
11. K. Kang, S. Xie, L. Huang, Y. Han, P.Y. Huang, K.F. Mak, C.-J. Kim, D. Muller, J. Park, High-mobility three-atom-thick semiconducting films with wafer-scale homogeneity, *Nature* 520 (2015) 656–660.
12. S. Wang, M. Pacios, H. Bhaskaran, J.H. Warner, Substrate control for large area continuous films of monolayer MoS_2 by atmospheric pressure chemical vapor deposition, *Nanotechnology* 27 (2016) 85604.
13. Z. Cai, B. Liu, X. Zou, H.-M. Cheng, Chemical vapor deposition growth and applications of two-dimensional materials and their heterostructures, *Chem. Rev.* 118 (2018) 6091–6133.
14. T. Roy, M. Tosun, J.S. Kang, A.B. Sachid, S.B. Desai, M. Hettick, C.C. Hu, A. Javey, Field-effect transistors built from all two-dimensional material components, *ACS Nano* 8 (2014) 6259–6264.
15. Y. Gao, Z. Liu, D.-M. Sun, L. Huang, L.-P. Ma, L.-C. Yin, T. Ma, Z. Zhang, X.-L. Ma, L.-M. Peng, H.-M. Cheng, W. Ren, Large-area synthesis of high-quality and uniform monolayer WS2 on reusable Au foils, *Nat. Commun.* 6 (2015) 8569.
16. Y. He, A. Sobhani, S. Lei, Z. Zhang, Y. Gong, Z. Jin, W. Zhou, Y. Yang, Y. Zhang, X. Wang, B. Yakobson, R. Vajtai, N.J. Halas, B. Li, E. Xie, P. Ajayan, Layer engineering of 2D semiconductor junctions, *Adv. Mater.* 28 (2016) 5126–5132.
17. J. Jeon, S.K. Jang, S.M. Jeon, G. Yoo, Y.H. Jang, J.-H. Park, S. Lee, Layer-controlled CVD growth of large-area two-dimensional MoS_2 films, *Nanoscale* 7 (2015) 1688–1695.
18. L. Jae-Hyun, L.E. Kyung, J. Won-Jae, J. Yamujin, K. Byung-Sung, L.J. Young, C. Soon-Hyung, A.S. Joon, A.J. Real, P. Min-Ho, Y. Cheol-Woong, C.B. Lyong, H. Sung-Woo, W. Dongmok, Wafer-scale growth of single-crystal monolayer graphene on reusable hydrogen-terminated germanium, *Science* 344 (2014) 286–289.
19. Q. Ji, Y. Zhang, J. Shi, J. Sun, Y. Zhang, Z. Liu, Morphological engineering of CVD-grown transition metal dichalcogenides for efficient electrochemical hydrogen evolution, *Adv. Mater.* 28 (2016) 6207–6212.
20. S. Wang, Y. Rong, Y. Fan, M. Pacios, H. Bhaskaran, K. He, J.H. Warner, Shape evolution of monolayer MoS_2 crystals grown by chemical vapor deposition, *Chem. Mater.* 26 (2014) 6371–6379.
21. S. Song, D.H. Keum, S. Cho, D. Perello, Y. Kim, Y.H. Lee, Room temperature semiconductor–metal transition of $MoTe_2$ thin films engineered by strain, *Nano Lett.* 16 (2016) 188–193.

22. S.J.R. Tan, I. Abdelwahab, Z. Ding, X. Zhao, T. Yang, G.Z.J. Loke, H. Lin, I. Verzhbitskiy, S.M. Poh, H. Xu, C.T. Nai, W. Zhou, G. Eda, B. Jia, K.P. Loh, Chemical stabilization of 1T′ phase transition metal dichalcogenides with giant optical kerr nonlinearity, *J. Am. Chem. Soc.* 139 (2017) 2504–2511.
23. Z. Wang, P. Liu, Y. Ito, S. Ning, Y. Tan, T. Fujita, A. Hirata, M. Chen, Chemical vapor deposition of monolayer $Mo_{1-x}W_xS_2$ crystals with tunable band gaps, *Sci. Rep.* 6 (2016) 21536.
24. C. Muratore, A.A. Voevodin, N.R. Glavin, Physical vapor deposition of 2D Van der Waals materials: A review, *Thin Solid Films* 688 (2019) 137500.
25. Y. Yamamura, H. Tawara, Energy dependence of ion-induced sputtering yields from monatomic solids at normal incidence, *At. Data Nucl. Data Tables* 62 (1996) 149–253.
26. J. Tao, J. Chai, X. Lu, L.M. Wong, T.I. Wong, J. Pan, Q. Xiong, D. Chi, S. Wang, Growth of wafer-scale MoS_2 monolayer by magnetron sputtering, *Nanoscale* 7 (2015) 2497–2503.
27. M.I. Ionescu, X. Sun, B. Luan, Multilayer graphene synthesized using magnetron sputtering for planar supercapacitor application, *Can. J. Chem.* 93 (2014) 160–164.
28. J. Deng, R. Zheng, Y. Zhao, G. Cheng, Vapor–solid growth of few-layer graphene using radio frequency sputtering deposition and its application on field emission, *ACS Nano* 6 (2012) 3727–3733.
29. X.P. Qiu, Y.J. Shin, J. Niu, N. Kulothungasagaran, G. Kalon, C. Qiu, T. Yu, H. Yang, Disorder-free sputtering method on graphene, *AIP Adv.* 2 (2012) 32121.
30. C.R. Serrao, A.M. Diamond, S.-L. Hsu, L. You, S. Gadgil, J. Clarkson, C. Carraro, R. Maboudian, C. Hu, S. Salahuddin, Highly crystalline MoS2 thin films grown by pulsed laser deposition, *Appl. Phys. Lett.* 106 (2015) 52101.
31. C.W. Tang, Two-layer organic photovoltaic cell, *Appl. Phys. Lett.* 48 (1986) 183–185.
32. T. Mirabito, B. Huet, J.M. Redwing, D.W. Snyder, Influence of the underlying substrate on the physical vapor deposition of Zn-phthalocyanine on graphene, *ACS Omega* 6 (2021) 20598–20610.
33. T. Ogikubo, H. Shimazu, Y. Fujii, K. Ito, A. Ohta, M. Araidai, M. Kurosawa, G. Le Lay, J. Yuhara, Continuous growth of germanene and stanene lateral heterostructures, *Adv. Mater. Interfaces* 7 (2020) 1902132.
34. J.A. Lipton-Duffin, O. Ivasenko, D.F. Perepichka, F. Rosei, Synthesis of polyphenylene molecular wires by surface-confined polymerization, *Small* 5 (2009) 592–597.
35. M. Bieri, M.-T. Nguyen, O. Gröning, J. Cai, M. Treier, K. Aït-Mansour, P. Ruffieux, C.A. Pignedoli, D. Passerone, M. Kastler, K. Müllen, R. Fasel, Two-dimensional polymer formation on surfaces: Insight into the roles of precursor mobility and reactivity, *J. Am. Chem. Soc.* 132 (2010) 16669–16676.
36. G. Franc, A. Gourdon, Covalent networks through on-surface chemistry in ultra-high vacuum: State-of-the-art and recent developments, *Phys. Chem. Chem. Phys.* 13 (2011) 14283–14292.
37. G. Otero, G. Biddau, C. Sánchez-Sánchez, R. Caillard, M.F. López, C. Rogero, F.J. Palomares, N. Cabello, M.A. Basanta, J. Ortega, J. Méndez, A.M. Echavarren, R. Pérez, B. Gómez-Lor, J.A. Martín-Gago, Fullerenes from aromatic precursors by surface-catalysed cyclodehydrogenation, *Nature* 454 (2008) 865–868.
38. G. Yang, S.-J. Park, Conventional and microwave hydrothermal synthesis and application of functional materials: A review, *Materials (Basel)* 12 (2019) 1177.
39. K.-J. Huang, J.-Z. Zhang, G.-W. Shi, Y.-M. Liu, Hydrothermal synthesis of molybdenum disulfide nanosheets as supercapacitors electrode material, *Electrochim. Acta* 132 (2014) 397–403.
40. B. Xie, Y. Chen, M. Yu, T. Sun, L. Lu, T. Xie, Y. Zhang, Y. Wu, Hydrothermal synthesis of layered molybdenum sulfide/N-doped graphene hybrid with enhanced supercapacitor performance, *Carbon N. Y.* 99 (2016) 35–42.

5 Types of Energy Devices and Working Principles

Yuyan Wang, Yang Liu, Linrui Hou, and Changzhou Yuan
University of Jinan

CONTENTS

5.1 SOLAR CELLS

As one kind of renewable energy, solar energy is widely distributed, abundant resources, environmentally friendliness, and sustainable used. And solar cells, which aim at collecting and converting the solar energy, have become a sustainable energy device with wide applications and promising prospects. By directly using solar

DOI: 10.1201/9781003178453-5

energy to convert into electrical energy through photoelectric effect or photochemical effect, it is also called solar photovoltaic technology. Generally, solar cells can be divided into crystalline silicon solar cells (SSCs; the first generation), thin-film solar cells (TFSCs; the second generation), and emerging solar photovoltaic cells (the third generation).

5.1.1 Silicon Solar Cells (SSCs)

Silicon (Si) is the second (the content only lower than oxygen) abundant reserves in the earth's crust, and the silicon-based solar cells with a stable, non-toxicity, abundant resources are currently the most mature solar cells, which have occupied about 90% of the solar cell market. And crystalline silicon (C-Si) with an energy bandgap of 1.12 eV is one of the most appropriate candidates for building multi-junction cells. Moreover, the current crystalline SSCs with the maximum conversion efficiency as high as 26.7%, which is very close to the theoretical maximum conversion efficiency of 31% [1–3], that is far higher than other types of solar cells in practical application. Therefore, crystalline SSCs have become widely used and the best commercialized photovoltaic power generation products.

The core of crystalline SSCs is the PN-junction, and the basic principle is to convert solar energy into electricity based on the photovoltaic effect of the p-n junction, the working principle is shown in Figure 5.1a. When the P-N junction is irradiated by sunlight, the photogenerated carriers gather at both ends of the P-N junction under the action of internal electric field and generate potential difference. When the external circuit is connected, the current flows through the external circuit under the effect of the potential difference, thus generating the output power.

5.1.2 Thin-Film Solar Cells (TFSCs)

TFSCs have received great attention because of their wide material sources, good flexibility, easy modulation function, simple preparation process, low cost, easy to prepare large area of flexible devices. They are usually formed by deposing a layer or multiple layers of photovoltaic materials on the substrate to form P-i-N nodes, whose thickness is usually tens of nanometers to tens of microns. According to the different deposition materials, the TFSCs can be divided into amorphous silicon (a-Si) and mono-crystalline silicon (c-Si), indium phosphide (InP), cadmium telluride (CdTe), copper indium gallium selenium (CIS or CIGS), and dye solar cells (DSCs).

TFSCs are a kind of photodiodes, under the dark and no illuminated conditions, the current dense-voltage curve satisfies the ***Shockley*** Eq. (5.1) [4].

$$J = J_{sc}\left\{\exp\left[\frac{q(V - IR_s)}{\eta k_B T}\right] - 1\right\} + \frac{(V - IR_s)}{R_p} \tag{5.1}$$

where $\boldsymbol{J_{sc}}$ is the reverse saturation current density, $\boldsymbol{k_B}$ is the Boltzmann constant, $\boldsymbol{q}$ is the elementary charge, $\boldsymbol{T}$ is the Kelvin temperature, and $\boldsymbol{\eta}$ is the ideal factor. However, there are series resistance ($\boldsymbol{R_s}$) and parallel resistance ($\boldsymbol{R_p}$) in practical devices. $\boldsymbol{R_s}$

includes the bulk resistance, interface contact resistance, etc., which are in series with load resistance; $\boldsymbol{R}_p$ comes from various leakage paths of the photocell.

When light irradiates the device, a current will be generated in the device, that is the photogenerated current I. The Shockley under illumination becomes Eq. (5.1):

$$J = J_{sc}\left\{\exp\left[\frac{q(V - IR_s)}{(\eta + cV)k_B T}\right] - 1\right\} + \frac{(V - IR_s)}{R_p} \tag{5.2}$$

The active absorbing layer is sandwiched between two carrier-selective interfacial layers that are deposited from solution or thermally evaporated. Finally, a reflective metal contact is evaporated to complete the device.

5.1.3 Dye Solar Cells (DSCs)

DSCs as a semiconductor photovoltaic device can directly convert solar radiation into electric current, which have been attracting researcher's attention due to their abundant sources, low cost, and easily manufacturing process [5]. Typically DSCs consists of a semiconductor oxide (TiO_2, ZnO, or SnO_2) film, a transparent conductive substrate, a dye, an electrolyte (iodide/triiodide), and a counter electrode (Pt or carbon-based materials) coated on transparent conductive substrate [6].

The working principle of DSCs is based on the injection of electrons into nanostructured metal-oxide (TiO_2, ZnO) photoanodes from chemically adsorbed dye molecules, regenerated by a redox pair in a liquid electrolyte [7]. As shown in Figure 5.1c, when the dye molecules adsorbed on the semiconductor film are photoexcited by the light, the electrons transition from the ground state to the excited state, and the excited state electrons are injected into the conduction band of the semiconductor, making the sensitizer in an oxidized state. When the electrons injected into the conduction band flow out from the photoanode to the counter electrode, the working current is generated through the external circuit.

5.1.4 Perovskite Solar Cells (PSCs)

Perovskite is a ceramic oxide with a regular octahedron structure and a general molecular formula of ABX_3. "A" usually refers to the monovalent alkaline or organic cations, such as $CH_3NH_3^+$, $CH_3(NH_2)_2^+$, or Cs^+; "B" is typically divalent metal ion Pb^{2+} or Sn^{2+}; and "X" represents halide ion Cl^-, Br^-, and I^-. Perovskite solar cells (PSCs) with small bandgaps, high extinction coefficients, and high carrier mobility have become a promising new photovoltaic technology [8].

In PSCs, the perovskite material acts as an absorbing layer that can absorb solar photons to produce both electrons and holes, and the electron–hole pair separation occurs under the internal electric field. The electrons transition from the valence band of perovskite to the conduction band and the corresponding holes are left in the valence band, and electrons and holes are collected at each electrode, respectively. Finally, generated the current in the external circuit and completed the conversion of light energy to electric energy. Its basic working principle is shown in Figure 5.1d.

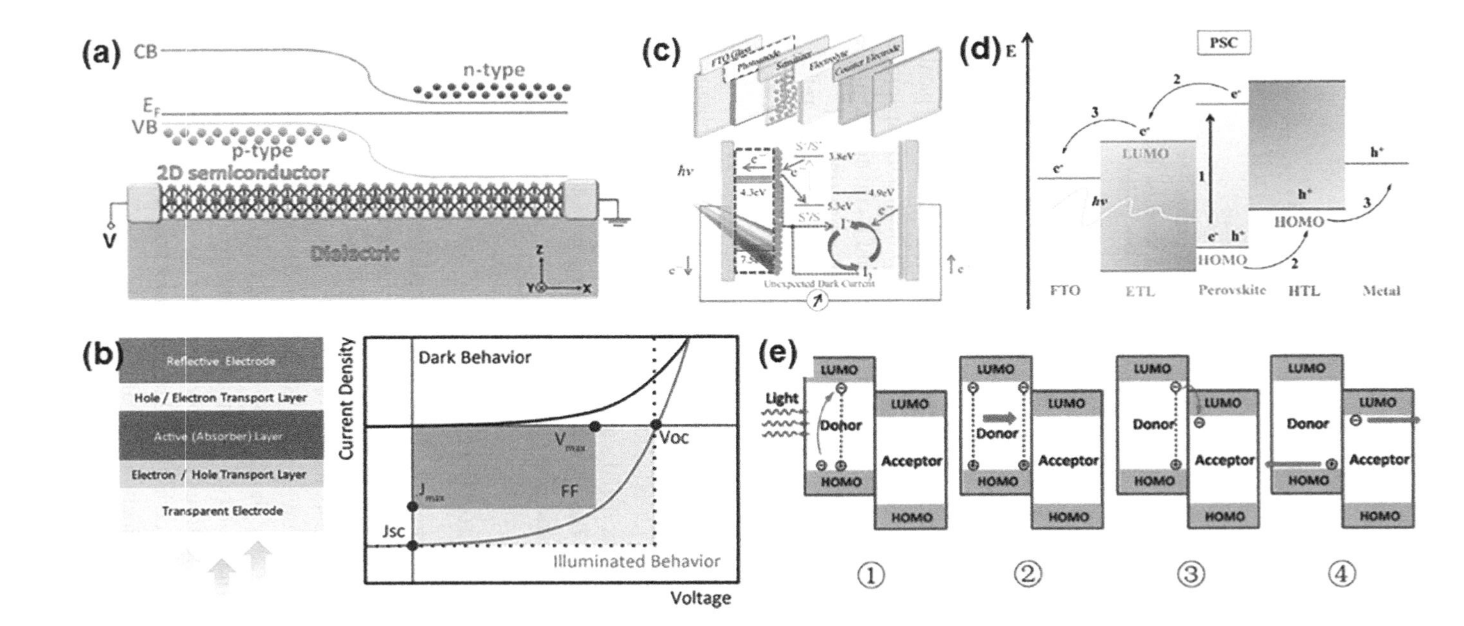

FIGURE 5.1 (a) Schematic diagram of PN-junction for the crystalline silicon solar cells. Adapted with permission from [1]. Copyright (2020) The Authors, some rights reserved; exclusive licensee [The Royal Society of Chemistry].) (b) The architecture and materials (left) and current density versus voltage curve (right) for the thin-film solar cells [11]. Copyright (2016) WILEY-VCH. (c) Structure and schematic diagram of charge transfer in dye solar cells [12]. (Copyright (2019) WILEY-VCH.) (d) Energy band and component diagram of a typical perovskite solar cells [13]. Copyright (2016) Elsevier. (e) Working mechanism of organic solar cells [14]. Copyright (2016) Elsevier.

5.1.5 Organic Solar Cells (OSCs)

Organic solar cells (OSCs) with easy preparation, lightweight, low cost, strong flexibility, environmental friendliness, etc. have been developed rapidly in recent years [9,10]. And it mainly uses photosensitive organic compounds as semiconductor materials to generate voltage and current by photovoltaic effect.

OSCs absorb light energy and convert it into electricity. As shown in Figure 5.1e, the main working principle can be divided as (a) light absorption, (b) exciton transport, (c) exciton dissociation, and (d) charge transport and collection [10]. When the OSC device is exposed to the sunlight, the donor material in the active layer absorbs photons and produces excitons. The exciton diffuses to the interface of the donor and dissociates, and the electron jumps into the acceptor material and the holes remain in the donor material. The electrons and holes then migrate to the cathode and anode, respectively, and eventually form a photocurrent in an external circuit.

5.2 FUEL CELLS

Fuel cells are a variety of galvanic cells with an open system, where the electrode (both anode and cathode) is just the charge transfer media and the active materials for the redox reaction are transported from outside. Using pure hydrogen as the fuel that successful converts the chemical energy into electric energy while causing little pollution. Usually, five major types of fuel cells can be distinguished by the type of electrolyte: alkaline fuel cell (AFC), phosphoric-acid fuel cell (PAFC), molten-carbonate fuel cells (MCFCs), solid-oxide fuel cells (SOFCs), and proton-exchange membrane fuel cell (PEMFC).

5.2.1 Alkaline Fuel Cells (AFCs)

The AFCs were the first type of fuel cells to be put into the practical service. By employing liquid alkaline solution as electrolytes (aqueous KOH), hydrogen as fuel, inexpensive Ni or Ag as catalysts, the AFCs can convert the H_2 gas directly into electricity [15,16]. As shown in Figure 5.2a, the H_2 directly reacts with OH^- to generate water and electrons (hydrogen oxidation reaction) on the anode, and the electrons generated at the anode are transferred to the cathode through an external circuit, where the oxygen reduction reaction of O_2 reacts with H_2O to generate OH^-. The direction of the reactions on the electrodes is opposite to that in alkaline water electrolysis.

The overall reactions are given by:

Anode: $H_2 + 2OH^- \rightarrow 2H_2O + 2e^-$
Cathode: $1/2O_2 + H_2O + 2e^- \rightarrow 2OH^-$
The overall reaction: $H_2 + 1/2O_2 \rightarrow H_2O$

5.2.2 Phosphoric Acid Fuel Cells (PAFCs)

PAFCs are one of the most practical fuel cells, which used concentrated phosphoric acid (95% ~ 100%) as electrolyte, Pt or Pt alloy supported on carbon as catalyst, natural gas or methanol conversion gas as raw material, and the working temperature is

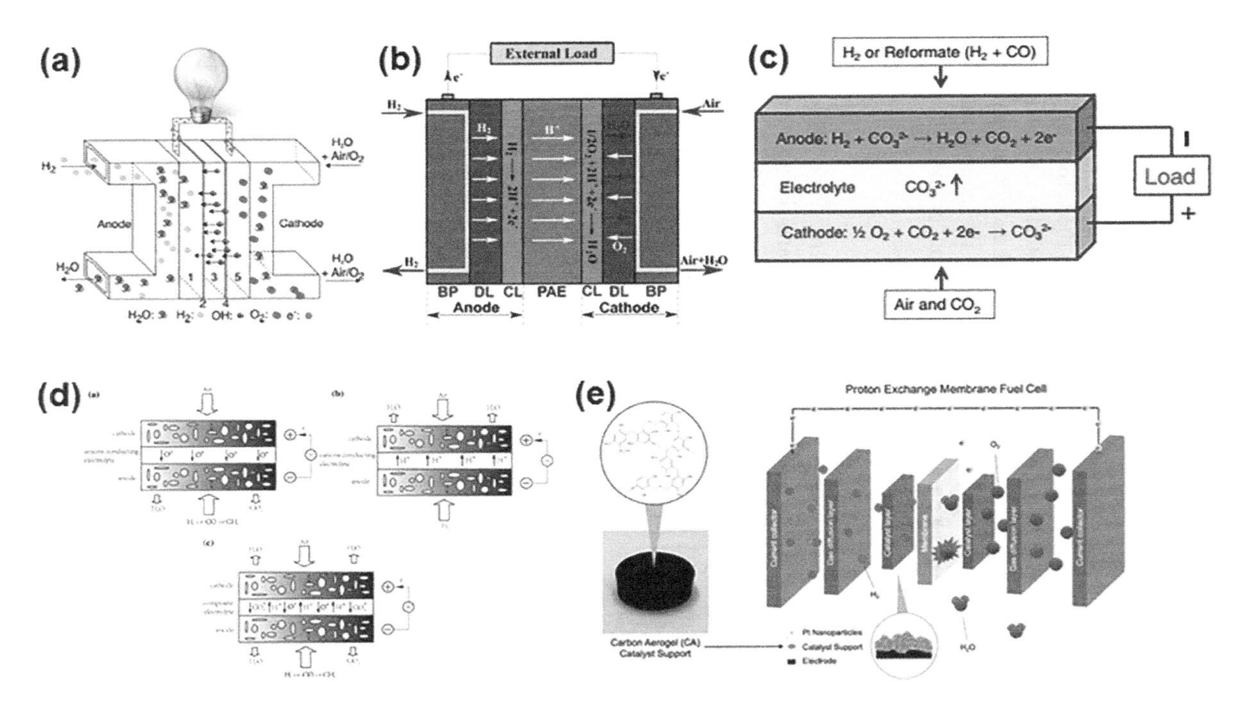

FIGURE 5.2 (a) The essential structures of alkaline-based fuel cells. Adapted with permission from [27]. Copyright (2021) The Authors, some rights reserved; exclusive licensee [Elsevier].) (b) The structure and working principle of the PAFCs [17]. Copyright (2020) Elsevier. (c) A schematic diagram of molten-carbonate fuel cells [20]. (Copyright (2012) Springer-Verlag.) (d) The schematic of three-type SOFCs according to the different types of the electrolyte ion conduction: (a) oxygen ion conductor (O-SOFC); (b) proton conduction (H-SOFC); and (c) mixed ion conductor [21]. Copyright (2020) Springer-Verlag. (e) The essentials structure of proton-exchange membrane fuel cells [28]. Copyright (2020) Elsevier.

150°C ~ 210°C, the power generation efficiency is about 40% [17,18]. Commonly, the PAFCs consist of bipolar plants, gas diffusion layers, catalyst layers, and a phosphate acid electrolyte layer (Figure 5.2b).

The working process is as follows: hydrogen fuel is first added to the anode, and then oxidized into protons under the catalyst. Two free electrons are released at the same time. Hydrogen protons combine with phosphoric acid to form phosphoric acid protons and move to the anode electrode. At the same time, electrons move to the anode electrode, while hydrated protons move to the cathode through phosphoric acid electrolyte. Therefore, on the anode electrode, electrons, hydrated protons, and oxygen generate water molecules under the action of catalyst [18]. The specific electrode reaction is expressed as follows:

Anode: $H_2 + 2OH^- \rightarrow 2H_2O + 2e^-$
Cathode: $1/2O_2 + H_2O + 2e^- \rightarrow 2OH^-$
The overall reaction: $H_2 + 1/2O_2 \rightarrow H_2O$

5.2.3 Molten Carbonate Fuel Cells (MCFCs)

Generally, the MCFCs are composed by a porous Ni/Al(Cr) as anode, NiO as cathode, and bicarbonate or alkaline carbonate as electrolyte. Compared with low-temperature fuel cells, the MCFCs operate at 600°C–700°C, which eliminates the need for the noble catalysts such as Pt [19].

The essential working process of MCFCs is the fuel oxidation and redox process as shown in Figure 5.2c [20]. CO_2 is used to produce carbonate ions in the cell, i.e., the ionic conductor in the electrolyte. The fuel and oxidant gases flow through the anode and cathode channels. The electrochemical reaction includes the formation of the carbonate ion $\left(CO_3^{2-}\right)$ at the cathode and transporting the carbonate ion to anode through the molten carbonate mixture.

At the cathode, O_2 and CO_2 in the oxidant react with electrons to produce CO_3^{2-}. CO_3^{2-} in the electrolyte moves directly from the cathode to the anode, and H_2 react with CO_3^{2-} at the anode to generate CO_2, H_2O, and electrons. The electrons are collected by the collector and flowed through the external circuit to complete the reaction.

Cathode: $1/2O_2 + CO_2 + 2e^- \rightarrow CO_3^{2-}$
Anode: $H_2 + CO_3^{2-} \rightarrow CO_2 + 2e^- + H_2O$
Overall reaction: $H_2 + 1/2O_2 \rightarrow CO_2 + H_2O$

5.2.4 Solid Oxide Fuel Cells (SOFCs)

SOFCs are a kind of electrochemical power generation device with all-ceramic structure. Generally, cermet oxides with fluorite structure and perovskite structure are used as electrolytes, which can avoid the possible loss or corrosion of the electrolyte [21,22]. The ions conducted by electrolytes can be oxygen ions and protons, and the working temperature of the SOFCs is as high as 400°C–1,000°C, and many fuels can be used directly as fuel without precious metals as catalysts, such as pure hydrogen, methane, alcohol, natgas, and marsh gas [23].

According to the different types of electrolyte ion conduction, SOFCs can be divided into three categories: oxygen ion conductor (O-SOFC), proton conduction (H-SOFC), and mixed ion conductor in Figure 5.2d [21,24]. And the electronic conductivity of the electrolyte needs to be negligible to avoid voltage loss and the possibility of short circuit.

For the O-SOFCs transportation mode, O_2 from the air is adsorbed on the cathode surface and electrons are reduced to oxygen ions, and O^{2-} is transferred to the anode through the electrolyte layer, while the fuel is oxidized at the anode to generate electrons, and then transferred to the cathode via the external circuit. When the fuel gas is hydrogen, the corresponding electrochemical reaction is:

At the anode:

$$H_2 + O^{2-} \rightarrow H_2O + 2e^-$$

At the cathode:

$$1/2O_2 + 2e^- \rightarrow O^{2-}$$

The overall reaction:

$$H_2 + 1/2O_2 \rightarrow H_2O$$

For H-SOFCs transportation mode, the protonation reaction of water and hydrogen occurs at the anode, and hydrogen ions are transferred to the cathode through the electrolyte, and react with oxygen ions to produce water. The corresponding electrochemical reaction is as follows:

At the cathode:

$$H_2 \rightarrow 2H^+ + 2e^-$$

At the anode:

$$1/2O_2 + 2H^+ + 2e^- \rightarrow H_2O$$

The overall reaction:

$$1/2O_2 + H_2 \rightarrow H_2O$$

For mixed ion conductor, the ions conduction occurs that depend on the different conducting medium.

5.2.5 Proton-Exchange Membrane Fuel Cell (PEMFCs)

PEMFCs have attracted significant attention as a promising clean energy source due to their excellent characteristics of low operating temperature (typically from 25°C to 90°C), fast start, no electrolyte loss, easy water discharge, long life, high specific power, and specific energy [25].

For the PEMFCs system, hydrogen is oxidized at the anode and protons enter into the electrolyte and are transported to the cathode [26]. As shown in Figure 5.2e, hydrogen gas flows and diffuses into the anode along with the collector channel, then split into two hydrogen protons and two electrons. Protons penetrate through the electrolyte membrane and reach the cathode, oxygen reacts with protons to form water, and electrons generate current in the process of arriving from the cathode to the anode. In this process, the electrolyte membrane provides the reaction place, both the effective contact between the electrode and the electrolyte membrane and the catalyst are the guarantee of the flowing of the proton and electron.

At the anode, hydrogen gas (H_2) is separated into two hydrogen protons (H^+) and two electrons:

$H_2 \rightarrow 2H^+ + 2e^-$

At the cathode, the supplied oxygen reacts according to:

$O_2 + 4e^- \rightarrow 2O_{2-}$

Electrons flow in the external circuit during these reactions. The oxygen ions react with protons to form water:

$O^{2-} + 2H^+ \rightarrow H_2O$

5.3 RECHARGEABLE BATTERIES

The increasing energy crisis and environmental pollution caused by fossil fuel combustion have prompted people to deeply explore efficient and renewable sources [29]. Clean and renewable energies (such as water, wind, or sunlight) play an important role in resolving the increasingly serious environmental problems. However, renewable energy is inherently intermittent, and limited by time and place. To improve the intermittency of renewable energy production, rechargeable secondary batteries as the green energy resources have been considered as a promising option for grid-level stationary energy storage systems and transportation technology [30,31].

5.3.1 Organic Rechargeable Metal Ion Batteries

5.3.1.1 Alkali Metal Ion Batteries

Alkali metal ion battery mainly includes lithium-ion battery, sodium-ion battery, and potassium ion battery. Compared with the traditional commercial batteries, such as lead-acid, metal hydride, and alkaline batteries, LIBs have the advantages of long cycle life, high charge–discharge voltage and energy density, and have been widely employed in portable electronic products, electric vehicles, and large energy storage power stations [32,33]. However, the uneven distribution and limited lithium resources have restricted the future application of lithium-ion battery technology. The content of sodium (2.74wt %) and potassium (2.09wt %) in nature is much higher than that of lithium (0.0017wt %), which reduces the production cost to a certain extent. Moreover, the standard potential of sodium (−2.71V vs. SHE) and potassium (−2.93V vs. SHE) is not very different from that of lithium (−3.04V vs. SHE), ensuring the high specific energy of sodium ion and potassium ion batteries [31]. Therefore, the concept of sodium ion battery and potassium ion battery has been put forward and studied as a potential substitute for lithium ion battery.

Similarly with Li-ion batteries, sodium-ion batteries/potassium-ion batteries (NIBs/KIBs) are also composed of positive and negative electrodes, electrolyte, and

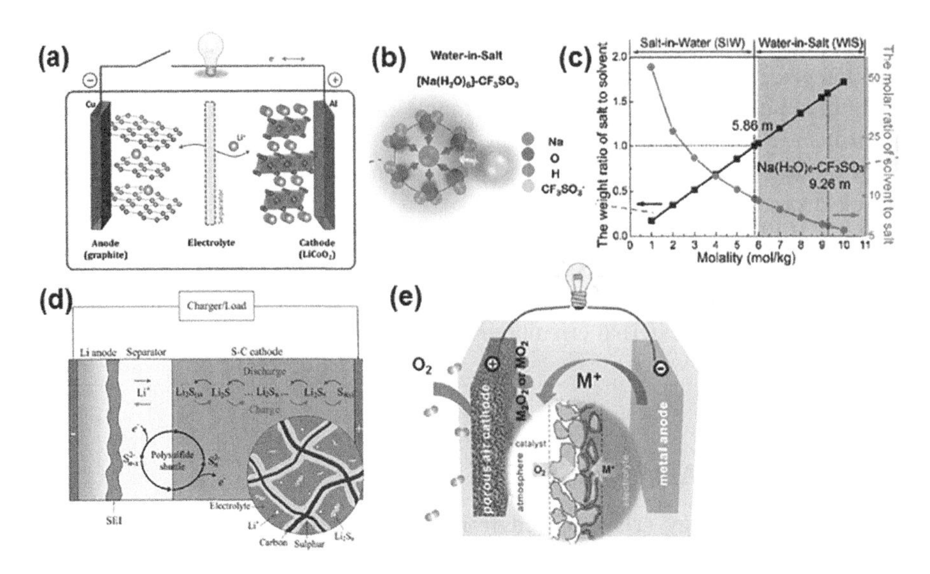

FIGURE 5.3 (a) Schematic illustration of the commercial Li-ion battery ($LiCoO_2$/graphite). Adapted with permission from [32]. Copyright (2013) The Authors, some rights reserved; exclusive licensee [American Chemical Society]. (b, c) The molar and weight salt/solvent ratios in NaOTF-H_2O binary system [39]. Copyright (2017) WILEY-VCH. (d) The typical schematic illustrations of Li-S batteries. Adapted with permission from [43]. Copyright (2015) The Authors, some rights reserved; exclusive licensee [The Royal Society of Chemistry]. (e) Schematic principle for metal-air batteries. Adapted with permission from [45]. Copyright (2017) The Authors, some rights reserved; exclusive licensee [American Chemical Society].

they use the same "rocking-chair"-type mechanism like LIBs do. Typically, a commercial LIB is using graphite anode and transition metal oxide-based cathodes and uses the "rocking-chair"-type mechanism [32]. And the corresponding process is shown in Figure 5.3a.

During the charging process, Li^+ was first escaped from the $LiCoO_2$, then flow through the electrolyte to the electrolyte at the other end, and finally inserted into the layered graphite anode material. To achieve charge balance, electrons must pass through an external circuit from the positive electrode to the negative electrode. At this time, a large number of lithium ions gathered on the anode material temporarily formed a lithium-rich state, while the positive electrode is in a lithium poor state.

The whole charge and discharge process can be summarized as follows:

Cathode: $LiCoO_2 \rightarrow Li_{1-x}CoO_2 + xLi^+ + xe^-$

Anode: $6C + xLi^+ + xe^- \rightarrow LixC_6$

Overall reaction: $LiCoO_2 + 6C \rightarrow Li_{1-x}CoO_2 + Li_xC_6$

5.3.1.2 Multivalent Rechargeable Batteries

The rapid development and urgent demand for consumer electronics, electric transportation technologies, and grid-scale energy storage have forced the enhanced improvement capacity storage for rechargeable batteries. Multivalent rechargeable batteries (Zn/Mg/Ca/Al) are considered as another promising alternative energy storage advice that has gained considerable attention due to their abundant resources, high volumetric energy density, and multielectron redox capability [34].

5.3.2 Aqueous Rechargeable Metal Ion Batteries (ARMIBs)

Aqueous rechargeable metal ion batteries (ARMIBs) have recently attracted extensive attention as promising energy storage systems due to their high safety, low cost, easy-to-assemble, high ionic conductivity, and environmental friendly [35]. These features ascribe to the use of aqueous electrolyte with non-flammability, high ionic conductivity, low interface resistance, and stable voltage window [36,37]. According to the different ionic valence states, the ARMIBs can be roughly divided into two types: aqueous monovalent metal ion batteries (M = Li^+, Na^+, K^+ and Cl^-) and aqueous polyvalent metal ion batteries (M = Zn^{2+}, Mg^{2+}, Ca^{2+}, Al^{3+}).

The fundamental mechanism of ARMIBs is similar to the commercial rechargeable organic metal ion battery systems, which transfers electrons through an external circuit between two electrodes and metal ions through the electrolytes [37]. However, the narrow electrochemical stability window of aqueous electrolyte restricts the selection of electrode materials, the side reactions between the active electrode materials and the aqueous solution to generate hydroxide and H_2 tremendous affect cycle ability of ARMIBs, and metallic dendrite formation also made the metal ion extraction/insertion process more complicated [37,38].

To solve conventional hydrogen evolution for the ARMIBs, the development of the "water-in-salt" electrolyte systems has successfully expanded the electrochemical voltage window up to 3 V by using high concentration electrolytes to avoid the direct interaction between electrode and water, and eliminates the free water molecules in the solution [38,39].

5.3.3 Metal-Sulfur Batteries (MSBs)

Metal-sulfur (Metal = Li, Na, K, Mg, Al) batteries (MSBs) are the most promising battery technologies due to high energy density, abundance of elements, low costs, and less toxic [40,41]. Sulfur is considered as a promising cathode material because of their earth abundant resources, a high theoretical specific capacity of 1,675 mAh g^{-1} (based on a two-electron-transfer reaction mechanism), which has promoted the rapid development of MSB systems [42].

Li-S batteries are the most studied field for a long time, in which the lithium metal anode offers a high energy density (3,681 mAh g^{-1}), and the sulfur cathode provides a theoretical energy density (1,675 mAh g^{-1}), respectively. So the theoretical capacity of Li-S battery is as high as 1,167 mAh g^{-1}. Here, we take Li-S batteries as example to explore the mechanism.

The mechanism of Li-S battery is between LIBs and Li-O_2 battery to some extent [42,43], which is based on a conversion or integration reaction of S_8 and Li^+ at the cathode interface. As shown in Figure 5.3d, in the discharging process, lithium ions and electrons are produced by oxidation reaction of lithium anode. The positive sulfur gains electrons and the reduction reaction forms lithium sulfide. In the positive electrode reduction process, the S_8 molecule first opens the ring long chain Li_2S_8 molecule. With the progress of reduction reaction, the long chain Li_2S_8 molecule is further reduced to the short chain polysulfide ion Li_2S_x ($4 \leq x \leq 6$), and finally is reduced to Li_2S_2/Li_2S. In the charging process, on the contrary, Li_2S_2/Li_2S goes through multiple steps of oxidation reaction to finally generate elemental sulfur and generate electrons and lithium ions. The electrons return to the lithium anode through external circuit, and the lithium ions are also transferred to the metal lithium anode through electrolyte, where they get electrons and are reduced to lithium elemental.

5.3.4 Metal-Air Batteries (MABs)

Metal-air batteries (MABs; metals such as Li, Na, K, and Zn) with an open cell structure, large storage capacity have received revived interest recently. These systems combine the design characteristics of conventional rechargeable secondary batteries and fuel cells in which used a metal as negative electrode, neutral or alkaline solvents as electrolytes, oxygen as cathode active materials that breathed from ambient atmosphere. As the cathode material is directly replaced by O_2 in the air, the total weight of the MABs can be greatly reduced, thus greatly improving the energy density of the battery. Different types of MABs can be found in recent researches, including Li-air, Na-air, K-air, Zn-air, Mg-air, and Al-air batteries [44,45].

Generally, the reaction for the MABs is considered as a redox process (Figure 5.3e). During the discharging process, the metal element is oxidized at the anode, and O_2 from the surrounding atmosphere is reduced on supported electrocatalysts in the cathode as illustrated in Figure 5.3d, and the corresponding reaction equations are as follows:

Anode: $M \leftrightarrow M^{n+} + ne^-$

Cathode: $nO_2 + 4ne^- + 4M^{n+} \leftrightarrow 2M_2O_n$

where M presents the metal and n is the oxidation number of the metal ion.

5.4 SUPERCAPACITORS

Supercapacitors as high electrochemical energy storage devices are widely used in heavy electric vehicles and other types of hybrid batteries, which require high power due to their fast charge/discharge rates (charging times ranging from seconds to minutes), excellent electrochemical performance, high reversibility, and long cycle life [46].

Like batteries, supercapacitor is also composed of two electrodes (an anode and a cathode), an electrolyte (usually is aqueous or organic solution), and a separator that allows ions to transfer and stores the charge through adsorption of ions from electrolytes on electrode surfaces. Generally, the supercapacitors can be divided into three types including electric double-layered capacitors (EDLCs), pseudocapacitors, and asymmetric supercapacitors (ASCs) [47].

5.4.1 Electric Double Layered Capacitors

EDLCs are the simplest and most commercial supercapacitor in which stores charge at the interface between electrode and electrolyte via electrostatic charge adsorption, thus forming the electrical double layers [48].

As we all know, the electrode materials play an important role in providing excellent electrochemical performance for supercapacitors. For EDLC electrodes, carbon-based materials (including the active carbon, carbon nanotubes, carbon aerogels, porous carbon, grapheme, and grapheme oxide) are the commonly used materials due to their high specific surface area, appreciable electrochemical stability, nontoxicity, open porosity, and high electrical conductivities [49]. The specific capacitance of an EDLC is strongly dependent on the accessible surface area of the electrode materials and surface properties of the carbon materials, which stored energy at the interphase between electrode and solution to form an electrical double-layer (Figure 5.4a) [49]. The capacitance of EDLCs can be estimated as following Eq. (5.3):

$$C = \frac{\varepsilon_r \varepsilon_0}{d} A \tag{5.3}$$

where $\boldsymbol{\varepsilon}_r$ is the relative electrolyte dielectric constant, $\boldsymbol{\varepsilon}_0$ is the permittivity of vacuum, $\boldsymbol{A}$ is the specific surface area of the electrode, and $\boldsymbol{d}$ is the effective thickness of charge separation distance layer (the Debye length). According to the physical electrostatic processes, the formation and relaxation of the electric double layer (EDL) occur within 10^{-8} s, which is shorter than that of in pseudocapacitor (10^{-2}–10^{-4} s). The charge/discharge process of EDLCs only store charge at the interface between electrode and electrolyte via charge rearrangement without any redox reactions (Faradaic reactions). Thus, the EDL can respond to change of potential immediately.

5.4.2 Pseudocapacitors

Pseudocapacitors (known as Faraday capacitors or gold capacitors) have a significant contribution in the energy storage fields, which can store charge by a fast and reversible redox reaction on the surface or near the surface of the material. Compared with EDLCs, pseudocapacitors perform higher capacitance due to the additional charges transferred within the defined potential. Generally, metal oxides and conductive polymers are the commonly used electrode materials for pseudocapacitors due to their high energy densities and high capacitance.

Commonly, the pseudocapacitors are classified into three types: (a) underpotential deposition, (b) redox pseudocapacitance (such as MnO_2, RuO_2, V_2O_5, or conductive polymer), and (c) intercalation pseudocapacitance (Nb_2O_5), corresponding processes are illustrated in Figure 5.4b [50]. The mechanism of pseudocapacitors can be summarized as a reversible redox reaction at the interface of the electrode/electrolyte, which is accompanied by the charge transfer, thus achieving the charge storage. During the energy storage process of the pseudocapacitor, a rapid redox reversible reaction will occur on the surface of the electrode active substance, and a large amount of charge or ion transfer can be carried out in a short time [51].

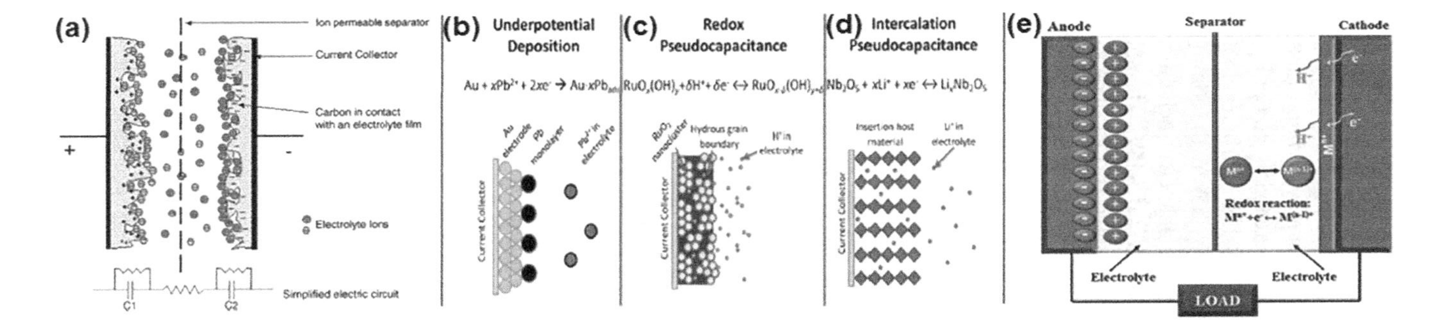

FIGURE 5.4 (a) The construction of the electric double-layer capacitors [56]. Copyright (2006) Elsevier B.V. (b–d) Three types of reversible redox mechanisms for the pseudocapacitance: (b) underpotential deposition, (c) redox pseudo-capacitance, and (d) intercalation pseudocapacitance. Adapted with permission from [50]. Copyright (2014) The Authors, some rights reserved; exclusive licensee [The Royal Society of Chemistry]. (e) The typical schematic construction of an asymmetric supercapacitor [55]. Copyright (2017) WILEY-VCH.

Different physical processes and with different types of materials determine these three different mechanisms, detailed processes are as below:

When metal cation is higher than its redox point, it forms a monolayer adsorption on the surface of another metal, which is the occurrence of underpotential deposition. And one typical example of underpotential deposition is the deposition of Pb^{2+} on the surface of Au [52]. Redox pseudocapacitance occurs when the electrolyte ions or electrons adsorbed on the surface or near the surface of materials, the metal oxides materials are often accompanied by redox reactions, while conductive polymer materials, charge storage is realized by absorbing N-P type doping or dedoping process through the abrupt conjugate effect of polymer chain [53]. The third intercalation pseudocapacitance can be accompanied by rapid Faraday charge transfer when ions are embedded on the surface, interlayer, or channels of an active material without phase transformation reactions.

5.4.3 Asymmetric Supercapacitors (ASCs)

ASCs (also known as hybrid capacitors) are devices composed of two electrical layers with complementary electrochemical windows and pseudocapacitor electrode material. It covers a wider working voltage window that can significantly improve the energy density [54,55].

According to the definition of pseudocapacitor behavior, the discussion between "asymmetric" and "hybridization" only involves devices, not electrodes [47,54]. The term "hybrid capacitor" is generally considered to describe a situation in which the electrodes have two different storage mechanisms: a capacitive storage mechanism and a battery-type storage mechanism (Figure 5.4e). Unlike traditional supercapacitors, ASCs assembled using two different electrode materials, namely battery-type Faradaic cathode electrode and capacitor-type anode electrode. Based on the different electrode materials, the charge storage behaviors can be divided into three types: EDLC//pseudocapacitive type, reversible redox type, and EDLC//battery type.

5.5 CONCLUSION

Electrochemical energy storage technologies can effectively reduce the traditional energy consumption of fossil fuels, avoid the restriction of renewable resources by geographical and environment, which can provide electricity directly when it is needed. Electrochemical energy devices (EEDs) are considered as an alternative energy resource in making renewable energies effective and efficient, while creating a more sustainable and environmentally friendly society. Commonly, the EEDs include solar cells, fuel cells, rechargeable batteries, and supercapacitors. The energy storage and conversion mechanisms are different: solar cells collecting and converting the solar energy into electric current, fuel cells and rechargeable batteries generated the electrical energy via redox reaction (fuel cells are operated in an open system, and batteries are closed system), and supercapacitors delivered the energy via adsorption or redox reaction. Currently, the solar cells are the most pollution-free in energy storage devices and have the largest market space, but so far limited by the energy conversion rate and stability. Rechargeable batteries are the most application markets' energy storage device with the widest market share and have established a certain market position. Whereas supercapacitors with their fast charge/discharge characteristics have occupied a certain market in some electronic devices, electric vehicles, shape memory applications, and light rails. Fuel cells are still in the development process and are looking for applications that are suitable for entering the market.

REFERENCES

1. Yoshikawa K., Yoshida W., Irie T., Kawasaki H., Konishi K., Ishibashi H., Asatani T., Adachi D., Kanematsu M., Uzu H., Yamamoto K. (2017) Exceeding conversion efficiency of 26% by heterojunction interdigitated back contact solar cell with thin film Si technology. *Sol. Energy Mat. Sol. C.* 173:37–42.
2. Tu Y., Wu J., Xu G., Yang X., Cai R., Gong Q., Zhu R., Huang W. (2021) Perovskite solar cells for space applications: Progress and challenges. *Adv. Mater.* 33:e2006545.
3. Zhou D., Liu D., Pan G., Chen X., Li D., Xu W., Bai X., Song H. (2017) Cerium and ytterbium codoped halide perovskite quantum dots: A novel and efficient downconverter for improving the performance of silicon solar cells. *Adv. Mater.* 29: 1704149.
4. Musolino M., Tahraoui A., Treeck D., Geelhaar L., Riechert H. (2016) A modified Shockley equation taking into account the multi-element nature of light emitting diodes based on nanowire ensembles. *Nanotechnology* 27:275203.
5. Ruba N., Prakash P., Sowmya S., Janarthana B., Prabu A.N., Chandrasekaran J., Alshahrani T., Zahran H.Y., Yahia I.S. (2021) Recent advancement in photo-anode, dye and counter cathode in dye-sensitized solar cell: A review. *J. Inorg. Organomet. P.* 31:1894–1901.
6. Raj C.C., Prasanth R. (2016) A critical review of recent developments in nanomaterials for photoelectrodes in dye sensitized solar cells. *J. Power Sources* 317:120–132.
7. Nazeeruddin M.K., Baranoff E., Grätzel M. (2011) Dye-sensitized solar cells: A brief overview. *Sol. Energy* 85:1172–1178.
8. Kim J.Y., Lee J.W., Jung H.S., Shin H., Park N.G. (2020) High-efficiency perovskite solar cells. *Chem. Rev.* 120:7867–7918.
9. Xu Y., Yao H., Ma L., Wang J., Hou J. (2020) Efficient charge generation at low energy losses in organic solar cells: A key issues review. *Rep. Prog. Phys.* 83:082601.
10. Feron K., Belcher W.J., Fell C.J., Dastoor P.C. (2012) Organic solar cells: Understanding the role of Forster resonance energy transfer. *Int. J. Mol. Sci.* 13:17019–17047.
11. Shastry T.A., Hersam M.C. (2017) Carbon nanotubes in thin-film solar cells. *Adv. Energy Mater.* 7:1601205.
12. Gao N., Huang L., Li T., Song J., Hu H., Liu Y., Ramakrishna S. (2019) Application of carbon dots in dye-sensitized solar cells: A review. *J. Appl. Polym. Sci.* 137:48433.
13. Marinova N., Valero S., Delgado J.L. (2017) Organic and perovskite solar cells: Working principles, materials and interfaces. *J. Colloid Interface Sci.* 488:373–389.
14. Zhao F., Wang C., Zhan X. (2018) Morphology control in organic solar cells. *Adv. Energy Mater.* 8:1703147.
15. Merle G., Wessling M., Nijmeijer K. (2011) Anion exchange membranes for alkaline fuel cells: A review. *J. Membrane Sci.* 377:1–35.
16. Wagner K., Tiwari P., Swiegers G.F., Wallace G.G. (2017) Alkaline fuel cells with novel gortex-based electrodes are powered remarkably efficiently by methane containing 5% hydrogen. *Adv. Energy Mater.* 8:17002285.
17. Chen W., Xu C., Wu H., Bai Y., Li Z., Zhang B. (2020) Energy and exergy analyses of a novel hybrid system consisting of a phosphoric acid fuel cell and a triple-effect compression-absorption refrigerator with [mmim]DMP/CH_3OH as working fluid. *Energy* 195:116951.

18. He Q., Mukerjee S., Zeis R., Parres-Esclapez S., Illán-Gómez M.J., Bueno-López A. (2010) Enhanced Pt stability in MO_2 (M=Ce, Zr or $Ce_{0.9}Zr_{0.1}$)-promoted Pt/C electrocatalysts for oxygen reduction reaction in PAFCs. *Appl. Catal. A-Gen.* 381:54–65.
19. Lan R., Tao S.W. (2016) A simple high-performance matrix-free biomass molten carbonate fuel cell without CO_2 recirculation. *Sci. Adv.* 2:e1600772.
20. Giddey A.K.S. (2012) Materials issues and recent developments in molten. *J. Solid State Electrochem.* 16:3123–3146.
21. Lyu Y.M., Xie J.T., Wang D.B., Wang J.R. (2020) Review of cell performance in solid oxide fuel cells. *J. Mater. Sci.* 55:7184–7207.
22. Duan C., Hook D., Chen Y., Tong J., O'Hayre R. (2017) Zr and Y co-doped perovskite as a stable, high performance cathode for solid oxide fuel cells operating below 500°C. *Energy Environ. Sci.* 10:176–182.
23. Kakac S., Pramuanjaroenkij A., Zhou X. (2007) A review of numerical modeling of solid oxide fuel cells. *Int. J. Hydrogen Energy* 32:761–786.
24. Shao Z.P., Haile S.M. (2004) A High Performance cathode for the next generation solid-oxide fuel cells. *Nature* 431:170–173.
25. Wu L., Zhang Z., Ran J., Zhou D., Li C., Xu T. (2013) Advances in proton-exchange membranes for fuel cells: An overview on proton conductive channels (PCCs). *Phys. Chem. Chem. Phys.* 15:4870–4887.
26. Jiao K., Xuan J., Du Q., Bao Z., Xie B., Wang B., Zhao Y., Fan L., Wang H., Hou Z., Huo S., Brandon N.P., Yin Y., Guiver M.D. (2021) Designing the next generation of proton-exchange membrane fuel cells. *Nature* 595:361–369.
27. Ferriday T.B., Middleton P.H. (2021) Alkaline fuel cell technology-A review. *Int. J. Hydrogen Energy* 46:18489–18510.
28. Gu K., Kim E.J., Sharma S.K., Sharma P.R., Bliznakov S., Hsiao B.S., Rafailovich M.H. (2021) Mesoporous carbon aerogel with tunable porosity as the catalyst support for enhanced proton-exchange membrane fuel cell performance. *Mater. Today Energy* 19:100560.
29. Etacheri V., Marom R., Elazari R., Salitra G., Aurbach D. (2011) Challenges in the development of advanced Li-ion batteries: A review. *Energy Environ. Sci.* 4:928–935.
30. Xu C., Chen Y., Shi S., Li J., Kang F., Su D. (2015) Secondary batteries with multivalent ions for energy storage. *Sci. Rep.* 5:14120.
31. Yang Z., Zhang J., Kintner-Meyer M.C., Lu X., Choi D., Lemmon J.P., Liu J. (2011) Electrochemical energy storage for green grid. *Chem. Rev.* 111:3577–3613.
32. Goodenough J.B., Park K.S. (2013) The Li-ion rechargeable battery: A perspective. *J. Am. Chem. Soc.* 135:1167–1176.
33. Larcher D., Tarascon J.M. (2015) Towards greener and more sustainable batteries for electrical energy storage. *Nat. Chem.* 7:19–29.
34. Xie J., Zhang Q. (2019) Recent Progress in multivalent metal (Mg, Zn, Ca, and Al) and metal-ion rechargeable batteries with organic materials as promising electrodes. *Small* 15:e1805061.
35. Liu T., Cheng X., Yu H., Zhu H., Peng N., Zheng R., Zhang J., Shui M., Cui Y., Shu J. (2019) An overview and future perspectives of aqueous rechargeable polyvalent ion batteries. *Energy Storage Mater.* 18:68–91.
36. Liu Y., He G., Jiang H., Parkin I.P., Shearing P.R., Brett D.J.L. (2021) Cathode design for aqueous rechargeable multivalent ion batteries: Challenges and opportunities. *Adv. Funct. Mater.* 31:2010445.
37. Ao H., Zhao Y., Zhou J., Cai W., Zhang X., Zhu Y., Qian Y. (2019) Rechargeable aqueous hybrid ion batteries: Developments and prospects. *J. Mater. Chem. A* 7:18708–18734.
38. Wang H.P., Tan R., Yang Z.X., Feng Y.Z., Duan X.C., Ma J.M. (2020) Stabilization perspective on metal anodes for aqueous batteries. *Adv. Energy Mater.* 11:2000962.

39. Suo L., Borodin O., Wang Y., Rong X., Sun W., Fan X., Xu S., Schroeder M.A., Cresce A.V., Wang F., Yang C., Hu Y.S., Xu K., Wang C.S. (2017) "Water-in-Salt" electrolyte makes aqueous sodium-ion battery safe, green, and long-lasting. *Adv. Energy Mater.* 7:1701189.
40. Salama M., Rosy, A.R., Yemini R., Gofer Y., Aurbach D., Noked M. (2019) Metal-sulfur batteries: Overview and research methods. *ACS Energy Lett.* 4:436–446.
41. Richter R., Häcker J., Zhao-Karger Z., Danner T., Wagner N., Fichtner M., Friedrich K.A., Latz A. (2021) Degradation effects in metal-sulfur batteries. *ACS Appl. Energy Mater.* 4:2365–2376.
42. Fang R., Xu J., and Wang D.-W. (2020) Covalent fixing of sulfur in metal–sulfur batteries. *Energy Environ. Sci.* 13:432–471.
43. Wild M., O'Neill L., Zhang T., Purkayastha R., Minton G., Marinescu M., Offer G.J. (2015) Lithium sulfur batteries, a mechanistic review. *Energy Environ. Sci.* 8:3477–3494.
44. Li L., Chang Z.-W., Zhang X.-B. (2017) Recent progress on the development of metal-air batteries. *Adv. Sustainable Syst.* 1:1700036.
45. Li Y., Lu J. (2017) Metal-air batteries: Will they be the future electrochemical energy storage device of choice? *ACS Energy Lett.* 2:1370–1377.
46. Zhu Q., Zhao D., Cheng M., Zhou J., Owusu K.A., Mai L., Yu Y. (2019) A new view of supercapacitors: Integrated supercapacitors. *Adv. Energy Mater.* 9:1901081.
47. Shao Y., El-Kady M.F., Sun J., Li Y., Zhang Q., Zhu M., Wang H., Dunn B., Kaner R.B. (2018) Design and mechanisms of asymmetric supercapacitors. Chem. Rev. 118:9233–9280.
48. Schütter C., Pohlmann S., Balducci A. (2019) Industrial requirements of materials for electrical double layer capacitors: Impact on current and future applications. *Adv. Energy Mater.* 9:1900334.
49. Cheng Q., Chen W., Dai H., Liu Y., Dong X. (2021) Energy storage performance of electric double layer capacitors with gradient porosity electrodes. *J. Electroanal Chem.* 889:115221.
50. Pandolfo A.G., Hollenkamp A.F. (2006) Carbon properties and their role in supercapacitors. *J. Power Sources* 157:11–27.
51. Liu L., Zhao H., Lei Y. (2019) Review on nanoarchitectured current collectors for pseudocapacitors. *Small Methods* 3:1800341.
52. Durst J., Siebel A., Simon C., Hasché F., Herranz J., Gasteiger H.A. (2014) New insights into the electrochemical hydrogen oxidation and evolution reaction mechanism. *Energy Environ. Sci.* 7:2255–2260.
53. Wang S., Gai L., Zhou J., Jiang H., Sun Y., Zhang H. (2015) Thermal cyclodebromination of polybromopyrroles to polymer with high performance for supercapacitor. *J. Phys. Chem. C* 119:3881–3891.
54. Wu N., Bai X., Pan D., Dong B., Wei R., Naik N., Patil R.R., Guo Z. (2020) Recent advances of asymmetric supercapacitors. *Adv. Mater. Interfaces* 8:2001710.
55. Choudhary N., Li C., Moore J., Nagaiah N., Zhai L., Jung Y., Thomas J. (2017) Asymmetric supercapacitor electrodes and devices. *Adv. Mater.* 29:1605336.
56. Pandolfo A.G., Hollenkamp A.F. (2006) Carbon properties and their role in supercapacitors. *J. Power Sources* 157:11–27.

6 Theoretical Considerations of 2D Materials in Energy Applications

Harsha Rajagopalan
Vellore Institute of Technology

Sumit Dutta and Sourabh Barua
Birla Institute of Technology

Pawan Kumar Dubey
Ariel University

Jyotsna Chaturvedi
Indian Institute of Science

Laxmi Narayan Tripathi
Vellore Institute of Technology

CONTENTS

DOI: 10.1201/9781003178453-6

6.1 INTRODUCTION

The field of two-dimensional (2D) materials got much attention after the discovery of graphene leading to a noble prize.[1] Bulk 2D materials can be used to prepare thin, mono/few-layer thick transition metal carbonitrides (MXenes), carbides, nitrides, oxides, and transition metal dichalcogenides (TMDCs).[2,3] TMDCs are a class of 2D materials with the chemical formula MX_2, where M represents a transition metal and X represents a chalcogen.[4,3] TMDCs are one of the most studied materials that have been isolated in monolayer form and show promising results like direct bandgap, which is a desired optoelectronic property.[5,6] Molybdenum- and tungsten-based TMDCs are quite popular as semiconductors, with bandgap ranging from visible to near-infrared.[7] Different types of 2D materials are widely used for the study of energy storage and energy conversion devices. They mainly include solar energy storage, photovoltaics, piezoelectric, and thermoelectric devices.[8,9] Nanostructured 2D materials possess unique properties like high surface areas, tunable bandgap leading to superior optoelectronic properties. The fact that the 2D materials can mechanically be exfoliated from bulk single crystal and transferred to any desired substrates[2,10] make them invaluable. Raman spectroscopy is an important tool to characterize the number of layers of 2D materials.[11] Graphene is one of the heavily used 2D materials for the energy application industry due to its unique properties like large surface area and a high electrochemical performance rate. The discovery of graphene, experimentally by Andre Geim and Konstantin Novoselov in 2004[12] highlighted the 2D materials for various applications. Since graphene doesn't have any bandgap, much effort has been made to find 2D materials that have a bandgap.[5]

In this work, we introduce several theoretical tools and theoretical principles for energy applications of 2D materials. To understand the electronic properties, we discuss the density functional theory or the first principle approach. We then discuss the Finite-Difference Time-Domain method (FDTD) or Yee's method (named after Kane S. Yee). The FDTD is simple yet conceptually elegant and frequently used full-wave techniques to solve problems in electromagnetics.[13–15] It incorporates the time-domain method technique, unlike the others, for a wide frequency range. Maxwell equations can characterize electromagnetic waves. Therefore, to solve an electrodynamical problem, we need to solve the four Maxwell equations. Finally, we review the piezoelectric and H_2 evolution reaction (HER) due to photoelectrochemical (PEC) applications of 2D materials. Under mechanical stress, piezoelectric materials undergo electrical polarization. We have illustrated this by an example of a piezoelectric device consisting of metal electrodes and a monolayer of MoS_2.[16] Thus, piezoelectric application of 2D materials opens new avenues for energy application.

Hydrogen is a source of clean energy and has tremendous potential to replace the fossil fuels. It could be a sustainable and environment-friendly source of energy. Recently, 1D and 2D nanostructures were used as a catalyst for the generation of H_2 from water or organic molecules. This process of generation of H_2 where the generation of H_2 is facilitated by 2D materials in the presence of UV/visible radiation is known as photocatalysis.[17] The hydrogen evolution can easily happen through water splitting. The separation of H_2 and O_2 in water requires a Gibbs free energy change, ΔG^0 of 237 kJ mol^{-1} or wavelength, 1,100 nm corresponding to energy 1.23 eV.[18] The photocatalytic water splitting has great potential as an up-scaled, cost-effective, and green source of energy in the form of H_2. The only requirement for the semiconductors to be used as photocatalysts is the appropriate bandgap (1.6–2.4 eV), corrosion resistance, stability, and recyclability.[19] A photocatalytic water splitting is important due to the presence of water as an abundant, clean, renewable, and natural source. The solar to chemical energy conversion efficiency in the water splitting can be enhanced significantly by the application of heterogeneous and non-precious metal electrocatalysts such as metal selenides and metal sulfides.[20] Furthermore, for upscaled usage, integrated PEC cell devices can be made scalable to large areas such as 64 cm^{-2}.[21]

6.2 CALCULATION OF ELECTRONIC PROPERTIES OF TRANSITION METAL DICHALCOGENIDES (TMDCs) FROM FIRST PRINCIPLES

Here we choose tungsten selenide (WSe_2) as a representative material to demonstrate how density functional theory (DFT) can be used to obtain important results, e.g., the evolution of the electronic band structure with thickness in TMDCs. For the DFT calculations, we have chosen the open-source code, Quantum ESPRESSO (QE),[22,23] which uses the plane wave method of DFT and pseudopotentials.

6.2.1 Density Functional Theory or First Principles Calculation

Ab initio/DFT calculations are based on the fact that the electron density contains all the necessary information about the ground state of any electronic system. This was shown by Hohenberg and Kohn in two theorems,[24] which form the basis of DFT. According to the two theorems, any many-body problem with N electrons, involving 3N spatial coordinates, can be reduced to a simple problem involving only three spatial coordinates when expressed in terms of the electron density. Thus, in DFT, rather than solving the practically unsolvable many-body Schrodinger equation, the single-particle Kohn–Sham equations are solved. Once these single-particle wave functions are found out, the electron density and from it, all properties of the solid can be known. The Kohn–Sham equations require an exchange-correlation functional to obtain the exact electric potential for which the equations are to be solved. This exchange-correlation functional is key to DFT and different methods exist to find the best possible exchange-correlation functional, like the Local Density Approximation and the Generalized Gradient Approximation (GGA). One such example of a GGA exchange-correlation functional is the Perdew-Burke-Ernzerhof (PBE) exchange-correlation functional,[25] which we have used here in our calculations.

Therefore, by using DFT, calculating the electronic properties of any solid, e.g., the Fermi energy, bandgap, and density of states, is possible, even before the solid is developed in the laboratory. So, as new materials such as TMDCs were discovered, scientists could study them extensively using DFT and arrive at important conclusions only by computation.

Using DFT, we can find the band structure of complex molecules and solids and account for the changes in the band structure while varying other physical properties associated with the system such as strain, electric field, and doping. For example, the effect of strain on the bandgap in WSe_2 and WS_2 films, and its conversion from indirect to direct bandgap due to wrinkle formation caused by strain have been reported.[26]

6.2.2 Computational Details

All the electronic structure calculations in this text are performed within the DFT framework, using the open-source code Quantum ESPRESSO.[22,23] The code uses the Projector Augmented Wave method along with pseudopotentials. We use the PBE exchange-correlation functional for the calculations. The energy cut-off for the wavefunction has been set to 40 Ry and 80 Ry for bulk and monolayer calculation, respectively. The k-point mesh size is set to $9 \times 9 \times 9$ for the bulk and $9 \times 9 \times 1$ for the monolayer self-consistent (SCF) calculations. For the monolayer calculations, a supercell with a vacuum of 15 Å is used. The vacuum is required so that interactions between monolayers of repeating units are minimized otherwise this can lead to artefacts in the calculated properties.

6.2.3 Results and Discussions

Figure 6.1a shows the layered structure of bulk WSe_2. The electronic band structure of bulk WSe_2 is calculated using the unit cell shown in Figure 6.1b. The bulk band structure is calculated from the first principles along the k-path Γ-M-K-Γ of the Brillouin zone and is shown in Figure 6.1c. The band diagram depicts the conduction band and the valence band separated by a significant bandgap. The bandgap is the energy gap between the minima in the conduction band and the maxima in the valence band. It is clear from the diagram that the minima in the conduction band and the maxima in the valence band are not at the same k-point. This implies that the bandgap is indirect in bulk WSe_2. Semiconductors with an indirect bandgap are not ideal for optoelectronic devices. In Figure 6.2a we show a monolayer of WSe_2, and in Figure 6.2b, we show the supercell used for the ab initio calculation of the properties of WSe_2. The electronic band structure of the monolayer WSe_2 is calculated along the *k*-path Γ-M-K-Γ of the Brillouin zone and is shown in Figure 6.2c. From the band diagram in the monolayer, the minima in the conduction band and the maxima of the valence band occur at the same point in *k*-space, which is the high symmetry K point in this case. Thus, monolayer WSe_2 is a direct bandgap semiconductor, which makes it suitable for optoelectronic devices. The bandgap in our calculation comes out to be 1.63 eV.

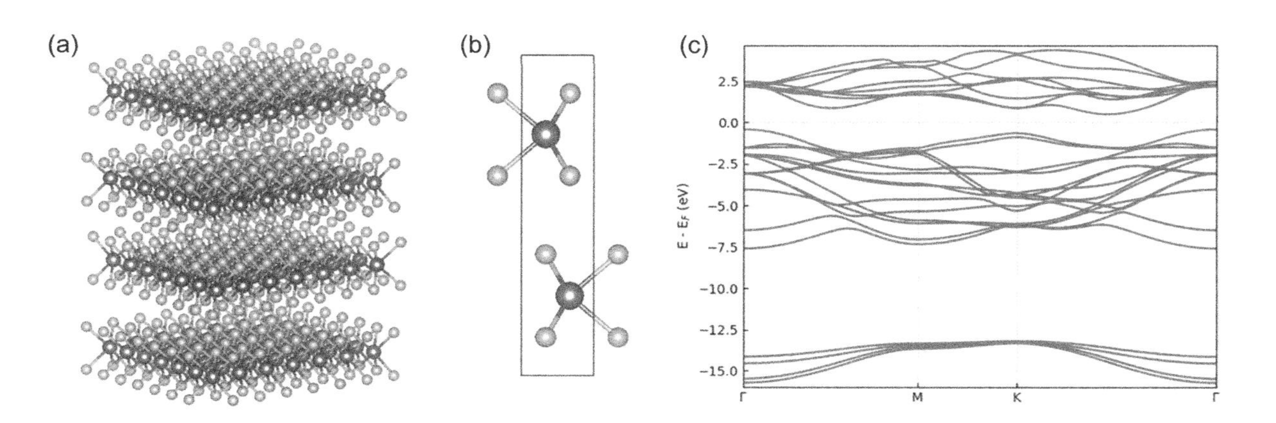

FIGURE 6.1 (a) Bulk WSe_2 showing multiple layers. The grey atoms are tungsten (W), and the green atoms are selenium (Se). (b) The unit cell of bulk WSe_2 used for the DFT calculations. Bulk WSe_2 has a 2H structure, i.e., the structure repeats after every two layers in the vertical (c-axis) direction. (c) Electronic band structure of bulk WSe_2. The Fermi level (FL) is set to zero in the band structure.

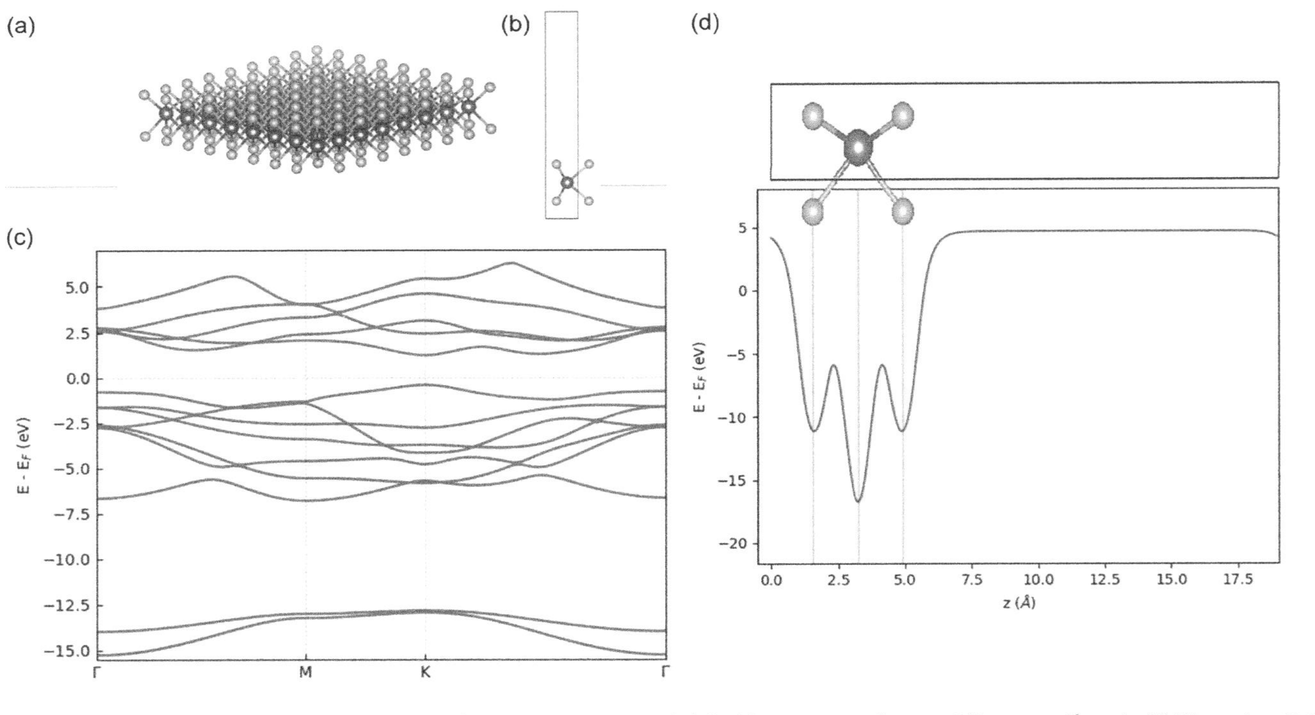

FIGURE 6.2 (a) Monolayer of WSe_2 showing a layer of W atoms (grey) sandwiched between two layers of Se atoms (green). (b) The unit cell (supercell) of monolayer WSe_2 used for the DFT calculation. A vacuum of 15 Å was added above the WSe_2 layer in the supercell to prevent artifacts arising from the interaction between two repeating monolayers. (c) Electronic band structure of monolayer WSe_2. The Fermi level (FL) is set to zero in the band structure. From the band diagram, the conduction band minima and valence band maxima are at the same k point, which is the K point here. (d) Plot of the electrostatic potential (ionic potential plus Hartree potential) inside the supercell of the monolayer WSe_2. The green line shows the fluctuation of the potential with position along the c-axis of the supercell. The vertical magenta lines indicate the location of the ions in the supercell, which are shown in the superposed image of the supercell at the top. The potential is plotted as energy and the Fermi energy is subtracted from it. Since the work function of a semiconductor is the difference between the potential energy in the vacuum and the Fermi level, it can be read directly from our plot.

The work function of a semiconductor material is the separation between the vacuum energy level and the Fermi energy in the semiconductor. The work function is a relevant quantity in semiconductors that are required for the calculation of various properties, e.g., Schottky barrier height. Using DFT it is possible to calculate the work function quite simply. After the self-consistent calculation, post-processing allows the plotting of the electrostatic potential (sum of the ionic potential plus the Hartree potential) in the supercell. We plot this for the monolayer WSe_2 supercell in Figure 6.2d.

The green line shows the variation of the potential with the position along the c-axis in the supercell. For a guide to the eye, the positions of the various ions in the supercell are marked by vertical magenta lines. Furthermore, we also superpose the supercell structure on this plot to help understand the variation of the potential with position better. Inside the cell, the potential varies periodically with the positions of the ions. In the vacuum region, it is flat as expected. The potential is plotted in terms of energy, and we subtract the Fermi energy from the potential energy in the plot. So, the zero lines on the energy scale signify the Fermi level. Since the work function of a semiconductor is defined as the difference between the vacuum potential energy and the Fermi level, this can be read from the graph in our case and comes to be 4.65 eV.

6.3 FINITE-DIFFERENCE TIME-DOMAIN (FDTD) AND APPLICATION FOR FIELD ENHANCEMENT IN TWO-DIMENSIONAL MATERIALS

6.3.1 The Finite-Difference Time-Domain Method

FDTD or Yee's method incorporates the time-domain method technique, which provides advantages over other methods because it can be used for a wide frequency range. Maxwell equations can characterize electromagnetic waves. Therefore, to solve electrodynamics, we need to solve the four Maxwell equations. The Maxwell equation is continuous, but there is no general way of solving the Maxwell equation; rather, one needs to be more precise about the problem and the inputs. In other words, one needs to solve the Maxwell equation on every point, even for the most straightforward geometry to study, how the field components are propagating with time (or how fast the fields are changing at each point). Thus, there is no closed-form solution of the Maxwell equation. Here FDTD simulation technique comes into the picture, which simplifies the equation solving methodology by discretizing the Maxwell equation using central-difference approximations to space and time partial derivatives. Thus, using the FDTD method for Maxwell's equations, any arbitrary media types (either in 2D or 3D) can be modeled, including inhomogeneous and lossy, frequency-dependent (dispersive), anisotropic, bi-anisotropic, chiral, or nonlinear media. Mathematically 1D Maxwell's equation can be written as[13–15]

$$\frac{\partial E_x}{\partial t} = \frac{1}{\epsilon}\frac{\partial H_y}{\partial z} \tag{6.1}$$

$$\frac{\partial H_y}{\partial t} = -\frac{1}{\mu}\frac{\partial E_x}{\partial z} \tag{6.2}$$

Here, symbols have their usual meaning. The two equations represent a plane wave traveling on the z-axis. Considering a central difference approximation of the derivatives and shifting E_x and H_y by half cell in space and half time in the time domain, we get:

$$\frac{E_x^{n+1/2}(k) - E_x^{n-1/2}(k)}{\Delta t} = -\frac{1}{\epsilon}\frac{H_y^n\left(k+\frac{1}{2}\right) - H_y^n\left(k-\frac{1}{2}\right)}{\Delta z} \tag{6.3}$$

$$\frac{H_y^{n+1}\left(k+\frac{1}{2}\right) - H_y^n\left(k+\frac{1}{2}\right)}{\Delta t} = -\frac{1}{\mu}\frac{E_x^{n+1/2}(k+1) - E_x^{n+1/2}(k)}{\Delta Z} \tag{6.4}$$

Here, superscript 'n' indicates time and means a time $t = \Delta t \cdot n$. The term '$n+1$' means a one-time step later. The term in parenthesis 'k' indicates distance with actual distance $z = \Delta x \cdot k$. These two equations assume that E and H are interleaved in both time and space. In particular, the left-hand side term in Eq. (6.3) says that the derivative of the E field at time $n \cdot \Delta t$ can be expressed as a central difference using E field values at times $\left(n + \frac{1}{2}\right) \cdot \Delta t$ and $\left(n - \frac{1}{2}\right) \cdot \Delta t$. The right-hand side term in Eq. (6.3) approximates the derivative of the H field at point $k \cdot \Delta x$ as a central difference using H field values at points $\left(k + \frac{1}{2}\right) \cdot \Delta x$ and $\left(k - \frac{1}{2}\right) \cdot \Delta x$. This approximation obliges us to calculate the value of E and H at the central and neighboring steps of time $(n \cdot \Delta t)$ and space $(k \cdot \Delta x)$. Thus, this algorithm by Allen Taflov and JJ Simpson[14] helps us calculate many values of E and H within the modeled geometry, and the result after the simulation is collected in data form, which can be visualized as some form of map or graph.

6.3.2 Field Enhancement in General in Nanogap Antennas

Characteristic properties of a single or array of nanoholes and nanoslit apertures in metals for microwave, terahertz, and infrared frequency regions have long been pursued subject, not only for its fundamental importance in spectroscopy (for instance, in surface-enhanced Raman and extreme-ultraviolet generation) but also for biomolecule detection and next-generation photovoltaics. Recently, due to innovation in nanotechnology, several metal nanostructures of different shapes and sizes have been studied for surface plasmons.[27] Phenomena like field enhancement due to the interaction of electromagnetic light with metallic nanostructures have gained attention from almost every aspect of the active research field related to light-matter interaction. Such metal nanoparticles are subwavelength scattering elements or can be used as nearfield antennas that can be efficiently coupled to the material due to increased effective absorption cross-section. Such phenomena can be observed in different nanostructure arrangements, consisting of adjacent Nobel metal (gold and silver) structures like nanorods or nanosphere separated by a nanoscale gap. Such a system can strongly scatter, absorb, and even confine or squeeze the optical field in the nanometer dimension. The study of such light-matter interaction, which deals with the generation, manipulation, and transmission of this excitation, is known as 'plasmonics'. It is crucial to study the fundamental mechanism and theoretical model of these interactions. One can explain the basic phenomena behind plasmonic by considering localized surface plasmon resonances (SPRs; in simple terms, charge-density oscillations) on the closed surfaces of the particles. Surface charge plasmons are oscillatory charge waves in a thin metallic film that can be excited by coupling light into this metal film at a suitable angle. When conduction electrons oscillate coherently in response to incident light, they displace the nuclei's electronic arrangement, giving rise to surface charge distribution. Here the Coulomb force between positive and negative charges is acting as a restoring force. The frequency of the oscillation is called the plasmon frequency.[28]

$$\omega_p = \left(\frac{4\pi N e^2}{\varepsilon_\infty m^*} \right)^{1/2} \tag{6.5}$$

Here, e is the electronic charge, N is the carrier density (number electron per unit volume), and m^* is the effective mass of the carrier. Each collective oscillation of different surface charge distribution has its resonant mode known as SPR. The electron density, effective mass, particle's shape, size, and the dielectric function of the material determine the number of modes and the frequency of the oscillation.[27] Under the quasi-static limit, we assume that the particle size is much smaller than the wavelength of the incident light.[27]

Here it is worth mentioning that some applications like enhanced nonlinear optical microscopy or near field Raman microscopy need sufficiently large field enhancement. For such cases, aperture-less and sharp tip nanostructures like nanorods tips or bow-tie-shaped structures are more valuable than structures like nanosphere.[30] An elongated structure like a metal nanorod has surface plasmons confined in two dimensions, one along the length and the other across the width. When the light of proper polarization is incident on it, the surface plasmons are excited in both directions, giving rise to a longitudinal and transverse mode of oscillation (Figure 6.3b). Hence, they have non-zero absorption coefficients at two different wavelengths. A shift in peaks was also observed if the aspect ratio of the nanorods was varied.[31] As evident from Eq. (6.5) $\omega_p \propto N^{1/2}$ surface plasmons absorption efficiency is affected by the nanoparticle size. In a study related to UV–Vis absorption spectra of colloidal solutions, it was observed that if we increase the particle size in the case of the gold nanosphere, the absorption is red-shifted.[32] Ward et al. studied the theoretical approach of variation of field enhancement with gap width and found that the intensity increases as the decrease in gap width 'd'.[33] This result was later confirmed by Garcia et al.,[34] in which they studied the intensity variation with the gap width between the tip of gold nanorods (similar to structure as shown in Figure 6.3d). Here we present a similar result in which we have observed that the field intensity decreases exponentially as we go away from the point of field enhancement. The simulation was performed using the FDTD package from Lumerical inc. Figure 6.3f(II) shows the variation

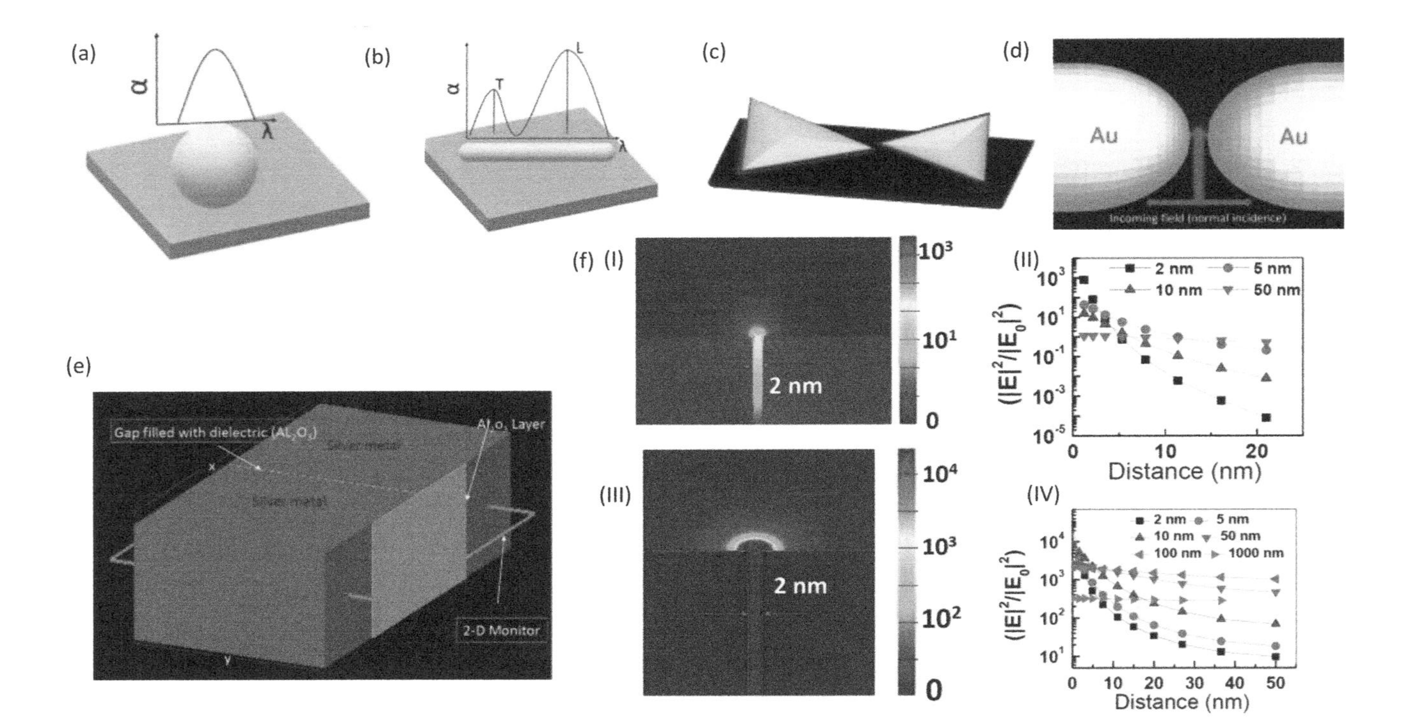

FIGURE 6.3 (a–d) represents a schematic diagram of the standard nanostructure shape used to study plasmons enhanced light-matter interaction. Figures (a) and (b) are adapted with permission from Dubey et al.,[29] Copyright (2019) Elsevier. (e) The schematic of simulated nanoslit to study the effect of field enhancement dependence on gap width. (f) Normalized intensity of incident light vs gap width for visible frequency range (I and II) and terahertz frequency range (III and IV).

of normalized field intensity with the distance away from the gap and for different gap widths (2, 5, 10, and 50 nm) between the metallic (silver) slit. Figure 6.3f(I) was simulated for the visible frequency range. On the other hand, Figure 6.3f(IV) was simulated for the terahertz frequency range. In both cases, for sufficiently larger gap width (50 nm and 1 micrometer), there is minimal field enhancement, and it is almost constant as we go away from the surface of the slit.

6.3.3 Application of Field Enhancement for Energy Application Using 2D Materials

Conventional solar cells were designed significantly thick to increase the absorption path of incident light in the material. Such designs have not successfully created efficient solar cells as the short minority carrier diffusion length compared to the overall thickness of material significantly reduced device efficiency. Modern technology leaped forward with the advent of 2D material like TMDCs, graphene as the photovoltaics (PV) material that has dramatically reduced the thickness of solar cell devices with improved efficiency. Utilizing Plasmonic nanoparticle for improved incident light absorption is another milestone scientist has achieved recently. The hybrid structure of 2D materials with metallic nanostructure uses phenomena like thermo-plasmonic effect, scattering, absorption enhancement, and near field enhancement for efficient light energy harvesting. Such light-matter interaction has the potential to extend the detection capabilities of semiconductors beyond their bandgap. For example, in the thermo-plasmonic effect, plasmonic nanostructure generates plasmon-generated hot electrons, moving in the semiconductor material, transferring their energy to other electrons. Suppose these electrons have enough energy to overcome the Schottky barrier of the metal-semiconductor boundary, in that case, the electron will inject into the semiconductor's conduction band to generate current for photodetection.[32] Authors have suggested several designs in which nanoparticles can be embedded on the material's surface or within the semiconducting material to increase the coupling of the incident light.[33] The nanoantenna embedded on the surface of the semiconducting PV material (Figure 6.4c) acts as a subwavelength scatterer that scatters the sunlight into the thin absorber layer of the PV cell at some angle, thus increasing the path length and absorption. Mathematically, we can illustrate this phenomenon using lamberts beers law as follows: Let I_{fe} be the intensity due to field enhancement and L_{fe} be the distance travelled by the light inside the semiconductor material after the scattering from the plasmonic nanoparticle, then the intensity I at the later distance is given by[29]

$$I = I_{fe} e^{-\alpha \cdot L_{fe}} \tag{6.6}$$

where α is the absorption coefficient of the material. Let us further assume that $I_{w,fe}$ is the intensity without any field enhancement (in the absence of any plasmonic nanoantenna). For the same final intensity I (assuming that most of the intensity has been absorbed in the end of the path) and $L_{w,fe}$ as the path length traveled by the incident light, we get:

$$I = I_{w,fe} \cdot e^{-\alpha \cdot L_{w,fe}} \tag{6.7}$$

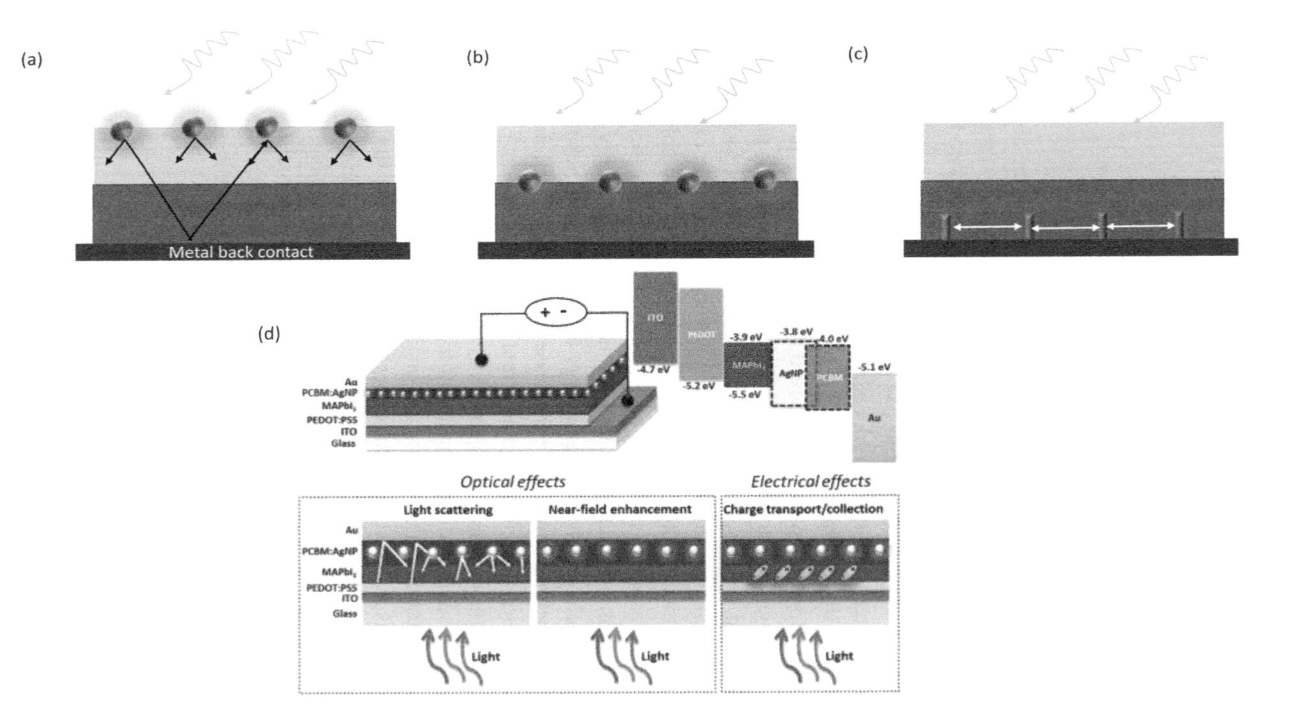

FIGURE 6.4 (a) The schematic diagram shows embedded silver nanoparticles on the material's surface. The nanoparticles scatter the light deep into the medium, effectively increasing the path length and absorption. (b) This structure utilizes the near field enhancement to couple light into the material due to an increase in effective absorption cross-section. (c) Here corrugated metal back surface has been used instead of spherical nanoparticle. When longer wavelength light penetrates the material, they interact with a tip-like metallic nanostructure giving comparatively more significant field enhancement and efficient coupling into the semiconductor material. (d) This figure depicts the layer integration of silver nanoparticles with the material and how the optical and electrical effects together improve the efficiency of solar cell technology. Figure (d) is adapted with permission from Higgins et al.[35] Copyright (2018) AIP publishing.

Dividing Eq. (6.6) by (6.7), we get:

$$L_{fe} - L_{w,fe} = \frac{1}{\alpha} \ln \left(\frac{I_{fe}}{I_{w,fe}} \right) \tag{6.8}$$

The term $L_{fe} - L_{w,fe}$ is the difference in path length. Since $I_{fe} > I_{w,fe}$ the right-hand side of the Eq. (6.8) is always greater than one, proving that plasmonic particles' use facilitates more significant absorption within the material.

In addition to the hybrid structure mentioned above, the plasmonic 'tandem' geometry-based solar cell has also proven to be very efficient.[36] In the plasmonic tandem geometry, the different semiconductor material is stacked on the top of each other separated by the metal film with the plasmonic nanostructure. Some other designs involve the coupling of sunlight into surface plasmon in quantum dot solar cells.[37] Such techniques are handy since it has flexibility in bandgap engineering of the semiconductor by controlling the size of the semiconducting particle. Plasmonic structures are becoming more and more involved in solar cell technology; hence, we should expect to see several other plasmonic facilitated efficient hybrid solar cell designs that may or may not be based on 2D materials.

6.4 APPLICATIONS OF 2D MATERIALS IN PIEZOELECTRIC DEVICES

Piezoelectric current and voltage can be generated in 2D materials such as MoS_2 after application of mechanical stress along x-direction.[16] The authors[16] found that the voltage and current were positive (negative) with an increase(decrease) of strain. When strain was applied in the x-direction with an increase of strain, positive voltage and current output are obtained. When the strain was decreasing negative output was observed. This is an example of the conversion of mechanical energy to electrical energy using 2D materials. The voltage response and current response were found to be increased with the increase of the magnitude of the applied strain. The peak current reached 27 pA at the applied strain of 0.64%. The output current and voltage as a function of load resistance show constant output current for a load resistance. Few kΩ till 1 MΩ while it decreases after a further increase of load resistance. The mechanical to electrical energy conversion was stable for cyclic loading up to 0.43% strain at 0.5 Hz for 300 minutes.[16] The piezoelectric output decreases with an increase in the number of layers. We can conclude that the piezoelectric output (current) is maximum for monolayers MoS_2.

The piezoelectric charge polarization strongly depends on the direction of applied strain in 2D materials. The coupling between polarization (P_i) and strain tensor (ε_{ijk}) can be quantified to the first order by the third-order piezoelectric tensor, $\boldsymbol{e}_{ijk}$[16]

$$e_{ijk} = \left(\partial P_i / \partial \varepsilon_{jk} \right) \tag{6.9}$$

Here i, j, k are 1, 2, 3, corresponding to x, y, z axes, respectively.

6.5 APPLICATIONS OF 2D MATERIALS IN HYDROGEN PRODUCTION: PHOTOELECTROCHEMICAL CELL AND PHOTOCATALYTIC WATER SPLITTING

Electrochemical water splitting is inspired by natural light-driven chemical reactions, i.e., photosynthesis (Figure 6.5). The evolution of O_2 through photosynthesis follows the Calvin–Benson cycle where light-driven (absorption of light (1.76–2.47 eV)) electron and proton transfer happen in the chloroplast (Figure 6.5b). Natural photosynthesis happens to be the first example of water splitting through light.

Figure 6.5 shows a comparative diagram for the mechanism of H_2 evolution using photochemical water splitting as opposed to O_2 generation in photosynthesis. Both processes occur in the presence of solar energy. In the classical photochemical cell (Figure 6.5a), used for water splitting by Honda and Fujishima,[38] a TiO_2 electrode was connected with a platinum electrode. When the surface of the TiO_2 was irradiated with visible light, a current was detected in the external circuit. Honda and Fujishima suggested that the water could be decomposed into hydrogen and oxygen by the visible light without any external voltage. They[39] suggested the following reaction pathways.

1. The generation of electron (e^-) and hole (p^+) pairs by absorption of photons

$$TiO_2 + 2h\nu \rightarrow 2e^- + 2p^+$$

2. The oxidation of water at the TiO_2 electrode

$$2p^+ + H_2O \rightarrow \frac{1}{2}O_2 + 2H^+$$

3. Reduction of Hydrogen at a platinum electrode

$$2e^- + 2H^+ \rightarrow H_2$$

The overall reaction was

$$H_2O + 2h\nu \rightarrow + \frac{1}{2}O_2 + H_2$$

A current of few mA flowed when the TiO_2 electrode was irradiated with a 500 W Xe lamp. The TiO_2 electrode surface area illuminated by the authors[38] was 1 cm^2. The estimated quantum efficiency, i.e., the number of electrons generated per photons absorbed was 10%. The free energy change that occurred during the change of one molecule of water to H_2 and ½ O_2 under standard conditions was given by $\Delta G = 237.2$ kJ/mol, which corresponds to an energy of 1.23 eV.[39] Therefore, to drive this reaction with the help of a semiconductor, the materials must absorb solar energy of more than 1.23 eV, i.e., wavelength less than or equal to 1,000 nm of light. This solar energy is further converted into H_2 and O_2. This process requires the generation of two electron–hole pairs per molecule of H_2, i.e., $2 \times 1.23\,eV = 2.46\,eV$ of energy.[39]

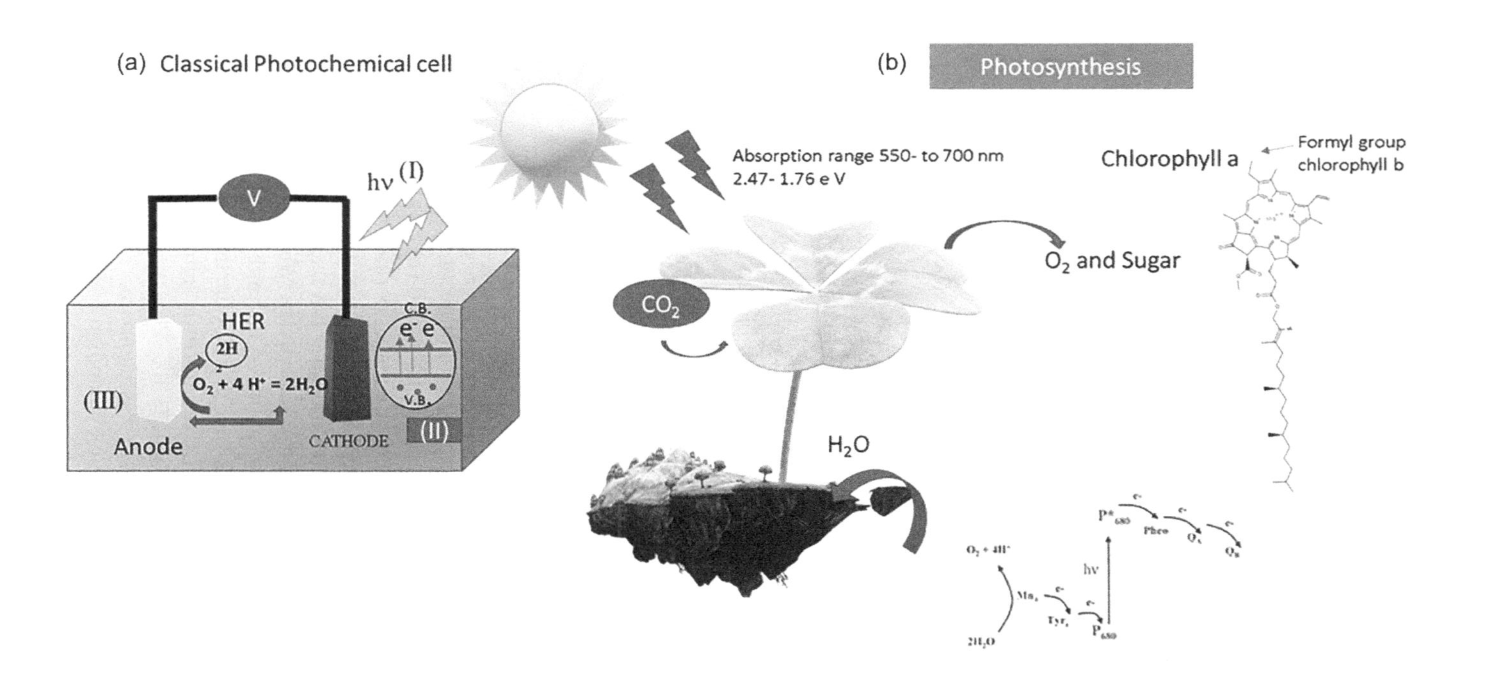

FIGURE 6.5 (a) Pictorial representation of a classical photochemical cell; (I) absorption of irradiation of light (UV/Visible) (II) Absorption of photons by the photocatalyst leads to the generation of electron/holes in the conduction and valence band (III). (b) Photosynthesis process in plants driven by Calvin–Benson Cycle. The light-induced electron/proton transfer inside the chloroplast.

The photo-induced charge carriers (electrons and holes) must travel in the liquid. The electron transfer process at the semiconductor/liquid junction suffers losses due to kinetic overpotential and concentration needed to drive the HER.[39] Therefore, the energy needed to drive the HER reaction due to photoelectrolysis at a semiconductor electrode is 1.6–2.4eV. Thus, a water-splitting reaction needs semiconductors having a bandgap of 1.6–2.4eV.

6.5.1 Solar Energy to Chemical Energy Efficiency, η without External Bias

The efficiency of conversion of solar energy to chemical energy[39] is given by, η

$$\eta = \frac{J_g \mu_{ex} \Phi_{\text{conv}}}{S}$$

where J_g is the absorbed flux, μ_{ex} is the excess chemical potential generated by absorption of light, Φ_{conv} is the quantum yield for absorbed photon, and S is the total incident solar radiation ($mWcm^{-2}$)

6.5.2 Solar Energy to Chemical Energy Efficiency, $\eta_{w/o}$ with External Bias

Some photoelectrode devices require an external bias to drive water electrolysis. In PEC water splitting, photoelectrode devices require an external bias for water electrolysis. The efficiency of such water splitting[39] is given by

$$\eta_{w/o} = \frac{J_{mp}\left(1.23V - V_{\text{app}}\right)}{P_{\text{in}}} \tag{6.10}$$

where J_{mp} is the externally measured current density and V_{app} is the applied voltage between the photoanode and the photoelectrode. The P_{in} is the power density during illumination. The two criteria for good evolutions of hydrogen are (a) the catalyst must generate more electrons and hole pairs and (b) catalyst must work for a longer time scale, i.e., must be robust over a longer period for commercial applications. HER catalyst such as TMDC,[40] which have been used as high-performance materials to satisfy three main criteria—(a) They have active sites (e.g., edges), (b) intrinsic catalytic activity, and (b) high conductivity. We can increase the number of active sites by an increase of a load of materials on the electrode and improved structuring exposes more active sites per gram.[40] Furthermore, the efficiency of H_2 evolution can also be enhanced by the intrinsic activity of each site.[41,42] The use of plasmonic nanoparticles enhances the PEC water-splitting reaction.[43,44] Relatively high overpotential of TMDC as compared to Pt electrodes possesses a serious problem. This was overcome by doping of Mn in $MoSe_2$ nanoflowers by Vasu et al.[45] The authors also claim that doping also enhanced the charge transfer kinetics.

6.6 CONCLUSION

2D materials have an immense potential for application in a wide variety of fields. DFT is very instrumental in understanding many of the properties of 2D materials. It can calculate a wide variety of properties like the thickness dependence of band structures, phase stability, catalytic properties for hydrogen evolution reactions, work function, Schottky barriers, and so on. Moreover, it allows calculating properties of new materials even before they are synthesized in the laboratory thus offering a unique perspective to material scientists in designing new materials. Besides, 2D materials are playing the lead role in modern thin and ultralight solar energy conversion devices. For the lowest per watt cost and lowest gram consumed per watt power generation, the commercial players are shifting their focus from crystalline solar cell material to 2D materials. Experimental designs incorporating metallic nanoparticles in these 2D materials have dramatically improved 2D material-based solar energy conversion device efficiency. Due to excellent carrier mobility and electron transport properties, not graphene but other direct bandgap monolayer materials such as TMDCs have tremendous potential for designing low-cost, flexible photovoltaic devices. The next-generation concept, such as tandem and hot carrier photovoltaic devices, has been made possible by the advent and utilization of 2D materials. On-chip integrated design using these concepts can utilize almost the entire energy of the incident photon. Such solar cell devices can be made very compact to be installed in miniature next-generation advanced miniature devices.

2D materials are the wonder material exhibiting high electrochemical performance and are very useful for photovoltaics, piezoelectric materials, and energy production in the form of hydrogen. 2D material-based energy devices can be improved by combining different 2D materials. The individual components in a composite can introduce the characteristics like high porosity, large surface area, increased active sites, enhanced cyclic stability. 2D materials are suitable for solar to chemical energy conversion and thus can become a nature-friendly, low-cost source of energy. In photocatalytic water splitting, the 2D/2D interface has an important role since the 2D cocatalysts on 2D semiconductors create a large interface that helps for the separating electron–hole pairs. It promotes the movement of electron–hole pairs between the 2D semiconductors and cocatalysts. Doping plays an important role in the formation of active sites in 2D materials. In the case of $MoSe_2$ Mn doping enhances the HER activity of $MoSe_2$.

ACKNOWLEDGEMENTS

Sourabh Barua acknowledges the Ministry of Human Resource Development (MHRD), Government of India for financial support through the NPIU TEQIP-III CRS scheme (CRS Application ID: 1-5743255881), and Birla Institute of Technology Mesra.

Ranchi for financial support under the NPIU TEQIP-III seed money scheme 2018. Jyotsna Chaturvedi acknowledges the department of science and technology (DST) for financial support through women scientist scheme A (WOSA) with file no: SR/WOSA-A/CS-26/2019.

REFERENCES

1. Novoselov, K. S.; Mishchenko, A.; Carvalho, A.; Neto, A. H. C. 2D Materials and van Der Waals Heterostructures. *Science* **2016**, *353*, aac9439.
2. Tripathi, L. N.; Iff, O.; Betzold, S.; Dusanowski, Ł.; Emmerling, M.; Moon, K.; Lee, Y. J.; Kwon, S.-H.; Höfling, S.; Schneider, C. Spontaneous emission enhancement in strain-induced WSe_2 monolayer-based quantum light sources on metallic surfaces. *ACS Photonics* **2018**, *5*, 1919–1926
3. Iff, O.; Lundt, N.; Betzold, S.; Tripathi, L. N.; Emmerling, M.; Tongay, S.; Lee, Y. J.; Kwon, S.-H.; Höfling, S.; Schneider, C. Deterministic coupling of quantum emitters in WSe_2 monolayers to plasmonic nanocavities. *Opt. Express* **2018**, *26*, 25944.
4. Tripathi, L. N.; Barua, S. Growth and characterization of two-dimensional crystals for communication and energy applications. *Prog. Cryst. Growth Charact. Mater.* **2019**, *65* (4), 100465.
5. Mak, K. F.; Lee, C.; Hone, J.; Shan, J.; Heinz, T. F. Atomically thin MoS_2 : A new direct-gap semiconductor. *Phys. Rev. Lett.* **2010**, *105*, 136805.
6. Jariwala, D.; Sangwan, V. K.; Lauhon, L. J.; Marks, T. J.; Hersam, M. C. Emerging device applications for semiconducting two-dimensional transition metal dichalcogenides. *ACS Nano* **2014**, *8*, 1102–1120.
7. Hu, W.; Sheng, Z.; Hou, X.; Chen, H.; Zhang, Z.; Zhang, D. W.; Zhou, P. Ambipolar 2D semiconductors and emerging device applications. *Small Methods* **2021**, *5*, 2000837.
8. Bonaccorso, F.; Colombo, L.; Yu, G.; Stoller, M.; Tozzini, V.; Ferrari, A. C.; Ruoff, R. S.; Pellegrini, V. Graphene, related two-dimensional crystals, and hybrid systems for energy conversion and storage. *Science* **2015**, *347*, 1246501.
9. Das, S.; Pandey, D.; Thomas, J.; Roy, T. The role of graphene and other 2D materials in solar photovoltaics. *Adv. Mater.* **2018**, *31*, 1802722.
10. Castellanos-Gomez, A.; Buscema, M.; Molenaar, R.; Singh, V.; Janssen, L.; van der Zant, H. S. J.; Steele, G. A. Deterministic transfer of two-dimensional materials by all-dry viscoelastic stamping. *2D Mater.* **2014**, *1*, 11002.
11. Mishra, P.; Tripathi, L. N. Characterization of two-dimensional materials from Raman spectral data. *J. Raman Spectrosc.* **2019**, *51*, 37–45.
12. Novoselov, K. S. Electric field effect in atomically thin carbon films. *Science* **2004**, *306*, 666–669.
13. Schneider, J. B. Understanding the Finite-Difference Time-Domain Method, www.eecs.wsu.edu/~schneidj/ufdtd/. **2010**.
14. Allen Taflove, S. C. *Hagness. Computational Electrodynamics: The Finite-Difference Time-Domain Method*; Archtech House Inc., Norwood, MA, **2005**.
15. Adak, S.; Tripathi, L. N. Nanoantenna enhanced terahertz interaction of biomolecules. *Analyst* **2019**, *144*, 6172–6192.
16. Wu, W.; Wang, L.; Li, Y.; Zhang, F.; Lin, L.; Niu, S.; Chenet, D.; Zhang, X.; Hao, Y.; Heinz, T. F.; Hone, J.; Wang, Z. L. Piezoelectricity of single-atomic-layer MoS_2 for energy conversion and piezotronics. *Nature* **2014**, *514*, 470–474.
17. Babu, V. J.; Vempati, S.; Uyar, T.; Ramakrishna, S. Review of one-dimensional and two-dimensional nanostructured materials for hydrogen generation. *Phys. Chem. Chem. Phys.* **2015**, *17*, 2960–2986.
18. Acar, C.; Dincer, I.; Naterer, G. F. Review of photocatalytic water-splitting methods for sustainable hydrogen production. *Int. J. Energy Res.* **2016**, *40*, 1449–1473.
19. Hisatomi, T.; Kubota, J.; Domen, K. Recent advances in semiconductors for photocatalytic and photoelectrochemical water splitting. *Chem. Soc. Rev.* **2014**, *43*, 7520–7535.
20. Zou, X.; Zhang, Y. Noble metal-free hydrogen evolution catalysts for water splitting. *Chem. Soc. Rev.* **2015**, *44*, 5148–5180.

21. Turan, B.; Becker, J.-P.; Urbain, F.; Finger, F.; Rau, U.; Haas, S. Upscaling of integrated photoelectrochemical water-splitting devices to large areas. *Nat. Commun.* **2016**, *7*, 12681.
22. Giannozzi, P.; Baroni, S.; Bonini, N.; Calandra, M.; Car, R.; Cavazzoni, C.; Ceresoli, D.; Chiarotti, G. L.; Cococcioni, M.; Dabo, I.; Dal Corso, A.; de Gironcoli, S.; Fabris, S.; Fratesi, G.; Gebauer, R.; Gerstmann, U.; Gougoussis, C.; Kokalj, A.; Lazzeri, M.; Martin-Samos, L.; Marzari, N.; Mauri, F.; Mazzarello, R.; Paolini, S.; Pasquarello, A.; Paulatto, L.; Sbraccia, C.; Scandolo, S.; Sclauzero, G.; Seitsonen, A. P.; Smogunov, A.; Umari, P.; Wentzcovitch, R. M. QUANTUM ESPRESSO: A modular and open-source software project for quantum simulations of materials. *J. Phys. Condens. Matter* **2009**, *21*, 395502.
23. Giannozzi, P.; Andreussi, O.; Brumme, T.; Bunau, O.; Buongiorno Nardelli, M.; Calandra, M.; Car, R.; Cavazzoni, C.; Ceresoli, D.; Cococcioni, M.; Colonna, N.; Carnimeo, I.; Dal Corso, A.; de Gironcoli, S.; Delugas, P.; DiStasio, R. A.; Ferretti, A.; Floris, A.; Fratesi, G.; Fugallo, G.; Gebauer, R.; Gerstmann, U.; Giustino, F.; Gorni, T.; Jia, J.; Kawamura, M.; Ko, H.-Y.; Kokalj, A.; Küçükbenli, E.; Lazzeri, M.; Marsili, M.; Marzari, N.; Mauri, F.; Nguyen, N. L.; Nguyen, H.-V.; Otero-de-la-Roza, A.; Paulatto, L.; Poncé, S.; Rocca, D.; Sabatini, R.; Santra, B.; Schlipf, M.; Seitsonen, A. P.; Smogunov, A.; Timrov, I.; Thonhauser, T.; Umari, P.; Vast, N.; Wu, X.; Baroni, S. Advanced capabilities for materials modelling with quantum ESPRESSO. *J. Phys. Condens. Matter* **2017**, *29*, 465901.
24. Hohenberg, P.; Kohn, W. Inhomogeneous electron gas. *Phys. Rev.* **1964**, *136*, B864–B871.
25. Perdew, J. P.; Burke, K.; Ernzerhof, M. Generalized gradient approximation made simple. *Phys. Rev. Lett.* **1996**, *77*, 3865–3868.
26. Dhakal, K. P.; Roy, S.; Jang, H.; Chen, X.; Yun, W. S.; Kim, H.; Lee, J.; Kim, J.; Ahn, J.-H. Local strain induced band gap modulation and photoluminescence enhancement of multilayer transition metal dichalcogenides. *Chem. Mater.* **2017**, *29*, 5124–5133.
27. Noguez, C. Surface plasmons on metal nanoparticles: the influence of shape and physical environment. *J. Phys. Chem. C* **2007**, *111*, 3806–3819.
28. Yu, H.; Peng, Y.; Yang, Y.; Li, Z.-Y. Plasmon-enhanced light–matter interactions and applications. *NPJ Comput. Mater.* **2019**, *5*, 45.
29. Dubey, P. K.; Tripathi, L. N. Hybrid metal nanoantenna 2D-material photovoltaic device. *Sol. Energy Mater. Sol. Cells* **2019**, *200*, 109918.
30. Krug, J. T.; Sánchez, E. J.; Xie, X. S. Design of near-field optical probes with optimal field enhancement by finite difference time domain electromagnetic simulation. *J. Chem. Phys.* **2002**, *116*, 10895–10901.
31. Jain, P. K.; Lee, K. S.; El-Sayed, I. H.; El-Sayed, M. A. Calculated absorption and scattering properties of gold nanoparticles of different size, shape, and composition: Applications in biological imaging and biomedicine. *J. Phys. Chem. B* **2006**, *110*, 7238–7248.
32. Link, S.; El-Sayed, M. A. Spectral properties and relaxation dynamics of surface plasmon electronic oscillations in gold and silver nanodots and nanorods. *J. Phys. Chem. B* **1999**, *103*, 8410–8426.
33. Atwater, H. A.; Polman, A. Plasmonics for improved photovoltaic devices. *Nat. Mater.* **2010**, *9*, 205–213.
34. García-Martín, A.; Ward, D. R.; Natelson, D.; Cuevas, J. C. Field enhancement in subnanometer metallic gaps. *Phys. Rev. B* **2011**, *83*, 193404.
35. Higgins, M.; Ely, F.; Nome, R. C.; Nome, R. A.; dos Santos, D. P.; Choi, H.; Nam, S.; Quevedo-Lopez, M. Enhanced reproducibility of planar perovskite solar cells by fullerene doping with silver nanoparticles. *J. Appl. Phys.* **2018**, *124*, 065306.

36. Fahr, S.; Rockstuhl, C.; Lederer, F. Metallic nanoparticles as intermediate reflectors in tandem solar cells. *Appl. Phys. Lett.* **2009**, *95*, 121105.
37. Pacifici, D.; Lezec, H. J.; Atwater, H. A. All-optical modulation by plasmonic excitation of CdSe quantum dots. *Nat. Photonics* **2007**, *1*, 402–406.
38. Fujihima, A.; Honda, K. Electrochemical photolysis of water at a semiconductor electrode. *Nature* **1972**, *238*, 37–38.
39. Walter, M. G.; Warren, E. L.; McKone, J. R.; Boettcher, S. W.; Mi, Q.; Santori, E. A.; Lewis, N. S. Solar water splitting cells. *Chem. Rev.* **2010**, *110*, 6446–6473.
40. Seh, Z. W.; Kibsgaard, J.; Dickens, C. F.; Chorkendorff, I.; Nørskov, J. K.; Jaramillo, T. F. Combining theory and experiment in electrocatalysis: insights into materials design. *Science* **2017**, *355*, 12.
41. Benck, J. D.; Hellstern, T. R.; Kibsgaard, J.; Chakthranont, P.; Jaramillo, T. F. Catalyzing the hydrogen evolution reaction (HER) with molybdenum sulfide nanomaterials. *ACS Catal.* **2014**, *4*, 3957–3971.
42. Chia, X.; Eng, A. Y. S.; Ambrosi, A.; Tan, S. M.; Pumera, M. Electrochemistry of nanostructured layered transition-metal dichalcogenides. *Chem. Rev.* **2015**, *115*, 11941–11966.
43. Linic, S.; Christopher, P.; Ingram, D. B. Plasmonic-metal nanostructures for efficient conversion of solar to chemical energy. *Nat. Mater.* **2011**, *10*, 911–921.
44. Mascaretti, L.; Dutta, A.; Kment, Š.; Shalaev, V. M.; Boltasseva, A.; Zbořil, R.; Naldoni, A. Plasmon-enhanced photoelectrochemical water splitting for efficient renewable energy storage. *Adv. Mater.* **2019**, *31*, 1805513.
45. Kuraganti, V.; Jain, A.; Bar-Ziv, R.; Ramasubramaniam, A.; Bar-Sadan, M. Manganese doping of $MoSe_2$ promotes active defect sites for hydrogen evolution. *ACS Appl. Mater. Interfaces* 2019, *11*, 25155–25162.

7 2D Nanomaterials Using Thin Film Deposition Technologies

Kwadwo Mensah-Darkwa and
Daniel Nframah Ampong
University of Science and Technology

Daniel Yeboah
Institute of Industrial Research - Council for Scientific and Industrial Research

Emmanuel Acheampong Tsiwah
University of Science and Technology Of China

Ram K. Gupta
Pittsburg State University

CONTENTS

DOI: 10.1201/9781003178453-7

7.1 INTRODUCTION

Over the last several decades, rapid economic growth has led to an increase in energy demand and consumption. The majority of this demand is met by non-renewable energy sources including petroleum, natural gas, coal, oil shale, and so on. Our over-reliance on traditional energy sources has prompted severe concerns about resource distribution inequity, intermittent fossil fuel supply, and environmental concerns. Because of the paucity and the depletion of non-renewable energy and the worsening environmental condition, environmentally friendly energy conversion/storage technologies are receiving more attention in research and development. In this regard, solar cells, supercapacitors, batteries, fuel cells, and other innovative sustainable energy conversions/storage technologies have evolved as a viable technology that has reduced our reliance on fossil fuels while also protecting the environment. Nevertheless, new technologies must be developed to synthesize and fabricate materials that have unique properties to enhance the performance of these technologies. Through extensive interdisciplinary research, new functional materials have been developed and applied to several emerging technologies, to develop environmentally clean and renewable energy conversion/storage systems.

Nanomaterials have demonstrated exceptional properties and have been accepted for application in energy conversion/storage systems, as well as a variety of other material discoveries. It is widely reported in the literature that, nanomaterials can exhibit a wide range of material properties based on particle size and shape. For example, a particular element or compound can exhibit unique properties depending on the crystal structure and composition. Depending on the manipulation of particle size and dimensions, nanomaterials can exhibit tunable physical and chemical properties. These materials' physical and chemical properties make them more attractive in applications such as photovoltaic cells, batteries, and supercapacitors. Nanomaterials are materials that have a length scale of 1–100 nm in at least one dimension. In regards to their dimensions, they are categorized as zero-dimensional (0D), one-dimensional (1D), two-dimensional (2D), and three-dimensional (3D) structures. 0D nanomaterials tend to have spherical shapes due to the restrictions in all dimensions (x, y, z) in the nanoscale (a dot). The restrictions limit it to atomic/molecular-scale structures such as quantum dots and nanoparticles. On the other hand, 1D nanomaterials have one dimension outside the nanoscale length scale. These usually exhibit needle-like-shaped structure, which includes nanotubes, nanorods, and nanowires.

2D nanomaterials, on the other hand, have two dimensions that aren't confined to the nanoscale. Nanocoatings, thin films, and nanolayers are all 2D nanomaterials. Finally, 3D nanomaterials (also known as bulk nanomaterials) are nanomaterials that are not constrained in all dimensions. This definition is ambiguous, leaving room for other interpretations; however, it is important to note that these materials must have a nanocrystalline structure as one of their main features at the nanoscale. Based on the definition, 3D nanomaterials can include dispersions of nanoparticles (0D), bundles of nanowires and nanotubes (1D), and multi-nanolayers (2D). The concept of 2D nanomaterials has inspired considerable research interest since Novoselov et al. [1] discovered graphene in 2004. When compared to other types of nanomaterials, 2D nanomaterials have been used in a variety of energy conversion/storage technologies due to their outstanding physical and chemical properties, including large surface to volume ratio versus shape and remarkable electrical and thermal characteristics. Many 2D materials have been fabricated and a lot more have been theoretically predicted. Because recent studies have focused on the unique features of 2D nanomaterials, we will begin by discussing 2D nanomaterials, thin films, and the properties that distinguish them for next-generation energy conversion and storage applications.

Thin-film nanomaterials have superior property stability due to their inherent structure and large surface area when compared to other types of nanomaterials. Most thin-film fabrication processes will necessitate the use of a substrate as support. This means that the substrate can be used as a seed to facilitate growth, or it can provide the necessary support to prevent aggregation during the fabrication process. This is especially problematic for 0D nanomaterials during the fabrication process, where aggregation/agglomeration can occur, causing the material to be unstable. Thin-film materials are commonly used in applications that require a large exposed surface area or interface. Because of these inherent properties, thin films are better suited for electrochemical energy conversion/storage applications that require more active sites and a larger contact surface area, such as batteries and supercapacitors. Recently, 2D TiO_2 [2], MXene/TiO_2 [3], reduced graphene oxide [4], Bi_2Te_3 [5], and $BiFeO_3$ [6] thin film have been widely reported in applications for energy conversion/storage.

Chen et al. [7] reported on the use of In_2S_3 nanosheet arrays fabricated on TiO_2 thin films using molecular precursor solution. The 2D nanosheet arrays were used for a heterojunction solar cell, which exhibited a power conversion efficiency (PCE) of 2.58% depending on the height of the array. In another work, 2D WO_3 nanosheets were prepared by a controlled hydrothermal method [8]. The fabricated CdTe solar cell using the nanosheets as back contact exhibited improved thermal stability and device efficiency. 2D nanomaterials exhibit more stable structure and performance compared to the other types of nanomaterials because of the lateral exposed area (sheet-like structure), which could also be tuned to vary its properties. For instance, Malek et al. [9] prepared MoS_2 nanosheets on indium tn oxide (ITO) substrate as an electron transfer layer for perovskite solar cells (PSCs) application. The PCE was found to increase with the decrease in the number of MoS_2 nanosheets. These are just a few examples of how 2D nanoparticles could improve the performance of materials in energy conversion/storage applications.

The most recent advancements in 2D nanomaterials for energy conversion/storage applications will be discussed in this chapter. First, we begin by reviewing the most common synthesis methods for 2D nanomaterials and then focus on methods for fabricating 2D materials using thin-films deposition, followed by a brief overview of characterization techniques for 2D materials fabrication. Second, we will look at 2D thin films and their applications in various energy conversion/storage systems.

7.2 SYNTHESIS OF 2D MATERIALS

7.2.1 Synthesis Approaches

2D nanomaterials have found their way into modern optoelectronic, catalytic, photovoltaic, imaging, sensing, and energy applications. Their use owes to the inherently high surface to volume ratio, structure, quantum size effect, and morphological (e.g., shape, size, and surface porosity of individual particles) attributes they possess. An attribute of particular importance for many of these applications is large surface area, as it facilitates fast (chemical and charge) transport phenomena. Another is, morphology that dictates how well they adhere, absorb, refract/reflect, and also, whether or not they are printable. These key attributes, among others mentioned, hinge on both synthesis and deposition techniques used. As a result, varied procedures and thus setups have been applied in synthesizing and depositing 2D nanomaterials as thin films for their required use in the applications of research and industry. Among the techniques include chemical vapor deposition (CVD), nanolithography, chemical/mechanical exfoliation, sol–gel, solvothermal, and others. The two types of synthesis/deposition procedures are top-down and bottom-up synthesis/deposition. Thus, entailed in this section is a discussion of such techniques explored in research and industry under these categories.

7.2.1.1 Top-Down Approach

Top-down methods involve disintegrating bulk material to obtain nanoscale forms of the material. It can be done by machining, etching, or breaking down successively the bulk material to atomic crystal lattices. For 2D materials, this is achieved by targeting and severing weak interlayer forces. Some examples are the mechanical exfoliation method, chemical exfoliation method, laser ablation method, and nanolithography (Figure 7.1).

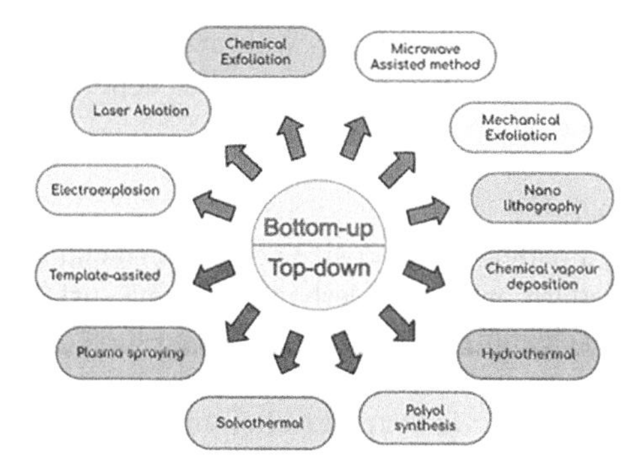

FIGURE 7.1 Examples of popular top-down and bottom-up approaches used to synthesize 2D nanomaterials.

7.2.1.1.1 Mechanical Exfoliation Method

The scotch tape approach is popular for mechanical exfoliation where the scotch tape is stuck onto the surface of bulk material and peeled off repeatedly to obtain thin sheets of material. It is a simple method and gives clean surfaces with minimum defects. The downside to this approach is that parameters cannot be controlled to produce specific quality. Also, it is not suited for mass production and has low yields. It has been used to synthesize graphene, antimonene, $NbSe_2$, TiS_2, WS_2, WSe_2, and $TaSe_2$ [10–13]. A shear exfoliation method has also been explored, which yielded graphene sheets in organic solvent without any defects.

7.2.1.1.2 Chemical Exfoliation Method

This involves, first of all, the intercalation of bulk material by chemical species followed by either a mechanical (e.g., ultrasonication and stirring) or electric (e.g., microwave irradiation and electric current probes) means of exfoliation. Chemical species like bromine, glycolic acid, hydrogen peroxide, and perchloric acid have been used [14–16]. The approach was successfully used to synthesize graphene, boron nitride nanosheets, and molybdenum disulfide having large lateral size (up to 256.8 μm^2) and high quality; few structural defects hence showing outstanding electrical conductivity [16,17]. Compared to mechanical exfoliation, this has higher yields, better quality, and takes less time. The downside is that it is not benign and requires high energy input if corrosive acids or high-boiling solvents are used, respectively.

7.2.1.1.3 Laser Ablation Method

In laser ablation, short wavelength, high fluence laser irradiation is used to remove bulk material and by severing Van der Waal forces. The laser irradiation can be continuous or pulsed. The method has wide use in processing ceramic, metallic, and polymeric materials. Noble metal nanoparticles have been synthesized in liquid solution through laser ablation [18]. In recent research, laser ablation, mainly pulsed laser deposition (PLD), has been explored in synthesizing 2D materials. WS_2 films have been grown on sapphire substrates using a pulsed KrF excimer laser [19]. Others like

Bi_2Te_3, SnSe, SnS, MoS_2, and $Mo_{0.5}W_{0.5}S_2$ with mono-/few layer(s) have also been prepared via PLD [20–23]. The thickness of 2D films deposited by PLD is precisely controlled by varying the number of laser pulses.

7.2.1.1.4 Nanolithography

It involves imprinting a pattern of nano-size onto a sacrificial substrate and thereafter, transferring the pattern onto an underlying substrate surface for its application. The process starts with cleaning a substrate, then deposition of resist onto its surface followed by baking. Afterward, the resist material is etched by light, electrons, or ions. The pattern is developed via the removal of unexposed resist portions. Finally, it is pressed onto an underlying substrate. Examples are photolithography, electron beam lithography, atomic force microscopy (AFM) manipulation, ion beam lithography, dip-pen lithography, and x-ray lithography. Graphene films were manipulated at the nanoscale through anodic oxidation to structure isolating trenches into single-layer graphene flakes of sizes less than 30 nm via the AFM technique [24].

7.2.1.2 Bottom-Up Approach

This method involves using chemical means to self-assemble smaller units, such as atoms or molecules, into bigger nanomaterials with desired chemical composition, size, shape, stacking order, crystal structure, and edge/surface defects. Examples in this category are solvothermal synthesis, CVD, and the sol–gel method.

7.2.1.2.1 Solvothermal Synthesis

It is a wet-chemical process that occurs at elevated temperatures and pressures. An organic or inorganic solvent is mixed with precursor materials and transferred into a reaction chamber. The solvent catalyzes the growth of nanoparticles. Hydrothermal synthesis is a similar process that uses water as a solvent. It is particularly used for the crystal growth of substances that are insoluble at normal temperatures yet soluble in water at high temperatures and pressure. This method has been used to fabricate 2D nanosheets such as metals, metal oxides, and metal chalcogenides. It is a low-cost process, can produce various morphologies, has high yields, and is green.

7.2.1.2.2 Chemical Vapor Deposition

In this technique, gas precursors are introduced into a chamber, and the substrate is heated to high temperatures, causing a reaction that results in the formation of the desired film. Low-pressure, atmospheric pressure, and plasma-enhanced CVD methods are all variations of this process. Parameters that affect the chemical and physical properties of thin-film material of varying sizes include substrate type, the concentration of the gas mixture, temperature, and pressure. This process yields a stable, homogeneous film with low porosity and high purity. It does, however, have significant drawbacks, such as the need for expensive equipment and the release of toxic gaseous by-products during the reaction process.

7.2.1.2.3 Sol–gel Method

Sol–gel method uses high chemically active compounds as precursors. They are mixed in the liquid phase, hydrolyzed, and condensed via chemical reactions to form a stable sol system in solution. The hydrolysis reaction causes an active monomer to polymerize forming a sol, and then a gel with a 2D spatial structure. The gel is dried, heat-treated, and prepared into required nanoparticular materials. It provides film homogeneity, facilitates uniform doping, synthesizes at lower temperatures, and is adapted to various new materials [25]. Synthesis of some 2D nanomaterials such as graphene and its composites via the sol–gel method is a challenge because graphene has low solubility in commonly used solvents.

7.2.2 2D Materials Using Thin Films Deposition

7.2.2.1 Physical Deposition

Physical vapor deposition happens in a highly evacuated chamber and coats substrate by transferring atoms from a source material via its evaporation or sputtering (Figure 7.2a–c).

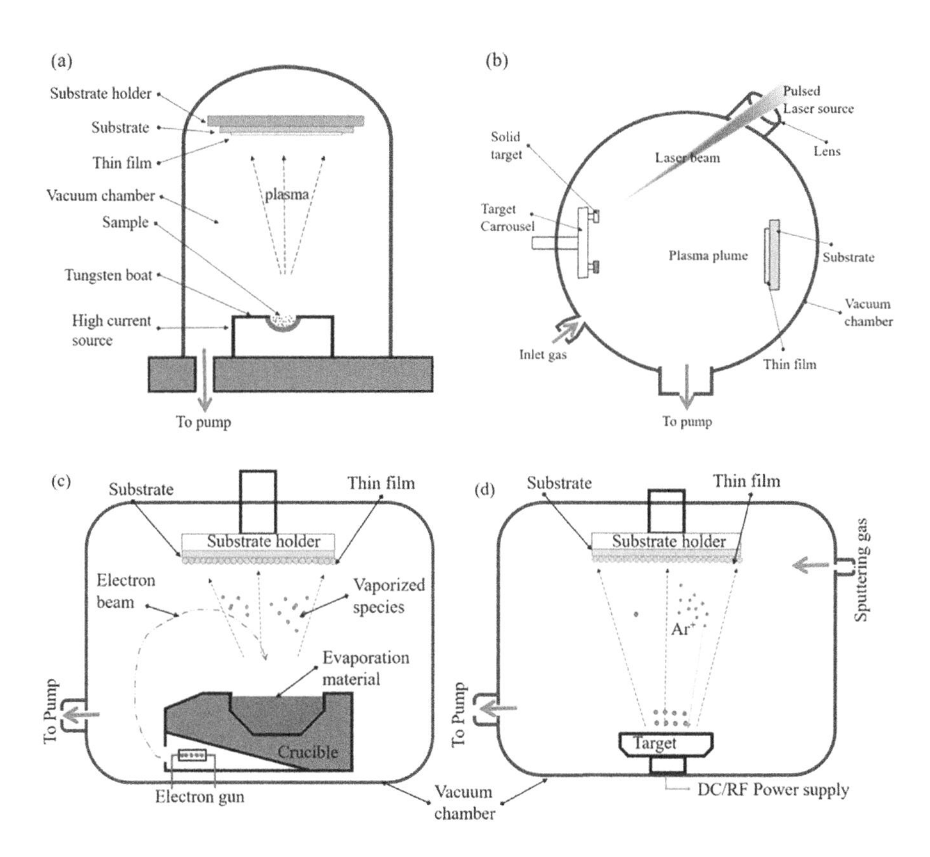

FIGURE 7.2 Schematic diagrams of deposition techniques for 2D thin films; (a) thermal evaporation, (b) pulsed-laser deposition, (c) electron beam evaporation, and (d) sputtering.

7.2.2.1.1 Evaporation Techniques

Deposition by evaporation takes place either in a vacuum or under controlled atmospheric conditions. The fundamental principle underlying this approach is to transform the material from a solid to a vapor phase and back to solid once it reaches the substrate surface [26]. Evaporation of the material can be achieved thermally, by electrons, laser, as well as ion-beam. CdSSe, MnS, and Ge-Te-Ga thin films were deposited by the vacuum thermal evaporation technique [26]. The thin-film quality depends on laser wavelength, energy, gas pressure ambient, pulsed duration, and the distance of the target to the substrate in the PLD evaporation technique. PLD deposits fast, use less time and are compatible with oxygen and other inert gases.

7.2.2.1.2 Sputtering Techniques

In a sputtering set-up, target material (acting as the cathode) supplied with either direct current or radio-frequency (RF) power is placed in a vacuum chamber at a distance from the substrate (acting as the anode). A sputtering gas (e.g., argon, neon, and krypton) is let into the chamber to cause a glow discharge, which releases ions from the target, bombarding the substrate thereby producing dense thin films (Figure 7.2d). By bleeding small amounts of oxygen with the sputtering gas, TiO_2 thin-films can be deposited from the metallic Ti. The method is suitable for high melting point materials and ultrahigh vacuum applications. Conversely, it is difficult to achieve uniform deposition on complex shapes [27].

7.2.2.2 Chemical Deposition

7.2.2.2.1 Chemical Vapor Deposition (CVD)

In conventional CVD, volatile gaseous precursors are allowed into a chamber containing a substrate. The reaction of the precursors takes place on the surface of the substrate, which is covered by a thin film. High-quality monolayer graphene films are easily grown by a self-limited growth mode employing a metal substrate as a catalyst in the process. Advantages of using this method include tunable physical and chemical properties and reduced porosity of films formed. CVD that makes use of liquid precursors also exists. Graphene-like films have been deposited on sapphire and SiO_2/Si from acetone and isopropanol liquid precursors and showed superior properties in terms of carrier mobility [28].

7.2.2.2.2 Sol–gel Technique

This is an appropriate method for producing 2D nanomaterials from the liquid phase. The method employs a low-temperature synthesis to allow the incorporation of organic molecules into the matrix while avoiding the degradation associated with a high-temperature synthesis. A challenge in using this synthesis approach is the uncontrolled aggregation side reactions observed in the precursor sol [29]. Graphene-titania nanocomposites have been prepared to harness the photocatalytic properties of graphene. A variety of techniques have been used to integrate graphene into mesoporous titania, silica, and other nanocomposites. Also, the graphene-silica sol–gel system was prepared as transparent nanocomposites and assessed for their optical applications [26].

7.3 CHARACTERIZATION OF 2D MATERIALS

After 2D nanomaterials are synthesized, peculiar properties such as film size, thickness, crystallinity, composition, and others are measured via characterization techniques. Based on the synthesis/deposition approach taken, different 2D materials tend to have different morphological and structural traits. Hence there is a need to accurately understand how the growth mechanism contributes to the structure, how the structure affects and relates to the functionality of the synthesized material, which could ultimately aid in the design of 2D materials with desired features for specific applications. As a result, a wide range of characterization techniques has been used in probing 2D materials to understand such correlations. This section explores the working principle, the strengths, and the limitations of some of these techniques.

7.3.1 Optical Microscopy

This technique is affordable and facilitates the study of 2D materials without the need for sophisticated imaging instrumentation. Generally, the interference of reflected light at the film/substrate interface is responsible for the generation of the image. Information on the location, thickness, and shape is obtained by generating a contrast between the substrate and the 2D nanomaterial. Mechanically exfoliated graphene has been probed via this technique with monochromic light [30]. The method is adapted to materials with good optical properties to obtain an estimated thickness of the 2D material according to color and contrast. The limitation this method has is that most insulating 2D nanomaterials have a small refractive index, as a result, a few layers might lack the contrast needed to be notable in reflected light [31].

7.3.2 Scanning Probe Microscopy

This consists collectively of methods for studying the surface morphology of nanomaterials at high resolutions. By using a probe to interact with the surface, the topography and electronic band structures of 2D materials can be obtained and interpreted.

7.3.2.1 Atomic Force Microscopy

This method, when combined with other techniques like photoluminescence spectroscopy, can measure the thickness of 2D materials. Thickness measured via this method can be affected by contaminants on substrate or chemical contrast between thin film and substrate leading to some discrepancies in thickness relative to theoretical values. For that reason, appropriate precautionary steps have to be taken [32]. Additionally, AFM can be used to monitor changes in morphology over time to reveal degradation activity in unstable 2D thin films. Nanosheets of black phosphorus, for example, have been found to degrade rapidly in ambient conditions due to a chemical reaction with oxygen. This observed dissolution of the reactants was studied with AFM [33].

7.3.2.2 Scanning Tunneling Microscopy

This method provides resolved atomic morphologies of materials. It can also give information on the localized electron density of states. Recently, it has been used to probe 2D materials to study band structure and the effects of doping. It reveals band alignment of Van der Waal heterostructures [34]. The particle bandgap of MoS_2 has been obtained through this method. Band structure of MoS_2-WS_2 heterofilms has also been reported. Aside from that, the position of Fermi level in a material can be deduced by this approach. It has been discovered that the Fermi levels of metallic substrate metals such as Ti, Au, or Pd exhibit intrinsic defects close to their conduction band [35].

7.3.2.3 Raman Spectroscopy

This method gives information about the structural and electronic properties of a material with high spatial and spectral resolution. It is non-destructive and fast. It can reveal crystalline orientation and phases, the number of layers, strain, and doping effects of 2D materials [36]. Raman spectroscopy has further been used to investigate perturbations like doping effect and strain in 2D materials. Changes in peak intensities and positions were observed for MoS_2 nanosheets during gradual stretching or via the application of gate voltages. In addition, slight shifts in Raman peaks can be linked to substrate effects that can show the effects of doping in 2D nanomaterial [37].

7.3.2.4 X-ray Photoelectron Spectroscopy

X-ray photoelectron spectroscopy is useful for qualitative and quantitative elemental and chemical analysis, respectively. This method can be used to identify the elements that comprise a material because each element is represented by a peak that also represents distinct binding energy. The electronic configuration within the atoms corresponds to the position and shape of the peak and hence can be used to extract parameters such as chemical composition and atomic bonding of the material. This technique measures information within 10 nm of the material surface thus, making it well suited for thin-film 2D materials as their thicknesses are usually <5 nm [38].

7.4 ENERGY APPLICATIONS OF THIN FILMS BASED 2D MATERIALS

The ever-rapid world development, industrialization, and increase in the standard of living of modern societies have necessitated the need for efficient, reliable, cost-effective, and eco-friendly energy conversion and storage systems. The development of alternative clean energy generation and storage sources is critical, given the growing concern about the global carbon footprint caused by the use of conventional fossil fuels. Batteries and supercapacitors are prominent energy storage devices, whereas fuel cells and solar cells provide clean energy generation; these technologies are discussed in the subsequent sections. To meet the demand for portable, lightweight, and energy-efficient requirements for certain device applications, thin-film nanostructures are crucial for this realization.

7.4.1 Thin Film-Based Supercapacitors

7.4.1.1 Graphene-Based Supercapacitors

Graphene-based electrodes are attractive for supercapacitor applications due to their improved charge mobility and surface area on the electrode surface. To study the performance of supercapacitors, thin films of graphene oxide/tungsten oxide (GO/WO_3) were fabricated using the chemical bath deposition (CBD) technique by employing different substrate materials [39]. The fabricated device recorded a maximum specific capacitance (C_{sp}) of 268.5 F g^{-1} and energy intensity of 52.2 W kg^{-1} on the Glass/GO/WO_3 with the specific capacitance increasing with decreasing scanning rate as shown in Figure 7.3a and b. This performance was attributed to the graphene oxide's large surface area. A simple hydrothermal process was adopted to produce thin-film binder-free GO-nickel sulfide nanoplates to investigate its pseudocapacitance behavior [40]. Using the three-electrode system, a maximum C_{sp} of 1,745.50 F g^{-1} at a current density of 5 mA cm^{-2} was recorded. This performance was due to the enhanced specific surface area and higher electrical conductivity.

Another cost-effective electrodeposition approach was adopted to produce a flexible thin-film RuO_2 electrode on a substrate of copper and graphene-coated Cu foil [41]. The electrochemical performance was studied for supercapacitor applications and a C_{sp} of 1,561 F g^{-1} was recorded at 5 mV s^{-1} scan rates for the RuO_2/Gr/Cu electrode, with 98% retention improvement under bending conditions (Figure 7.3c and d). Thin films of reduced graphene oxide (GO) used in supercapacitor applications with a solid-state electrolyte recorded an optimum electrode capacitance of 8.55 mF cm^{-2}

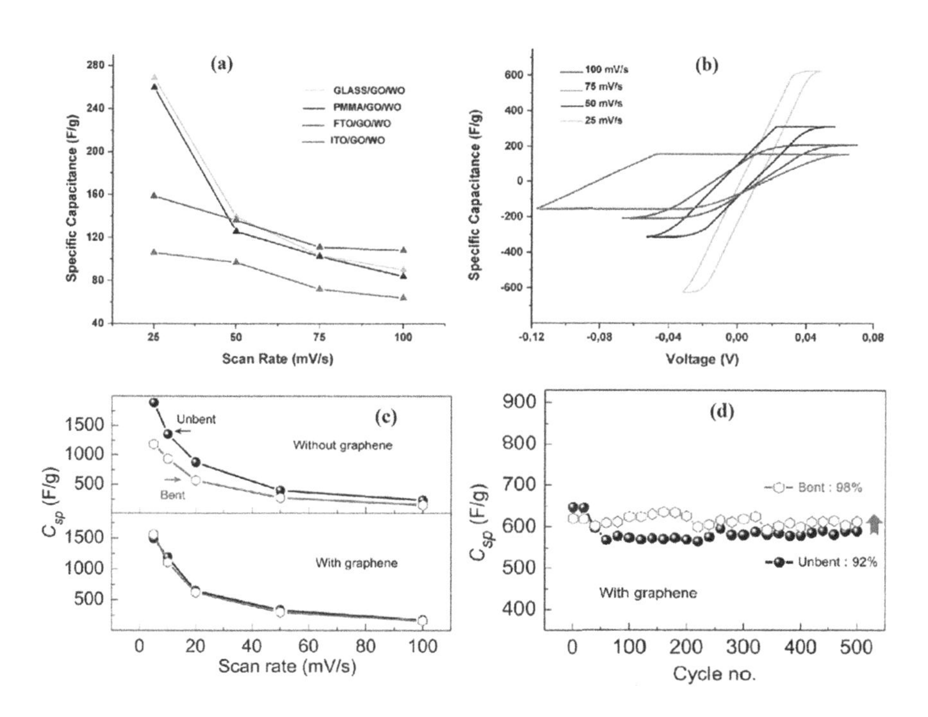

FIGURE 7.3 (a) C_{sp}-R curves for GO/WO_3 on different substrates. (b) C_{sp}-V curves for Glass/GO/WO_3. Adapted with permission from [39]. Copyright (2020) Elsevier. (c) C_{sp} of the RuO_2/Cu and RuO_2/Gr/Cu electrodes at different scan rates in the bent and unbent states. (d) Capacitance retention for the RuO_2/Gr/Cu electrode for 500 cycles in the bent and unbent states. Adapted with permission from [41]. Copyright (2017) Elsevier.

as compared to the liquid-based Na_2SO_4 electrolyte (10.67 mF cm^{-2}). Here, the GO thin films fabricated on ITO substrates were reduced electrochemically to obtain the reduced GO (rGO). This performance is encouraging for the possible integration of solid-state thin-film rGO supercapacitor electrodes in portable electronic devices [42].

7.4.1.2 Transition Metal Oxides and Hydroxides-Based Supercapacitors

Transition metal oxides and transition metal hydroxides offer great pseudocapacitive performance for supercapacitors due to enhanced charge separation at the interface of the electrode and electrolyte. Thin films of cobalt hydroxide [$Co(OH)_2$]-based electrode for symmetric supercapacitor was successfully fabricated utilizing 1 M KOH as the electrolyte onto stainless steel (SS) substrate by potentiodynamic electrodeposition process (Figure 7.4a and b) [43]. The performance of the device was evaluated and maximum specific energy and specific power of 3.96 Wh kg^{-1} and 42 kW kg^{-1} at 5 mV s^{-1} scan rates were recorded respectively, as shown in Figure 7.4c. A C_{sp} of 1,727 F g^{-1} at 5 mA cm^{-2} has been reported with about 91% capacitance retention after 2,000 cycles for NiO/multiwalled CNTs (MWCNTs) flexible SS electrode using ionic layer adsorption and reaction [44].

Temperature is critical in supercapacitor applications if the energy storage device is to be employed in operations that need high-temperature variations. The device's electrochemical performance strongly depends on the operating temperatures. The performance of the thin-film silver oxide (Ag_2O) electrode in an aqueous solution was evaluated in a temperature range of 35°C–75°C for supercapacitor applications [45]. The capacitance increased with an increase in temperature with higher stability as shown in Figure 7.5a and b. This improved electrochemical performance was attributed to the resistance reduction in charge transfer, an enhancement in ionic conductivity, and quick adsorption/desorption of ions at elevated temperatures. The effect of temperatures on the electrochemical performance of nanostructured hydrous copper oxide/hydroxide [$CuO/(OH)_2$] deposited onto SS substrate by CBD method has been reported [46]. The binder-free solid-state electrode used for supercapacitors recorded a maximum C_{sp} of up to ~340 F g^{-1} at 1 mA cm^{-2} for 105°C (Figure 7.5c). Figure 7.5d shows the capacitance retention of the ($CuO/(OH)_2$)-based supercapacitor electrode up to about 91% after 1,000 cycles.

7.4.1.3 Transition Metal Dichalcogenides (TMDs)-Based Supercapacitors

Higher specific area and better conductivity of 2D nanostructure materials are necessary for enhanced supercapacitor performance. Several 2D transition metal dichalcogenides (TMDs), for example, MoS_2, $MoSe_2$, and WS_2 have been studied for electrochemical energy conversion/storage applications due to good metal ion coordination and high electrical conductivity. Electrochemical intercalation of ions such as H^+, K^+, Li^+, and Na^+ is possible in nanosheets of MoS_2, which contains a greater concentration of metallic 1T phase. This results from the chemical exfoliation of the material to achieve excellent efficiency and capacitance values. This material also showed good power density values as well as better cycling stability [47]. A simple and resizable spray-painting technique and a further laser-patterning can be adopted to fabricate and optimize thin-film 2D MoS_2 nanosheets for micro-supercapacitors to obtain high area capacitance (8 mF cm^{-2}) with improved cycle stability [48].

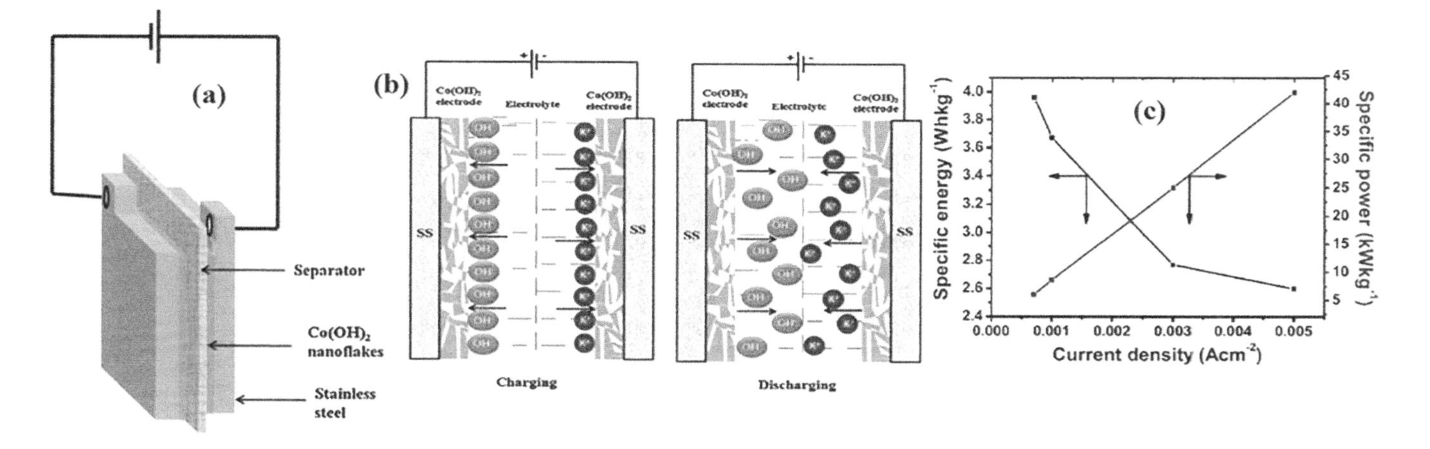

FIGURE 7.4 Symmetric $Co(OH)_2$-$Co(OH)_2$ supercapacitor (a) test cell and (b) charging–discharging mechanism, (c) Variations of specific energy and power as the function of current densities. Adapted with permission from [43]. Copyright (2013) Elsevier.

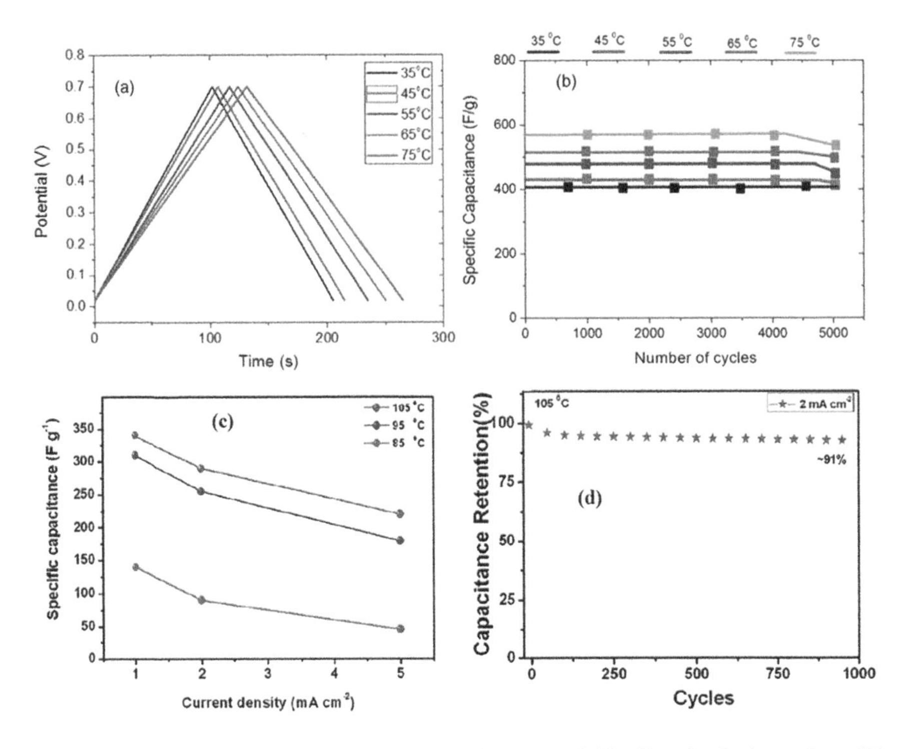

FIGURE 7.5 (a) Charge–discharge and (b) stability test of thin-film Ag_2O electrode at different temperatures. Adapted with permission from [45]. Copyright (2021) Elsevier. (c) Graph of C_{sp} with different current densities of CuO/$(OH)_2$ electrode at different temperatures. (d) Cyclic stability of CuO/$(OH)_2$ electrode (prepared at 105°C) at 2 mA cm^{-2} charging current density over 1,000 cycles. Adapted with permission from [46]. Copyright (2017) Elsevier.

2D-tungsten disulfide (WS_2) possesses enormous active sites that enhance charge accumulation. Incorporating WS_2 with rGO nanosheets (Figure 7.6a) as the electroactive material through a simple molten salt technique improves the electrochemical performance for supercapacitor applications. A higher C_{sp} of 2,508 F g^{-1} at 1 mV s^{-1} scan rate can be achieved, with excellent stability of 98.6% retention after 5,000 charge/discharge cycles as shown in Figure 7.6b [49]. TMDs can be engineered with defects to create active sites for enhanced electrochemical performance. A facile moderate solution reduction technique was adopted by a group of Chinese researchers to fabricate high-performance supercapacitors comprising of reduced $CoNi_2S_4$ (r-$CoNi_2S_4$) nanosheets as shown in Figure 7.6c [50]. A specific capacitance of 1,117 C/g at 2 A/g was obtained as compared to 882 C/g for the pristine $CoNi_2S_4$ (Figure 7.6d) due to the generation of sulfur vacancies of the reduced nanosheets, which enhanced conductivity.

7.4.1.4 MXenes-Based Supercapacitors

2D MXene ($Ti_3C_2T_x$) films can be derived from MXene inks for higher electron transfer for purposes of manufacturing electrodes for portable thin-film supercapacitors. The main problems hindering its real-life application are the dismantling of MXene components and oxidative instabilities when contacted with water. To control this,

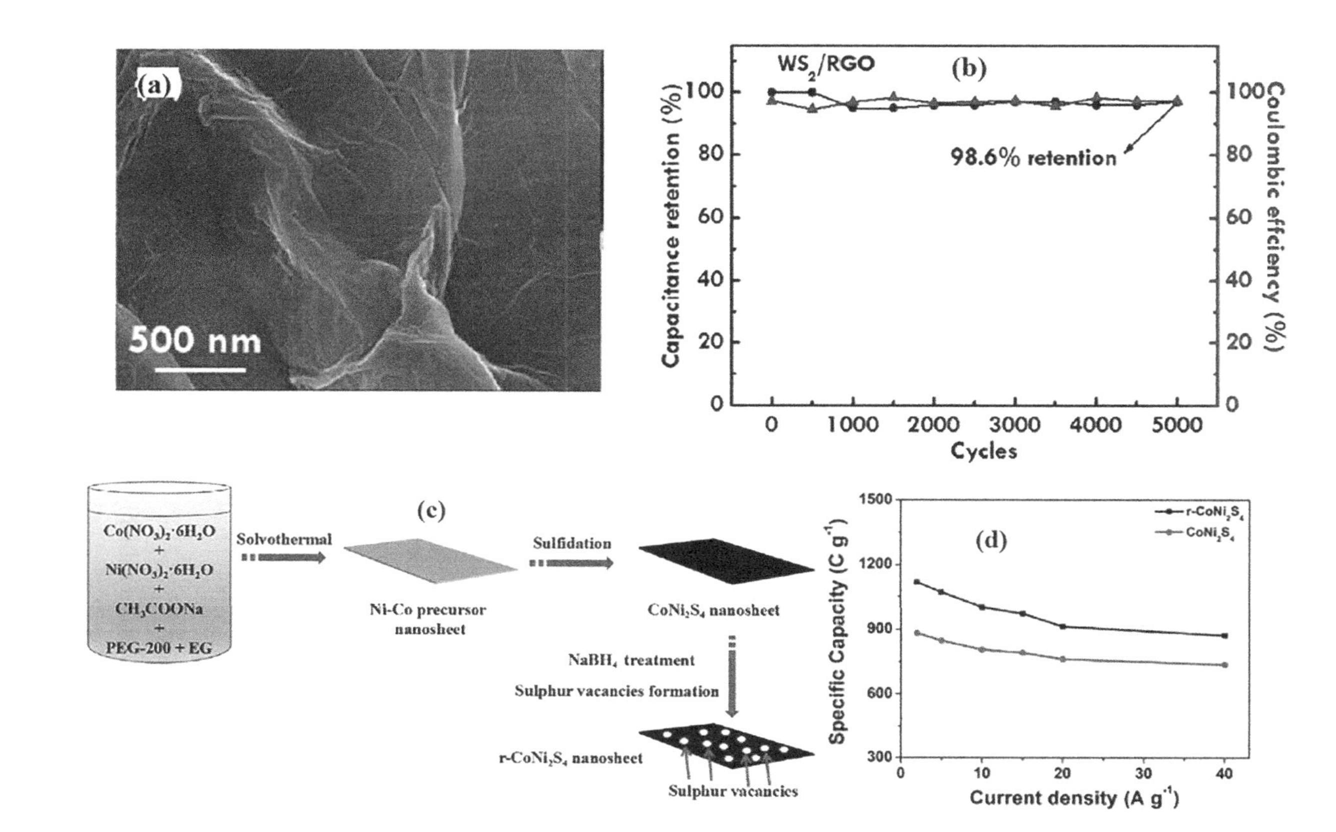

FIGURE 7.6 (a) SEM images of WS_2/RGO hybrid. (b) The capacitance retention and the Coulombic efficiency as the function of the charge/discharge cycle for the SC electrodes with WS_2/RGO. Adapted with permission from [49]. Copyright (2016) Elsevier. (c) Schematic illustration of the synthesis of r-$NiCo_2S_4$ nanosheets. (d) Specific capacity at different current densities for the r-$CoNi_2S_4$ and $CoNi_2S_4$ nanosheets electrodes. Adapted with permission from [50]. Copyright (2018) Elsevier.

tartaric acid can be added to the MXene composite to impede oxidation and firmly hold the assembly units together [51]. Incorporating TMOs such as Fe_3O_4 [52] into MXene interlayers to form sandwiched electrodes (MXene/Fe_3O_4/MXene) by laser crystallization technique improves the areal capacitance and cycling stability of the device. MXene-based electrode films (PANI/MXene) can be fabricated by simple blade-coating technique by intercalating PANI into interlayers of MXene nanosheets for improved electrochemical performance, flexibility, and structural stability. By so doing, the operating potential window, volumetric energy density, and volumetric capacitance can also be enhanced [53]. Annealing treatment can also be used to modify 2D MXene thin films, which affect the electrochemical and structural properties [54]. Direct annealing techniques help to improve the strength of the structures, cycling stability, and electrochemical performance due to structural and chemical variations on the surface of the film. Polystyrene microspheres can be incorporated into MXene films to generate porous interconnected structures (3D) to promote ion transport and obtain higher electrochemical performance. The 3D channels promote electrolyte accessibility for high-performance thin-film supercapacitors [55].

7.4.2 Thin Films-Based Batteries

7.4.2.1 Lithium (Li)-Based Batteries

The high capacity and lightweight requirement of many microelectronics require thin-film power sources and lithium-ion batteries (LIBs) have been promising in that regard. The electrode materials have a significant impact on the performance of these devices. Thin films of RuO_x nanocrystals deposited (sputter deposition) at room temperature on SS were employed as cathodes for use in thin-film micro batteries (MBs). The device showed enhanced cycling stability and excellent electrochemical properties. The oxygen content in the sputter gas contributed significantly to this performance, which makes it possible for integration into microsystems [56]. There is a need to understand the diffusion barriers of Li in thin-film batteries to enable its integration on silicon substrates since it can cause damage to surrounding units, device separation, and loss of battery capacity. For the first time, the thermal atomic layer deposition (ALD) technique was used to fabricate ultrathin films of titanium nitride (TiN) as a current collector and bifunctional Li-ion diffusion barrier. The device recorded the lowest specific resistance and high electrical conductivity, which is a requirement for current collectors [57]. PLD was also employed to produce a thin film of TiO_2 nanoparticles to be used as Li-ion MBs' negative electrodes. Post-deposition annealing improved the electrochemical activity whereas additional carbon layer PLD onto the films enhanced the Li^+ transport properties and battery capacity retention [58].

7.4.2.2 Zn-MnO_2 Battery

The units of LIBs are poisonous and highly flammable, making them unfeasible power sources for implantable and wearable electronics close to the body. Zn-MnO_2 batteries have several qualities such as high flexibility, low cost, eco-friendliness, high electrochemical performance, and abundance, making them desirable candidate materials for making wearable storage systems. Free-standing electrode materials are crucial to attaining higher device performance. Figure 7.7 shows the mechanism of charge/discharge

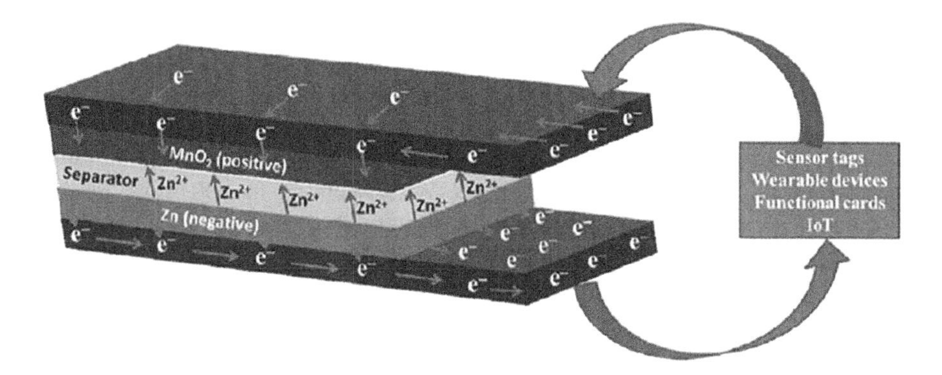

FIGURE 7.7 Charge-transfer mechanism during battery operation of a unit cell. Adapted with permission from [59]. Copyright (2021) Elsevier.

cycles during the operation of the Zn-MnO_2unit battery. The thickness of the current collector, Ag coating, and the amount of carbon contents in the carbon current collector have proven to be the major factors that contribute to the device's maximum discharge capacities and conductivity. These factors were optimized in the current collector for its application in thin-film Zn-MnO_2 battery with high flexibility [59]. A one-step electrodeposition technique was employed by Huang et al. [60] to deposit nanoflakes of MnO_2 onto CNT film, which enhanced the electrochemical kinetics when used in Zn-MnO_2 battery. The device recorded high capability rate of 105.6 mAh g^{-1} at 3 mA cm^{-2} outstanding cycling stability even after 1,000 cycles, and mechanical robustness during bending. Recently, Liang and colleagues [61] integrated nanosheets of graphite with little surface defects in MnO_2 to suppress Zn-MnO_2 batteries in situ electrodeposition to investigate the factors that affect the system's stability. The fabricated electrode device showed enhanced cycling stability over 600 cycles with high columbic efficiency (~99.8%).

7.4.2.3 Nickel-Metal Hydride

Thin-film nickel-metal hydride (Ni-MH) batteries are advantageous over the LIBs due to their competitive price and power density making them ideal candidates for power supply in microsystems. Surface modification of the MH electrode used in Ni-MH batteries can improve the conductivity and reduce the ohmic resistance of the battery. Yang et al. [62] employed a vacuum evaporation plating technique to plate thin Ag film on both sides of the electrode and observed appreciable durability, improved efficiency, and stability of the battery. Tsai et al. [63] also utilized a cost-effective and efficient process to fabricate thin film and flexible buckypapers derived from MWCNTs to be used for Ni-MH batteries anode. An improved electrochemical performance was observed, thus, the buckypaper could be used as a replacement for the metal substrate in Ni-MH battery anode for miniature storage systems.

7.4.2.4 Flow Batteries

Redox flow batteries (RFBs) demonstrating a lot of advantages such as long cycle life, low cost, high energy efficiency, and safety are considered as one of the best electrochemical energy storage devices. To obtain higher performance of the device, thin-film electrodes are required, aside from the membrane being the major component. A two-layer thin-film electrode was developed by Wu et al. to be used for vanadium RFB [64]. The electrode comprised of electrospun fiber mat (EFM) as catalyst layer and carbon cloth, carbon paper, and graphite felt as the backing layer as shown in Figure 7.8a. From Figure 7.8b, the energy efficiency of 76.1% at 240 mA cm^{-2} was recorded for the carbon cloth/EFM electrode. After operating for 800 cycles, no significant decay was observed and hence, the battery showed higher stability. Teng et al. [65] used interfacial polymerization to prepare and optimize thin-film composite (TFC) of polypyrrole (PPy) to enhance the porous membrane for vanadium RFB. A coulombic efficiency of 98.1% was recorded for the device with minimal deterioration upon cycling tests. The enhanced performances of these TFC materials suggest a great potential for large-scale energy storage systems.

7.4.3 Thin Film-Based Solar Cells

7.4.3.1 Amorphous Silicon

A well defect passivated c-Si surface by intrinsic amorphous silicon (i-a-Si:H) layer prior to deposition of the n/p-doped layers is critical to achieve high efficiency for silicon heterojunction (SHJ) solar cells, as the i-a-Si:H layer can significantly reduce the surface defect related to recombination losses. Ru et al. introduced a low deposition rate, approximately 2 Å s^{-1}, i-a-Si:H (i_1) buffer layer grown prior to the bulk i-a-Si:H layer, both deposited with RF-PECVD to improve c-Si surface passivation and thus the efficiency of SHJ solar cell was improved 25.11%. The fabrication process and device performance have been shown in Figure 7.9 [66]. Ruan et al. also deposited a high CH buffer layer (i_1) with pure silane plasma, followed by deposition of intrinsic bulk layer. The i_1 layer suppressed the epitaxial growth at interface, while low-defect i_2 bulk layer further promoted passivation quality [67]. Cho et al. then used plasma-enhanced chemical vapor deposition (PECVD) to create bifacial and semitransparent a-Si:H thin-film solar cells with bifacial transparent conducting oxide (TCO) contacts, obtaining PCE values of 6.39%–6.62% and 4.95%–5.15% using front and rear lighting, respectively [68]. Wu et al. reported the implementation of trimethyl boron $B(CH_3)_3$ (TMB) doped p-type a-Si:H (a-Si:H(p)) film as hole transport layer

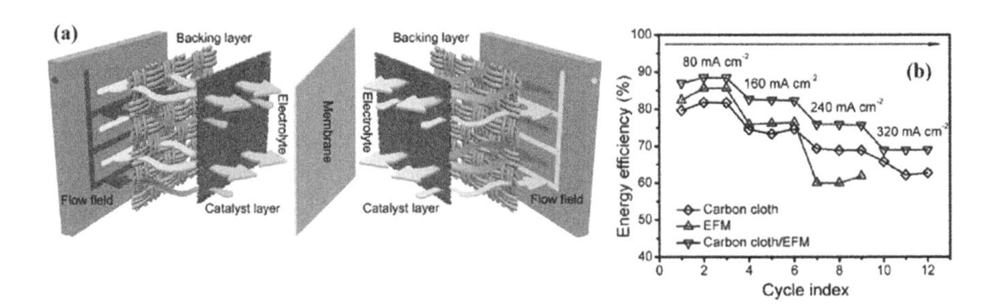

FIGURE 7.8 (a) Schematics of the dual-layer electrode structured flow cell. (b) Energy efficiency of batteries as a function of cycle index. Adapted with permission from [64]. Copyright (2019) Elsevier.

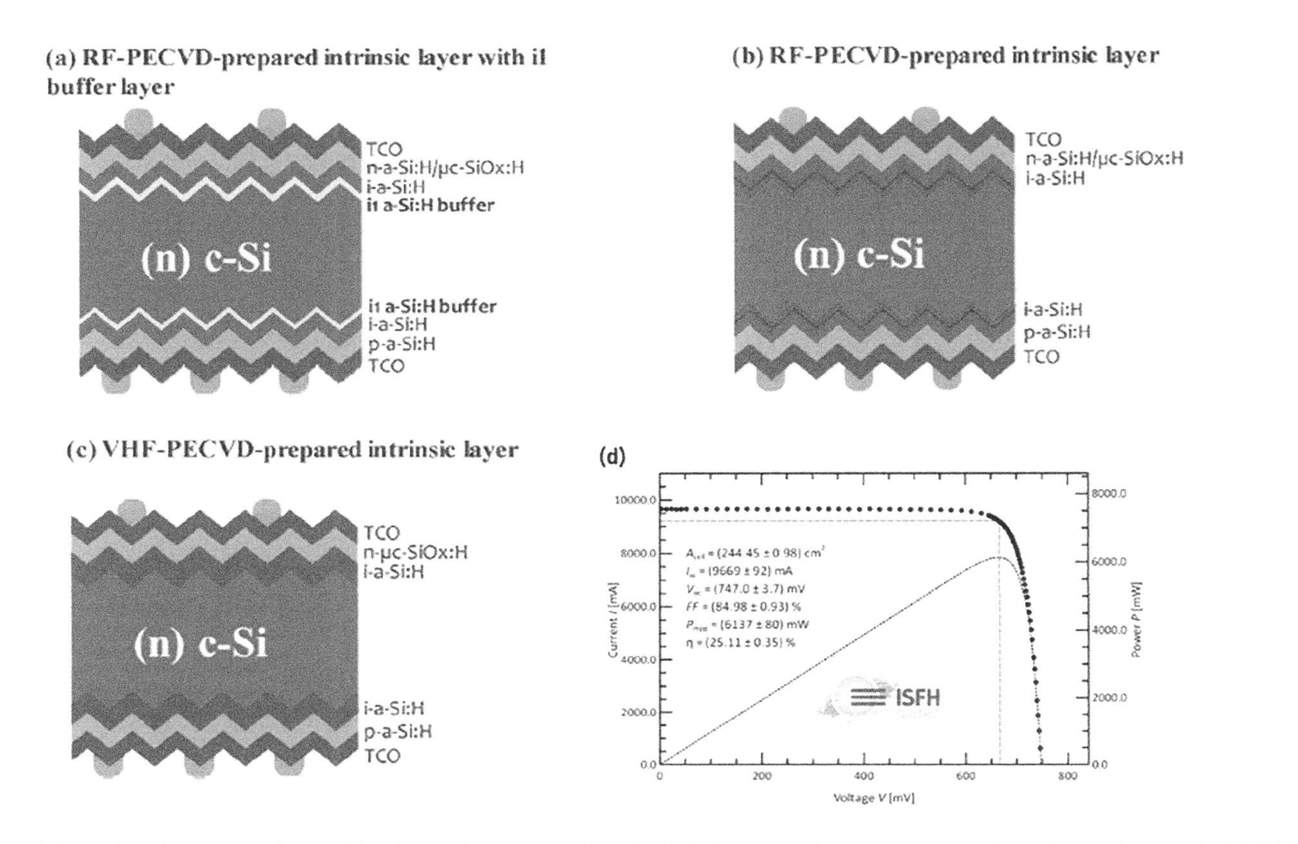

FIGURE 7.9 Schematic illustration of the SHJ solar cell structure: (a) RF-PECVD-prepared intrinsic layer with i_1 buffer layer, (b) RF-PECVD-prepared normal intrinsic layer, (c) VHF-PECVD-prepared intrinsic layer. (d) I–V characteristics of a 25.11% efficiency SHJ solar cell on a total area of 244.5 cm^2. Adapted with permission from [66]. Copyright (2020) Elsevier.

contacting with indium-free aluminium-doped zinc oxide (AZO) in SHJ solar cells. The optimized low-resistivity a-Si:H (p)/AZO contact enables a fill factor above 81% and an efficiency of 23.6% for M2 SHJ solar cells [69]. Sedani et al. also studied the solid-phase crystallization of boron-doped non-hydrogenated amorphous Si films fabricated by electron beam evaporation equipped with effusion cells (e-Beam EC) on silicon nitride-coated glass substrates [70].

7.4.3.2 Cadmium Telluride (CdTe) Solar Cells

Because of its high absorption coefficient, nearly optimum optical bandgap, and ease of manufacturing, CdTe is an excellent candidate for thin-film PV solar cell applications. To deposit the CdTe absorber layers, several growth methods have been developed. The efficiencies of CdTe cells grown on commercial fluorine doped tin oxide (FTO)-coated substrates were significantly lower. A lot of methods have been developed but their results are significantly below the results from world-record cells grown on non-commercial TCO glass substrates. Paudel et al. recently focused on optimizing the fabrication and characterization of high-efficient CdTe solar cells grown on commercial TCO glass substrates using the close space sublimation (CSS) system developed in their lab, and efficiencies of greater than 15% were achieved [71]. Hu et al. attempted to minimize absorption loss from the CdS:O window layer, which is one of the challenges in the development of CdTe-based thin-film solar cells, by using a post-annealing process to further reduce the CdS:O thickness to 30 nm while still achieving 18% efficiency. The device performance is shown in Figure 7.10 [72]. Major et al. incorporated ZnO NWs deposited by a low-temperature electrochemical deposition route, which achieved up to 9.5% efficiency [73].

7.4.3.3 Dye-Sensitized Solar Cells (DSSCs)

Many materials, including TiO_2, ZnO, Al_2O_3, MgO, and others, have been deposited using various thin-film deposition techniques, including sputtering, ALD, spin coating, thermal oxidation, spray pyrolysis, and electrochemical methods for use in dye-sensitized solar cells (DSSCs). Pham et al. recently addressed the use of SnO_2-based photoanodes in aqueous dye-sensitized PV solar cells, as well as a new liquid deposition strategy for forming SnO_2@TiO_2 surfaces. The device achieved an overall

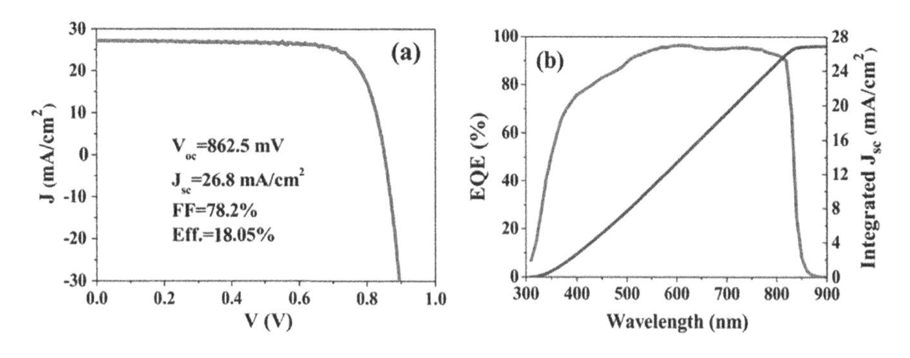

FIGURE 7.10 (a) J–V and (b) EQE curves of the cell with 18% efficiency (with AR coating). Adapted with permission from [72]. Copyright (2021) Elsevier.

solar conversion efficiency of 0.66% [74]. Fu et al. also proposed a simple method for fabricating efficient panchromatic DSSCs with a photoelectrode composed of bi-layered TiO_2 thin films coated with one or two dyes. A screen-printing process was used to create the bottom layer of the TiO_2 thin film on an FTO glass electrode [75]. RF-sputtered ZnO and TiO_2/ZnO play an important role in the charge dynamics.

7.4.3.4 Perovskite Solar Cells (PSCs)

For high-performing PSCs, the quality of the fabricated perovskite film is very essential. The PCEs of PSCs have been improved over the past decade with lead-based perovskite having the highest of them (25.5%) as shown in Figure 7.11 [76]. To provide a full picture of realizing high-performance PSCs, the focus has been on the strategies for preparing high-quality perovskite films (including antisolvent, additive engineering, scalable fabrication, bandgap adjustment, and strain engineering). A Gallium-doped zinc oxide (GZO) layer was used as TCO film for low-cost $Cu_2O/CH_3NH_3PbI_3/SiO_2$ structure to improve the heterojunction planar PSCs [77]. Ayad et al. used the well-known SCAPS-1D software to simulate a planar structure ($CuI/CH_3NH_3PbI_3$-xClx/ZnO/TCO). Also, CuI like transport layer of the holes was used, which can be deposited by simple techniques at low temperature. A three-step fabrication method at ambient atmospheric conditions was presented by Zanca et al. involving an initial step to fabricate $MAPbI_3$ thin films directly on conductive and optically transparent substrates [78]. Chen et al. reported the thin-film deposition of a representative Ti-based HP, Cs_2TiBr_6, and provided insights into the film-formation mechanism. The optical bandgap of the as-deposited Cs_2TiBr_6 HP thin films was 1.8 eV, making them highly suitable for tandem PV applications [79].

7.4.4 Thin Film-Based Fuel Cells

7.4.4.1 Alkaline Fuel Cells

A significant attempt is being made to design high-utilization, low-loading electrodes for target performance. The electrode fabrication method and substrate material have a significant impact on this development. Electron beam physical vapor deposition was used to create thin-film anodes with low catalyst loading on both Ni foam and carbon paper as electrode substrates. [80]. Although Ni foam performed better than carbon paper as an electrode substrate, it was less efficient. The anode was made entirely of Ni foam, which produced a peak power density of 174 mW cm^{-2} at 60°C. As a result, nickel foam had both a positive (electro-oxidation of $NaBH_4$) and a negative (hydrolysis) effect on the performance of the fuel cells. Liu et al. fabricated Cu_2O-Cu composite catalyst by a facile laser-irradiation method. The addition of Cu_2O-Cu composite in activated carbon air-cathode greatly improved the performance of the cathode [81]. Takahashi et al. then investigated the catalytic activity of the Pt oxide thin film for the ethanol oxidation reaction (EOR) in alkaline media. The Pt oxide thin film showed high activity for the EOR in alkaline media, and it was possible to approach the realization of direct ethanol fuel cells (DEFCs) that use a safe fuel with high energy density. The Pt oxide thin film and Pt thin film were prepared by reactive sputtering and the electrocatalytic activity of the ethanol oxidation reaction and it was investigated in a KOH solution for developing the alkaline direct ethanol fuel cells [82].

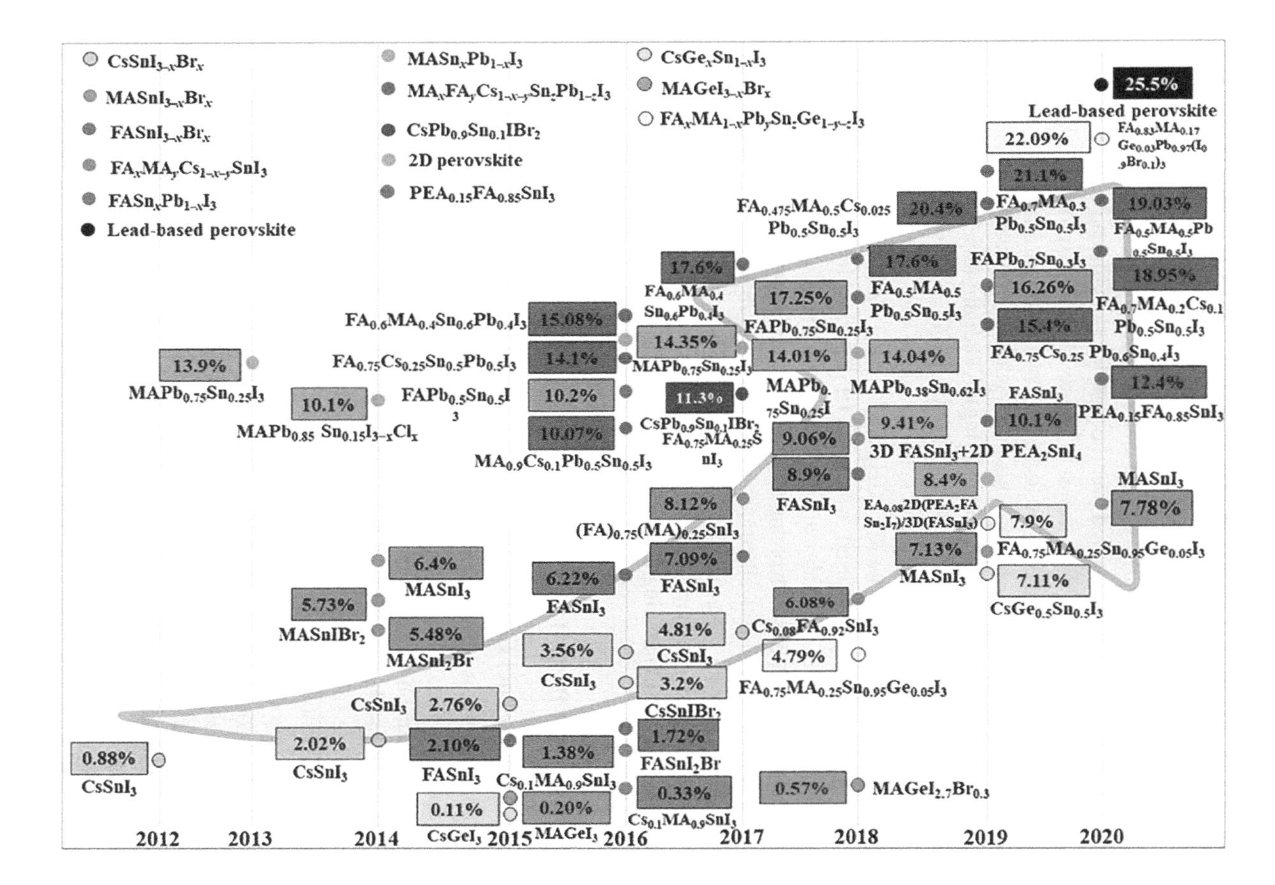

FIGURE 7.11 Improvement of PCE of various PSCs in the last decade. Adapted with permission from [76]. Copyright (2021) Elsevier.

7.4.4.2 Solid Oxide Fuel Cells (SOFCs)

A solid oxide fuel cell (SOFC) is a form of high-temperature fuel cell constructed of solid oxides, such as ceramic materials (generally over 700°C). Fuel flexibility, high efficiency, and high power density/specific power are all advantages of this method. A unit cell was successfully manufactured using a 2-inch sputtering system, indicating that significantly larger cells can be fabricated using larger sputtering equipment. Despite these accomplishments, there have been concerns raised because the technology used to fabricate the multiscale-architecture SOFC was PLD, which has well-known commercialization constraints. Kang et al. introduced the anode functional layer on the surface of the substrate to stably deposit the thin-film electrolyte and additional design parameters including the roughness and density were also introduced. The fabrication process and device performance are shown in Figure 7.12 [83]. Seo et al. also demonstrated how Al_2O_3 layers coated through ALD effectively suppress the degradation of nanoporous Pt thin-film electrodes [84]. These results suggested that a simple and scalable coating strategy enables the implementation of TF-SOFCs with ideal performance and durability outcomes. Normal butane was selected as the fuel by Thieu et al. and multiscale-architectured thin film-based SOFCs were operated in direct butane utilization mode at a temperature of 600°C. They tried to secure fuel flexibility at lower operating temperatures (≤600°C), to employ low temperature solid oxide fuel cells (LTSOFCs) as portable and mobile power sources [85].

7.4.4.3 Microbial Fuel Cells (MFCs)

A microbial fuel cell (MFC) is a device that converts chemical energy to electrical energy through the catalytic reaction of microorganisms. MFCs open up new opportunities for generating sustainable energy from biodegradable compounds. MFCs have also gotten a lot of attention because of their inherent advantages in both fundamental studies (particularly of anodes) and applications as high-throughput platforms (compared to the cultivation and hydrogen production of algae). There has been a lot of research on MFCs. Recently, Lee et al. modified the anodic electrodes of MFCs with Rhodobacter sphaeroide-imprinted polymers by using various polymer concentrations to perform micro-contact imprinting of bacteria [86]. Park et al. also investigated electricity generation in MFCs using e-beam deposited Pt electrodes to improve efficiency and minimize Pt loading [87]. Narayanasamy et al. then reported a carbon cloth-based composite material with effectively improved electrochemical properties resulting from surface modifications with two redox species, namely nickel cobaltite ($NiCo_2O_4$) and a conductive polymer (PANI), through a hydrothermal method and the electro-polymerization technique, respectively [88]. Ying et al. prepared a TiO_2 thin film-modified SS mesh (SSM/TiO_2-H) by an electrochemical method and applied it as an anode in an MFC to enhance their power output. A maximum projected current density of 69.5 ± 1.2 A m^{-2} was obtained from the SSM/TiO_2-H electrode [89].

7.4.4.4 Direct Methanol Fuel Cells (DMFCs)

Membranes commonly used in a direct methanol fuel cell (DMFC) are expensive and show a great permeability to methanol which reduces fuel utilization. Santiago et al. developed sulfonated styrene-ethylene-butylene-styrene-modified membranes with zirconia silica phosphate sol–gel phase and studied them to evaluate their potential

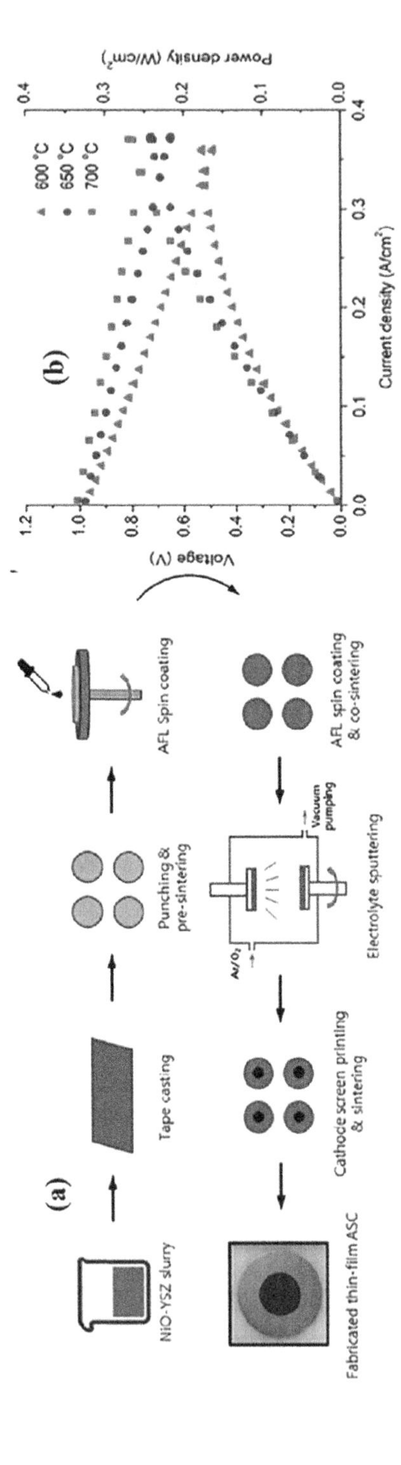

FIGURE 7.12 (a) Schematic of the developed anode-supported thin-film SOFC fabrication process. (b) Current–voltage curves of the fabricated large-sized thin-film SOFC. Adapted with permission from [83]. Copyright (2020) Elsevier.

use in DMFC applications [90]. Nanocellulose composite membrane was successfully synthesized by the phase inversion method and investigated its properties for DMFC [91]. Suhaimin et al. synthesized SPEEK/GO membrane using a simple and facile method. The carbonaceous filler, GO was incorporated into the SPEEK matrix for improving the physical and chemical properties of the fabricated membrane [92]. A novel flexible gas diffusion layer was prepared by vacuum filtration of a suspension of carbon fibers and highly-dispersed multiwall carbon nanotubes in a polytetrafluoroethylene binder and water repellent [93].

7.5 CONCLUSION

To meet the rising energy demand, it is critical to creating energy generation and storage systems that are sustainable, efficient, eco-friendly, and low-cost. Faced with the massive energy demands against the increasing carbon footprints and an impending energy crisis, innovative ways to develop clean and renewable energy materials and systems are extremely eminent. This is possible with microfabrication techniques. The merits of utilizing microfabrication techniques such as thin film deposition for 2D nanomaterials stern from the ease to control the product quality to the creating of enormous electron confinements in the 2D planes, which enhance electronic properties. Thin-film 2D materials feature excellent electrical, thermal, and optical properties as compared to their bulky ones, which render them flexible and make them suitable for wearable electronics. The efficiency, behavior, and response of a material in service can be influenced by its surface characteristics. It is believed that these surface features can be modified and tuned to fulfill a specific application for greater performance, and this method has been widely used in a variety of fields. The focus of this chapter is on the most recently evolved thin film deposition techniques for high-tech energy applications, particularly in advanced microelectronic device production which necessitates the most enhanced performance for its applications. Attention is paid to the materials, synthesis methods, and characterization techniques employed in thin-film 2D nanomaterials. The advantages and limitations of using the various synthesis methods and characterization techniques have been reported.

REFERENCES

1. Novoselov K.S., Geim A.K., Morozov S.V., Jiang D., Zhang Y., Dubonos S.V., Grigorieva I.V., Firsov A.A. (2004) Electric field effect in atomically thin carbon films. *Science* 306:666–669.
2. Adewinbi S.A., Buremoh W., Owoeye V.A., Ajayeoba Y.A., Salau A.O., Busari H.K., Tijani M.A., Taleatu B.A. (2021) Preparation and characterization of TiO_2 thin film electrode for optoelectronic and energy storage Potentials: Effects of Co incorporation. *Chem Phys Lett* 779:138854.
3. Yu J., Zeng M., Zhou J., Chen H., Cong G., Liu H., Ji M., Zhu C., Xu J. (2021) A one-pot synthesis of nitrogen doped porous MXene/TiO_2 heterogeneous film for high-performance flexible energy storage. *Chem Eng J* 426:130765.
4. Korkmaz S., Meydaneri Tezel F., Kariper İ.A. (2021) Reduced graphene oxide/molybdenum oxide thin films and its' capacitance properties: Different substrates effect. *J Energy Storage* 40:102736.

5. Buathet S., Simalaotao K., Reunchan P., Vailikhit V., Teesetsopon P., Raknual D., Kitisripanya N., Tubtimtae A. (2020) Electrochemical performance of Bi_2Te_3 heterostructure thin film and Cu_7Te_4 nanocrystals on undoped and In3+-doped WO3 films for energy storage applications. *Electrochim Acta* 341:136049.
6. Lamichhane S., Sharma S., Tomar M., Kumar A., Gupta V. (2020) Influence of laser fluence in modifying energy storage property of $BiFeO_3$ thin film capacitor. *J Energy Storage* 32:101769.
7. Chen W., Wan Z., Cao W., Dong C., Chen J., Liu R., Zhu L., Zhang X., Wang M. (2021) Solution-processed In_2S_3 nanosheet arrays for $CuInS_2$ thin film solar cells. *Mater Lett* 290:129490.
8. Masood H.T., Anwer S., Rouf S.A., Nawaz A., Javed T., Munir T., Zheng L., Deliang W. (2021) Back contact buffer layer of WO_3 nanosheets in thin-film CdTe solar cell. *J Alloys Compd* 887:161367.
9. Abd Malek N.A., Alias N., Md Saad S.K., Abdullah N.A., Zhang X., Li X., Shi Z., Rosli M.M., Tengku Abd Aziz T.H., Umar A.A., Zhan Y. (2020) Ultra-thin MoS2 nanosheet for electron transport layer of perovskite solar cells. *Opt Mater (Amst)* 104:109933.
10. Geim A.K., Novoselov K.S. (2007) The rise of graphene. *Nat Mater* 6:183–191.
11. Ares P., Palacios J.J., Abellán G., Gómez-Herrero J., Zamora F. (2018) Recent progress on antimonene: A new bidimensional material. *Adv Mater* 30:1703771.
12. Zeng Z., Tan C., Huang X., Bao S., Zhang H. (2014) Growth of noble metal nanoparticles on single-layer TiS_2 and TaS_2 nanosheets for hydrogen evolution reaction. *Energy Environ Sci* 7:797–803.
13. Staley N.E., Wu J., Eklund P., Liu Y., Li L., Xu Z. (2009) Electric field effect on superconductivity in atomically thin flakes of $NbSe_2$. *Phys Rev B - Condens Matter Mater Phys* 80:1–6.
14. Anderson S.H., Chung D.D.L. (1983) Exfoliation of single crystal graphite and graphite fibers intercalated with halogens. *Synth Met* 8:343–349.
15. Wong L.N.W., Chung D.D.L. (1986) Electromechanical Behaviour of Graphite Intercalated with Bromine. Carbon N Y.
16. Wang P., Guo B., Ma H., Wu W., Zhang Z., Liu X., Cui M., Zhang R. (2020) Green and eco-friendly chemical exfoliation of two-dimensional nanosheets in the $C_2H_4O_3$-based system under the assistance of microwave irradiation. *Chem Eng J* 399:125758.
17. Wu W., Liu M., Gu Y., Guo B., Ma H.X., Wang P., Wang X., Zhang R. (2020) Fast chemical exfoliation of graphite to few-layer graphene with high quality and large size via a two-step microwave-assisted process. *Chem Eng J* 381:122592.
18. Prasad S., Kumar V., Kirubanandam S., Barhoum A. (2018) Engineered nanomaterials: Nanofabrication and surface functionalization. In: Ahmed Barhoum and Abdel Salam Hamdy Makhlouf (eds.) *Emerging Applications of Nanoparticles and Architectural Nanostructures: Current Prospects and Future Trends.* Elsevier, Cambridge, United States.
19. Tian K., Baskaran K., Tiwari A. (2018) Growth of two-dimensional WS_2 thin films by pulsed laser deposition technique. *Thin Solid Films* 668:69–73
20. Barvat A., Prakash N., Satpati B., Singha S.S., Kumar G., Singh D.K., Dogra A., Khanna S.P., Singha A., Pal P. (2017) Emerging photoluminescence from bilayer large-area 2D MoS_2 films grown by pulsed laser deposition on different substrates. *J Appl Phys* 122:15304.
21. Yao J., Zheng Z., Yang G. (2017) All-Layered 2D optoelectronics: A high-performance UV–vis–NIR broadband SnSe photodetector with Bi_2Te_3 topological insulator electrodes. *Adv Funct Mater* 27:1701823.
22. Yao J., Yang G. (2018) Flexible and high-performance all-2D photodetector for wearable devices. *Small* 14:1704524.
23. Yao J., Zheng Z., Yang G. (2016) Promoting the performance of layered-material photodetectors by alloy engineering. *ACS Appl Mater Interfaces* 8:12915–12924.

24. Giesbers A.J.M., Zeitler U., Neubeck S., Freitag F., Novoselov K.S., Maan J.C. (2008) Nanolithography and manipulation of graphene using an atomic force microscope. *Solid State Commun* 147:366–369.
25. Wang X. (2020) Preparation, synthesis and application of Sol-gel method University Tutor : Pr. Olivia GIANI Internship Tutor : Mme. WANG Zhen.
26. Innocenzi P., Malfatti L., Lasio B., Pinna A., Loche D., Casula M.F., Alzari V., Mariani .A (2014) Sol-gel chemistry for graphene-silica nanocomposite films. *New J Chem* 38:3777–3782.
27. Jilani A., Abdel-wahab M.S., Hammad A.H. (2017) Advance deposition techniques for thin film and coating. In: N.N. Nikitenkov (ed.) *Modern Technologies for Creating the Thin-film Systems and Coatings*. InTech, London.
28. Sedlovets D.M., Knyazev M.A., Zotov A.V., Naumov A.P., Korepanov V.I. (2021) Acetone and isopropanol – a new liquid precursor for the controllable transfer- and lithography-free CVD of graphene-like films. *J Mater Res Technol* 14:1339–1346.
29. Innocenzi P., Brusatin G. (2001) Fullerene-based organic–inorganic nanocomposites and their applications. *Chem Mater* 13:3126–3139.
30. Jung I., Pelton M., Piner R., Dikin D.A., Stankovich S., Watcharotone S., Hausner M., Ruoff R.S. (2007) Simple approach for high-contrast optical imaging and characterization of graphene-based sheets. *Nano Lett* 7:3569–3575.
31. Li H., Lu G., Yin Z., He Q., Li H., Zhang Q., Zhang H. (2012) Optical identification of single- and few-layer MoS_2 sheets. *Small* 8:682–686.
32. Xu K., Cao P., Heath J.R. (2010) Graphene visualizes the first water adlayers on mica at ambient conditions. *Science* 329:1188–1191.
33. Island J.O., Steele G.A., Van Der Zant H.S.J., Castellanos-Gomez A., Wood J.D., Wells S.A., Jariwala D., Chen K., Cho E., Sangwan V.K., Liu X., Lauhon L.J., Marks T.J., Hersam M.C. (2014) Effective passivation of exfoliated black phosphorus transistors against ambient degradation. *Nano Lett* 14:6964–6970.
34. Du Y., Zhuang J., Liu H., Xu X., Eilers S., Wu K., Cheng P., Zhao J., Pi X., See K.W., Peleckis G., Wang X., Dou S.X. (2014) Tuning the band gap in silicene by oxidation. *ACS Nano* 8:10019–10025.
35. McDonnell S., Addou R., Buie C., Wallace R.M., Hinkle C.L. (2014) Defect-dominated doping and contact resistance in MoS_2. *ACS Nano* 8:2880–2888.
36. Zhang X., Qiao X.-F., Shi W., Wu J.-B., Jiang D.-S., Tan P.-H. (2015) Phonon and Raman scattering of two-dimensional transition metal dichalcogenides from monolayer, multilayer to bulk material. *Chem Soc Rev* 44:2757–2785.
37. Chakraborty B., Bera A., Muthu D.V.S., Bhowmick S., Waghmare U.V., Sood A.K. (2012) Symmetry-dependent phonon renormalization in monolayer MoS_2 transistor. *Phys Rev B* 85:161403.
38. Tan C., Cao X., Wu X.-J., He Q., Yang J., Zhang X., Chen J., Zhao W., Han S., Nam G.-H., Sindoro M., Zhang H. (2017) Recent advances in ultrathin two-dimensional nanomaterials. *Chem Rev* 117:6225–6331.
39. Korkmaz S., Tezel F.M., Kariper A. (2020) Facile synthesis and characterization of graphene oxide/tungsten oxide thin film supercapacitor for electrochemical energy storage. *Phys E Low-Dimens Syst Nanostruct* 116:113718.
40. Iqbal M.F., Yousef A.K.M., Hassan A., Hussain S., Ashiq M.N., Mahmood-Ul-Hassan, Razaq A (2021) Significantly improved electrochemical characteristics of nickel sulfide nanoplates using graphene oxide thin film for supercapacitor applications. *J Energy Storage* 33:102091.
41. Cho S., Kim J., Jo Y., Ahmed A.T.A., Chavan H.S., Woo H., Inamdar A.I., Gunjakar J.L., Pawar S.M., Park Y., Kim H., Im H. (2017) Bendable RuO_2/graphene thin film for fully flexible supercapacitor electrodes with superior stability. *J Alloys Compd* 725:108–114.

42. Giannakopoulou T., Todorova N., Erotokritaki A., Plakantonaki N., Tsetsekou A., Trapalis C. (2020) Electrochemically deposited graphene oxide thin film supercapacitors: Comparing liquid and solid electrolytes. *Appl Surf Sci* 528:146801.
43. Jagadale A.D., Kumbhar V.S., Dhawale D.S., Lokhande C.D. (2013) Performance evaluation of symmetric supercapacitor based on cobalt hydroxide [$Co(OH)_2$] thin film electrodes. *Electrochim Acta* 98:32–38.
44. Gund G.S., Dubal D.P., Shinde S.S., Lokhande C.D. (2014) Architectured morphologies of chemically prepared NiO/MWCNTs nanohybrid thin films for high performance supercapacitors. *ACS Appl Mater Interfaces* 6:3176–3188.
45. Oje A.I., Ogwu A.A., Oje A.M. (2021) Effect of temperature on the electrochemical performance of silver oxide thin films supercapacitor. *J Electroanal Chem* 882:115015.
46. Patil U.M., Nam M.S., Lee S.C., Liu S., Kang S., Park B.H., Chan Jun S. (2017) Temperature influenced chemical growth of hydrous copper oxide/hydroxide thin film electrodes for high performance supercapacitors. *J Alloys Compd* 701:1009–1018.
47. Acerce M., Voiry D., Chhowalla M. (2015) Metallic 1T phase MoS_2 nanosheets as supercapacitor electrode materials. *Nat Nanotechnol* 10:313–318.
48. Cao L., Yang S., Gao W., Liu Z., Gong Y., Ma L., Shi G., Lei S., Zhang Y., Zhang S., Vajtai R., Ajayan P.M. (2013) Direct laser-patterned micro-supercapacitors from paintable MoS_2 films. *Small* 9:2905–2910.
49. Tu C.C., Lin L.Y., Xiao B.C., Chen Y.S. (2016) Highly efficient supercapacitor electrode with two-dimensional tungsten disulfide and reduced graphene oxide hybrid nanosheets. *J Power Sources* 320:78–85.
50. Li Z., Zhao D., Xu C., Ning J., Zhong Y., Zhang Z., Wang Y., Hu Y. (2018) Reduced $CoNi_2S_4$ nanosheets with enhanced conductivity for high-performance supercapacitors. *Electrochim Acta* 278:33–41.
51. Zhang M., Héraly F., Yi M., Yuan J. (2021) Multitasking tartaric-acid-enabled, highly conductive, and stable MXene/conducting polymer composite for ultrafast supercapacitor. *Cell Reports Phys Sci* 2:100449.
52. Li H., Liu Y., Lin S., Li H., Wu Z., Zhu L., Li C., Wang X., Zhu X., Sun Y. (2021) Laser crystallized sandwich-like MXene/Fe_3O_4/MXene thin film electrodes for flexible supercapacitors. *J Power Sources* 497:229882.
53. Wang Y., Wang X., Li X., Bai Y., Xiao H., Liu Y., Yuan G. (2021) Scalable fabrication of polyaniline nanodots decorated MXene film electrodes enabled by viscous functional inks for high-energy-density asymmetric supercapacitors. *Chem Eng J* 405:126664.
54. Zhao X., Wang Z., Dong J., Huang T., Zhang Q., Zhang L. (2020) Annealing modification of MXene films with mechanically strong structures and high electrochemical performance for supercapacitor applications. *J Power Sources* 470:228356.
55. Yao M., Chen Y., Wang Z., Shao C., Dong J., Zhang Q., Zhang L., Zhao X. (2020) Boosting gravimetric and volumetric energy density via engineering macroporous MXene films for supercapacitors. *Chem Eng J* 395:124057.
56. Perego D., Heng J.S.T., Wang X., Shao-Horn Y., Thompson C.V. (2018) High-performance polycrystalline RuOx cathodes for thin film Li-ion batteries. *Electrochim Acta* 283:228–233
57. Speulmanns J., Kia A.M., Kühnel K., Bönhardt S., Weinreich W. (2020) Surface-dependent performance of ultrathin TiN films as an electrically conducting Li diffusion barrier for li-ion-based devices. *ACS Appl Mater Interfaces* 12:39252–39260.
58. Curcio M., De Bonis A., Brutti S., Santagata A., Teghil R. (2021) Pulsed laser deposition of thin films of TiO_2 for Li-ion batteries. *Appl Surf Sci Adv* 4:100090.
59. Tan Thong P., Sadhasivam T., Kim N.I., Kim Y.A., Roh S.H., Jung H.Y. (2021) Highly conductive current collector for enhancing conductivity and power supply of flexible thin-film Zn–MnO_2 battery. *Energy* 221:119856.
60. Huang A., Chen J., Zhou W., Wang A,. Chen M., Tian Q., Xu J. (2020) Electrodeposition of MnO_2 nanoflakes onto carbon nanotube film towards high-performance flexible quasi-solid-state Zn-MnO_2 batteries. *J Electroanal Chem* 873:114392.

61. Liang R., Fu J., Deng Y.P., Pei Y., Zhang M., Yu A., Chen Z. (2021) Parasitic electrodeposition in Zn-MnO_2 batteries and its suppression for prolonged cyclability. *Energy Storage Mater* 36:478–484.
62. Yang K., Wu F., Chen S., Zhang C. (2007) Effect of surface modification of metal hydride electrode on performance of MH/Ni batteries. *Trans Nonferrous Met Soc China (English Ed* 17:200–204.
63. Tsai P.J., Chiu T.C., Tsai P.H., Lin K.L., Lin K.S., Chan S.L.I. (2012) Carbon nanotube buckypaper/$MmNi_5$ composite film as anode for Ni/MH batteries. *Int J Hydrogen Energy* 37:3491–3499.
64. Wu Q., Lv Y., Lin L., Zhang X., Liu Y., Zhou X. (2019) An improved thin-film electrode for vanadium redox flow batteries enabled by a dual layered structure. *J Power Sources* 410–411:152–161.
65. Teng X., Wang M., Li G., Dai J. (2020) Polypyrrole thin film composite membrane prepared via interfacial polymerization with high selectivity for vanadium redox flow battery. *React Funct Polym* 157:104777.
66. Ru X., Qu M., Wang J., Ruan T., Yang M., Peng F., Long W., Zheng K., Yan H., Xu X. (2020) 25.11% efficiency silicon heterojunction solar cell with low deposition rate intrinsic amorphous silicon buffer layers. *Sol Energy Mater Sol Cells* 215:110643.
67. Ruan T. Qu M., Qu X., Ru X., Wang J., He Y., Zheng K., Lin B.H.H., Xu X., Zhang Y., Yan H. (2020) Achieving high efficiency silicon heterojunction solar cells by applying high hydrogen content amorphous silicon as epitaxial-free buffer layers. *Thin Solid Films* 711:138305.
68. Cho J.S., Seo Y.H., Choi B.H., Cho A., Lee A., Shin M.J., Kim K., Ahn S.K., Park J.H., Yoo J., Shin D., Jeong I., Gwak J. (2019) Energy harvesting performance of bifacial and semitransparent amorphous silicon thin-film solar cells with front and rear transparent conducting oxide contacts. *Sol Energy Mater Sol Cells* 202:110078.
69. Wu Z., Duan W., Lambertz A., Qiu D., Pomaska M., Yao Z., Rau U., Zhang L., Liu Z., Ding K. (2021) Low-resistivity p-type a-Si:H/AZO hole contact in high-efficiency silicon heterojunction solar cells. *Appl Surf Sci* 542:148749.
70. Sedani S.H., Yasar O.F., Karaman M., Turan R. (2020) Effects of boron doping on solid phase crystallization of in situ doped amorphous Silicon thin films prepared by electron beam evaporation. *Thin Solid Films* 694:137639.
71. Paudel N.R., Yan Y. (2013) Fabrication and characterization of high-efficiency CdTe-based thin-film solar cells on commercial SnO_2:F-coated soda-lime glass substrates. *Thin Solid Films* 549:30–35.
72. Hu A., Zhou J., Zhong P., Qin X., Zhang M., Jiang Y., Wu X., Yang D. (2021) High-efficiency CdTe-based thin-film solar cells with unltrathin CdS:O window layer and processes with post annealing. *Sol Energy* 214:319–325.
73. Major J.D., Tena-Zaera R., Azaceta E., Bowen L., Durose K. (2017) Development of ZnO nanowire based CdTe thin film solar cells. *Sol Energy Mater Sol Cells* 160:107–115.
74. Pham B., Willinger D., Mcmillan N.K., Roye J., Burnett W., Achille A.D., Coffer J.L., Sherman B.D. (2021) Tin (IV) oxide nanoparticulate films for aqueous dye-sensitized solar cells. *Sol Energy* 224:984–991.
75. Fu G., Cho E.J., Luo X., Cha J., Kim J.H., Lee H.W., Kim S.H. (2021) Enhanced light harvesting in panchromatic double dye-sensitized solar cells incorporated with bilayered TiO_2 thin film-based photoelectrodes. *Sol Energy* 218:346–353.
76. Sun H., Dai P., Li X., Ning J., Wang S., Qi Y. (2021) Strategies and methods for fabricating high quality metal halide perovskite thin films for solar cells. *J Energy Chem* 60:300–333.
77. Tseng C.C., Chen L.C., Chang L.B., Wu G.M., Feng W.S., Jeng M.J., Chen D.W., Lee K.L. (2020) Cu_2O-HTM/SiO_2-ETM assisted for synthesis engineering improving efficiency and stability with heterojunction planar perovskite thin-film solar cells. *Sol Energy* 204:270–279.

78. Zanca C., Piazza V., Agnello S., Patella B., Ganci F., Aiello A., Piazza S., Sunseri C., Inguanta R.(2021) Controlled solution-based fabrication of perovskite thin films directly on conductive substrate. *Thin Solid Films* 733:138806.
79. Chen M., Ju M.G., Carl A.D., Zong Y., Grimm R.L., Gu J., Zeng X.C., Zhou Y., Padture N.P. (2018) Cesium Titanium(IV) bromide thin films based stable lead-free perovskite solar cells. *Joule* 2:558–570.
80. Ma J., Sahai Y. (2013) Effect of electrode fabrication method and substrate material on performance of alkaline fuel cells. *Electrochem Commun* 30:63–66.
81. Liu P., Liu X., Dong F., Lin Q., Tong Y., Li Y., Zhang P. (2018) Electricity generation from banana peels in an alkaline fuel cell with a Cu_2O-Cu modified activated carbon cathode. *Sci Total Environ* 631–632:849–856.
82. Takahashi H., Sagihara M., Taguchi M. (2014) Electrochemically reduced Pt oxide thin film as a highly active electrocatalyst for direct ethanol alkaline fuel cell. *Int J Hydrogen Energy* 39:18424–18432.
83. Kang S., Lee J., Cho G.Y., Kim Y., Lee S., Cha S.W., Bae J. (2020) Scalable fabrication process of thin-film solid oxide fuel cells with an anode functional layer design and a sputtered electrolyte. *Int J Hydrogen Energy* 45:33980–33992.
84. Seo H.G., Ji S., Seo J., Kim S., Koo B., Choi Y., Kim H., Kim J.H., Kim T.S., Jung W.C. (2020) Sintering-resistant platinum electrode achieved through atomic layer deposition for thin-film solid oxide fuel cells. *J Alloys Compd* 835:155347.
85. Thieu C.A., Ho-Il J., Kim H., Yoon K.J., Lee J.H., Son J.W. (2019) Palladium incorporation at the anode of thin-film solid oxide fuel cells and its effect on direct utilization of butane fuel at 600°C. *Appl Energy* 243:155–164.
86. Lee M.H., Thomas J.L., Chen W.J., Li M.H., Shih C.P., Lin H.Y. (2015) Fabrication of bacteria-imprinted polymer coated electrodes for microbial fuel cells. *ACS Sustain Chem Eng* 3:1190–1196.
87. Park H.I., Mushtaq U., Perello D., Lee I., Cho S.K., Star A., Yun M. (2007) Effective and low-cost platinum electrodes for microbial fuel cells deposited by electron beam evaporation. *Energy and Fuels* 21:2984–2990.
88. Narayanasamy S., Jayaprakash J. (2021) Carbon cloth/nickel cobaltite (NiCo2O4)/polyaniline (PANI) composite electrodes: Preparation, characterization, and application in microbial fuel cells. *Fuel* 301:121016.
89. Ying X., Shen D., Wang M., Feng H., Gu Y., Chen W. (2018) Titanium dioxide thin film-modified stainless steel mesh for enhanced current-generation in microbial fuel cells. *Chem Eng J* 333:260–267.
90. Santiago O., Mosa J., Escribano P.G., Navarro E., Chinarro E., Aparicio M., Leo T.J., del Río C. (2020) $40SiO_2$–$40P_2O_5$–$20ZrO_2$ sol-gel infiltrated sSEBS membranes with improved methanol crossover and cell performance for direct methanol fuel cell applications. *Int J Hydrogen Energy* 45:20620–20631.
91. Priyangga A., Pambudi A.B., Atmaja L., Jaafar J. (2021) Synthesis of nanocellulose composite membrane and its properties for direct methanol fuel cell. *Mater Today Proc* 46:1998–2003.
92. Suhaimin N.S., Jaafar J., Aziz M., Ismail A.F., Othman M.H.D., Rahman M.A., Aziz F., Yusof N. (2021) Nanocomposite membrane by incorporating graphene oxide in sulfonated polyether ether ketone for direct methanol fuel cell. *Mater Today Proc* 46:2084–2091.
93. Shu Q.Z., Xia Z.X., Wei W., Xu X.L., Wang S.L., Zhao H., Sun G.Q. (2021) A novel gas diffusion layer and its application to direct methanol fuel cells. *Xinxing Tan Cailiao/New Carbon Mater* 36:409–419.

8 Wafer-Scale Growth and High-Throughput Characterization of Ultrathin 2D Transition Metal Dichalcogenides (TMDCs) for Energy Applications

Shraddha Ganorkar
Sungkyunkwan University

Mangesh Diware
Seoul National University

CONTENTS

DOI: 10.1201/9781003178453-8

8.1 INTRODUCTION

An imminent threat for exhausting natural fossil fuels, climate change, and global warming is lingering over our heads. A new era of 'green energy' is already started where various innovative techniques have been tried and tested for sustainable energy generation and developed strategies for its wise use. The challenge with energy utilization is to use it efficiently and consistently through smart and low-energy devices with high performance. For example, massive data centers run by big tech giants use nearly 1.15% of total global electricity consumption and are predicted to increase up to 1.89% by 2030, which is ~ 658 TWh [1]. The design of energy-efficient devices is a long-waited quest, and billions of dollars are being invested every year searching for novel materials that can fulfill this task. The discovery of graphene, a two-dimensional (2D) array of carbon atoms in a hexagonal pattern, surged a massive interest in exfoliable layered materials, searching for a new degree of freedom whose control can give the new exciting properties. 2D electronic materials have gained momentous attention with the rise of graphene, realizing the lack of bandgap making it unsuitable for smart electronics devices. Graphene has led the path for 2D materials, and transition metal dichalcogenide (TMDC) has been born. TMDC is a family of 2D layered materials with the general formula MX_2, where M is transition metal and X is chalcogenide. There are more than 40 TMDC possible combinations with 2D structure and increasing.

A typical 2D layered TMDC structure comprises a hexagonal plane of metal (M) sandwiched between two hexagonal planes of chalcogenide (X) atoms with a valance of +4 and –2, respectively [2]. Like graphene, each 2D layer with a thickness of 6–7 Å has a strong M-X-M in-plane covalent bond and weak out-of-plane Van der Waals interactions. The 2D TMDC stabilizes in two structural polytypes – trigonal prismatic (2H) (D_{h3} point group) and octahedral (1T) (D_{3d} point group) in the honeycomb and centered-honey comb motif, respectively, based on the position of M relative to X in the M-X-M structure [2,3]. The irreversible transition of 2H to 1T polytype in a few TMDC is tunable via chemical processing [4]. Owing to the confined movement of electrons in 2D layers, TMDC has remarkable properties such as direct bandgap in the visible-near infrared (IR) range, which can be engineered with the number of layers, high carrier mobility, and high on/off ratio, spin–orbit coupling (SOC), and ultra-strong Coulomb interaction linked to the valley degree of freedom [5]. The semiconducting monolayer of TMDC is successfully molded in green energy applications [6] such as energy generators, energy storage, energy generator catalysis, electronic devices, and solar energy harvesting. Strong interactions with light give rise to a series of emerging electronic and optoelectronic device applications [7], not limited to strong piezoelectric coupling, field-effect transistors, heterostructure junctions, photodetectors, photovoltaics, and sensors with the advantage of ultrafast responses, compact, light-weighted and energy-efficient. 2D TMDC has revolutionized the electronic industry. However, most of the applications mentioned above are in the lab-based research phase. These devices need to reach the batch-processing capability for practical applications.

8.2 WAFER-SCALE GROWTH OF ULTRATHIN TMDC FILMS

Large area with a uniform thickness of ultrathin TMDC films are prerequisites for batch processing to meet the commercial application standards. The pivotal top-down synthesis methods such as mechanical exfoliation for layered TMDC materials give very high-quality pristine 2D sheets, useful for lab-scale research purposes but suffer spatial and structural uniformity, including intrinsic defects. The exfoliated size is limited to the quality of 3D precursor and impossible to achieve wafer-scale 2D single crystal. The bottom-up approach facilitates large-area uniformity by layer engineering via parametric control of vapor-phase deposition. However, the conventional epitaxial thin films suffer from lattice mismatch due to the strong bonding nature of two surfaces (layer–substrate (LS)) resulting in intrinsic strain in the thin films, deteriorating the quality of the thin films. On the contrary, due to the high anisotropic atomic bonding and advantage of dimensionality, 2D materials are privileged with weak substrate interactions; hence, lattice mismatch and growth mechanism have an insignificant correlation. Theoretical studies by density function theory show the typical strength of interlayer interaction is 0.03 eV atom^{-1} (layer–layer, LL) and LS is 0.2eV atom^{-1} for most of the 2D materials [8,9], much lower than typical chemical bonds strength (2–8 eV atom^{-1}). According to the theory of thin-film growth [10], the LL and LS interactions lead to three different mechanisms. The Frank–van der Merwe (FM) growth facilitates layer-by-layer growth through surface nucleation due to stronger LS and weaker LL interaction. On the other hand, in Volmer–Weber (VW) growth, the clusters form without nucleation due to stronger LL interaction comparatively weaker LS interactions leading to island formation. Stranski–Krastanov growth is transitional through surface nucleation similar to FM followed by VW type over the critical thickness of the film [11]. Following the growth models, 2D materials with very weak LL interaction and weak LS interaction must follow the FM growth mechanism enabling selective growth of mono or few layers of 2D materials [12]. After nucleation in the FM mechanism, there are three possible scenarios for large-scale growth of 2D material: first, forming a single large nucleus on a substrate, which then grows to wafer-scale. Second, the formation of multi-nucleus orientated in the same direction grows bigger flawlessly to form a large-scale single crystal. Third, differently oriented nano-crystallites are grown into a large-scale polycrystalline 2D film with unrestrained grains and grain boundaries. Enormous efforts have been dedicated to the synthesis of large-area ultrathin TMDC films, including chemical vapor deposition (CVD) [13], metal-organic chemical vapor deposition (MOCVD) [12], and micromechanical cleavage [5]. In this section, the most effective wafer-scale synthesis methods for ultrathin TMDC are described.

8.2.1 Metal-Organic Chemical Vapor Deposition

The MOCVD technique uses the metal-organic sources (i.e., $Mo(CO)_6$, $W(CO)_6$, $(CH_3)_2Se$, $(C_2H_5)_2S$), and H_2S) that are generally preferred due to cost-effectiveness and precise control over the concentration of gases. MOCVD facilitates several degrees of freedom such as temperature, pressure, carrier gas, and chalcogen/transition (X/M) metal ratio to precise control of grain size, nucleation density, structure, etc. In showerhead-type MOCVD, the precursor gases are inserted into the reactor through the showerhead across the entire substrate area (wafer). The gas inlet is equipped with an array of multi-tube assemblies to spread the individual gases. The gases are introduced in the reactor separately for even distribution during the reaction. The substrate (wafer) is mounted on a rotating susceptor, which is heated by the resistance heater. The advantage of MOCVD over CVD is the temperature for the process that can be as low as 250°C.

Figure 8.1a shows schematics of a typical MOCVD with a mass flow controller connected to various gases. The top inset of Figure 8.1a shows an SEM image of MoS_2 island that is eventually grown to 8-inch MoS_2 ultrathin polycrystalline film using MOCVD, with 30 sccm H_2S gas, chamber pressure of 10 Torr, growth temperature of 400°C, and S/Mo, Ar/H_2S, and H_2/H_2S molar ratios of 200, 5, and 5, respectively, with an optimum growth time of 9 hours [14]. The growth mechanism for the polycrystalline large-scale ultrathin film follows the FM growth. The parametric studies of MOCVD grown MoS_2 show a critical balance of control parameters for wafer-scale growth [14]. Ar flow rate has a significant influence on MOCVD. Figure 8.1b indicates the effect of growth pressure on nucleation density. The nucleation density increases with an increasing flow rate, which could be due to the effective temperature of the substrate, as the higher flow rate will lose the temperature equilibrium in the reactor. The ambient gas, H_2, acts as a catalyst for the nucleation site and prevents inter-grain coupling. This results in a higher lateral growth rate by blocking bilayer formation. Comparing the effect of the flow rate of two gases, higher H_2 facilitates the growth of quasi monolayer formation of MoS_2. In contrast, a higher Ar flow rate deteriorates the quality of MoS_2 crystal and leads to the formation of multilayer, confirmed by Raman spectra [14]. The growth pressure is the most important parameter to produce wafer-scale MoS_2 via optimizing the higher lateral growth rate. The parametric variation study under the growth pressure 0.3–9 Torr results in decreasing nucleation site and increasing grain size (with constant parameters: T = 400°C, X/M = 300) due to competing driving forces of crystallization with increasing the pressure. The grain size is inversely proportional to nucleation density for 2D TMDC and increases with the X/M ratio due to the higher diffusivity of M compared to X. The MOCVD is one of the promising techniques for industrial production of wafer-scale 2D TMDC, enabling the easy and precise control of growth parameters. The major drawback of MOCVD is that it is a time-consuming process. The temperature equilibrium of the chamber is very sensitive to growth and needs to optimize in the future.

8.2.2 Vertical-Ostwald Ripening Method

The MOCVD method is a promising technique to grow wafer-scale TMDC ultrathin films, as mentioned above. The major drawback is its slow growth rate due to the use of needed low precursor pressure to avoid the secondary nucleation over

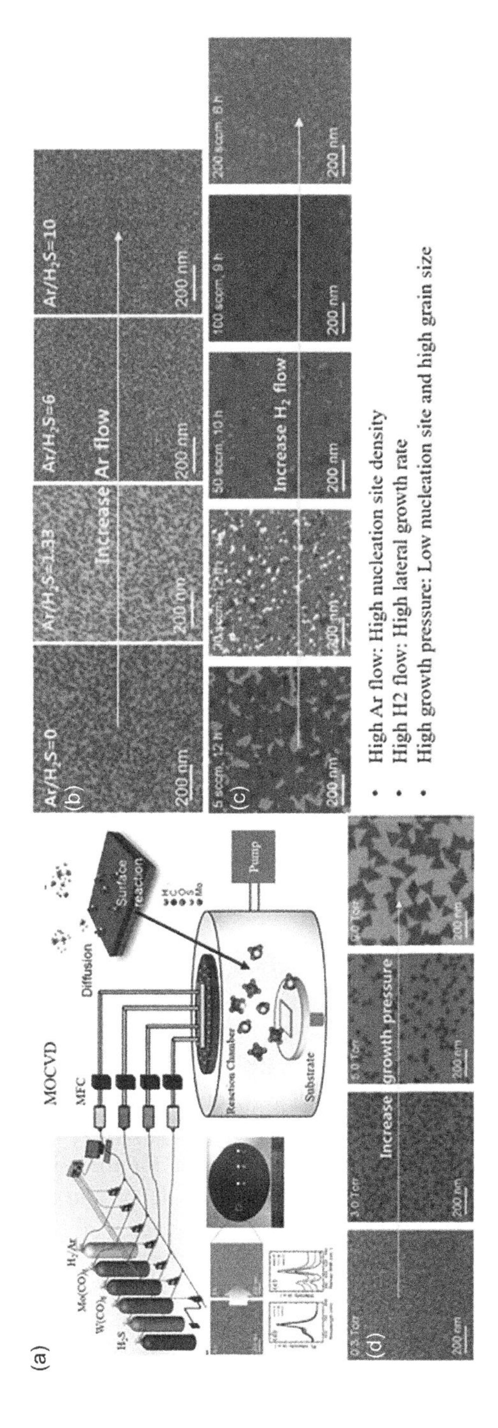

FIGURE 8.1 (a) Schematics of MOCVD and (b-d) SEM images for parametric study MoS_2 ultrathin film growth. Inset of Figure (a) and Figure (b-d) are adapted with permission from Ref. [14]. Copyright 2017 Institute of Physics Publishing.

pre-deposited TMDC. Large-area growth of monolayer will require several hours, not a cost-effective way. The higher partial pressure of precursors is essential to increase the growth rate, but it favors the island-type growth. Secondary nucleation over pre-grown monolayer should be regulated to achieve fast wafer-scale monolayer growth. The modified-MOCVD method introduced by Seol et al. is called pulsed MOCVD for the rapid growth of wafer-scale MoS_2 and WS_2 monolayer [15]. Schematic of TMDC monolayer growth via the vertical-Ostwald ripening process is illustrated in Figure 8.2a. Nano-TMDC (e.g., MoS_2) crystallites grow over the entire wafer surface during the initial stage of the growth process. Secondary nucleation could start on these pre-deposited MoS_2 nanocrystals unless suddenly precursor supply cuts off, stopping the MoS_x cluster from reaching critical nuclei size. As shown in Figure 8.2b, adatoms energetically favor the edge than the basal plane of the pre-deposited MoS_2 nanocrystals. Both adatoms and unstable MoS_x clusters diffuse from surface to edge of the pre-deposited MoS_2 nanocrystals, called vertical-Ostwald ripening. Seol et al. observed that Mo adatoms can survive for 5.9 μs and travel up to 65 nm at 900°C before desorption [15]. So, cutting precursor supply provides enough time for adatoms to diffuse from surface to edge of the pre-deposited MoS_2 nanocrystals. Accordingly, pulsating precursor injection, shown in Figure 8.2d, subdues the vertical growth and helps to achieve wafer-scale MoS_2 monolayer.

Seol et al. used the showerhead-type cold wall reactor to facilitate the uniform and controlled precursor supply over the entire substrate surface, as shown in Figure 8.2c-i. Figure 8.2c-ii shows the reaction cycles; gas phase ($Mo(CO)_6$) or ($W(CO)_6$), ($(C_2H_5)_2S_2$), and H_2 precursors were injected for 2 minutes using N_2 carrier gas followed by purging the chamber with N_2 for 1 minute after stopping precursor supply. Thus, four of such reaction cycles were enough to grow MoS_2 and WS_2 monolayer over a 6-inch quartz wafer. The total growth time was only 12 minutes. Raman line-scan across the wafer and photoluminescence peak intensity map confirms the uniformity of the grown MoS_2 monolayer, shown in Figure 8.2e and f, respectively [15].

8.2.3 Self-limiting Growth Method Using Atomic Layer Deposition

In the previous sections, we have discussed vapor depositions of ultrathin 2D TMDC films that indicate the sensitivity of deposition parameters on the growth of the large-area ultrathin film. The change in the control parameters or the thermodynamics equilibrium in the reactor can lead to multilayer deposition. The self-limiting growth of ultrathin films is an effective and efficient way to grow wafer-scale 2D TMDC materials. Atomic layer deposition (ALD) is known for the self-limiting growth technique for 2D TMDC films using a layer-by-layer deposition approach via alternate precursor exposure. ALD consists of four major steps – (a) precursor exposure, (b) purge/evacuating precursor/byproducts, (c) precursor/reactant exposure, and (d) purge/evacuate reactants/byproducts. Contrary to CVD, the growth in ALD does not depend on the flow rate; rather, it depends on the kinetics of surface reaction leading to self-limiting growth. The self-limiting growth technique is important for industrial production since critical control of precursor gases is no longer required to avoid multilayer formation. Eventually, wafer-scale growth time can be dramatically

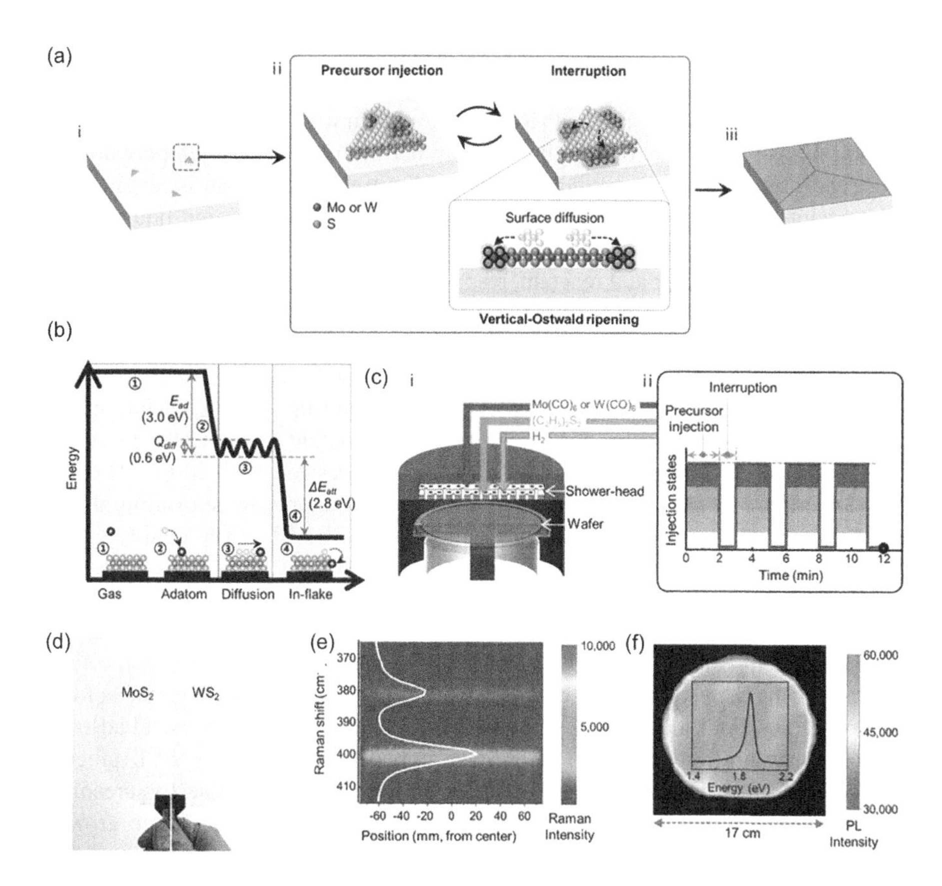

FIGURE 8.2 Pulsed MOCVD facilitates the vertical-Ostwald ripening process to grow wafer-scale TMDC monolayer. Growth of wafer-scale monolayer TMDCs. a) Schematic of the growth mechanism by pulsed MOCVD. After nucleation in the early stage (i), lateral growth is induced by repeating the precursor injection and interruption steps (ii), which results in continuous monolayer TMDCs (iii). b) Calculated reaction energy diagram of Mo adsorbed on the basal plane of MoS_2. c) Schematic illustration of the reactor geometry (i) and injection sequence of precursors (ii) for growing wafer-scale monolayer TMDCs. A shower-head-type, cold-wall reactor system was used with MO precursors. d) Photograph of wafer-scale monolayer MoS_2 and WS2 grown on 6-inch quartz glass wafers. e) Color-coded rendering of the spatially resolved Raman spectra for MoS_2 films. The data were collected from 300 positions (with an interval of 500 μm along the horizontal direction of the entire 6-inch wafer). Inset: Raman spectrum of the MoS_2 film. f) PL peak intensity mapping of the MoS_2 film on the wafer scale. The spatial resolution is 1 cm. Inset: PL spectrum of the MoS_2 film. Adapted with permission from Ref. [15]. Copyright 2020 Wiley-VCH GmbH.

shortened. The first step in the self-limiting process is the adsorption of precursor molecules on the surface, achieved by the ligands bonded to the metal atoms in the precursors (halogen or organic ligands), resulting in the growth of the first monolayer. Consequently, after forming a fully covered monolayer, no further catalytic surface is available for the additional decomposition of ligands. This process limits the adsorption of the precursor by passivating the adsorption sites after the saturation coverage is reached, i.e., *self-limiting*. Excess exposure of precursor gases will not take part in the formation of layers. Large-scale 2D MoS_2 and WS_2 are successfully grown by this self-limiting method [16,17]. The self-limiting technique can be tuned to control layer thickness by varying the number of ALD cycles, temperature, or flow concentration [16]. Nevertheless, the self-limiting method can be engineered to be completely temperature-dependent (cycle independent) by using liquid-phase organic precursor in place of the gaseous H_2S reactant. Park et al. [18] used WCl_2 and $C_4H_{10}Se$ as precursor and reactant, respectively, with temperature as an efficient parameter for controlling the number of layers of WSe_2. The pre-heated precursor was carried into the tubular furnace by a pure Ar carrier gas. H_2 gas was used with $C_4H_{10}Se$ for reducing WCl_6.

Figure 8.3a shows a schematic of the ALD self-limiting process, including the four steps: precursor exposure for 4 s, Ar purge 5 s, reactant exposure for 3 s, and a final 5 s Ar purge. The ultrathin WSe_2 films are synthesized with 100 ALD cycle at a temperature ranging from 600°C to 800°C with the precise controlling thickness from mono- to penta-layer as shown in Figure 8.3b and c. The quality of the films is confirmed by photoluminescence, X-ray photoelectron spectroscopy, and Raman spectroscopy, shown in Figure 8.3d–j. The uniformity of tri-layer WSe_2 on the 6-inch wafer strip was confirmed by measuring Raman spectra line-scan along the wafer length (shown in Figure 8.3h–j). The temperature sensitivity of the self-limiting mechanism can be understood from the surface potential difference of the used precursor. First, the surface of TMDC is inherently inert for physical adsorption over the substrate. Second, the surface potential depth of the WCl_6 precursor decreases with the increasing thickness of WSe_2 2D layers due to screening from the electric field of the pre-deposited layer [18]. Hence at the given growth pressure, the number of layers can be terminated by lowering the potential depth of WCl_6 via manipulating the growth temperature. This method has the potential for industrial-scale production of wafer-scale 2D TMDC films using liquid-phase organic reactants.

8.2.4 Layer-Resolved 2D Material Splitting Technique

Exfoliation of 2D layers is one of the pioneer methods, which gives pristine 2D layer materials. The vapor deposition methods can lead to contamination due to precursors, reactants, and the carrier gas used during the process, thereby affecting the properties of 2D TMDC. Nevertheless, the exfoliation with adhesive tape is time-consuming, and wafer-scale exfoliation is impossible. The size of exfoliated gains highly depends on the quality of the 3D crystal used. Recently, Shim et al. [19] introduced the layer-resolved spitting (LRS) technique by integrating with preliminary wafer-scale deposition methods such as MOCVD and vapor phase epitaxy. The steps involved in the LRS technique are illustrated in Figure 8.4a. A thick 2D TMDC film is deposited on a wafer using vapor phase deposition, followed by a capping layer of 600 nm thick Ni. A thermal releasing tape is used as a handler to peel off 2D TMDC/Ni layer completely from the 5 cm diameter wafer to giving a TMDC/Ni/Tape stack. Another Ni layer is deposited on the bottom of the TMDC/Ni/Tape stack to form strong adhesion to the 2D TMDC layers; these give the Ni/

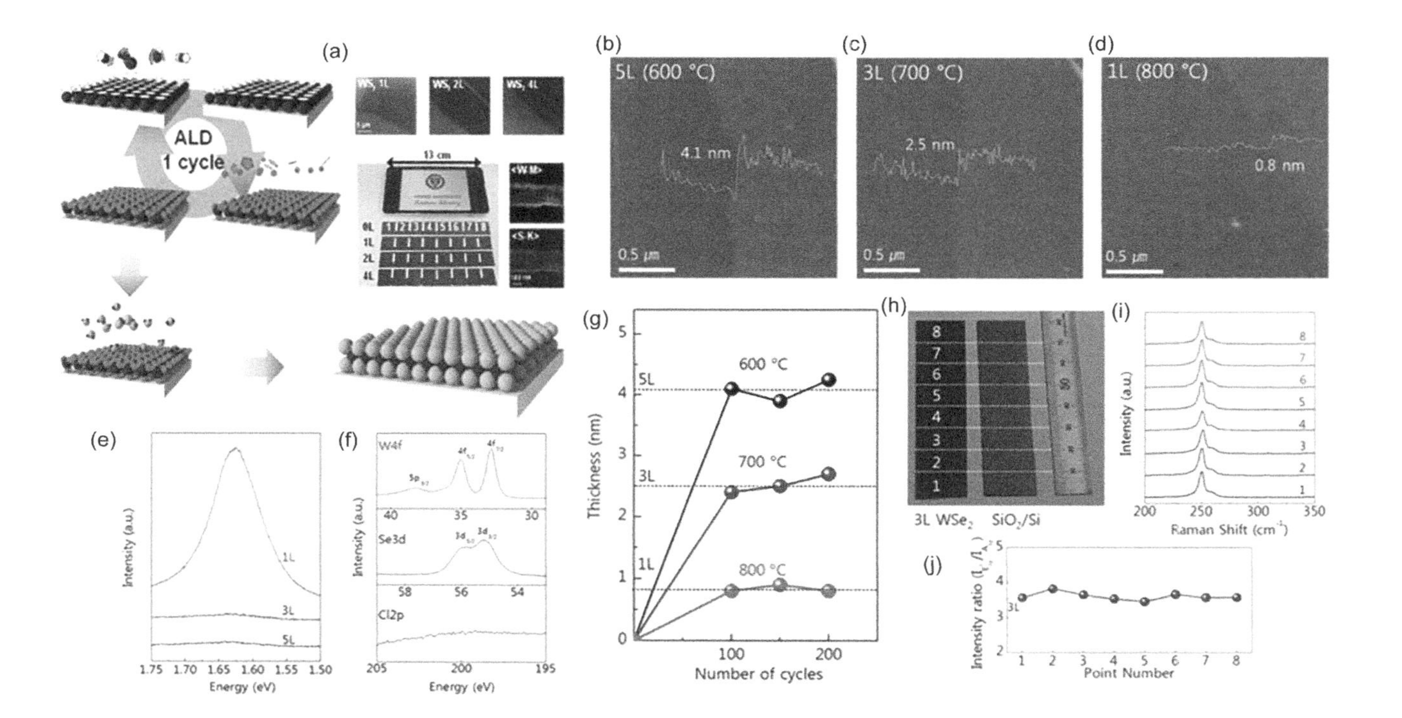

FIGURE 8.3 (a) Schematic of self-limiting process, (b-d) AFM images of 1L, 3L, and 5L of WSe_2 films showing their thickness with growth temperatures of 600C,700C, and 800C, respectively. (e) Photoluminescence as a function of a number of layers and (f) XPS spectra show the quality of the grown layers, (g) number of ALD cycles with temperature, and (h-j) uniformity investigation of tri-layer WSe_2 deposited SiO_2/Si substrate by Raman spectroscopy. The intensity ratio of characteristic Raman peaks is used to quantify thickness uniformity. ((a) Adapted with permission from Ref. [17] Copyright 2013 American Chemical Society and (b-j) adapted with permission from Ref. [16] Copyright 2016 Springer Nature.)

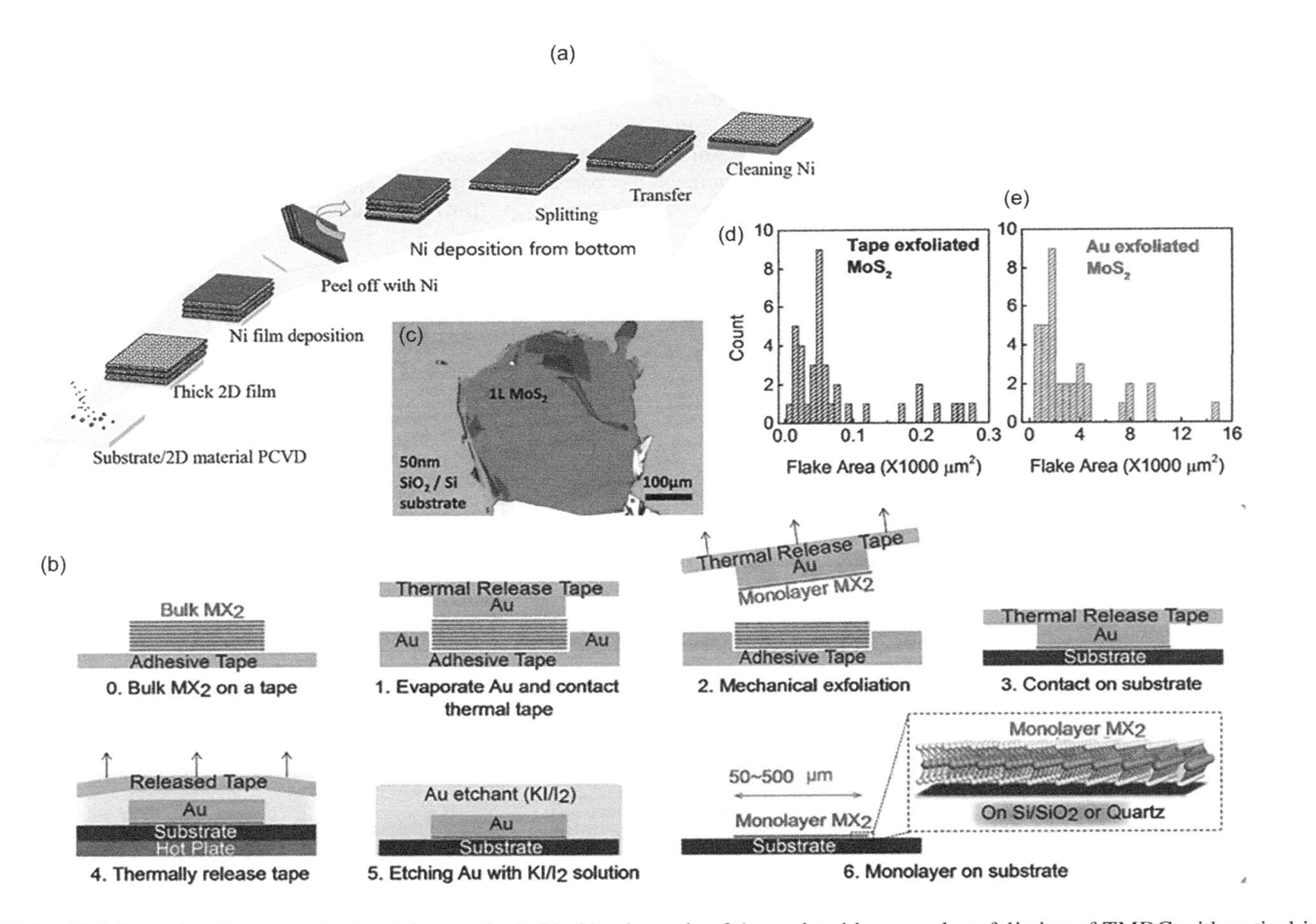

FIGURE 8.4 (a) Schematic of layer-resolved spitting method, (b)–(e) schematic of Au-assisted large-scale exfoliation of TMDC with optical image and fleck area count. Figures (b)–(e) adapted with permission from Ref [20]. Copyright 2016, John Wily & Sons.

TMDC/Ni/Tape stack. A second thermal tape handler was used on the bottom Ni. The stack is then split, followed by transfer on Si/SiO_2 substrate (Si/SiO_2/TMDC/Ni/Tape). The transferred stacking was then heated to 110°C to release the tape. The Ni film was etched in the solution of $FeCl_3$ followed by rinsing with DI water. The above steps can be used for multiple splitting. The key parameter of the LRS process is the interfacial toughness (Γ) between Ni-TMDC-substrate and TMDC interlayer. The Γ_{TMDC} - Γ_{TMDC} < Γ_{TMDC} - $\Gamma_{sapphire}$ < Γ_{TMDC} - $\Gamma_{Ni,}$ (0.26, 0.45, and 1.4 Jm^{-2}), respectively. Shim et al. claim that LSR is the universal method applicable to all TMDC. The underlying mechanism is explained on the basis of external force triggered bending moment torque applied across a small distance in the sample that initiates spalling mode fracture and cracks formation. The LSR method is a modified version of metal-assisted TMDC exfoliation described by Desai et al. [20,21] as depicted in Figure 8.4b. The large-scale Au-assisted exfoliated MoS_2 flake is shown in Figure 8.4c. The metal-assisted exfoliation has dramatically enhanced the flack area compared in Figure 8.4d and e.

8.3 HIGH-THROUGHPUT CHARACTERIZATION OF WAFER-SCALE ULTRATHIN MoS_2 USING SPECTROSCOPIC ELLIPSOMETRY

Mechanical exfoliation of TMDC single crystals can give defect-free TMDC mono- and multilayers very easily but limited to a small area, suitable for research purposes. On the other hand, commercial applications demand wafer-scale production of ultrathin TMDCs films with precise thickness control for batch processing. Also, there should be easy access to crucial information about these large-area film quality and thickness uniformity. Various attempts were made to grow wafer-scale TMDCs films [12,22–25]; poor thickness control and limited deposition scale are a few limitations of these methods. Recently, Kim et al. [14], Park et al. [18], Shim et al. [19], and Seol et al. [15] developed efficient recipes using MOCVD and ALD methods to grow wafer-scale TMDC monolayer films with precise thickness control by single deposition parameter; details are in the above section. Usually, the thickness and quality of TMDC ultrathin films are characterized by atomic force microscopy (AFM), tunneling electron microscope, and X-ray photoelectrons spectroscopy, which are inadequate for large-area samples. However, Raman spectroscopy and photoluminescence are being used for large area samples, but an intense laser beam could damage the underlying material. Therefore, a fast and non-destructive method is needed to quantify thickness uniformity and quality of wafer-scale TMDC ultrathin films.

Ellipsometry can provide that option; it has come a long way since the first appearance of the commercial instrument in the first half of the 19th century due to the advancement of automation and computation power and expanded in a wide application area. Specifically, it became a valuable measurement technique and real-time feedback control in the semiconducting industry. This section will discuss spectroscopic ellipsometry (SE) basics, data analysis, and application for wafer-scale characterization of ultrathin TMDC materials.

8.3.1 Basics of Spectroscopic Ellipsometry

Details about the ellipsometry technique are given in the standard books [26,27]. However, essential terms needed for discussion are explained in brief. Figure 8.5a illustrates the working principle of ellipsometry. SE measures the change in a polarization state of incident light after interaction with the sample surface. Spectroscopic means several wavelengths of light are used. The polarization state of light is defined in terms of coordinates of the waves propagating along two orthogonal axes: parallel (*p*-) and perpendicular (*s*-) to the plane of incidence. For example, linear polarized light is defined in Figure 8.5a where incident light is in the first quadrant, oriented at −45° with-respect-to $E_{ip} = E_{is}$; is incident electric field vector for *p*- and *s*- polarization. It means the amplitude of electric vectors for both the polarization is the same, and the phase difference between them is zero. In the same way, other polarization states can be defined from amplitude and phase differences between *p*- and *s*-polarization vectors. The most common polarization state is elliptical hence the name. The *p*- and *s*-polarized components of the incident beam behave differently after interacting with the sample, and amplitude reflection coefficients are different. Therefore, change in polarization state is measured from modified reflected light in terms of two ellipsometric angles Ψ: amplitude ratio and Δ: phase difference between reflected *p*- and *s*-polarizations. The complex reflectance ratio is given below:

$$\rho = \frac{(E_{rp} / E_{ip})}{(E_{rs} / E_{ip})} = \frac{r_p}{r_s} = \tan\Psi \exp(-i\Delta) \tag{8.1}$$

$$\Psi = \frac{|r_p|}{|r_s|}, \quad \Delta = \delta_{rp} - \delta_{rs} \tag{8.2}$$

where *E* is the electric field vector, '*r*' is for reflection, and '' for incidence.

SE instruments in various configurations are available depending on optical components and the modulation type used [26,27]. Rotating-analyze type SE (RASE) with simple optical configuration and automated data acquisition was first devised by Aspens et al. in 1975 [28]. To overcome the limitations of RASE, which is it cannot measure the complete range of Ψ and Δ, more complicated versions are invented. The most popular and accurate configuration is the rotating compensator SE (RCSE). The compensator is a quarter-wave plate that generates a phase difference between *p*- and *s*-components of the electric vector of the light. The RCSE measures the full range of Ψ and Δ ($0° \leq \Psi \leq 90°$ and $0° \leq \Delta \leq 360°$). Complete Stokes parameter can be measured, allowing uniform measurement sensitivity and depolarization from a sample.

Figure 8.5a shows the schematics of rotating compensator type SE (RCSE), consisting of light source – polarizer (*P*) – rotating compensator (*C*) – sample – analyzer (*A*) – detector. A combination of *P* and *C* convert the unpolarized incident beam into circularly polarized light. Therefore, the light of known polarization is incident on the sample surface, generating elliptically polarized reflected beam, passed through the

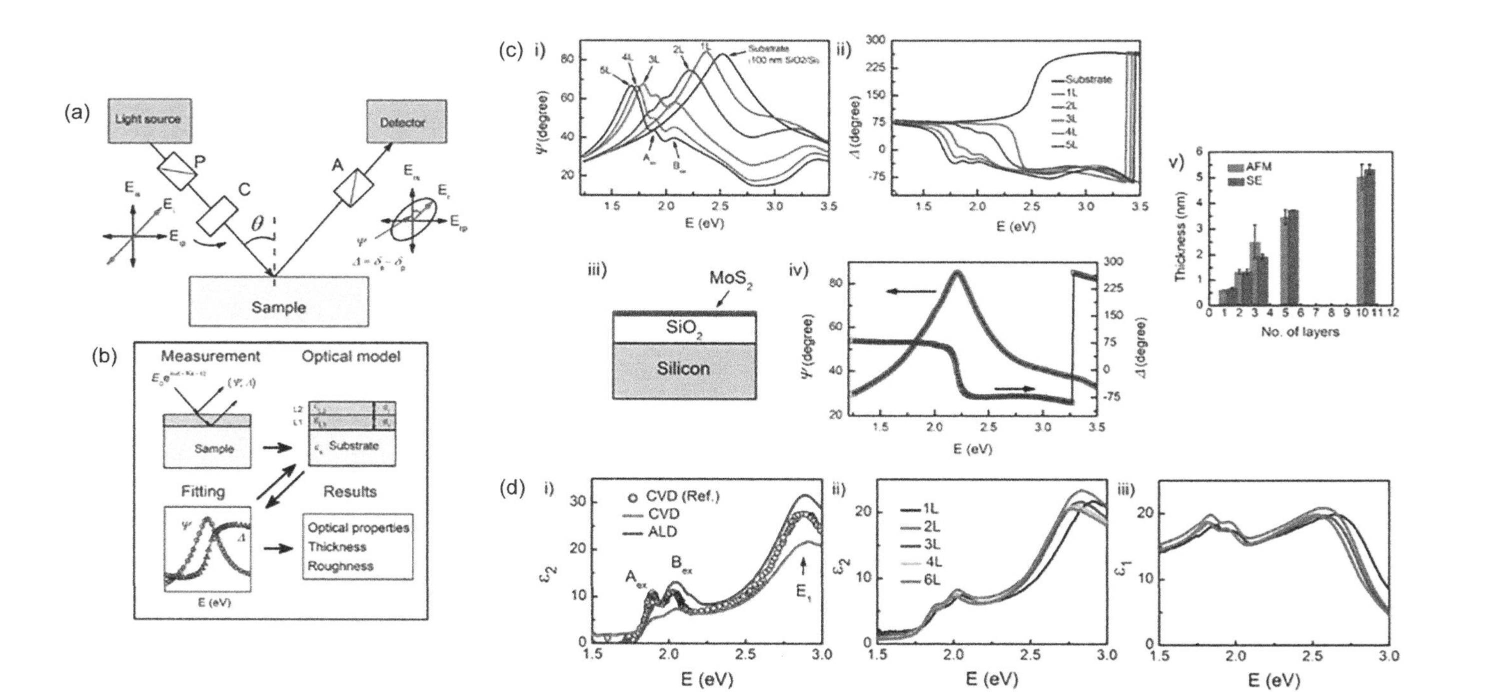

FIGURE 8.5 (a) Schematic of rotating compensator spectroscopic ellipsometer (SE) explaining the basic principles of ellipsometry technique. (b) Analysis procedure of SE data. (c) SE data analysis of MoS_2 ultrathin films: (c-i) Ψ and (c-ii) Δ spectra of mono-, bi-, tri-, tetra-, and hexa-layer MoS_2 films on SiO_2 (100 nm)/Si substrate. (c-iii) Four-layer (ambient/MoS_2/SiO_2/Si) optical model used for data analysis. (c-iv) Calculated (lines) and measured (symbols) data for MoS_2 monolayer show the good fit. (c-v) Thicknesses of MoS_2 thin films obtained from SE analysis, compared with values obtained using AFM. (d) Extracted dielectric functions of MoS_2 ultrathin films. (d-i) Imaginary part of the dielectric function of monolayer MoS_2 grown by MOCVD and ALD, compared with CVD grown Ref. data [29]. The (d-ii) ε_2 and (d-iii) ε_1 of MOCVD grown mono-, bi-, tri-, tetra-, and hexa-layer MoS_2 films.

analyzer and finally to the detector to measure the polarization change. It is advised that measurements should be done at multiple angles around the Brewster angle for ex situ measurements. For an in situ application, suppose an ellipsometer is attached to the deposition chamber, there is a restriction of fix angle. In that case, it should be aligned at the Brewster angle where p- and s-polarization have the highest sensitivity. Proper calibration should be performed to initialize the optical components before measurements. Also, chromatic errors from optical elements should be corrected for spectroscopic measurements.

8.3.2 Data Analysis

The bitter truth about ellipsometry is that if you don't have a complete description of a sample to construct an optical model as close as an actual sample, you cannot figure out what you have measured despite being the most surface-sensitive technique known today. Measured SE data contain the information of the whole sample structure accessed by the entire penetration depth of the probing light. Therefore, to make sense of measured data, a proper analysis procedure should be followed, as shown in Figure 8.5b. The first step is the construction of an optical model, which can regenerate measured data and separate the optical response of the thin film from substrates using Fresnel's equations [26]. An optical model is the structure of stacked flat layers representing an actual sample as close as possible. Each layer used in the optical model is defined by thickness and dielectric function $(\varepsilon = \varepsilon_1 + i\varepsilon_2)$. For example, to represent the measured SE data for silicon wafer requires a three-layer optical model: the first layer is silicon, the second layer is SiO_2, which grows naturally over silicon, and the third layer is ambient. Analytical oscillator models (AOMs) are used to represent the ε of material if it is not known. Various AOMs are available and can be used concerning the properties of the material. For example, the Cauchy model is for transparent material, Drude model represents the free-electron response. For absorbing materials, Lorentz or Tauc-Lorentz models can be used [30]. An important point to remember is that the ε of the substrate should be known in advance to analyze a film. So measurement of the substrate is advisable before film deposition.

The next step is to match the calculated data with experimental data via fitting some parameters representing the layers in the optical model. We can fit any SE data if many fitting parameters are used, but results may be unphysical. So, we need to use a minimum number of fitting parameters that show no correlations. The microstructure and thickness of the films should be checked using AFM, SEM, or TEM; their advanced knowledge can ease the fitting process and may lead to accurate extraction of thin film's dielectric function. Multi-wavelength, multi-angle, and multi-sample measurements generate an extensive data set, which will help to extract reliable results from the analysis. Finally, fitting results can be improved by correcting the model appropriate to the real sample, i.e., adding interference effect, surface roughness, and an interface layer. Physical properties like dielectric function, fundamental bandgap, and thin-film thickness can be extracted by achieving an excellent fit between calculated and measured data. Dielectric function is related to complex reflectance ratio as

$$\varepsilon = \sin^2\theta\left[1+\tan^2\theta\left[\frac{1-\rho}{1+\rho}\right]^2\right] \tag{8.3}$$

where θ is the angle of incidence. The ε is a dielectric function of a whole sample accessed by a probing beam. The ε of the thin films is obtained from Eq. (8.3) using Fresnel's equations. Other optical functions can be easily obtained as $\varepsilon_1 = n^2 - k^2$ and $\varepsilon_2 = 2nk$, where n is the refractive index, k is absorption coefficient, ε_1 and ε_2 are real and imaginary parts of the ε. The thickness of a thin film is obtained from phase thickness as $\beta = 2\pi dN\cos\theta/\lambda$, where β is phase thickness, d is a film thickness, $N = n + ik$ is a complex refractive index, and λ is the wavelength of probing light. For a detailed derivation of these quantities, please refer to these books [26,27].

8.3.3 Uniformity of Wafer-Scale MoS_2 Monolayer

In this investigation, MoS_2 ultrathin films with varying thickness (mono- (1L), bi- (2L), tri- (3L), tetra- (4L), and hexa-layer (6L)) were grown on $2\times 2\,cm^2$ SiO_2(100 nm)/Si substrate to optimize the MoS_2 thickness extraction process from SE data. Actual large-area MoS_2 monolayer and multilayers samples were grown on an 8-inch SiO_2(300 nm)/Si wafer. SiO_2 was grown from wet thermal oxidation of Si wafer. Therefore, SE data are represented as [31]:

$$\rho = f\left(\varepsilon_{air}, \varepsilon_{MoS_2}, \varepsilon_{SiO_2}, \varepsilon_{Si}, d_{SiO_2}, d_{MoS_2}, E, \theta\right) \tag{8.4}$$

Measured Ψ and Δ spectra are the function of eight parameters: dielectric functions of air (ε_{air}), MoS_2 (ε_{MoS_2}), SiO_2 (ε_{SiO_2}), Si (ε_{si}), thicknesses of SiO_2 (d_{SiO_2}), MoS_2 (d_{MoS_2}), the energy of probing beam (E), and angle of incidence (θ). We need to reduce the unknowns from Eq. (8.4) to avoid the complications in the analysis due to correlations among fitting parameters and for accurate extraction of the properties of ultrathin MoS_2 like thickness and ε_{MoS_2} from SE data. The E, θ, and $\varepsilon_{air} = 1$ are known. Si and SiO_2 are well-known materials, and their ε is available with good precision. The d_{SiO_2} value can be obtained by prior measurement of a substrate before deposition. The ε_{MoS_2}, d_{MoS_2} are the only remaining unknown parameters. The ε_{MoS_2} is thickness dependent [32], so its knowledge is essential for analyzing wafer-scale MoS_2 ultrathin films and mapping the thickness.

8.3.3.1 SE Measurement

Ellipsometric measurements were performed using M200D RCSE from J.A Woollam Co. inc. The M2000D model is equipped with an automated goniometer for variable angle measurement, XY stage for precise multi-sport measurement, and a laser system for sample alignment. This automation is necessary to measure wafer-scale samples. The measured energy range was 1.24–3.50 eV. Angles of incidence were 60°, 65°, 70°, and 75°. The 1L, 2L, 3L, 4L, and 6L MoS_2 films were measured at multiple spots. This extensive data set for every sample is helpful to accurately obtain the ε_{MoS_2}. Similar measurements were performed at 950 spots over an 8-inch wafer of MOCVD grown samples and at 51 spots for $1.5\times 20\,cm^2$ of ALD grown samples to map the thickness uniformity.

8.3.3.2 Thickness-Dependent Dielectric Functions Ultrathin MoS_2 Films

Figure 8.5c-i,ii show the measured Ψ and Δ spectra of 1L, 2L, 3L, 4L, and 6L MoS_2 films, clearly shows the effect of the thickness variation of these ultrathin films. Raw SE data can be used to detect the difference between a number of layers of MoS_2. A_{ex} and B_{ex} optical features are excitonic transitions from split valence band due to SOC to excitonic ground state. Four-layer (ambient/MoS_2/SiO_2/Si) optical model, shown in Figure 8.5c-iii, was used for data analysis. Thicknesses of the ambient and Si are taken as infinite and d_{SiO_2} =~ 100 nm, measured previously. The ε_{SiO_2} and ε_{Si} are taken from the literature [33]. The d_{MoS_2} was measured using AFM, and used as the initial guess and refined during fitting. Only unknown parameter is ε_{MoS_2}. Wavelength-by-wavelength (w-b-w) fitting approach was used to match the measured data. Measured and calculated Ψ and Δ spectra of monolayer MoS_2 on Si substrate is plotted in Figure 8.5c-iv. Used known parameters in Eq. (8.4) are available with good accuracy, results in that simple w-b-w fit provide better results [34]. Otherwise, we had to use AMO to represent the ε_{MoS_2}. Care should be taken that ε_{MoS_2} obtained from w-b-w fit is not necessarily Kramer–Kronig (KK) consistent. In our case, ε_1 and ε_2 follow KK relations, suggest the accuracy of our model. The obtained d_{MoS_2} values from SE analysis were compared with measured using AFM, shown in Figure 8.5c-v.

Extracted ε_{MoS_2} from the above procedure is plotted in Figure 8.5d. The ε_2 spectra of ALD and MOCVD grown MoS_2 1L are compared with reported data by Lee et al. [29], which were measured on CVD grown 1L MoS_2. Line shapes are nearly similar. Peaks A_{ex} and B_{ex} are the excitonic transitions from split valence band minimum due to SOC to the excitonic ground state at the K point of the Brillouin zone. The SOC splitting is 0.12 ± 0.02 eV. High energy absorption peak E_1 is from high energy inter-band transitions due to the presence of the Van Hove singularities in joint density of states at high-symmetry points and saddle-points from band nesting. It is observed that the ε_{MoS_2} depends on the thickness of the ultrathin films, shifts to the lower energy with an increasing number of layers, plotted in Figure 8.5c-ii,iii. Therefore, it is critical to know accurate ε_{MoS_2} to calculate other dependent properties like device performance.

8.3.3.3 Thickness Mapping of MOCVD and ALD Grown Wafer-Scale MoS_2 Ultrathin Films

As we mentioned above, if we want to extract one desired property from Eq. (8.4), the remaining seven parameters should be known with the required accuracy. An 8-inch wafer was used for MOCVD growth, and a 1.5 cm × 20.0 cm strip was used for ALD due to instrument restrictions. Commercially available 8-inch silicon wafers are with thermally grown 300 nm SiO_2 that was used to deposit MoS_2 films. Si wafers from two different manufacturers were used for MOCVD and ALD, their provided value $d_{SiO_2} = 300 \pm 5\,\text{nm}$. We have observed that d_{SiO_2} is not uniform over an 8-inch wafer, should be measured before deposition. Si wafers were measured with the same procedure mentioned above and obtained the map of the SiO_2 thickness using the following analysis. A three-layer (ambient/SiO_2/Si) optical model was used, as shown in Figure 8.6a. Figure 8.6b and c show the ε_{SiO_2} and ε_{Si}, respectively, used as the fix parameter and only d_{SiO_2} was varied during fitting.

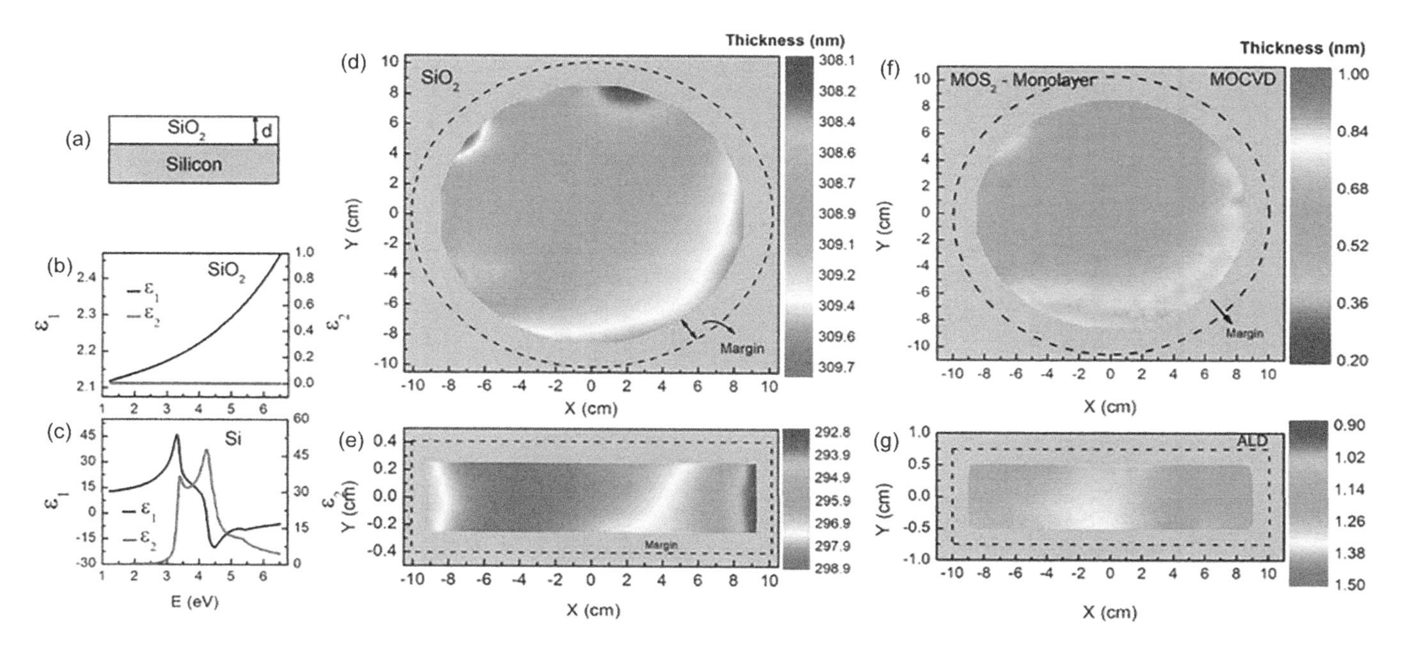

FIGURE 8.6 Characterization of substrate (SiO_2 (300nm)/Si) before MoS_2 deposition. (a) Three-layer optical model, and (b) ε_{SiO_2} and (c) ε_{Si}. Thickness map of SiO_2 over (d) 8-inch wafer used in MOCVD and (e) 1.5 cm × 20.0 cm strip used in ALD growth of MoS_2 ultrathin films. Thickness map of wafer-scale MoS_2 (f) monolayer grown by MOCVD and (g) bilayer grown by ALD.

The SiO_2 thickness map of both the wafers is given in Figure 8.6d and e. The 8-inch Si wafer has 308.9 ± 1 nm SiO_2 layer (Figure 8.6d), which suggests good quality wafer. While Si wafer used for ALD shows the non-uniform SiO_2 formation, edge to the center has ~ 5 nm thickness difference (Figure 8.6e), which may cause an error in d_{MoS_2} estimation. Therefore, it is highly advisable to measure substrates before deposition.

Finally, we have collected all the necessary parameters in Eq. (8.4) to estimate the thickness uniformity of wafer-scale MoS_2. We limit the use of energy range between 1.3 and 3.0 eV where the substrate is transparent, which will be helpful for easy converge of fitting data. Also, the high energy optical response of MoS_2 is unpredictable due to the complicated nature of its band structure; two samples with the same thickness show slightly different values of ε, which may induce an error in uniformity analysis. The same four-layer was used and plugged all the obtained parameters in Eq. (8.4), only d_{MoS_2} was the unknown parameter. Thickness maps are obtained from the above procedure for wafer-scale MoS_2 film grown over 8-inch wafer by MOCVD and 1.5 cm × 20.0 cm wafer strip by ALD, plotted in Figure 8.6f and g, respectively. It is observed that MOCVD-grown MoS_2 film has an average thickness of 0.65 ± 0.05 nm (monolayer), and ALD-grown film's average thickness is 1.25 ± 0.05 nm (bilayer). This precise thickness control growth of MoS_2 was achieved by controlling the partial pressure of precursor gases and growth temperature, explained in detail in the synthesis section.

Thickness uniformity (U) was quantified using the following expression [31],

$$U = \left(d_{avg} - d_{\sigma}\right) / d_{avg} \times 100 \tag{8.5}$$

where d_{avg} is the average thickness of ultrathin MoS_2 films, d_σ is the standard deviation in the measurement. Wafer-scale MoS_2 monolayer grown using MOCVD shows ~ 92%, and ALD grown bilayer shows ~ 96% thickness uniformity.

8.4 CONCLUSION

This chapter introduces the different methods for wafer-scale growth, which have the potential of mass production, and a non-destructive way to characterize the wafer-scale samples to quantify the uniformity and quality of ultrathin TMDC films. Wafer-scale production of high-quality 2D TMDC materials is necessary for batch processing to meet the criteria of practical commercial applications. Four different approaches, namely MOCVD, pulsed MOCVD, ALD with self-limiting approach, and layered resolved splitting, are summarized, which showed uniform wafer-scale growth of ultrathin TMDC films. The first three methods use vapor phase precursors and control the thickness of films via vapor pressure, precursors injection time, growth temperatures, and reactant gas ratio. However, the fourth method uses metal-assisted controlled crack initiations and propagation in wafer-scale artificially grown thick TMDC films. Pulsed MOCVD with the vertical-Ostwald ripening process to control the secondary nucleation over pre-deposited TMDC nanocrystals and surface diffusion to cover the entire surface is the most promising method for mass production. However, further investigations are required to reach that goal.

SE, a non-destructive optical technique, is a powerful tool that can optimize the growth process of wafer-scale ultrathin TDMCs films, leading to practical device development. This chapter explains the basics of SE and the data analysis procedure. It is observed that rotating compensator SE is a good choice for high-speed data acquisition and can be used for a variety of materials with high accuracy and precision. The optical properties of sensitive materials like TMDCs strongly depend on growth techniques and substrates. Therefore, we first build the database of thickness-dependent dielectric functions (ε) of MoS_2 ultrathin films, which will be used as a fixed parameter or as the initial guess during accurate thickness mapping. We build the extensive SE data set for each measured point using 370 wavelengths and 5 angles of incidences, which helps to reduce any correlation in fitting parameters. Using pre-measured parameters, which define the SE data of MoS_2 monolayer on SiO_2(300 nm)/Si substrate, the thickness map of MoS_2 was obtained, shows 96% uniformity. The capability of SE to measure the high-speed thickness and the parameter that could define the material's quality help to bring the proposed technologies from lab to production lines.

REFERENCES

1. M. Koot, F. Wijnhoven, Usage impact on data center electricity needs: A system dynamic forecasting model, *Appl. Energy.* (2021) 291, 116798.
2. N.V. Podberezskaya, S.A. Magarill, N.V. Pervukhina, S.V. Borisov, Crystal chemistry of dichalcogenides MX_2, *J. Struct. Chem.* (2001) 42, 654–681.
3. M. Chhowalla, H.S. Shin, G. Eda, L.-J. Li, K.P. Loh, H. Zhang, The chemistry of two-dimensional layered transition metal dichalcogenide nanosheets, *Nat. Chem.* (2013) 5, 263–275.
4. G. Eda, H. Yamaguchi, D. Voiry, T. Fujita, M. Chen, M. Chhowalla, Photoluminescence from Chemically Exfoliated MoS 2, *Nano Lett.* (2011) 11, 5111–5116.
5. H. Morgan, C. Rout, D.J. Late, *Fundamentals and Sensing Applications of 2D Materials*, 2019. Elsevier, Duxford, UK.
6. H. Li, Y. Shi, M.-H. Chiu, L.-J. Li, Emerging energy applications of two-dimensional layered transition metal dichalcogenides, *Nano Energy.* (2015) 18, 293–305.
7. D. Jariwala, V.K. Sangwan, L.J. Lauhon, T.J. Marks, M.C. Hersam, Emerging device applications for semiconducting two-dimensional transition metal dichalcogenides, *ACS Nano.* (2014) 8, 1102–1120.
8. Y. Qi, N. Han, Y. Li, Z. Zhang, X. Zhou, B. Deng, Q. Li, M. Liu, J. Zhao, Z. Liu, Y. Zhang, Strong adlayer–substrate interactions "Break" the patching growth of h -BN onto graphene on Re(0001), *ACS Nano.* (2017) 11, 1807–1815.
9. S. Deng, E. Gao, Z. Xu, V. Berry, Adhesion energy of MoS_2 thin films on silicon-based substrates determined via the attributes of a single MoS_2 wrinkle, *ACS Appl. Mater. Interfaces.* (2017) 9, 7812–7818.
10. E. Bauer, Phänomenologische Theorie der Kristallabscheidung an Oberflächen. I, *Zeitschrift Für Krist. - Cryst. Mater.* (1958) 110, 372–394.
11. I.V. Markov, *Crystal Growth for Beginners*, 2017. Wrold Scientific, Singapore.
12. L. Zhang, J. Dong, F. Ding, Strategies, status, and challenges in wafer scale single crystalline two-dimensional materials synthesis, *Chem. Rev.* (2021) 121, 6321–6372.
13. S. Ganorkar, J. Kim, Y.-H. Kim, S.-I. Kim, Effect of precursor on growth and morphology of MoS2 monolayer and multilayer, *J. Phys. Chem. Solids.* (2015) 87, 32–37.
14. T. Kim, J. Mun, H. Park, D. Joung, M. Diware, C. Won, J. Park, S.-H. Jeong, S.-W. Kang, Wafer-scale production of highly uniform two-dimensional MoS_2 by metal-organic chemical vapor deposition, *Nanotechnology* (2017) 28, 18LT01.

15. M. Seol, M. Lee, H. Kim, K.W. Shin, Y. Cho, I. Jeon, M. Jeong, H. Lee, J. Park, H. Shin, High-throughput growth of wafer-scale monolayer transition metal dichalcogenide via vertical ostwald ripening, *Adv. Mater.* (2020) 32, 2003542.
16. Y. Kim, J.-G. Song, Y.J. Park, G.H. Ryu, S.J. Lee, J.S. Kim, P.J. Jeon, C.W. Lee, W.J. Woo, T. Choi, H. Jung, H.-B.-R. Lee, J.-M. Myoung, S. Im, Z. Lee, J.-H. Ahn, J. Park, H. Kim, Self-limiting layer synthesis of transition metal dichalcogenides, *Sci. Reports* (2016) 6, 1–8.
17. J.-G. Song, J. Park, W. Lee, T. Choi, H. Jung, C.W. Lee, S.-H. Hwang, J.M. Myoung, J.-H. Jung, S.-H. Kim, C. Lansalot-Matras, H. Kim, Layer-controlled, wafer-scale, and conformal synthesis of tungsten disulfide nanosheets using atomic layer deposition, *ACS Nano.* (2013) 7, 11333–11340.
18. K. Park, Y. Kim, J.G. Song, S. Jin Kim, C. Wan Lee, G. Hee Ryu, Z. Lee, J. Park, H. Kim, Uniform, large-area self-limiting layer synthesis of tungsten diselenide, *2D Mater.* (2016) 3, 014004.
19. J. Shim, S.-H. Bae, W. Kong, D. Lee, K. Qiao, D. Nezich, Y.J. Park, R. Zhao, S. Sundaram, X. Li, H. Yeon, C. Choi, H. Kum, R. Yue, G. Zhou, Y. Ou, K. Lee, J. Moodera, X. Zhao, J.-H. Ahn, C. Hinkle, A. Ougazzaden, J. Kim, Controlled crack propagation for atomic precision handling of wafer-scale two-dimensional materials, *Science* (2018) 362, 665–670.
20. S.B. Desai, S.R. Madhvapathy, M. Amani, D. Kiriya, M. Hettick, M. Tosun, Y. Zhou, M. Dubey, J.W. Ager, D. Chrzan, A. Javey, Gold-mediated exfoliation of ultralarge optoelectronically-perfect monolayers, *Adv. Mater.* (2016) 28, 4053–4058.
21. H. Sun, E.W. Sirott, J. Mastandrea, H.M. Gramling, Y. Zhou, M. Poschmann, H.K. Taylor, J.W. Ager, D.C. Chrzan, Theory of thin-film-mediated exfoliation of van der Waals bonded layered materials, *Phys. Rev. Mater.* (2018) 2, 94004.
22. J. Mun, Y. Kim, I.-S. Kang, S.K. Lim, S.J. Lee, J.W. Kim, H.M. Park, T. Kim, S.-W. Kang, Low-temperature growth of layered molybdenum disulphide with controlled clusters, *Sci. Rep.* (2016) 6, 21854.
23. A.T. Hoang, K. Qu, X. Chen, J.-H. Ahn, Large-area synthesis of transition metal dichalcogenides via CVD and solution-based approaches and their device applications, *Nanoscale.* (2021) 13, 615–633.
24. A. Kozhakhmetov, R. Torsi, C.Y. Chen, J.A. Robinson, Scalable low-temperature synthesis of two-dimensional materials beyond graphene, *J. Phys. Mater.* (2020) 4, 012001.
25. J. Yu, X. Hu, H. Li, X. Zhou, T. Zhai, Large-scale synthesis of 2D metal dichalcogenides, *J. Mater. Chem. C.* (2018) 6, 4627–4640.
26. H. Fujiwara, *Spectroscopic Ellipsometry: Principles and Applications*, 2007. John Wiley & Sons, Ltd, Chichester, UK.
27. H.G. Tompkins, E.A. Irene, eds., *Handbook of Ellipsometry*, 2005. Springer-Verlag GmbH & Co., Heidelberg.
28. D.E. Aspnes, A.A. Studna, High precision scanning ellipsometer, *Appl. Opt.* (1975) 14, 220.
29. W. Li, A.G. Birdwell, M. Amani, R.A. Burke, X. Ling, Y.H. Lee, X. Liang, L. Peng, C.A. Richter, J. Kong, D.J. Gundlach, N.V. Nguyen, Broadband optical properties of large-area monolayer CVD molybdenum disulfide, *Phys. Rev. B.* (2014) 90, 1–8.
30. M.S. Diware, S.P. Ganorkar, J. Kim, S.N. Bramhe, H.M. Cho, Y.J. Cho, W. Chegal, Dielectric function of polycrystalline α -Ag_2S by spectroscopic ellipsometry, *Appl. Phys. Lett.* (2015) 107, 171905.
31. M.S. Diware, K. Park, J. Mun, H.G. Park, W. Chegal, Y.J. Cho, H.M. Cho, J. Park, H. Kim, S.-W. Kang, Y.D. Kim, Characterization of wafer-scale MoS_2 and WSe_2 2D films by spectroscopic ellipsometry, *Curr. Appl. Phys.* (2017) 17, 1329–1334.

32. Y. Yu, Y. Yu, Y. Cai, W. Li, A. Gurarslan, H. Peelaers, D.E. Aspnes, C.G. Van de Walle, N.V. Nguyen, Y.-W. Zhang, L. Cao, Exciton-dominated dielectric function of atomically thin MoS_2 films, *Sci. Rep.* (2015) 5, 16996.
33. C.M. Herzinger, B. Johs, W.A. McGahan, J.A. Woollam, W. Paulson, Ellipsometric determination of optical constants for silicon and thermally grown silicon dioxide via a multi-sample, multi-wavelength, multi-angle investigation, *J. Appl. Phys.* (1998) 83, 3323–3336.
34. M.S. Diware, S.P. Ganorkar, K. Park, W. Chegal, H.M. Cho, Y.J. Cho, Y.D. Kim, H. Kim, Dielectric function, critical points, and Rydberg exciton series of WSe_2 monolayer, *J. Phys. Condens. Matter.* (2018) 30, 235701.

9 Morphological Aspects of 2D Nanomaterials for Energy Applications

Jing Ning, Maoyang Xia, Dong Wang, Jincheng Zhang, and Yue Hao
Xidian University

CONTENTS

DOI: 10.1201/9781003178453-9

9.1 INTRODUCTION

In the information age of the rapid development of the Internet of Things connection, self-developed smart chips for various applications have become a problem to be solved in all countries. The energy supply system integrated with microelectronic devices is required for fast charging and discharging, high power density, long cycling lives, and good stability. Selecting the appropriate electrode material is the key to ensuring high energy density for the micro-energy storage system. So far, two-dimensional (2D) nanomaterials have shown promising prospects in the field of energy storage due to their excellent physical and chemical properties.

Since the preparation of graphene, many atomic-layer materials including 2D TMDCs and MXene have also been implemented in laboratories. These materials present layered structures with a dimensional constraint in one direction. The micron-scale lateral size and nanoscale thickness endow the 2D nanomaterials with ultra-high specific surface areas and exposure of the surface atoms, which make them suitable for surface modification. Due to their lower dimension, the open channels for ion diffusion and the presence of active sites enable the fast transport and efficient storage of ions. The strong covalent bonds in the 2D nanosheets result in prominent mechanical strength and flexibility for individual sheets. Therefore, excellent and unique physicochemical properties of 2D materials lead to high mobility and high energy density, making them very promising candidates for energy storage devices.

Although 2D materials have a larger theoretical capacity than other traditional bulk materials, their self-pseudocapacitance is limited because of poor processability. During micro-energy devices fabrication process, 2D materials nanosheets are easy to aggregate and restack, which seriously impedes rapid diffusion of electrolyte ions and affects the full utilization of the active surface of the electrodes. Converting the planar structure of 2D materials nanosheets into an interpenetrating network with open or porous structures is an effective way to construct high-performance micro-electrodes. In the past few years, a large number of reviews are on the application of 2D materials as electrode materials for energy storage devices. This chapter will focus on the morphological aspects of 2D nanomaterials and their energy applications.

9.2 GRAPHENE

9.2.1 Morphology and Synthesis Method

9.2.1.1 Graphene Nanosheet

Graphene nanosheets are generally prepared by mechanical stripping of graphite, chemical vapor deposition (CVD), or chemical oxidation or stripping of graphite. Each method has advantages and disadvantages in terms of the quality and yield of the prepared graphene.

Liquid-phase exfoliation (LPE). Graphite is dispersed in organic solvents or specific surfactants. Single or multiple layers of graphene are peeled off from the graphite surface by ultrasound. The graphene dispersion was obtained by centrifugation. Finally, graphene can be deposited on different substrates. Telkhozhayeva Madina's group [1] reported that monolayer graphene flakes are directly exfoliated from graphite using ethanol as a solvent by bath sonication. A total of 77% of the graphene flakes have a thickness below three layers with an average lateral size of 13 μm.

Chemical vapor deposition (CVD). Using carbon-containing gas (such as methane) as raw material, in an environment above 1,000°C, through chemical reaction, carbon is precipitated and grown on copper and other substrates to form graphene flakes. Kim et al. [2] developed a CVD device for producing high-quality (about 252 Ω per square) and large area (over 16 inches) graphene.

Redox method. Natural graphite is oxidized by strong acids and strong oxidants such as sulfuric acid, potassium permanganate, and hydrogen peroxide to obtain graphene oxide. It is then dispersed by physical stripping and high-temperature expansion to obtain graphene oxide. Finally, it is reduced to obtain reduced graphene oxide. Wang et al. [3] used hydroquinone to reduce graphene oxide in an aqueous solution to prepare graphene nanosheets.

9.2.1.2 Porous Graphene

At present, the preparation methods of porous graphene include the template method and colloid drying method.

Template-assisted methods are common methods for preparing the structure and morphology of specific nanomaterials. The template-assisted method mainly includes three steps: (a) Combine the precursors of the reaction to impregnate or infiltrate the template; (b) Form a solid state on the template through reaction, nucleation, and growth; and (c) The final product can be obtained after removing the template.

Metal Template-Assisted Chemical Vapor Deposition. Graphene is directly grown using nickel foam as a catalytic metal and template. Self-supporting porous graphene can be obtained by etching the nickel foam. Ning's group [4] reported that pore-adjustable copper-nickel alloy was prepared on nickel foam by electrochemical deposition and etching. Compared with commercially available Cu or Ni foams, graphene grown on alloy foams also has smaller and variable pore size and larger surface area. The specific surface area of graphene can reach $0.096\,m^2cm^{-3}$, which is four times larger than that of commercial templates.

Colloidal sphere synthesis. The photoresist method using polystyrene or silicon dioxide spheres as templates has been used to prepare various porous graphene. The structure prepared by this method has ordered aperture and controllable size. Self-assembled polystyrene micro-sphere templates and chemically modified graphene (CMG) have been reported to prepare self-supported network graphene [5]. In this method, self-supported polystyrene spheres/graphene are mixed with hydrocolloid suspended graphene and polystyrene spheres for vacuum filtration, and then polystyrene sphere templates are removed with toluene to obtain porous graphene. The resulting graphene pore diameter is the diameter of the polystyrene ball, so it is easy to control its pore size.

Interferometric lithography template method. The multilayer graphene prepared by converting the prepared 3D pyrolysis photoresists [6]. Its specific surface area is two orders of magnitude smaller than that of graphene grown directly on nickel foam. It consists of three steps: first, fabricating a 3D porous carbon face-centered cubic structure by interferometric lithography; then, 3D porous carbon was annealed at 750°C for 50 minutes and converted into 3D graphite material. Finally, the 500-nm pore size is obtained after etching the nickel foam.

Hydrothermal template method. Porous graphene was prepared by combining hydrothermal synthesis with nickel foam template method. The main method is to deposit chemically peeled graphene oxide onto nickel foam by hydrothermal synthesis. The deposited nickel foam was then reduced by vapor etching in ethylene to obtain porous graphene. This method is simpler than the foam nickel template CVD method. However, graphene has many stacking defects and its thickness is difficult to control [7].

Hydrothermal gel synthesis method. Self-supporting porous graphene can be prepared by graphite oxide or organic oxide graphite. But the stability and uniformity of graphene depend on gel molecules and polymers. When the gel and polymer are dried, small holes will form on the surface of the porous graphene. These small holes communicate with each other to form a 3D structure, not just connected on the surface. Shi et al. [8] reported that graphene oxide (GO) sheets could form complex hydrogels with poly(vinyl alcohol) (PVA). The role of GO sheets is similar to that of 2D macromolecules. The formation of the hydrogels depends on the assembly of GO sheets and the cross-linking effect of PVA chains.

Freeze-dry method. Freeze the graphene solution below the freezing point, convert the water to ice, and then remove the ice by converting it to steam under a higher vacuum. Material can be frozen in the freezer before drying. But it can also be frozen directly in a drying chamber by pumping it into a vacuum quickly. The water vapor generated by sublimation is removed by means of a condenser. The voids that remain form a porous material. Xu et al. [9] organized GO into 3D networks with a cylindrical structure by hydrothermal treatment of a GO suspension (1 or 2 mg ml^{-1}) at 180°C.

9.2.2 Application of Graphene as Electrode

9.2.2.1 Supercapacitor

When used in supercapacitors, graphene stores energy through an electric double-layer capacitance. Due to its high conductivity and high specific surface area, graphene can provide high-performance energy storage. The theoretical specific capacitance is 21 μF cm^{-2} (550 F g^{-1}).

Ruoff et al. [10] reported the work of CMG as electrode material for supercapacitors in 2008. CMGs are made from 1-atom thick carbon sheets by Hummer method. The specific capacitances in aqueous (5.5 M KOH) and organic electrolytes ($TEABF_4$/AN) are 135 and 99 F g^{-1}, respectively. Further, Ruoff et al. [11] also reduced GO by microwave heating and activated the product to obtain porous graphene. The surface area achieves 3,100 $m^2 g^{-1}$. The specific capacity in $TEABF_4$/AN increases to 166 F g^{-1}, and the energy density increases to 70 Wh kg^{-1}.

Kim et al. [12] prepared micron-scale graphene nanolayers (GNMs) with nanomanipulation by catalytic carbon gasification. The pore size and density distribution of GNMs are controlled by adjusting the size and fraction of metal oxides. GNM electrodes show high capacitance (253 F g^{-1} at 1 A g^{-1}).

Tang et al. [13] used polystyrene balls as templates to assemble graphene nanosheets into spherical shells. Compared with stacked planar graphene, the prepared graphene spherical shell has a larger free space between the spheres. Graphene hollow spheres show high specific capacitance 273 F g^{-1} and excellent electrochemical stability.

9.2.2.2 Lithium-Ion Batteries

Graphene has lithium storage activity in the low potential range (<1.5 V) and is usually used as a negative electrode material for lithium-ion batteries (LIBs). Lithium can be stored on both surfaces and edges of graphene, resulting in two layers of lithium per graphene sheet, with a theoretical capacity of 744 mAh g^{-1} by forming Li_2C_6. At the same time, graphene has a large interlayer spacing and abundant pores, and Li-ion can quickly diffuse through the interlayer spacing, which has good multiplication performance.

Wang et al. [14] prepared 2–3 layered graphene nanosheets from bulk graphite via Hummer method. The graphene nanosheets as anodes of LIBs exhibit an enhanced lithium storage capacity (650 mAh g^{-1}). Yoo et al. [15] reported that graphene layer spacing was extended from 0.365 to 0.42 nm by the incorporation of macromolecules of CNT and C_{60}. The capacity was increased from 540 to 784 mAh g^{-1}. Zhao et al. [16] synthesized mesoporous graphene nanosheets by a controlled low concentration monomicelle close-packing assembly approach. Mesoporous graphene provides high surface area for Li-ion adsorption and intercalation and enables efficient ion transport. Mesoporous graphene exhibited an excellent reversible capacity of 1,040 mAh g^{-1} at 100 mA g^{-1}.

9.2.2.3 Sodium-Ion Batteries

The size of Na-ion is large (102 pm) and the bonding to the graphite surface is weak, so graphite cannot be directly used as negative electrode material of sodium-ion batteries (SIBs). The replacement of graphite by 2D graphene as a cathode material for SIBs is considered to be an effective strategy.

Wen et al. [17] reported expanded graphene as an SIB anode. Expanded graphene has a 4.3 Å expanded interlayer lattice distance, but retains a similar long-range ordered layered structure. Expanded graphene can provide a high reversible capacity of 284 mAh g^{-1} at a current density of 20 mA g^{-1}.

The sodium storage properties of porous graphene depend on appearance, surface defects, and pore size. Lee et al. [18] prepared activated wrinkled graphene by CVD. For sodium storage, the reversible capacity is 280 mAh g^{-1} at 40 mA g^{-1}. The enhanced electrochemical performance stems from the adsorption of ions on various defects such as the Stone–Wales defect.

9.2.2.4 Potassium-Ion Batteries

For potassium-ion batteries (PIBs), the electrochemical properties of pure graphene materials are still unsatisfactory. One of the optimization principles of graphene-based electrodes is to design novel structures that can introduce more point defects, edges, grain boundaries, and doped atoms [19].

Share et al. [20] developed several layers of N-doped graphene, which provides a higher potassium storage capacity (more than 350 mAh g^{-1}) than KC_8 (278 mAh g^{-1}). N-doped uniformly distributed in the carbon matrix serves as the active sites for local storage of K-ions. Ju et al. [21] synthesized nitrogen-doped graphene from dicyandiamide and coal tar pitch. The electrode shows a high specific capacity (320 mAh g^{-1} after 60th cycle at 50 mA g^{-1}), and long-term cycling capability (over 150 mAh g^{-1} after 500 cycles at 500 mA g^{-1}) in PIBs. Ju et al. [22] using polyvinylidene difluoride (PVDF) as a single source reactant, prepared several F-doped graphene foams with a thickness of about 4 nm and high surface area (874 $m^2 g^{-1}$) by a high-temperature solid-state method. The F-doped graphene shows a high initial capacity of 863.8 mAh g^{-1}.

9.3 TRANSITION METAL DICHALCOGENIDES (TMDCs)

9.3.1 Morphology and Synthesis Method

"Top-down" and "Bottom-up" are used in the synthesis of 2D materials [23]. "Top-down" is the preparation of large-size materials into nanostructures through physical or chemical means, including mechanical stripping and LPE. "Bottom-top" is to self-assemble small structural units (such as molecules, atoms, and nanoparticles) into nano-sized structures through weak interactions, including CVD and hydrothermal/solvent thermal synthesis methods.

9.3.1.1 TMDC Nanoflakes

TMDC nanoflake is generally prepared by the mechanical exfoliation and LPE of bulk TMDC, and CVD.

Mechanical exfoliation is the simplest physical method to synthesize 2D TMDCs nanoflake in a short time. This method of preparing ultra-thin TMDCs by mechanical exfoliation is fast and simple. It does not need the conditions of high temperature and high pressure. Ultra-thin TMDCs can be prepared only through special tape, and the yield of stripped products is high. However, the sample size is small, the yield is low, and the thickness of sample layer cannot be controlled. Late et al. [24] successfully prepared single-layer and few layer WS_2 thin films on SiO_2/Si substrate by mechanical stripping of large WS_2 crystals and transparent tape technology at room temperature.

LPE method is to disperse TMDC crystal powder into liquid medium and provide external force assistance for stripping through ultrasonic, microwave, shear force, thermal stress, and centrifugation. Solvent-assisted stripping and ion intercalation-assisted stripping are commonly used in LPE. Coleman et al. [25] placed bulk TMDC in organic solvent for ultrasonic treatment to obtain monolayer nanosheets. The experimental results show that 1-methyl-2-pyrrolidone (NMP) is the most effective organic solvent for removing bulk TMDC. Yu et al. [26] placed the bulk WSe_2 in NMP solvent for ultrasonic treatment, and WSe_2 thin films were prepared. The treatment of TMDCs by solvent exfoliation method will not significantly interfere with its crystal structure. Fan et al. [27] found that ultrasonic treatment can help *n*-butyl lithium intercalate bulk TMDC in hexane solution. MoS_2 nanosheets can fall off after ultrasonic treatment in aqueous solvent for 5 minutes.

Through CVD technology, single layer or several layers of TMDCs can be grown on a variety of substrates, which is also one of the most practical methods for large-area synthesis of materials. Lee et al. [28] used MoO_3 and S powder as reactants and N_2 gas as protective gas to form large-area MoS_2 films on SiO_2/Si substrates through CVD for the first time.

9.3.1.2 TMDC Nanorod and Nanoflower

Hydrothermal/solvothermal method is a common wet chemical synthesis method, which has low cost and large amount of products. By adjusting the ion concentration, temperature, and reaction time in the hydrothermal process, the morphology and crystal structure of the products can be further regulated and a large number of active sites can be provided.

Ning et al. [29] used a one-step hydrothermal method to form sodium intercalation nanoflower 1T-2H $MoSe_2$-graphene. The insertion of Na-ions not only expands the distance between layers but also provides space for electrolyte Na-ions.

9.3.1.3 TMDC Nanofiber

Electrospinning depends on the electrostatic repulsion between surface charges to continuously prepare viscous fluids into nanofibers with diameters up to tens of nanometers. In Kumuthini et al. [30] MoS_2/CNFs composite was successfully fabricated by electrospinning technology. The diameter of the composite was 200–300 nm.

9.3.1.4 Porous TMDC

Hydrothermal gel synthesis method. Li et al. [31] prepared a polysulfide molybdenum gel (MoS_{12}) with $(NH_4)_2Mo_3S_{13}$ as a precursor and then obtained porous MoS_2 through critical point drying technology. The porous MoS_2 has a wide pore size. A specific surface area is 315 $m^2 g^{-1}$ and a pore volume is 1.9 $cm^3 g^{-1}$.

9.3.2 Application of TMDCs as Electrode

9.3.2.1 Supercapacitor

The two common structural phases of TMDC are 2H and 1T, which are characterized by trigonal prismatic and octahedral coordination of the transition metal atoms, respectively. The 2H Phase TMDC is semiconductor, which shows double electric layer energy storage properties in aqueous electrolyte. Conversely, 1T phase TMDC is metallic material, which shows pseudocapacitance energy storage properties. In addition, the electron mobility of 1T phase TMDC is higher than that of 2H phase TMDC. Therefore, 1T phase TMDC is widely used in the field of energy storage.

Habib et al. [32] reported CVT technique is used to form a layered single crystal of WS_2 and WSe_2. The TMDC electrodes exhibit excellent cycle stability and high capacitance retention of 80 and 99% for WS_2 and WSe_2 after 20,000 cycles, respectively. Ning et al. [33] synthesized gypsophila-like 1T-WSe_2/graphene via ammonia-assisted hydrothermal treatment. The 1T-WSe_2/graphene electrode shows a large specific capacitance (1,735 F g^{-1} at 1 A g^{-1}), and the energy density of all-solid-state supercapacitor reaches 48.2 Wh kg^{-1} at 250 W kg^{-1}. Tu et al. [34] used molten salt process to prepare WS_2 reduced graphene oxide (rGO) nanosheets. WS_2/rGO electrodes reach a high specific capacitance of 2,508.07 F g^{-1} at 1 mV s^{-1}. The hybrid WS_2/rGO achieves an excellent cycling stability, and the capacitance retention is as high as 98.6% after 5,000 cycles.

9.3.2.2 Lithium-Ion Batteries

The layered structure of TMDCs is conducive to the rapid diffusion of Li-ions in the electrode. At the same time, due to the very loose stacking structure, few layer TMDCs can greatly adapt to the structural changes before and after Li-ion insertion, and the volume expansion rate is low. It is considered an ideal cathode material for LIBs.

Guo et al. [35] synthesized MoS_2 nanospheres with a diameter of about 200 nm by hydrothermal method as the cathode electrode of LIBs, and the first discharge specific capacity reached 1,272 mAh g^{-1}. Du et al. [36] obtained restacked MoS_2 by Li-ion chemical exfoliation and hydrothermal treatment. Through stripping and restacking process, the c lattice parameter and surface area of MoS_2 were increased. Restacked MoS_2 shows high reversible lithium storage capacity (750 mAh g^{-1} after 50 cycles) as anode for LIBs.

9.3.2.3 Sodium-Ion Batteries

The large interlayer distance in TMDCs favors the adaptation of Na-ion. However, because of their low electrical conductivity, the overall performance, especially the high-rate performance, still far exceeds satisfaction. Therefore, opening TMDCs and introducing agents and good conductivity are expected to solve the problem.

Bai et al. [37] developed a layered yarn spherical MoS_2 nanosphere structure coated with an N-doped carbon layer. As the anode material for SIB, the MoS_2-PVP@NC electrode exhibits an excellent cycling performance of 410.2 mAh g^{-1} at 1 A g^{-1} after 200 cycles. Ye et al. [38] reported that metal-semiconductor mixed-phase twinned hierarchical (MPTH) MoS_2 nanowires. The expanded interlayer spacing accelerates Na-ion insertion/extraction kinetics, and the metal-semiconductor mixed-phase enhances electron transfer ability. The MPTH MoS_2 electrode delivers high reversible capacities of 200 mAh g^{-1} at 0.1 A g^{-1} for 200 cycles.

9.3.2.4 Potassium-Ion Batteries

In non-aqueous electrolytes, the standard redox potential of K is lower than that of Na (or even Li). Compared with SIBs and LIBs, K can be converted into a potentially higher battery voltage for PIBs.

Guo et al. [39] reported single crystalline metallic graphene-like VSe_2 ultrathin nanosheets as anode materials. The large-sized ultrathin wrinkle-like nanosheets show a large surface-area-to-volume ratio, ultrafast electron/K-ion transport, limited self-aggregation, and excellent structural stability. The ultrathin VSe_2 nanoflakes exhibit high reversible capacity (366 mAh g^{-1} at 100 mA g^{-1}).

9.4 MXene

9.4.1 Morphology and Synthesis Method

9.4.1.1 MXene Nanoflake

MXene is usually obtained by etching MAX materials with different concentrations of HF, NH_4HF_2, and the mixture of fluoride and concentrated HCl. Single layer or few layers MXene is usually obtained from multi-layer MXene etched by liquid phase stripping HF in various polar organic solvents. Hydrazine, urea, dimethyl sulfoxide, isopropylamine, or organic base molecules can be embedded between MXene layers, and single or few layers of MXene is obtained by mechanical vibration such as ultrasound and ball milling. Gene et al. [40] achieved one-step etching exfoliation of MAX pretreated with low concentration HF using TMAOH to obtain a single layer of MXene with $Al(OH)^{4-}$ functional groups on the surface. Li et al. [41] prepared large monolayer MXenes without F functional groups by etching and exfoliation at 180°C for 24 hours with inorganic alkali KOH solution.

9.4.1.2 Porous MXene

After the porogen is removed by a specific method, the interconnected scaffold can be retained and a porous MXene material is formed. These methods mainly include freeze-drying, spray drying, hard template strategy, and also chemical reduction.

Freeze-drying has been widely used to create porous 2D materials by discharging nanoflake precursors into the boundaries of ice crystals. Wang et al. synthesized porous MXene with a large surface area by freeze-drying and used the product as the substrate to prepare porous MXene/Fe_2O_3 composites.

Spray drying is an alternative method to prepare porous 2D materials by capillary forced crumpling and self-assembly. Recently, Qiu et al. used spray drying to prepare porous MXenes. First, MXene dispersion is transferred into aerosol droplets. Second, directly calcined at a high temperature to remove solvents. Then, the inward capillary force leads to the isotropic compression of MXene nanoflakes and rapid assembly into 3D structures with fluffy shape after the solvent completely evaporated.

Hard templating strategy is another effective method for preparing porous materials, which depends on the pre-deposition of 2D materials on the surface of the hard template and then removing the hard template. For instance, Gogotsi et al. synthesized MXene-coated polymethyl methacrylate (PMMA) spherical composite material based on the interactions between their surface hydroxyl groups and then vacuum filtered them into independent films. After removing the PMMA hard template, the hollow MXene spheres and 3D macroporous MXene films are obtained.

The chemical reduction method can produce pores in the dense MXene film through the foaming process induced by the removal of surface functional groups. For example, Yu et al. treated dense MXene films via hydrazine to prepare MXene foam. Hydrazine reacts with oxygen-containing groups to quickly release a large amount of gaseous substances, such as CO_2 and H_2O, which generate high pressure between MXene sheets and overcome the van der Waals force that holds the flakes together. Therefore, the dense hydrophilic Ti_3C_2Tx film is transformed into a hydrophobic porous Ti_3C_2Tx film with a honeycomb structure [42].

9.4.2 Application of MXene as Electrode

9.4.2.1 Supercapacitor

In neutral aqueous electrolyte or organic electrolyte, the energy storage mechanism of $Ti_3C_2T_x$ may behave as electric double-layer capacitance. When Ti_3C_2Tx is in the H_2SO_4 electrolyte, the storage mechanism of pseudocapacitance is dominant. The embedding of hydrogen ions protonated the oxygen functional groups on the surface to form hydroxyl groups, and the oxidation state of Ti changed. The reversible change of Ti oxidation valence from +3 to +4 follows the bonding and bond breaking of oxygen functional groups, respectively. Similar to acidic electrolytes, KOH can provide ion intercalation without the reaction of surface functional groups.

Zhang et al. [44] fabricated the free-standing, flexible 3D porous MXene $Ti_3C_2T_x$/CNTs film via freeze-drying MXene $Ti_3C_2T_x$-based water membranes without post-treatment. The 3D porous $Ti_3C_2T_x$/CNT electrode shows the excellent specific capacitance of 372 F g^{-1} at a scan rate of 5 mV s^{-1}.

9.4.2.2 Lithium-Ion Batteries

MXenes can be a very suitable anode material for LIBs because of their large interlayer spacing (0.7–1.1 nm), excellent ionic and electronic conductivity, and adjustable elemental composition.

In 2012, Gogotsi et al. used a 2D Ti_2C-based material with an oxidized surface (prepared from Ti_2AlC by using HF etching Al at room temperature) as anode material for LIBs. When current rate is 0.1 C, the specific capacity stabilizes at 160 mAh g^{-1} after five cycles [45]. In addition, Gogotsi et al. fabricated porous MXenes with in-plane pores and then used carbon nanotubes (CNTs) filter it into an independent flexible film [46]. When used as an anode for LIB, the activated P-Ti_3C_2/CNT film provides high capacity approximately 1,250 mAh g^{-1} at 0.1 C.

9.4.2.3 Sodium-Ion Batteries

Layered MXene is considered an ideal electrode material due to its obvious interlayer spacing, which can promote the insertion and removal of sodium ions. Constructing 2D MXene nanosheets into 3D structure can inhibit the stacking between MXene sheets, improve the utilization of surface active sites, and then improve the electrochemical performance.

Zhang et al. [47] synthesized 3D carbon-coated $Ti_3C_2T_x$ by realizing self-polymerization of dopamine on the surface of original $Ti_3C_2T_x$ nanosheets followed by freeze-drying and carbonization under an inert air atmosphere. The 3D architecture exhibits a high capacity of 257.6 mAh g^{-1} at 0.05 A g^{-1} after 200 cycles for SIBs.

9.4.2.4 Potassium-Ion Batteries

The radius of K-ion is much larger than that of Li-ion and Na-ion, which leads to slow reaction kinetics and large volume expansion of active substance during charge and discharge. Therefore, expanding the MXene layer spacing and constructing a 3D structure is an effective strategy to improve the performance of PIBs.

Zhao et al. [48] designed PDDA-NPCN/Ti_3C_2 hybrids as PIBs anodes via an electrostatic attraction self-assembly approach, while N-rich porous carbon nanosheets (NPCNs) came from metal-hexamine frameworks. The hybrids possess a synergetic effect, resulting to a reversible capacity of 358.4 mAh g^{-1} after 300 cycles at 0.1 A g^{-1}.

9.5 CONCLUSIONS AND PERSPECTIVES

In this chapter, we discussed the preparation of 2D materials from exfoliated layer materials, assembling them into 2D films/films or 3D powdered samples/macrostructures, and the latest applications of solution-treated 2D materials in energy applications (including supercapacitors and batteries).

REFERENCES

1. Telkhozhayeva M., Teblum E., Konar R., Girshevitz O., Perelshtein I., Aviv H., Tischler Y. R., and Nessim G. D. 2021. Higher ultrasonic frequency liquid phase exfoliation leads to larger and monolayer to few-layer flakes of 2D layered materials. *Langmuir* 37:4504–4514.
2. Kim S. M., Kim J. H., Kim K. S., Hwangbo Y., Yoon J. H., Lee E. K., Ryu J., Lee H. J., Cho S., and Lee S. M. 2014. Synthesis of CVD-graphene on rapidly heated copper foils. *Nanoscale* 6:4728–4734.
3. Wang G., Yang J., Park J., Guo X., Wang B., Liu H., and Yao J. 2008. Facile synthesis and characterization of graphene nanosheets. *Journal of Physical Chemistry C* 112:8192–8195.
4. Li W., Xu X., Liu C., Tekell M. C., Ning J., Guo J., Zhang J., and Fan D. 2017. Ultralight and binder-free all-solid-state flexible supercapacitors for powering wearable strain sensors. *Advanced Functional Materials* 39:1702738.
5. Choi B. G., Chang S. J., Kang H. W., Park C. P., Kim H. J., Hong W. H., Lee S. G., and Huh Y. S. 2012. High performance of a solid-state flexible asymmetric supercapacitor based on graphene films. *Nanoscale* 4:4983.
6. Xiao X. Y., Beechem T. E., Brumbach M. T., Lambert T. N., Davis D. J., Michael J. R., Washburn C. M., Wang J., Brozik S. M., Wheeler D. R., Burckel D. B., and Polsky R. 2012. Lithographically defined three-dimensional graphene structures. *ACS Nano* 6:3573–3579.
7. Deng Y. F., Wan L. N., Xie Y., Qin X., and Chen G. 2014. Recent advances in Mn-based oxides as anode materials for lithium ion batteries. *RSC Advances* 4:23914–23935.
8. Bai H., Li C., Wang X. L., and Shi G. 2010. A pH-sensitive graphene oxide composite hydrogel. *Chemical Communications* 46:2376–2378.
9. Xu Y., Sheng K., Li C., and Shi G. 2010. Self-assembled graphene hydrogel via a one-step hydrothermal process. *ACS Nano* 4:4324–4330.
10. Stoller M. D., Park S., Zhu Y., An J., and Ruoff R. S. 2008. Graphene-based ultracapacitors. *Nano Letters* 8:3498–3502.
11. Zhu Y., Murali S., Stoller D. M., Ganesh K. J., Cai W., Ferreira P. J., Pirkle A., Wallace R. M., Cychosz K. A., Thommes M., Su D., Stach E. A., and Ruoff R. S. 2011. Carbon-based supercapacitors produced by activation of graphene. *Science* 332:1537–1541.
12. Kim H. K., Bak S. M., Lee S. W., Kim M. S., Park B., Lee S. C., Choi Y. J., Jun S. C., Han J. T., Nam K. W., Chung K. Y., Wang J., Zhou J., Yang X. Q., Roh K. C., and Kim K. B. 2016. Scalable fabrication of micron-scale graphene nanomeshes for high-performance supercapacitor applications. *Energy & Environmental Science* 9:1270–1281.
13. Shao Q., Tang J., Lin Y., Zhang F., Yuan J., Zhang H., Shinyaa N., and Qin L. C. 2013. Synthesis and characterization of graphene hollow spheres for application in supercapacitors. *Journal of Materials Chemistry A* 1:15423.
14. Wang G., Shen X., Yao J., and Park J. 2009. Graphene nanosheets for enhanced lithium storage in lithium ion batteries. *Carbon* 47:2049–2053.
15. Yoo E. J., Kim J., Hosono J., Zhou H. S., Kudo T., and Honma I. 2008. Large reversible li storage of graphene nanosheet families for use in rechargeable lithium ion batteries. *Nano Letter* 8:2277–2282.
16. Fang Y., Lv Y., Che R., Wu H., Zhang X., Gu D., Zheng G., and Zhao D. 2013. Two-Dimensional mesoporous carbon nanosheets and their derived graphene nanosheets: Synthesis and efficient lithium ion storage. *Journal of the American Chemical Society* 135:1524–1530.
17. Wen Y., He K., Zhu Y., Han F., Xu Y., Matsuda I., Ishii Y., Cumings J., and Wang C. 2013. Expanded graphite as superior anode for sodium-ion batteries. *Nature Communications* 5:4033.

18. Lee B., Kim M., Kim S., Nanda J., Kwon S. J., Jang H. D., Mitlin D., and Lee S. W. 2020. High capacity adsorption-dominated potassium and sodium ion storage in activated crumpled graphene. *Advanced Energy Materials* 10:1903280.
19. Wu X., Chen Y., Xing Z., Lam C. W. K., Pang S. S., Zhang W., and Ju Z. 2019. Advanced carbon-based anodes for potassium-ion batteries. *Advanced Energy Materials* 9:201900343.
20. Share K., Cohn A. P., Carter R., Rogers B., and Pint C. L. 2016. Role of nitrogen-doped graphene for improved high-capacity potassium ion battery anodes. *ACS Nano* 10:9738–9744.
21. Ju Z., Li P., Ma G., Xing Z., Zhuang Q., and Qian Y. 2018. Few layer nitrogen-doped graphene with highly reversible potassium storage. *Energy Storage Materals* 11:38–46.
22. Ju Z., Zhang S., Xing Z., Zhuang Q., Qiang Y., and Qian Y. 2016. Direct synthesis of few-layer F-doped graphene foam and its lithium/potassium storage properties. *ACS Applied Materials & Interfaces*, 8:20682–20690.
23. Cai X., Luo Y., Liu B., and Cheng H. M. 2018. Preparation of 2D material dispersions and their applications. *Chemical Society Reviews* 47:6224–6266.
24. Thripuranthaka M., and Late D. J. 2014. Temperature dependent phonon shifts in single-layer WS2. *ACS Applied Materials Interfaces* 6:1158–1163.
25. Coleman J. N., Lotya M., O'Neill A., Bergin S. D., King P. J., Khan U., Young K., Gaucher A., De S., Smith R. J., Shvets I. V., Arora S. K., Stanton G., Kim H. Y., Lee K., Kim G. T., Duesberg G. S., Hallam T., Boland J. J., Wang J. J., Donegan J. F., Grunlan J. C., Moriarty G., Shmeliov A., Nicholls R. J., Perkins J. M., Grieveson E. M., Theuwissen K., McComb D. W., Nellist P. D., and Nicolosi V. 2011. Two-dimensional nanosheets produced by liquid exfoliation of layered materials. *Science* 331:568–571.
26. Yu X. Y., Guijarro N., Johnson M., and Sivula K. 2018. Defect mitigation of solution-processed 2D WSe_2 nanoflakes for solar-to-hydrogen conversion. *Nano Letters* 18:215–222.
27. Fan X. B., Xu P. T., Zhou D. K., Sun Y. F., Li Y. C., Nguyen M.A.T., Terrones M., and Mallouk T. E. 2015. Fast and efficient preparation of exfoliated 2H MoS_2 nanosheets by sonication-assisted lithium intercalation and infrared laser-induced 1T to 2H phase reversion. *Nano Letters* 15:5956–5960.
28. Lee Y. H., Zhang X. Q., Zhang W. J., Chang M. T., Lin C. T., Chang K. D., Yu Y. C., Wang J. T. W., Chang C. S., Li L. J., and Lin T. W. 2012. Synthesis of large-area MoS_2 atomic layers with chemical vapor deposition. *Advanced Materials* 24:2320–2325.
29. Guo H. B., Ning J., Wang B. Y., Feng X., Xia M. Y., Wang D., Jia, Y. Q., Zhang J. C., and Hao Y. 2020. Sodium ion-intercalated nanoflower 1T-2H $MoSe_2$-graphene nanocomposites as electrodes for all-solid-state supercapacitors. *Journal of Alloys and Compounds* 853:157116.
30. Kumuthini R., Ramachandran R., Therese H. A., and Wang F. 2017. Electrochemical properties of electrospun MoS_2@C nanofiber as electrode material for high-performance supercapacitor application. *Journal of Alloys and Compounds* 705:624–630.
31. Li N., Chai Y. M., Dong B., Liu B., Guo H. L., and Liu C. G. 2012. Preparation of porous MoS2 via a sol-gel route using (NH4)(2)Mo3S13 as precursor. *Materials Letters* 88:112–115.
32. Habib M., Khalil A., Muhammad Z., Khan R., Wang C., Rehman Z., Masood H. T., Xu W., Liu H., Gan W., Wu C. Q., Chen H., and Song L. 2017. WX2(X=S, Se) single crystals: A highly stable material for supercapacitor applications. *Electrochimica Acta* 258:71–79.
33. Xia M. Y., Ning J., Wang D., Feng X., Wang B. Y., Guo H. B., Zhang J. C., and Hao Y. 2021. Ammonia-assisted synthesis of gypsophila-like 1T-WSe_2/graphene with enhanced potassium storage for all-solid-state supercapacitor. *Chemical Engineering Journal* 405:126611.

34. Tu C. C., Lin L. Y., Xiao B. C., and Chen Y. S. 2016. Highly efficient supercapacitor electrode with two-dimensional tungsten disulfide and reduced graphene oxide hybrid nanosheets. *Journal of Power Sources* 320:78.
35. Guo G. H., Hong J. H., Cong C. J., and Zhou X. W. 2005. Molybdenum disulfide synthesized by hydrothermal method as anode for lithium rechargeable batteries. *Journal of Materials Science* 40:2557.
36. Du G. D., Guo Z. P., Wang S. Q., Zeng R., Chen Z. X., and Liu H. K. 2010. Superior stability and high capacity of restacked molybdenum disulfide as anode material for lithium ion batteries. *Chemical Communications* 46:1106–1108.
37. Bai J., Zhao B. C., Wang X., Ma H. Y., Li K. Z., Fang Z. T., Li H., Dai J. M., Zhu X. B., and Sun Y. 2020. Yarn ball-like MoS_2 nanospheres coated by nitrogen-doped carbon for enhanced lithium and sodium storage performance. *Journal of Power Sources* 465:228282.
38. Ye W., Wu F. F., Shi N. X., Zhou H., Chi Q. Q., Chen W. H., Du S. Y., Gao P., Li H. B., and Xiong S. 2020. Metal-semiconductor phase twinned hierarchical MoS_2 nanowires with expanded interlayers for sodium-ion batteries with ultralong cycle life. *Small* 16:1906607.
39. Xia H., Xu Q., and Zhang J. 2018. Recent progress on two-dimensional nanoflake ensembles for energy storage applications. *Nano-Micro Letters* 10:66.
40. Xuan J., Wang Z., Chen Y., Liang D., Chen L., Yang X., Liu Z., Ma R., Sasaki T., and Geng F. 2016. Organic-base-driven intercalation and delamination for the production of functionalized titanium carbide nanosheets with superior photothermal therapeutic performance. *Angewandte Chemie International Edition* 55:14569–14574.
41. Li G., Tan L., Zhang Y., Wu B., and Li L. 2017. Highly efficiently delaminated single-layered MXene nanosheets with large lateral size. *Langmuir* 33:9000–9006.
42. Bua F., Zagho M. M., Ibrahimb Y., Ma B., Elzatahry A., and Zhao D. 2019. Porous MXenes: Synthesis, structures, and applications. *Nano Today* 30:100803.
43. Zang X., Wang J., Qin Y., Wang T., He C., Shao Q., Zhu H., and Cao N. 2020. Enhancing capacitance performance of $Ti_3C_2T_x$ MXene as electrode materials of supercapacitor: From Controlled preparation to composite structure construction. *Nano-Micro Letters* 12:77.
44. Zhang P., Zhu Q., Soomro R. A., He S., Sun N., Qiao N., and Xu B. 2020.In situ ice template approach to fabricate 3D flexible MXene film-based electrode for high performance supercapacitors. *Advanced Functional Materials* 30:20200922.
45. Naguib M., Come J., Dyatkin B., Presser V., Taberna P. L., Simon P., Barsoum M. W., and Gogotsi Y. 2012. MXene: A promising transition metal carbide anode for lithium-ion batteries. *Electrochemistry Communications* 16:61–64.
46. Ren C. E., Zhao M. Q., Makaryan T., Halim J., Boota M., Kota S., Anasori B., Barsoum M. W., and Gogotsi Y. 2016. Porous two-dimensional transition metal carbide (MXene) flakes for high-performance Li-ion storage. *ChemElectroChem* 3:689–693.
47. Zhang P., Soomro R. A., Guana Z., Sun N., and Xu B. 2020. 3D carbon-coated MXene architectures with high and ultrafast lithium/sodium-ion storage. *Energy Storage Materials* 29:163–171.
48. Zhao R. Z., Di H. X., Hui X. B., Zhao D. Y., Wang R. T., Wang C. X., and Yin L. W. 2020. Self-assembled Ti_3C_2 MXene and N-rich porous carbon hybrids as superior anodes for high-performance potassium-ion batteries. *Energy Storage Materials* 13:246–257.

10 Effect of Exfoliation on Structural and Electrochemical Properties

Gibin George
SCMS School of Engineering and Technology

Deepthi Panoth
Kannur University

Brijesh K
National Institute of Technology Karnataka (NITK) Surathkal

Anjali Paravannoor
Kannur University

Nagaraja Hosakoppa
National Institute of Technology Karnataka (NITK) Surathkal

Yu-Hsu Chang
National Taipei University of Technology

Sreejesh Moolayadukkam
Centre for Nano and Soft Matter Sciences (CeNS)

CONTENTS

DOI: 10.1201/9781003178453-10

10.1 INTRODUCTION

The term exfoliation represents a process during which the layered bulk materials are expanded through a chemical or physical method to overcome the weak inter-layer forces that hold the layers together. Generally, the stacked layered materials seized together by van der Waals forces can be easily intercalated or exfoliated by solution methods or simple physical means such as shear or ultrasonic vibrations to form 2D nanosheets. The exfoliated 2D nanosheets are often composed of single or few layers of atoms, and most importantly several of their properties are largely deviated from the bulk. Such materials find applications in electronics, photonics, catalysis, supercapacitors, fuel cells, batteries, etc. [1]. The success of graphene triggered the development of other 2D structured nanomaterials, especially by the exfoliation of layered bulk inorganic materials. Unlike bulk materials, 2D nanosheet counterparts exhibit unique electron and phonon transport characteristics, which leads to several fascinating properties such as thermal conductivity, ion transport, and charge carrier concentration, besides the structural and mechanical properties.

Many of the 2D nanosheets are non-toxic and can be handled easily, and they can be cast to any substrate as a thin film for device fabrication [2]. Over the years, exfoliated 2D nanolayers have become an essential part of electrochemistry, mainly in sensing, energy, and environmental applications. 2D carbon allotropes such as graphene and 2D porous carbon are not electrochemically active by themselves; therefore, they are often doped/modified by heteroatoms such as B, P, and N or transition metals. The high charge conductivity of the 2D carbon materials is highly favorable for several electrochemical applications such as batteries, supercapacitors, sensors, and catalysis. The stability of several inorganic 2D nanosheets in acidic and basic media makes them attractive for the aforesaid applications and they are considered as the immediate replacement for expensive noble metal electrocatalysts [3].

MXenes are 2D nanolayers of metal carbides, carbo-nitrides, and nitrides, an important class of electroactive 2D nanomaterials that are developed lately. $Ti_3C_2T_x$ is the first MXene discovered in 2011. So far about 50 different types of MXenes with wide chemical and structural variations are synthesized by exfoliating MAX phases by selective etching and mechanical shearing. MAX phases represent a family of ternary carbides and nitrides. MXenes are unstable in oxygen-containing environments. The hydrophilic nature and high surface charge of MXene nanosheets make them stable in polar solutions for device printing. The ability of MXenes to intercalate various cations including multivalent ions and polar organic molecules between its 2D layers makes them apt for non-lithium-ion batteries and supercapacitors [4]. Alike graphene, MXene exhibits excellent electronic conductivity and can be functionalized, hybridized, and doped for tuning the properties to meet the requirements of a specific application.

Many non-noble metal electrocatalysts are inactive and unstable in acidic mediums. The reaction in an acidic medium is highly efficient at a high current density. Transition metal dichalcogenides (TMDs) are highly active electrocatalysts for sensing, batteries, supercapacitor, water splitting, etc., especially in acidic and harsh environments. TMDs have the general formula MX_2, where M is the transition metal and X is the chalcogen (X = S, Se, and Te), having a similar layered structure to those of graphene. Alike any other 2D layered nanosheets, TMDs can be doped, functionalized, and hybridized for improving various operating parameters such as selectivity, sensitivity, and affordability in sensing and efficiency, stability, and life span in catalysis. Additionally, TMDs have good electronic and mechanical properties favorable for electrode materials [5].

2D nanosheet of layered hydroxides (LDHs) and oxides are also an important class of electrochemical materials, starting from sensing to fuel cells. The presence of oxyl and hydroxyl groups allows the efficient transport of ions when they are used as electrodes in energy storage. The possibility of intercalation of ions other than Li^+ makes them a promising candidate for non-lithium-ion batteries. The electronic conductivity of LDHs and oxides are poor, therefore these materials are often hybridized with carbon-containing conductive materials as an effective strategy to increase the intrinsic catalytic activity. In this chapter, the electrochemical applications of the exfoliated 2D nanosheets in batteries, supercapacitors, biological sensing, and water splitting are discussed concisely. The underlying mechanism of electrochemical activity of different classes of 2D layered nanosheets is different. Such unique characteristics of different classes of 2D nanosheets favorable for the respective applications are also explored in this chapter.

10.2 ELECTROCHEMICAL SENSORS

A large number of sensors are used in our daily life to monitor and modify ourselves and our surroundings in a positive way. Electrochemical sensors have the largest share among all the chemical sensors, which use an electrochemical reaction (parameters such as a change in current and impedance) of the analyte to quantify the concentrations. Analytes electrochemically interact with the active material to produce signals and the sites on which such interactions happen are known as electrochemically active sites. Usually, the concentration of electrochemically active surface area increases with the surface area of the active material. Interestingly, exfoliation of 2D materials increases the surface area and exposes active sites, which may not be active otherwise. Often, exfoliated materials take part in the electrochemical reaction or act as a host to molecules such as enzymes that catalyze the reaction. Exfoliation, being a top-down approach results in defects that can also have a positive influence on the electrochemical reactions because of their very high activity. Apart from this, the extend of exfoliation, lateral size, etc. is also critical in deciding exfoliated material's electrochemical activity [6].

Graphene, which is a carbon allotrope, is the first known material to be exfoliated into atomically thin layers from its bulk counterpart graphite. Graphite can be easily exfoliated by mechanical cleaving. This can be used as an advantage in sensing where the fouling of the electrode material is a serious concern. The detection of material like bisphenol-A involves the polymerization of the analyte molecules and results in the deposition of the material on the surface of the electrode, which results in the electrode fouling. Exfoliated graphite helps in tackling this issue wherein a mild polishing results in the removal of the polymerized products from the surface as described by Ndlovu et al. [7]. Figure 10.1 schematically represents how exfoliation acts as a tool to challenge the fouling issues in electrochemical sensing.

Graphite oxide samples are usually exfoliated using thermal shock to achieve high quality and are electrochemically active for the detection of hydrogen peroxide and this is extensively reported by many researchers. Moolayadukkam et al. in 2020, in detail, explained the effect of solar exfoliation on the H_2O_2 sensing performance. Exfoliated graphene sheet has more defect concentration, which acts as the electrocatalytically active sites by adsorbing the analyte molecules. These adsorbed analyte molecules are electrocatalytically oxidized and corresponding signals can be recorded with a technique such as chronoamperometry. Figure 10.2 schematically shows graphene layers with defects/pores and their activity in adsorbing H_2O_2 molecules (analyte) [8].

Non-carbonaceous materials are electrocatalytically more active and their exfoliation has revolutionized electrochemical sensing research and developments. Layered 2D TMDs offer a wide variety of materials that can be exfoliated and having electrical properties varying from metallic to semiconducting nature. The peculiar arrangement of each atomic layer in TMDs offers a variety of active sites for the analyte adsorption in each layer after the exfoliation process. This property is widely used in the efficient detection of biomolecules. MoS_2 is one of the most widely used TMDs for sensing and other applications. Ashwathi et al. studied the relation between the analyte affinity and the active material by taking MoS_2 and Hg (II) ions as an example. In this particular example, Hg (II) ions have a high affinity toward S-containing groups. Exfoliation leaves S on both the surfaces of nanosheets exposed while the Mo layer at the center acting as the backbone. This arrangement of atoms improves the sensitivity by many folds clearly showing exfoliation of 2D TMDs could be used as an effective method for fine-tuning sensing capabilities [9].

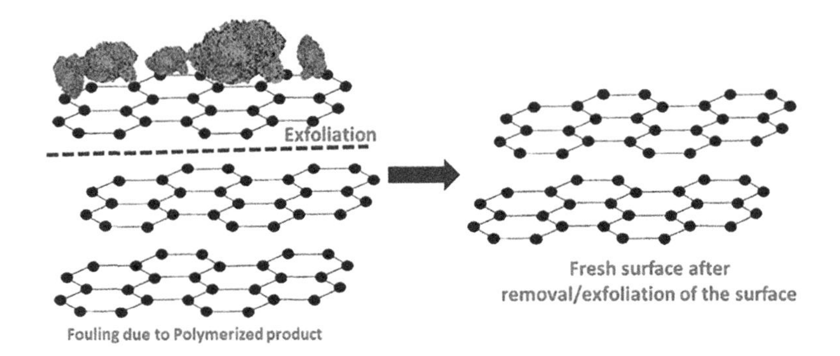

FIGURE 10.1 Shows how exfoliation of the material helps tackle the fouling issues in sensing.

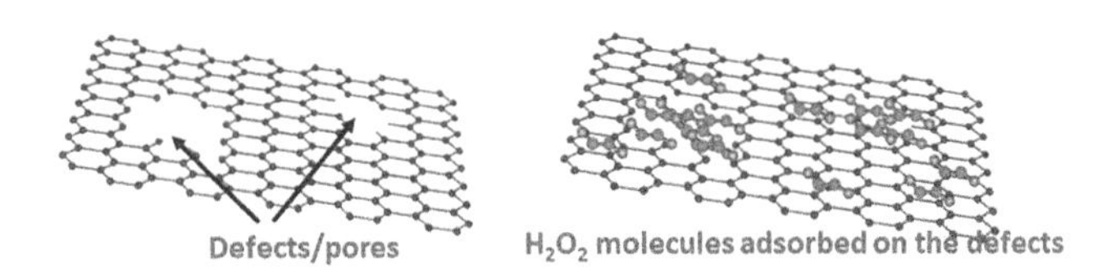

FIGURE 10.2 Schematic representation showing the importance of defects in adsorbing the analyte molecules on the graphene surface.

LDHs are another class of materials that can be exfoliated to form molecular layers with metal as the center layer. Compared to TMDs, LDHs have the advantage that there may be more than one metal in the metallic center layer, and varying the ratios of metals at the center and the metals themselves can tweak the sensing properties [10]. Sahoo et al. studied the sensing properties of ultrasonically exfoliated Ni_2Co-LDH with dopamine, an important biomolecule. The electron transfer rates are reported to be improved on moving from bulk to the monolayers of the LDH. Going from bulk to monolayers could help decrease the electron scattering at the active material, which can have a positive impact on the sensing properties [11]. Strong dependence of the exfoliation on the sensing properties is also reported by Chia et al., Authors explained the effect of exfoliation using enzymatic glucose sensing as a tool. Exfoliated 2D sheets show better sensing properties because of the high surface area and thin nature. Thinner sheets result in a decreased distance between adsorbed enzyme and the electrode, which facilitates efficient electron transfer. Polymeric 2D material, graphitic carbon nitride also shows similar sensing properties upon exfoliation. Kesavan et al. exfoliated graphitic carbon nitride using ultrasonication technique and demonstrated the flutamide (FLT) sensing properties. With the help of impedance spectroscopic studies, they have shown that active sites and conductivity are increased as a result of exfoliation. Along with this, the affinity of FLT and nitrogen on the graphitic carbon nitride played an important role in improving the sensing properties [12].

Irrespective of the layered material, exfoliation is observed to have a significant influence on the sensing properties. Exfoliation results in exposing active sites and the reduction in thickness resulting in better absorption of the analyte molecules and better electron transfer characteristics. Apart from this, the method of exfoliation induces different types of defects on the 2D crystal, the electron density on these defects such as edges and pores have an impact on the electrocatalysis of the analyte molecule. Carefully altering the method of exfoliation, sensing capabilities of the materials could be extended.

10.3 WATER SPLITTING AND FUEL CELLS

Water is an abundant source of energy and splitting water in a most economic route is a serious research concern in recent years for the production of hydrogen and oxygen. Hydrogen is considered the most advantageous renewable source of energy and the availability of oxygen is critical for the treatment of patients affected with COVID 19. Oxygen is also important for the complete combustion of any fuel, including hydrogen. The commercial electrocatalysts containing noble metals are currently used in fuel cells as hydrogen evolution reaction (HER), oxygen evolution reaction (OER), and oxygen reduction reaction (ORR) catalysts. The involvement of noble metals in crucial energy-related applications such as a fuel cell increase the installation and operation cost tremendously. Recently, several non-noble metal electrocatalysts are introduced as a replacement for noble metals and their derivatives. Several 2D layered nanosheets prepared by intercalation/exfoliation are subjected to HER and OER/ORR. Many are identified as potential replacements for noble metals in their respective applications. A list of widely studied exfoliated 2D materials as electrocatalysts are discussed in this session.

10.3.1 Hydrogen Evolution Reaction (HER)

The evolution of hydrogen by electrochemical water splitting can be a feasible way of storing hydrogen for energy-related applications, especially for fuel cells. Over the years, noble metals are broadly used as an efficient catalyst for HERs. However, the high cost of noble metals limits their extensive use as a catalyst at a large scale. To overcome the high cost of noble metal catalysts, electroactive materials that are available in abundance are proposed as catalysts. However, the major challenges of most non-noble materials used in HER are (1) the low efficiency, well below the thermodynamic limits of the water-splitting reaction and (2) the short lifetime [13]. Materials containing transitions metals are very active for HER. Though HER can be performed either in an acid $\left(2H^{+} + 2e^{-} \leftrightarrow H_2\right)$ or basic $\left(2H_2O + 2e^{-} \leftrightarrow H_2 + 2OH^{-}\right)$ medium, a basic medium is commonly preferred due to the short-term stability of many materials in the acid medium. Similarly, due to stability issues, pure metals are avoided for HER reactions. To improve the performance of electroactive materials par to the noble metals several strategies are adopted. The suitability of a nanostructured material as an electrocatalyst depends on the surface area, presence of defects such as oxygen vacancies, availability of active sites on the surface, and dopants. The surface area plays an important role in HER since HER is a surface-active reaction.

Materials with the layered structure are identified as a suitable candidate for the HER since the layered materials are often characterized by the presence of multivalent transition metals in their crystal structure and the synergic interaction of these elements can augment the catalysis by offering many active sites for catalyzing the reaction. Interestingly, the conductive flexible 2D nanosheets enable the easy access of the electron from the catalyst substrate to the surface through intimate contact. As a result, the interfacial electron transfer resistance can be reduced and electrons can circulate through the external circuit efficiently [14]. The most active sites of exfoliated 2D materials for HER are located along the edges of the layers, but its performance is currently limited by the density and reactivity of active sites. The unprecedented HER activity of the layered materials is observed when they are exfoliated by intercalating a charged ion such as Li and Na, and thereby surface area is increased enormously in addition to the increased electrical conductivity. The overall HER activity is determined by how well hydrogen atoms can be adsorbed on the catalyst surface [14].

Among the layered materials, the introduction of exfoliated TMDs is a breakthrough in the history of non-noble metal catalysts for HER. Chemically exfoliated layers of dichalcogenides such as MoS_2, WS_2, CoS_2, VS_2, and NiS_2 are extensively studied as a promising electroactive HER catalyst. The above materials exhibit a low overpotential in the range of 100–250 mV vs reversible hydrogen electrode in an acidic medium. Overpotential is the measure of the efficiency of a material for a water-splitting reaction and it represents the loss of the applied voltage. The overpotential of platinum/carbon commercial electrodes are ~30–50 mV. The layered materials without exfoliation or intercalation are often inactive as in the case of MoS_2. The ultra-thinning and 2D nanosheet formation create an abundance of HER active sites at the edges [15]. Moreover, the planar mobility of electrons along the 2D layer guarantees rapid electron transfer from the substrate to active sites. The exfoliated transition metal selenides and tellurides are also reported as electroactive materials for HERs.

For instance, exfoliated WSe_2, $MoSe_2$, $MoS_{2(1-x)}Se_{2x}$, $MoTe_2$, WTe_2, $MoSe_2/WSe_2$, VSe_2, etc. 2D nanosheets exhibited a superior performance than the bulk counterparts. Doping noble metals such as Pt and Ru to the 2D chalcogenides can increase the catalytic activity tremendously. $MoSe_2$ is an n-type semiconductor, converting $MoSe_2$ to a p-type semiconductor by Nb or Ta doping reduces its activity toward HER [16].

MXene (layered metal nitrides and carbides) is a new family of exfoliated materials and potential electrocatalyst for HER, MXene adopts a general formula of $M_{n+1}X_nT_x$ ($n = 1$–3), where M is a transition metal such as Mo, V, or Ti, X is C and/or N, and Tx represents surface functional groups such as H or OH. Despite the high surface area, MXenes are characterized by excellent hydrophilicity and conductivity. Interestingly, the active HER sites for MXene are located on the O^* basal plane, which makes them ideal for HER [17]. The HER activity of MXene is enhanced by modifying the transition metal, during which the Gibbs free energy for hydrogen adsorption is improved, subsequently, one can obtain a decreased barrier energy for hydrogen production [18]. MXene combined with nanostructured platinum is widely used as the electrocatalyst. $Mo_2TiC_2T_x$, $Ti_3C_2T_x$, $V_4C_3T_x$, $Mo_2TiC_2T_x$, etc. are some representative MXene electrocatalysts for HER.

Layered carbon allotropes such as graphene and its oxide exhibit poor adsorption toward hydrogen; therefore, they are not efficient catalysts for HER. However, these materials are extensively used as supporting materials for electroactive elements and nanostructures. The graphene decorated with electroactive nanostructures of Pt, Ni-Mo-N, Ni, CoP, MoS_2, $ReSe_2$, WS_2, etc. is identified as excellent catalysts for HER in a basic medium. In addition to the large surface area, the high conductivity of the graphene/graphene oxide significantly reduces the interfacial electron transfer resistance between the catalyst support and the active sites, which ultimately improves the efficiency toward HER.

10.3.2 Oxygen Evolution Reaction (OER) and Oxygen Reduction Reaction (ORR)

The electrochemical generation of oxygen through water splitting is critical in metal-air batteries and fuel cells. The electrochemical OER ($2H_2O \rightarrow O_2 + 4H^+ + 4e^-$) and ORR ($O_2 + 4H^+ + 4e^- \rightarrow 2H_2O$) are four-electron transfer reactions. Due to the complicated multi-electron transfer steps, the ORR/OER suffers from sluggish kinetics. Similar to HER, noble metals and their derivatives exhibit low overpotential for both ORR (e.g., Pt) and OER (e.g., IrO_2 and RuO_2) applications. 2D nanolayers are unique due to a large number of surface atoms as compared to the internal atoms, which makes them highly electroactive for a variety of applications. Exfoliated 2D materials like graphene and graphene oxide, inorganic monolayer materials such as metal oxides, TMDs, LDHs, MXenes, diatomic hexagonal boron nitride, and black phosphorous (BP or phosphorene) are studied as potential candidates for OER and ORR applications. In addition to the planar strength, exfoliated 2D materials are flexible with an atomic or few-layer thickness. Interestingly, the most single or few layers of graphene, carbon nanosheets (CNS), TMDs, LDHs, and MXenes are exfoliated from their bulk, and these are the most extensively studied 2D materials for OER application.

To overcome the scarcity of OER electrocatalysts for acid medium the transition metal dichalcogenides (TMD) are proposed. The exfoliated 2D nanosheets of MoS_2, TaS_2, WS_2, $MoSe_2$, etc. either in 1T and 2H polymorphic forms are the common electroactive catalysts for OER. The performance of the above materials for OER is par to stable IrO_2. Liquid phase and ion intercalation are the most common routes for the exfoliation of TMDs nanosheets from the bulk by overcoming the weak van der Waals interaction among layers. The step-by-step exfoliation of bulk TMDs using Isopropyl alcohol and the preparation of electrodes using exfoliated nanolayers are shown in Figure 10.3. Alike HER, the dominant active sites of TMDs for OER are on the edges rather than the surface [19]. The dichalcogenides of noble metals such as Rhenium-, Ruthenium-, and Iridium- exhibit exceptional activity toward OER and ORR.

Unlike in the HER, MXenes themselves are not directly active for ORR or OER electrocatalysis; however, they serve as excellent supports for various electroactive materials. MXenes are better catalyst support for Pt nanoparticles or Pt/Pd atoms than carbon as in the commercial Pt/C electrode for OER due to the strong interaction between Pt and the respective MXene layers. Likewise, other electroactive materials such as metal-organic frameworks, carbon nitride, LDHs, oxides, borate, sulfides, and metals bound to the surface of MXenes also exhibited superior OER activity par to the commercial noble metal catalysts. Hybrid TMD–MXene-like materials are recently introduced as OER catalysts. The heterostructure of the above hybrids allows the synergistic interactions between TMDs and MXenes and one can achieve a significant improvement in the OER activity.

Carbon allotropes themselves are not active for OER or ORR, though when doped with heteroatoms (B, S, N, P, F, and O) or transition metals (Ni, Co, Fe, etc.), they become excellent ORR and OER catalysts. The conductivity of graphene, 2D porous

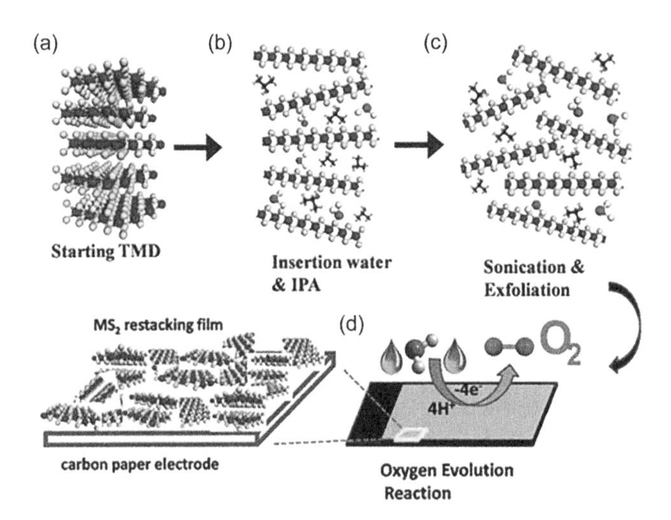

FIGURE 10.3 Schematic representation of step-by-step electrode fabrication process using exfoliated TMDs. (a) Starting TMD, (b) Insertion water and IPA, (c) Sonication and exfoliations, and (d) application for OER (Wu et al. 2016). Reprinted with permission from © 2016 John Wiley and Sons.

carbon, and graphitic carbon nitride (g-C_3N_4) layers can significantly reduce the interfacial resistance between the electroactive materials or the active sites and the current-carrying substrate. Additionally, as discussed in the case MXenes, the exfoliated 2D carbon layers are commonly used as a support for nanosized or atomic catalytic materials. MoS_2, Fe_3O_4, FeP, Ni_2P, CoP_2, CoO_x, NiO, etc. are some representative nanoparticles grown on 2D carbon materials for OER. Nevertheless, the long-term stability of carbon-based electrocatalysts is inferior to MXenes. Both MXenes and 2D carbon allotropes are mostly sought for OER and ORR in a basic medium.

Among the OER catalysts, layered double hydroxides (LDHs) are extensively studied as a potential replacement for noble metal catalysts due to their compositional and structural flexibility in addition to the simple preparation routes. Often LDHs adopt a formula either $M_x^{2+}M_{1-x}^{3+}(OH)_2\left(A^{n-}\right)_{\frac{x}{n}}\cdot yH_2O$ or $M_x^{1+}M_{1-x}^{4+}(OH)_2\left(A^{n-}\right)_{\frac{x}{n}}\cdot yH_2O$; where M is a metal and A is the intercalating anion. In LDHs, every single layer is composed of edge-sharing octahedral MO_6 moieties (M stands for metal) as shown in Figure 10.4. The color code used in the figure are: purple for metals, red for oxygen, and grey for inter-layer anions and water molecules. If d_1 is the inter-layer distance before intercalation, the inter-layer distance increases after intercalation to d_2 and $d_2 > d_1$. One can observe the change in interlayer spacing under an electron microscope and the subsequent change in the crystal structure from X-ray diffraction. The transition metal oxides (TMOs) with *d*-orbitals can effectively bind oxygen species on its surface, which is an essential requirement for OER/ORR catalyst. The substitution of elements in M^{2+} and M^{3+} sites can fine-tune the electronic as well as the catalytic properties of LDHs. Exfoliated LDHs formed by a combination of the transition metals, Ni-Co, Ni-Fe, Co-Fe, Co-Co, Ni-Mn, Co-Mn, etc. are some representative low overpotential electrocatalysts for OER in a basic medium among the non-noble metal catalysts.

Exfoliated layered perovskite with the general formula ABO_3 (A and B can be occupied by a large number of elements in the periodic table) and delafossite with the general formula AMO_2 is also studied as potential OER catalysts [20]. The above oxides with transition metals such as Co, Ni, and Fe at one of the sites are excellent OER catalysts. Such oxides are stable than the carbon-containing catalysts under oxidative environments and offer a competitive catalytic property comparable to noble metals.

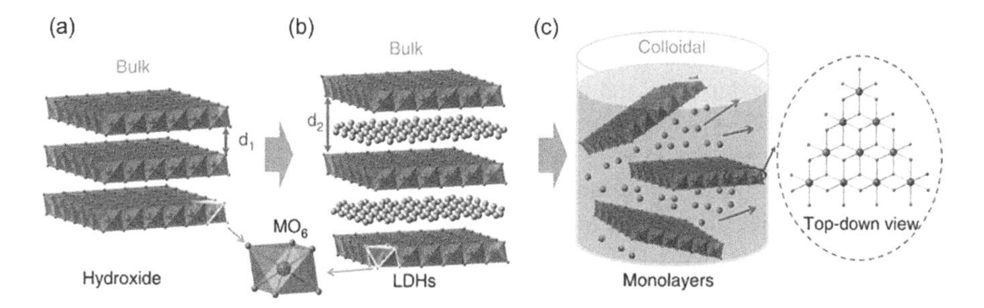

FIGURE 10.4 (a) Structure of layered hydroxides, (b) LDHs intercalated with a layer of anions and water molecules, and (c) exfoliated LDH monolayers in a colloidal solution (Song and Hu 2014). Reprinted with permission from © 2014 Springer Nature.

10.4 SUPERCAPACITORS

Supercapacitors bridge the gap between rechargeable batteries and conventional capacitors. But one of the major restrictions of supercapacitors is their lower energy density than the rechargeable batteries. There are several reported attempts to improve and enhance the energy density of supercapacitors. Supercapacitors mainly consist of electrodes, electrolytes, current collectors, sealants, and separators. The selection and design of the electrode materials have a major role in the overall performance of a supercapacitor as it determines the ionic conductivity, surface area, and chemical and thermal stability [21].

Supercapacitors are categorized mainly into two, based on their charge storage mechanism, one is electric double-layer (EDLC) or faradaic capacitor where energy is stored via non-Faradaic electrostatic interaction and the other one is pseudocapacitor where the energy storage is accomplished through Faradaic redox charge transfer reactions [22]. When 2D layered nanomaterials are used as electrodes in both Faradaic and non-Faradaic storage systems, the charge is mainly stored at the basal plane of the layered nanosheet, i.e., with the larger planar area. Additionally, the presence of active edge sites and the weak van der Waals gap between the nanosheet layers of 2D nanomaterials offer enhanced and suitable electrochemical performance in supercapacitors. Here in this section, the most commonly used exfoliated 2D nanosheets of both carbon-based and non-carbon-based are discussed in detail.

Graphene is one of the most common 2D layered carbon sheets with a hexagonal lattice structure, widely investigated for supercapacitor applications. The kinetics of an electrode material mainly depends on the transportation and diffusion of electrolyte ions. Due to the lack of enough edge planes and surface charges, monolayer graphene is considered one of the most chemically and electrochemically inert materials [23]. During the charge storage process, graphene acts as a superior active material as the electrolyte ions like Na^+, K^+, etc., can be stored electrostatically on the electrode. But the agglomeration of graphene nanosheets due to the strong van der Waals interaction limits the full utilization of graphene surface for ion adsorption. The agglomerated structure extremely limits the direct access to the charge-storage surfaces, which finally leads to the increase in ionic resistance at the electrode [24]. Higher agglomeration, hydrophobicity, and the random orientation of graphene nanosheets restrict the availability of ions on the active surface. Thus, the morphology of the electrode materials plays a vital role in the charge storage mechanism of supercapacitors.

Stoller et al. developed chemically modified graphene (CMG) electrodes with good electrical conductivity and a specific surface area of $705\,m^2g^{-1}$, by chemical functionalization of monolayer graphene. The CMG electrode materials exhibited a specific capacitance of 135 F g^{-1} in aqueous electrolyte (5.5 M KOH) and 99 F g^{-1} in the organic electrolyte [25]. Most reported graphene-derived electrode materials exhibited lower specific surface area than their theoretical value ($2{,}630\,m^2g^{-1}$). But the Ruoff group reported KOH-activated thermally exfoliated graphene oxide and microwave exfoliated graphene oxide (MEGO) electrode material, which exhibited an ultrahigh specific surface area value of $3{,}100\,m^2g^{-1}$, a high electrical conductivity (~500 $S.m^{-1}$), high content of sp^2-bonded carbon, and low hydrogen content. The KOH-activated MEGO electrode exhibited a notable high energy density (~70 Wh kg^{-1})

and power density (~250 kW kg^{-1}) at a current density of 5.7 A g^{-1} [26]. El-Kady et al. fabricated a graphene-based supercapacitor via laser irradiation of a graphene oxide film coated on a flexible substrate mounted in a LightScribe DVD optical drive. The graphene oxide sheets stacked in the film were reduced and exfoliated simultaneously upon laser irradiation and this structure restricts the agglomeration of graphene sheets and also the open pores in them facilitate the easy accessibility of electrolyte on the electrode surface. The resultant laser-scribed graphene sheets exhibited a high specific surface area of 1,520 m^2g^{-1}, good mechanical flexibility, and high electrical conductivity (1,738 S.m^{-1}) [27]. Miller and his group fabricated supercapacitor electrodes using radio frequency plasma-enhanced chemical vapor deposition in which vertically oriented graphene nanosheets were deposited on a heated Ni-substrate. They showed a specific surface area of ~ 1,100 m^2g^{-1} and effective filtering of 120 Hz current with a resistance-capacitance time constant value less than 0.2 ms. With the exposed edge planes the vertically aligned graphene nanosheets showed enhanced charge storage as compared to the flat graphene nanosheets [28]. The exceptional properties and promising application of graphene in energy storage devices have triggered a remarkable interest in exploring other non-carbon 2D layered nanostructures with versatile properties.

Non-carbon-based 2D layered nanomaterials have been considered as a potential candidate for supercapacitor electrodes owing to their unique physical and chemical properties such as high electronic conductivity, tunable surface chemistry, more surface-active sites, dual non-faradaic and faradaic electrochemical performances, and larger mechanical strength. 2D non-CNSs include TMDs (MoS_2, WS_2, TiS_2, ZrS_2, $MoSe_2$, WSe_2, etc.), layered metal-oxides, hexagonal boron nitride (h-BN), LDHs, graphitic carbon nitride (g-C_3N_4), and MXenes (Ti_3C_2, V_2C, Ti_2AlC, TiAlC, Ti_3CN) [29]. Among TMDs, 2D MoS_2 nanosheets are a potential supercapacitor electrode material that exhibits large electrical double layer capacitance (EDLC) owing to their stacked sheet-like structure, and large pseudocapacitance due to the different Mo oxidation states (+2 to +6). Tour and his co-workers developed vertically aligned/edge-oriented MoS_2 nanosheets that offer a high capacitive property with more van der Waals gaps and rendered reactive dangling bonds sites for the electrolyte ions. Areal Capacitance of 12.5 mF cm^{-2} was obtained for sponge-like vertically aligned MoS_2 electrodes [30]. Layered 2D TMOs exhibit exceptionally high surface area and high conductivity as they are capable of holding charged ions on their surface without intermixing. Supercapacitors based on layered TMOs feature superior cyclic stability, high energy density, and high discharge currents. Commonly used 2D layered TMOs include MnO_2, NiO, Co_3O_4, and RuO_2. MnO_2 possesses low conductivity and thus they require a conductive matrix of graphene or metal foam. Peng et al. fabricated a supercapacitor electrode integrating 2D graphene and 2D MnO_2 into a planar capacitor design that was highly flexible [31].

2D LDH sheets are a class of multi-metal clay materials that consist of metal cations brucite layers octahedrally surrounded by hydroxyls forming $M^{2+}(OH)_6/M^{3+}/M^{4+}(OH)_6$ octahedra. Their high redox activities can be attributed to their unique properties like cations, easy tenability in their host layers and they are capable of exchanging anions without disturbing the structure. In NiAl-LDH, its electrochemical property is due to a mixed mechanism comprising of 'electron hopping' along with the layers of LDH and the migration of protons from the host layer to the solution [32]. MXenes have become a widely accepted supercapacitor electrode material with their impressive electrochemical properties due to their unique 2D structure and well-defined geometry. MXenes are one of the fast-growing materials among 2D materials, which include metal carbides, nitrides, and carbonitrides. One of the promising features of MXene is the exceptionally large interlayer spacing, which helps in the de/intercalation of ions like Na^+, Li^+, etc. Mainly hydrogen bonding and van der Waals bonding interactions act between the MXene layers. To produce MXene single flake suspensions, water, cations, tetrabutylammonium hydroxide (TBAOH), dimethylsulfoxide (DMSO), etc. are intercalated into the MXene interlayer spacing followed by the sonication process. In the H_2SO_4 electrolyte, $Ti_3C_2T_x$ shows a high volumetric capacitance of ~1,500 F cm^{-3} (380 F g^{-1}), and the conductive, transparent $Ti_3C_2T_x$ films are used to fabricate solid-state transparent supercapacitors [33].

10.5 LITHIUM-ION BATTERY

Lithium-ion batteries (LIBs) are the answer to many of the energy storage-related challenges. LIBs become an essential part of everyday life. LIBs work by the rocking chair mechanism wherein the lithium ions are moved between the anodes and cathodes. The electrodes play an important role in storing the lithium ions by the intercalation and deintercalation reactions. Historically, layered materials have played an important role in the development of LIBs by allowing the layered structures of the electrodes like graphite to intercalate lithium ions. Currently, LIBs use a wide variety of electrodes having mechanisms such as insertion, alloying, and conversion reactions [34]. Electrodes with higher rate capability, higher charge capacity, and (for cathodes) sufficiently high voltage can improve the energy and power densities of Li batteries and make them smaller and cheaper. The fast-paced life around the globe is forcing researchers to focus on materials that can be charged faster and hold more energy per volume and weight. Layered materials are often helpful in achieving faster lithium-ion diffusion and have a higher capacitive contribution. Owing to compelling electrochemical and mechanical properties, exfoliated 2D nanomaterials have been propelled to the forefront in investigations of electrode materials in recent years.

Exfoliated 2D nanomaterials are exceedingly desirable as anodes and cathodes. As anodes, the famous candidates are graphene and graphene-based composite materials, including carbon nanotubes/graphene, nonmetal/graphene, TMOs/graphene, sulfide/graphene, and salts/graphene. As cathodes, exfoliated 2D nanomaterials have remarkable electron transport velocity, high theoretical capacity, and excellent structural stability. The exfoliation of bulk material and Li^+ insertion was represented in Figure 10.5, which shows easier paths for lithium-ion storage.

Graphite is the most traditionally used anode in LIBs, which is a layered material. Expansion and exfoliation of the graphite are well reported by various researchers. Graphite as such shows a theoretical capacity of 372 mAh g^{-1}. Due to good electrical conductivity, high surface area, and greater mechanical flexibility, graphite exfoliation has attracted the most attention for fabricating high-performance electrode material for LIB. Lithium may bond both graphene sheet sides as well as edges and

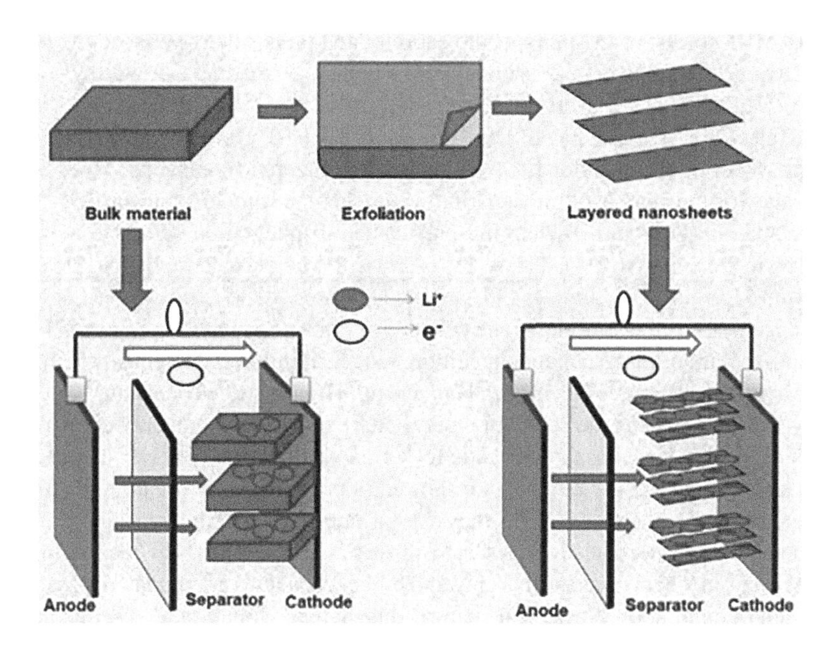

FIGURE 10.5 Sketch of the exfoliated layered material as a cathode for Li^+ insertion in LIBs.

covalent sites. Recent studies demonstrated that the small lateral sizes of narrow graphene nano-ribbons can accommodate Li^+ ions at the edges sites more efficiently than basal sites, thus leading to maximum Li-storage in the form of Li_4C_6. The probable defects formed during the exfoliation process become an advantage in such cases. Apart from this, graphene nanosheets are widely used to make composites with other electrode materials. In materials like silicon, graphene sheets are also used to give cushioning effect to accommodate the high volume change during the lithium uptake.

Exfoliated carbons are trustable electrodes for lithium battery electrodes but lack high capacity, which restricts the overall capacity of the batteries. Exfoliated 2D group V nano-crystals have a greater theoretical capacity than graphite. Exfoliation of these metallic electrodes is challenging because of the stability issues. Among them, layered 2D antimony has the potential as electrode material for LIBs, owing to their large interlayer distance in their layered structure, high capacity, long mean free path, and environmental friendliness. The theoretical capacity of antimony is moderate therefore other 2D materials are also explored as LIB electrodes. The layered transition metal oxides (LTMOs) require a special mention in LIBs. The exceptional feature of these materials is the presence of an interlayer region that serves as the host for ion intercalation. The extensive interlayer spacing and weak interlayer bonding of LTMOs permit the intercalation of an enormous variety of guest species, like cations, polymers, and anions. LTMO has excellent electronic and ionic conductivity, the attainability of interlayer sites for the intercalation of cations from the electrolyte, and the ability to undergo redox reaction property for high energy density LIB. Several mechanisms are possible when LTMO is in contact with an electrolyte like intercalation, conversion, double layer capacitance, conversion, and pseudocapacitance [35].

2D TMDs consist of greater specific capacity and larger interlayer spacing, which permit a quick Li^+ insertion/extraction process without persuading noteworthy volume changes [36]. Exfoliated layers of chalcogenides such as MoS_2, $NbSe_2$, WS_2, $MoSe_2$, $TaSe_2$, and $MoTe_2$ nanosheets are widely used for the LIB. Among them, MoS_2 is an exciting electrode material for LIBs due to its high theoretical capacity. MoS_2 nanolayers allow intercalation of Li^+ ions into the structure without noteworthy volume change and charging and discharging prevent the disintegration of active material. Based on the reaction $MoS_2 + 4Li^+ + 4e^- \leftrightarrow 2Li_2S + Mo$, the electrochemical reaction of Li with MoS_2 involves 4 moles of Li per mole of MoS_2. The main concern of MoS_2 layered nanomaterial is low electronic conductivity and poor cyclic stability [37].

Another class of materials that is gaining recent attention is MXenes, which possess 2D layered structure. The main advantage of MXene as electrode material for the energy storage device is the separation between MXene layers that can be controlled systematically. MXenes usage as anode for LIBs was first reported by Naguib et al. [38]. The MXenes prepared by Naguib showed improved surface area by ~10-fold as compared with graphene since MXenes exhibit improved specific capacity. Layered morphology of the electrodes always had a positive impact on LIBs by facilitating the rocking chair mechanism. Fast charging and higher capacities are repeatedly reported as a result of exfoliation. Structural changes during the exfoliation are usually acting in favor of the intercalation of more lithium ions to the electrodes. Therefore, exfoliated 2D materials are going to have a large impact on the future development of LIBs.

10.6 CONCLUSIONS

Exfoliated 2D nanosheets have gained considerable attention from the research community in recent years. The development of various 2D nanosheets of different origins allows the researchers to resolve numerous bottlenecks associated with many electrochemical devices, especially in sensors, fuels cells, supercapacitors, and batteries. Though exfoliation is a top-down approach, it can produce reasonably good quality nanosheets in large quantities, which is essential for device fabrication at a large scale. Interestingly the defects generated during exfoliation favor electrochemical activity than the ones prepared by chemical vapor deposition with fewer defects. The exfoliated 2D materials are expected to play an important role in the further advancements in electrochemical devices in the coming years.

REFERENCES

1. Guo B., Xiao Q., Wang S., Zhang H. (2019) 2D layered materials: Synthesis, nonlinear optical properties, and device applications. *Laser Photon Rev* 13:1800327.
2. Alzakia F.I., Tan S.C. (2021) Liquid-exfoliated 2D materials for optoelectronic applications. *Adv Sci* 8:2003864.
3. Zhou Y., Pondick J.V., Silva J.L., Woods J.M., Hynek D.J., Matthews G., Shen X., Feng Q., Liu W., Lu Z., Liang Z., Brena B., Cai Z., Wu M., Jiao L., Hu S., Wang H., Araujo C.M., Cha J.J. (2019) Unveiling the interfacial effects for enhanced hydrogen evolution reaction on MoS_2/WTe_2 hybrid structures. *Small* 15:1900078.
4. Anasori B., Lukatskaya M.R., Gogotsi Y. (2017) 2D metal carbides and nitrides (MXenes) for energy storage. *Nat Rev Mater* 22(2):1–17.

5. Kumar S., Pavelyev V., Mishra P., Tripathi N., Sharma P., Calle F. (2020) A review on 2D transition metal di-chalcogenides and metal oxide nanostructures based NO_2 gas sensors. *Mater Sci Semicond Process* 107:104865.
6. Huo C., Yan Z., Song X., Zeng H. (2015) 2D materials via liquid exfoliation: A review on fabrication and applications. *Sci Bull* 60:1994–2008.
7. Ndlovu T., Arotiba O.A., Sampath S., Krause R.W., Mamba B.B. (2012) An exfoliated graphite-based bisphenol a electrochemical sensor. *Sensors* 12:11601–11611.
8. Sreejesh M., Huang N.M., Nagaraja H.S. (2015) Solar exfoliated graphene and its application in supercapacitors and electrochemical H_2O_2 sensing. *Electrochim Acta* 160:94–99.
9. Aswathi R., Sandhya K.Y. (2018) Ultrasensitive and selective electrochemical sensing of Hg(II) ions in normal and sea water using solvent exfoliated MoS_2: Affinity matters. *J Mater Chem A* 6:14602–14613.
10. Moolayadukkam S., Thomas S., Sahoo R.C., Lee C.H., Lee S.U., Matte H.S.S.R. (2020) Role of transition metals in layered double hydroxides for differentiating the oxygen evolution and nonenzymatic glucose sensing. *ACS Appl Mater Interfaces* 12:6193–6204.
11. Sahoo R.C., Moolayadukkam S., Thomas S., Asle Zaeem M., Matte H.S.S.R. (2021) Solution processed Ni_2Co layered double hydroxides for high performance electrochemical sensors. *Appl Surf Sci* 541:148270.
12. Kesavan G., Chen S.M. (2020) Sonochemically exfoliated graphitic-carbon nitride for the electrochemical detection of flutamide in environmental samples. *Diam Relat Mater* 108:107975.
13. Strmcnik D., Lopes P.P., Genorio B., Stamenkovic V.R., Markovic N.M. (2016) Design principles for hydrogen evolution reaction catalyst materials. *Nano Energy* 29:29–36.
14. Di J., Yan C., Handoko A.D., Seh Z.W., Li H., Liu Z. (2018) Ultrathin two-dimensional materials for photo- and electrocatalytic hydrogen evolution. *Mater Today* 21:749–770.
15. Gao M.-R., Chan M.K.Y., Sun Y. (2015) Edge-terminated molybdenum disulfide with a 9.4-Åinterlayer spacing for electrochemical hydrogen production. *Nat Commun* 6:1–8.
16. Chua X.J., Luxa J., Eng A.Y.S., Tan S.M., Sofer Z., Pumera M. (2016) Negative electrocatalytic effects of p-doping niobium and tantalum on MoS_2 and WS_2 for the hydrogen evolution reaction and oxygen reduction reaction. *ACS Catal* 6:5724–5734.
17. Handoko A.D., Fredrickson K.D., Anasori B., Convey K.W., Johnson L.R., Gogotsi Y., Vojvodic A., Seh Z.W. (2017) Tuning the basal plane functionalization of two-dimensional metal carbides (MXenes) to control hydrogen evolution activity. *ACS Appl Energy Mater* 1:173–180.
18. Li P., Zhu J., Handoko A.D., Zhang R., Wang H., Legut D., Wen X., Fu Z., Seh Z.W., Zhang Q. (2018) High-throughput theoretical optimization of the hydrogen evolution reaction on MXenes by transition metal modification. *J Mater Chem A* 6:4271–4278.
19. Hemanth N.R., Kim T., Kim B., Jadhav A.H., Lee K., Chaudhari N.K. (2021) Transition metal dichalcogenide-decorated MXenes: Promising hybrid electrodes for energy storage and conversion applications. *Mater Chem Front* 5:3298–3321.
20. George G., Ede S.R., Luo Z. (Professor of materials science) (2020) *Fundamentals of Perovskite Oxides: Synthesis, Structure, Properties and Applications.* CRC Press, Boca Raton, FL.
21. Forouzandeh P., Pillai S.C. (2021) Two-dimensional (2D) electrode materials for supercapacitors. *Mater Today Proc* 41:498–505.
22. Gholamvand Z., McAteer D., Harvey A., Backes C., Coleman J.N. (2016) Electrochemical applications of two-dimensional nanosheets: The effect of nanosheet length and thickness. *Chem Mater* 28:2641–2651.
23. Huang X., Zeng Z., Fan Z., Liu J., Zhang H. (2012) Graphene-based electrodes. *Adv Mater* 24:5979–6004.

24. Dong Y., Wu Z.S., Ren W., Cheng H.M., Bao X. (2017) Graphene: A promising 2D material for electrochemical energy storage. *Sci Bull* 62:724–740.
25. Stoller M.D., Park S., Zhu Y., An J., Ruoff R.S. (2008) Graphene-based ultracapacitors. *Nano Lett* 8:3498–3502.
26. Zhu Y., Murali S., Stoller M.D., Ganesh K.J., Cai W., Ferreira P.J., Pirkle A., Wallace R.M., Cychosz K.A., Thommes M., Su D., Stach E.A., Ruoff R.S. (2011) Carbon-based supercapacitors produced by activation of graphene. *Science* 332:1537–1541.
27. El-Kady M.F., Strong V., Dubin S., Kaner R.B. (2012) Laser scribing of high-performance and flexible graphene-based electrochemical capacitors. *Science* 335:1326–1330.
28. Miller J.R., Outlaw R.A., Holloway B.C. (2010) Graphene double-layer capacitor with ac line-filtering performance. *Science* 329:1637–1639.
29. Yang Y., Hou H., Zou G., Shi W., Shuai H., Li J., Ji X. (2018) Electrochemical exfoliation of graphene-like two-dimensional nanomaterials. *Nanoscale* 11:16–33.
30. Yang Y., Fei H., Ruan G., Xiang C., Tour J.M. (2014) Edge-oriented MoS_2 nanoporous films as flexible electrodes for hydrogen evolution reactions and supercapacitor devices. *Adv Mater* 26:8163–8168.
31. Peng L., Peng X., Liu B., Wu C., Xie Y., Yu G. (2013) Ultrathin two-dimensional MnO_2/Graphene hybrid nanostructures for high-performance, flexible planar supercapacitors. *Nano Lett* 13:2151–2157.
32. Li X., Du D., Zhang Y., Xing W., Xue Q., Yan Z. (2017) Layered double hydroxides toward high-performance supercapacitors. *J Mater Chem A* 5:15460–15485.
33. Hu M., Zhang H., Hu T., Fan B., Wang X., Li Z. (2020) Emerging 2D MXenes for supercapacitors: Status, challenges and prospects. *Chem Soc Rev* 49:6666–6693.
34. Pender J.P., Jha G., Youn D.H., Ziegler J.M., Andoni I., Choi E.J., Heller A., Dunn B.S., Weiss P.S., Penner R.M., Mullins C.B. (2020) Electrode degradation in lithium-ion batteries. *ACS Nano* 14:1243–1295.
35. Augustyn V. (2017) Tuning the interlayer of transition metal oxides for electrochemical energy storage. *J Mater Res* 32:2–15.
36. Wu S., Du Y., Sun S. (2017) Transition metal dichalcogenide based nanomaterials for rechargeable batteries. *Chem Eng J* 307:189–207.
37. Choi W., Choudhary N., Han G.H., Park J., Akinwande D., Lee Y.H. (2017) Recent development of two-dimensional transition metal dichalcogenides and their applications. *Mater Today* 20:116–130.
38. Naguib M., Mochalin V.N., Barsoum M.W., Gogotsi Y. (2014) 25th anniversary article: MXenes: A new family of two-dimensional materials. *Adv Mater* 26:992–1005.

11 Tuning of Bandgap and Electronic Properties for Energy Applications

Maoyang Xia, Jing Ning, Dong Wang, Jincheng Zhang, and Yue Hao
Xidian University

CONTENTS

DOI: 10.1201/9781003178453-11

11.1 INTRODUCTION

In the information age of the rapid development of the Internet of Things connection, the large demand for mobile electronic devices has stimulated intensive research on energy-storage devices. With the wearability of mobile electronics, the matching energy systems need not only intelligence, miniaturization, and stretchability but also fast charging and discharging, high power density, long cycling lives, and good stability. Selecting the appropriate electrode material is the key to ensuring high energy density for the micro-energy storage system. So far, 2D nanomaterials have shown broad application prospects in the field of energy storage because of their outstanding physical and chemical properties.

Since its discovery, 2D materials have attracted extensive research attention because of their excellent and controllable physical and chemical properties. The 2D materials are mostly bonded by strong ionic or covalent bonds in layers, and the layers are stacked together by van der Waals forces, so that few or single-layer 2D materials can be stripped. The weak interaction between 2D materials layers also makes it possible to use van der Waals gaps to adjust the electronic structure of the system and further optimizing the material properties. The introduction of intercalator in van der Waals gap can significantly change the layer spacing of host materials and regulate the van der Waals coupling between layers, so as to optimize the physical and chemical properties of 2D materials.

In recent years, a large number of studies have proposed many effective methods to prepare 2D materials, and studied its electronic structure and energy band regulation, in order to solve the bottleneck problem in the application. This chapter systematically summarizes the research progress in the electronic structure and energy band regulation of 2D materials and further discusses the energy application fields and prospects of 2D materials.

11.2 GRAPHENE

11.2.1 Structure and Electronic Properties

Graphene is a 2D crystalline ultra-thin material with 2D sp^2-hybridized carbon nanostructure and a honeycomb lattice. Graphene exhibits a large theoretical specific surface area ($2{,}600\,m^2 g^{-1}$), high electrical conductivity (106 S cm^{-1}), and low bandgap (0eV). Because of these advantages, graphene is selected as one of the most promising candidates in energy storage devices electrode [1].

11.2.2 Control Method of Bandgap and Electronic Properties

The electronic structure and energy band of graphene determine its applications in the fields of electronics, optoelectronics, and energy storage. Generally speaking, the methods to control the bandgap structure of graphene include doping, preparation of 1D graphene with quantum effect, and chemical modification.

11.2.2.1 Non-Metallic Doping of Graphene

Doping non-metallic atoms in graphene is an effective way to change their properties [2]. The introduction of nitrogen (N) or boron (B) atoms into graphene to replace the position of carbon atoms can enable graphene to exhibit the properties of N- or P-type semiconductors. Because N, B, and C atoms differ by only one valence electron and have similar size and electronic structure, N and/or B doping of graphene can avoid large stress, maintain the integrity of graphene structure, and make the intrinsic graphene change from metallic to semiconductor through changing the energy bandgap structure.

At present, the main methods to prepare N-doped graphene (NDG) are chemical vapor deposition (CVD), ion-nitriding, arc discharge, and flame method.

CVD method. Graphene was prepared and doped via high-temperature decomposition of ammonia and methane, mutual solubility of catalysts, and cooling precipitation. Wei et al. [3] deposited a layer of Cu nanocrystalline film with a thickness of about 25 nm on the Si wafer and then prepared N-doping graphene in a mixed atmosphere of CH_4, H_2, and NH_3 at 800°C by the CVD method. N atoms are embedded in graphene by introducing NH_3 during the growth process. The incorporation of N atoms transforms graphene from a conductor to an n-type semiconductor.

Ion-nitriding method. Carbon ring structure in graphene is destroyed by nitrogen ion bombardment, and N atom is introduced to fill the vacancy defect of C atom. Dai et al. [4] treated graphene oxide with NH_3 between 300°C and 900°C to achieve reduction and doping of graphene oxide at same time.

Arc discharge method. In carbon-nitrogen-hydrogen atmosphere, the carbon source and nitrogen source are decomposed by arc discharge to prepare NDG. Hydrogen is used to avoid the curling of graphene nanosheets to form carbon nanotubes. Shi et al. [5] prepared a large number of NDG sheets by DC arc evaporation graphite electrode in NH_3 atmosphere. The doping concentration of N element is adjusted by controlling the concentration of NH_3.

Flame method. The mixed fuel is used as a fuel to provide reaction energy, and as a material to provide reactants, with Ni as a catalyst. According to the different solubility of C and N atoms in Ni, NDG was prepared. Zhang et al. [6] mixed amine and ethanol in 7:3 ratio as fuel, coated a layer of Ni nanocrystalline film on Si wafer, putted it on flame for 1–3 minutes. NDG nanosheet layer is obtained on Ni surface.

11.2.2.2 One-Dimensional Graphene Nanoribbons

The electronic structure and energy band of graphene are related to its width and boundary geometry. The 2D graphene sheet is cut into 1D graphene nanoribbons with finite size in the transverse direction. The quantum confinement effect and boundary effect will limit the movement of electrons in the transverse direction of graphene nanoribbons so that the energy gap is opened. At the same time, the width of 1D graphene nanoribbons can be accurately controlled to tune its energy gap. So far, the preparation technologies of 1D graphene nanoribbons mainly include carbon nanotube slitting, organic synthesis, and mask etching.

Carbon nanotube slitting method. Carbon nanotubes replace graphite as the starting material, break the bonding on the surface of carbon nanotubes through sulfuric acid and potassium permanganate oxidation treatment or plasma etching treatment, and cut longitudinally to form graphene nanoribbons.

Organic synthetic method. Cai et al. [7] reported atomically precise graphene nanoribbons with different topologies and widths, which utilize surface-assisted coupling of molecular precursors to linear polyphenylenes and their subsequent cyclodehydrogenation. The topology, width, and edge periphery of the graphene nanoribbon products are determined by the structure of the precursor monomers, which can be designed for various graphene nanoribbons.

Mask etching method. Bai et al. [8] fabricated graphene nanoribbons with a width of less than 10 nm via using chemically synthesized nanowires as the physical protection mask for oxygen plasma etching. Banded or branched or crossed graphene nanostructures can be produced.

11.2.2.3 Chemical Modification

Graphene has a large specific surface area and strong adsorption capacity. The energy gap of multilayer graphene can be opened by adsorbing metal atoms on the one side. Chemical modification method is to place graphene in a space rich in adsorbents. Chemical modification adjusts the size of the energy gap by altering the electronic structure of graphene. Chemical modification of graphene can also destroy the symmetry of graphene to open its energy gap. For example, restore graphene prepared by graphite oxide has a certain bandgap because of the large number of oxygen-containing functional groups and defects on the surface. Its bandgap can be adjusted with the change of oxygen content. Mathkar et al. [9] found that the bandgap of graphene is adjusted from 1 to 3.5 eV via changing the degree of oxidation and oxidation region.

11.2.3 Application of Graphene as Electrode

11.2.3.1 Supercapacitor

Graphene is the most promising electrode material in supercapacitor because of its high electronic conductivity and outstanding mechanical stability. The self-doping defect structure is controllably introduced on the surface of graphene to improve the energy storage performance by doping and introducing functional groups.

Wu et al. [10] synthesized the 3D structure of graphene aerogel and assembled all-solid-state supercapacitors. The power density and energy density reach 62 Wh kg^{-1} and 1,600 W kg^{-1}, respectively. Chen et al. [11] deposited the graphene hydrogel inside the nickel foam to open the pore structure of the electrode and allow the electrolyte ions to pass smoothly, so it has excellent rate performance.

Ogata et al. [12] reported a flexible solid-state reduced graphene oxide (rGO)/graphene oxide (GO)/rGO supercapacitor. High ionic concentration and fast ion conduction in the H_2SO_4-intercalated GO electrolyte/separator and abundant CH defects, which serve as pseudocapacitive sites on the rGO electrode, were responsible for the high capacitance (14.5 mF cm^{-2}) of the device. Moon et al. [13] fabricated 3D porous stretchable supercapacitor with wrinkled-structure poly(3,4-ethylenedioxythiophene)- poly(styrenesulfonate) on a 3D porous reduced graphene oxide-coated elastic substrate under pre-strained conditions via solution process. The stretchable supercapacitor exhibits an electrochemical energy capacity up to 82.4 F g^{-1} and 85.1% capacitance retention at 300% strain.

Dong et al. [14] reported defect-enriched graphene block with a low specific surface area of 29.7 $m^2 g^{-1}$ and high packing density of 0.917 g cm^{-3} performs high gravimetric, volumetric, and areal capacitances of 235 F g^{-1}, 215 F cm^{-3}, and 3.95 F cm^{-2} at 1 A g^{-1}, respectively. Zhang et al. [15] fabricated N-doping graphene paper using GO-ethanol dispersion filtration. NDG papers maintain good foldability with improved electric conductivity and porous structure. When the NDG paper is fabricated as flexible device, this device also demonstrates a high charge/discharge capacitances 312.5 F g^{-1}.

11.2.3.2 Lithium-Ion Batteries

Carbon/graphite is a commercially available lithium-ion battery (LIB) anode material because of low cost and stable performance. The theoretical specific capacity is 376 mAh g^{-1}. Every six carbon atoms are combined with one lithium atom to store energy in LiC_6. The doping and modification of graphene can increase the active sites and improve the energy storage performance of LIBs.

Wei et al. [16] reported all-solid-state LIB is made of monolayer graphene grown by CVD directly onto Cu foil. The total thickness of the resulting battery was 50 mm. Graphene LIBs showed the highest energy density of 10 W h L^{-1} and the highest power density of 300 W L^{-1}. Yin et al. [17] reported a self-assembly strategy to create bioinspired hierarchical structures composed of functionalized graphene sheets. The electrodes with multilevel architectures simultaneously optimize ion transport and capacity, leading to a high performance of reversible capacity of up to 1,600 mAh g^{-1}, and 1,150 mAh g^{-1} after 50 cycles.

Liu et al. [18] reported graphene paper is made from a graphene aerogel, which is prefabricated by freeze-drying a GO aqueous dispersion and subsequent thermal reduction. The graphene paper is free-standing and flexible. When used as electrodes for a LIB, this graphene paper can reach 557 mAh g^{-1}. Mukherjee et al. [19] reported photoflash and laser-reduced free-standing graphene paper as high-rate capable anodes for LIBs. Photothermal reduction of graphene oxide yields an expanded structure with micrometer-scale pores, cracks, and intersheet voids. At charge/discharge rates of 40 C, the anodes delivered a steady capacity of 156 mAh g^{-1}.

Liu et al. [20] synthesized sandwich architecture of N-carbon/rGO by simple pyrolysis of a polypyrrole/GO nanosheet precursor. The N-carbon/rGO exhibits the N-doping of 15.4% and a specific surface area of 327 $m^2 g^{-1}$. The N-carbon/rGO shows a high initial reversible capacity of 1,100 mAh g^{-1} at 100 mA g^{-1}. Mo et al. [21] reported the synthesis of nitrogen-doped, mesoporous graphene particles through CVD with magnesium-oxide particles as the catalyst and template. Such particles possess structural and electrochemical stability, electronic and ionic conductivity, enabling their use as anodes with reversible capacity (1,138 mAh g^{-1} at 0.2 C).

11.2.3.3 Sodium-Ion Batteries

Carbon-based materials have significant advantages such as low price, convenient operation, low toxicity, and high efficiency. However, due to the uniform shrinkage of C = C bond, narrow layer spacing, and weak surface bond cooperation, layered graphite still has a bottleneck in the application of sodium-ion battery (SIB). Therefore, the modification of graphene and the increase of layer spacing are conducive to improve the performance.

Lee et al. [22] reported the ion storage capacity of activated-crumbled graphene (A-CG) for fast charging and extended cyclability. A-CG was synthesized by low-temperature spraying of graphene oxide slurry, reduction annealing, and air activation. For Na storage, the reversible capacities are 280 mAh g^{-1} at 0.04 A g^{-1}.

Wang et al. [23] produced boron-functionalized reduced graphene oxide (BF-rGO), with expended interlayer spacing and defect-rich structure, thereby effectively accommodating to sodiation/desodiation and providing more active sites. The Na/BF-rGO half cells exhibit unprecedented long cycling stability, with 89.4% capacity retained after 5,000 cycles at current density of 1,000 mA g^{-1}.

Dan et al. [24] produced the NC/RGO by pyrolyzing the cellulose, chitosan, and GO. Nitrogen doping not only improved the conductivity but also increased layer spacing and promoted the insertion of Na-ions. The NC/RGO shows a sodium storage capacity of 395 mAh g^{-1} at 0.1 A g^{-1}.

11.2.3.4 Potassium-Ion Batteries

In K-insertion/extraction and the corresponding potassium storage mechanism, KC_{36} (Stage III) was first observed between 0.3 and 0.2 V, followed by KC_{24} (Stage II) between 0.2 and 0.1 V. Finally, the pure KC_8 (Stage I) was decomposed near 0.01 V. No KC_{24} (Stage II) was detected during potassium-depotassiation, and KC_8 (Stage I) was replaced by KC_{36} (Stage III) until around 0.3 V.

Ju et al. [25] prepared few-layer nitrogen-doped graphene (FLNG) by bottom-up synthesis using Dicyandiamide and Coal tar pitch as raw materials. The FLNG with high surface area (479.21 $m^2 g^{-1}$) and nitrogen content (14.68 at%) exhibits excellent K-ion storage performances (320 mAh g^{-1} at 50 mA g^{-1}).

Dong et al. [26] synthesized carbon dots-modified reduced graphene oxides (LAP-rGO-CDs) via a microwave-assisted method. Due to the introduction of carbon dots, the interlayer space of LAP-rGO-CDs increase and ion transfer rate become faster. As a potassium-ion battery (PIB) electrode, LAP-rGO-CDs showed a specific capacity of 299 mAh g^{-1} at 1 A g^{-1}.

Qiao et al. [27] synthesized columnar graphene nanosheets with expanded interlayer spacing via an in situ growth strategy. It is confirmed by kinetics that the extended interlayer spacing can accelerate ion transport. The graphene nanosheet/NiO-500 composite anode provides ultrahigh reversible capability 416.1 mA h g^{-1} at 0.1 A g^{-1}.

11.3 TRANSITION METAL DICHALCOGENIDES (TMDCs)

11.3.1 Structure and Electronic Properties

Transition metal dichalcogenide (TMDC) is a layered compound similar to graphene, and its molecular formula is generally MX_2, where M represents a transition metal atom (Mo, W, etc.) and X represents a chalcogen atom (S, Se, etc.). Single-layer X-M-X has a sandwich structure. Similar to other 2D materials, TMDC layers are connected by van der Waals force. TMDCs exist in several structural phases according to different coordination modes of the transition metal atoms. The three common structural phases are 3R, 2H, and 1T, which are characterized by rhombic, trigonal prismatic, and octahedral coordination of the transition metal atoms, respectively. Different crystal forms also give different properties to multilayer TMDCs: the conductivity of 1T phase is 10^7 times that of 2H phase. Both 3R phase and 2H phase have trigonal prismatic structure, but their stacking order is different, and their application range is narrow. Due to their similar crystal structure, TMDCs are easy to construct heterostructures. The heterogeneous interface not only exposes more active sites of electrochemical reaction but also helps to improve the properties of ion transport and electron conduction. The combination of different 2D TMDCs provides an opportunity for the design of artificial layered materials and widens the application of TMDCs in the field of electrochemistry [28].

11.3.2 Control Method of Bandgap and Electronic Properties

11.3.2.1 Phase Control Engineering

2D TMDCs are composed of more than 40 compounds. Complex metal TMDCs assume a 1T phase in which transition metal atoms are coordinated as octahedrons. The 2H phase is stable in semiconductor TMDCs, where the coordination of metal atoms is triangular prism. Phase transitions in TMDCs involve transformations by chemistry at room temperature and pressure. 2H phase 2D TMDCs can be converted to 1T phase, or 1T phase can be patterned locally on 2H phase.

1T phase was realized in alkali metal (Li and K) intercalated TMDC decades ago. The embedding of TMDC leads to a significant expansion of layer spacing, so that the solvation and the reduction of embedding agent lead to stripping into a single layer. These monolayer nanosheets consist of most of the 1T phase. The conversion of 2H phase to 1T phase in the embedding process is attributed to the charge transfer from alkali atoms to nanosheets. This additional charge leads to the density of states of Fermi level, which makes the material metallic. In addition to alkali doping by intercalation, substitution doping with elements with more valence electrons than transition metal ions (e.g., Re and Mn) in TMDC can also lead to the formation of 1T phase.

Although the transition from 2H to 1T phase requires additional electrons from dopants, the presence of such impurities is undesirable. The removal of dopant will destroy the stability of 1T phase. In lithium-embedded TMDCs (e.g., Li_xMoS_2), butyl lithium is used. Organic butyl and Li-ions can be removed from MoS_2 nanosheets by careful washing with hexane and water. Surprisingly, the 1T phase still exists after the removal of organic and alkali impurities. The "Dry" films of the chemically pure 1T phase TMDCs in multilayer and monolayer forms have been proved. The stability of these films is attributed to the presence of protons or other immobile, positively charged ions on the surface of the nanosheets, which counteract the additional charge provided by the dopants. After annealing to ~300°C in a controlled environment, the 1T phase relaxes to 2H phase, which supports the existence of positively charged counter ions adsorbed on the nanosheets [29,30].

11.3.2.2 Heteroatoms Doping

Doped heteroatoms are an effective way to change electron structure especially the d-band of TMDCs-based material. The layer spacing of 2D materials can be effectively changed by doping heteroatoms and intercalation with metal ions, small molecules, metal atoms, and organic molecules. Heteroatom doping usually causes charge transfer or other interactions with the matrix material while changing the interlayer coupling.

The presence of heavy atoms will also increase the spin–orbit coupling, resulting in changes in superconducting properties, thermoelectric properties, or spin polarization. It is also found that the interlayer coupling of multilayer 2D materials can be effectively decoupled by inserting other 2D monolayers or organic molecules into the layers of 2D materials to form superlattices [31].

11.3.3 Application of TMDCs as Electrode

11.3.3.1 Supercapacitor

TMDC-based supercapacitors can store energy through three different mechanisms; in addition to EDLC and pseudocapacitance (caused by the redox reaction of metal atoms), the large layer spacing of 2D TMDC can also accelerate the rapid reversible insertion of electrolyte ions between layers, which can help some embedded pseudocapacitance.

Cao et al. [32] fabricated 2D MoS_2 film-based micro-supercapacitors via spraying MoS_2 nanosheets on Si/SiO_2 chip and subsequent laser patterning. The best MoS_2-based micro-supercapacitor exhibits electrochemical performance for energy storage with aqueous electrolytes, with a high area capacitance of 8 mF cm^{-2} (volumetric capacitance of 178 F cm^{-3}).

Acerce et al. [33] have shown that chemically exfoliated nanosheets of MoS_2 containing a high concentration of the metallic 1T phase can electrochemically embed ions with extraordinary efficiency and achieve capacitance values ranging from similar to 400 to similar to 700 F cm^{-3} in a variety of aqueous electrolytes.

Ning et al. [34] synthesized W-doped $MoSe_2$/graphene heterojunctions by a one-step hydrothermal catalysis. The insertion of a small amount of W (~ 5%) into the $MoSe_2$/graphene heterostructure leads to the change of its lattice structure. The W-doped $MoSe_2$/graphene composite achieved excellent capacitance of 444.4 mF cm^{-2} at 1 mV s^{-1}.

11.3.3.2 Lithium-Ion Batteries

TMDCs have many advantages, such as high theoretical specific capacity, high natural reserves, and environmental friendliness. It has attracted people's attention and is widely used as cathode material for LIBs. However, TMDCs also have some shortcomings, such as low conductivity and structure instability. Therefore, it is necessary to design the TMDC structure to improve the lithium storage performance.

Dong et al. [35] synthesized tremella-like nitrogen-doped carbon encapsulated few-layer MoS_2 (MoS_2@NC) hybrid via a strategy with simultaneously polydopamine carbonization, and molybdenum oxide specifies sulfurization. The MoS_2@

NC exhibits enhanced high rate performance with a specific capacity of 208.7 mAh g^{-1} at a current density of 10 A g^{-1}.

Chan et al. [36] synthesized MoS_2/PANI-rGO composite by self-assembling 1T-MoS_2 2D sheets on PANI-rGO through an electrostatic interaction. The flower-like aggregate of 1T-MoS_2 with increased interlaminar distance favors diffusion and intercalation of Li-ions. The electrode has specific capacities of 812 mAh g^{-1} at 0.1 A g^{-1}.

11.3.3.3 Sodium-Ion Batteries

Based on multi-electron conversion reaction, layered TMDCs are considered to be a promising Na-ion anode material, but they usually show poor cycle and rate performance due to their low conductivity and large volume changes during charge and discharge. Therefore, how to design and prepare metal chalcogenide cathode materials with stable structure and excellent sodium storage performance is still a challenge.

David et al. [37] reported layered free-standing papers composed of few-layer MoS_2 and rGO flakes exfoliated by acid, used as a self-supporting flexible electrode in SIBs. The electrode showed good Na cycling ability with a stable charge capacity of approximately 230 mAh g^{-1} with Coulombic efficiency reaching approximately 99%.

Yousaf et al. [38] pointed a 3D three-layer design that sandwiched $MoSe_2$ as the TMDC layer between the internal CNT core and external carbon layer to form a CNT/$MoSe_2$/C framework. The heterostructure produces a capacity of 347 mAh g^{-1} at 500 mA g^{-1}.

11.3.3.4 Potassium-Ion Batteries

Potassium and lithium have similar standard redox potentials, however, potassium has an abundant storage in the crust (1.5 wt.%), which leads to low cost of PIBs. At the same time, K-ion has fast ion transport kinetics in electrolyte, which makes PIB have excellent prospects.

Jiang et al. [39] showed that creating sufficient exposed edges in MoS_2 via constructing ordered mesoporous architecture is beneficial to improving kinetics. The engineered MoS_2 with edge-enriched planes has the characteristics of 3D bicontinuous frameworks and shows a high reversible charge capacity of 506 mAh g^{-1} at 0.05 A g^{-1}.

Soares et al. [40] studied electrochemical performance of WTe_2 as a working electrode for a PIB half-cell. WTe_2 has a high first cycle-specific charge capacity—each WTe_2 molecule stores up to 3.3 K-ion, a stable capacity of 143 mAh g^{-1} at 10th cycle number.

11.4 MXene

11.4.1 Structure and Electronic Properties

Metal carbides/nitrides (MXenes) appear as a new 2D materials family with a general formula $M_{n+1}X_nT_x$ (n = 1,2,3), where M stands for an early transition metal (Ti, Zr, Hf, V, Nb, etc.), X stands for the C and/or N, and T_x stands for a large number of surface-terminating moieties (–O, –F and –OH), due to its excellent inherent physical and chemical properties, such as superb intrinsic electrical conductivity, tunable surface chemistry (various functional groups), and tunable layer structure, have shown great potential to increase capacitance of micro-supercapacitors [41].

11.4.2 Control Method of Bandgap and Electronic Properties

11.4.2.1 Surface Control

Conventional etching methods involving HF or fluorine-containing salts will result in the termination of F and oxygen-containing functional groups. The content of functional groups can be adjusted due to the concentration of the etchant, the etching time, or by different post-treatments (such as annealing in different atmospheres and hydrazine treatment). Therefore, the engineering of MXenes surface terminals with various properties has attracted great attention.

Physical properties of MXenes highly rely on the surface terminations and the electronic properties of $Ti_3C_2T_x$. Through annealing at 380°C in vacuum, the work function of sample would increase from 3.9 to 4.8 eV, which is mainly due to desorption of water, contaminations, and OH species. With desorption of fluorine at higher temperatures from 500°C to 750°C, the work function of $Ti_3C_2T_x$ decreased to 4.1 eV again. In addition, the tunable electronic structure of MXenes turned by surface terminations shown broad prospects in energy storage. If possible, surface terminations of -H, -F, and -OH should be avoided to improve the energy storage performance and M_2C with non-functional groups (extremely difficult to achieve) or -O termination would provide the most promising performance [42].

11.4.3 Application of MXene as Electrode

11.4.3.1 Supercapacitor

The electrochemical energy storage performance of MXene is greatly affected by surface functional groups. On the one hand, the F-/OH- and other groups adsorbed on the surface are not conducive to the rapid migration of electrolyte ions between layers. On the other hand, they affect the redox reaction of active site transition metal M in MXene layer. Therefore, F-/OH- and other groups adsorbed on the surface are one of the main reasons affecting the performance of MXene material as supercapacitor electrode material.

Wu et al. [43] reported SA-MXene was dispersed via in situ synthesis and with sodium ascorbate (SA) terminated of Ti_3C_2Tx MXenes. The in situ synthesis process increases the interlayer spacing of SA-MXene sheets and improves their energy storage efficiency. The all-solid micro-supercapacitor exhibits capacitance of 108.1 mF cm^{-2}.

Levitt et al. [44] reported that MXene flakes are added into PAN solutions at a weight ratio of 2:1 (MXene:PAN) in spinning coatings, resulting in fiber mats with up to 35 wt% MXene. The electrodes have high areal capacitance, up to 205 mF cm^{-2} at 50 mV s^{-1}.

Cao et al. [45] synthesized Ti_3C_2/TiO_2-nanowires via room temperature oxidation methods. When used as electrodes for supercapacitors, Ti_3C_2/TiO_2-nanowires exhibit enhanced supercapacitive properties. The specific capacitance of Ti_3C_2/TiO_2-

nanowires electrode is 143 F g^{-1} at 2 mV s^{-1}. After 6,000 cycles, the 80% of the initial capacitance is retained for Ti_3C_2/TiO_2-nanowires.

11.4.3.2 Lithium-Ion Batteries

MXene is a 2D material with excellent electronic conductivity, low Li-ion diffusion barrier, and good mechanical properties. It is widely used for energy storage.

Naguib et al. [46] reported Li insertion into a 2D layered Ti_2C-based material (MXene) with an oxidized surface, formed by etching Al from Ti_2AlC in HF at room temperature. The steady state capacity was 225 mAh g^{-1} at C/25.

Liu et al. [47] synthesized an N-doped 2D Nb_2CTx MXene. The introduction of nitrogen into MXene nanosheets increases c-lattice parameter of Nb_2CTx MXene from 22.32 Å to 34.78 Å. The N-doped Nb_2CTx shows an increased reversible capacity of 360 mAh g^{-1} at 0.2 C.

11.4.3.3 Sodium-Ion Batteries

Ti_3C_2 has high rate capacity and good long-term cycle stability, which is most suitable for SIBs. The pillaring effect of Na-ion captured by MXene can effectively maintain the stability of layer spacing, which is conducive to the realization of rapid reversible sodium denaturation process.

Wei et al. [48] studied Mn^{2+} intercalation strategy to optimize the sodium storage performance of V_2C MXene. The intercalated Mn^{2+} formed a V-O-Mn covalent bond, which was beneficial to stabilize the structure of V_2C. The V_2C@Mn electrode showed a high specific capacity of 425 mAh g^{-1} at 0.05 A g^{-1}.

Zhu et al. [49] reported the use of Ti_3CN as anode material in SIBs. Introducing more electronegative nitrogen into the lattice grid of Ti_3C increases the electron count of MXenes. The Ti_3CN electrode provides a high specific capacity of 211.5 mAh g^{-1} at 20 mA g^{-1}.

11.4.3.4 Potassium-Ion Batteries

Because of the low cost, high element abundance and intrinsic safety, PIBs have attracted a surge of interest in recent years. Currently, the key challenge to the development of PIBs is to find suitable anode materials with large capacity, high rate capability, and small lattice changes during the charge/discharge process.

Zhao et al. [50] designed PDDA-NPCN/Ti_3C_2 hybrids as PIBs anodes via an electrostatic attraction self-assembly approach. The hybrids afford enlarged interlayer spacing and 3D interconnected conductive networks to accelerate the ionic/electronic transport rates. The hybrids exhibit a high capacity of 358.4 mAh g^{-1} at 0.1 A g^{-1}.

11.5 CONCLUSIONS AND PERSPECTIVES

In this chapter, we systematically describe a series of works on adjusting the bandgap and electronic properties of 2D materials. The electronic and bandgap structures of 2D materials are regulated by doping, atom insertion, and chemical modification, so as to change its electrical properties. It is of great significance to the development and application of 2D materials, especially in the fields of energy storage. However, there are still some problems in the study of 2D materials band regulation. For example, how to effectively and accurately control the bandgap width of 2D TMDC materials so that it will not cause a significant reduction in electron mobility in the process of changing from semiconductor attribute to metallicity, and how to control the stability of semiconductor TMDC. We believe that, while using intercalation to control the properties of materials, further application of heteroatom doping and strain control to optimize the properties of materials may break through the limitation of a single method to control the properties of materials, avoid or compensate for some performance degradation that may be caused by intercalation, resulting in better or more comprehensive materials.

REFERENCES

1. Seman R. N. A. R., Azam M. A., and Ani M. H. 2018. Graphene/transition metal dichalcogenides hybrid supercapacitor electrode: Status, challenges, and perspectives. *Nanotechnology* 29:502001.
2. Czerw R., Terrones M., Charlier J. C., Blase X., Foley B., Kamalakaran R., Grobert N., Terrones H., Tekleab D., Ajayan P. M., Blau W., Ruhle M., and Carroll D. L. 2001. Identification of electron donor states in N-doped carbon nanotubes. *Nano Letters* 9:457–460.
3. Wei D., Liu Y., Wang Y., Zhang H. L., Huang L. P., and Yu G. 2009. Synthesis of N-doped graphene by chemical vapor deposition and its electrical properties. *Nano Letter* 9:1752–1758.
4. Li X., Wang H., Robinson J. T., Sanchez H., and Diankov G. 2008. Simultaneous nitrogen-doping and reduction of graphene oxide. *Journal of American Chemistry Society* 131:15939–15944.
5. Li N., Wang Z., Zhao K., Shi Z., Gu Z., and Xu S. 2010. Large scale synthesis of N-doped multi layered graphene sheets by simple arc-discharge method. *Carbon* 48:255–259.
6. Zhang Y., Cao B., Zhang B., Qi X., and Pan C. 2012. The production of nitrogen-doped graphene from mixed amine plus ethanol flames. *Thin Solid Films* 520:6850–6855.
7. Cai J., Ruffieux P., Jaafar R., Bieri M., Branun T., Blankenburg S., Muoth M., Seitsonen A. P., Saleh M., Feng X., Mullen K., and Fasel R. 2010. Atomically precise bottom-up fabrication of graphene nanoribbons. *Nature* 466:470–473.
8. Bai J., Duan X., Huang Y. 2009. Rational fabrication of graphene nanoribbons using a nanowire etch mask. *Nano Letters* 9:2083–2087.
9. Mathkar A., Tozier D., Cox P., One P., Galande C., Balakrishnan K., Reddy A. L. M., and Ajayan P. M. 2012. Controlled, stepwise reduction and band gap manipulation of graphene oxide. *Journal of Physical Chemistry Letters* 3:986–991.
10. Wu Z. S., Winter A., Chen L., Sun Y., Turchanin A., Feng X., and Müllen K. 2012. Three-dimensional nitrogen and boron co-doped graphene for high-performance all-solid-state supercapacitors. *Advanced Materials* 24:5130–5135.
11. Chen J., Sheng K., Luo P., Li C., and Shi G. 2012. Graphene hydrogels deposited in nickel foams for high-rate electrochemical capacitors. *Advanced Materials* 24(33):4569–4573.
12. Ogata C., Kurogi R., Awaya K., Hatakeyama K., Taniguchi K., Koinuma M., and Matsumoto Y. 2017. All-graphene oxide flexible solid-state supercapacitors with enhanced electrochemical performance. *ACS Applied Materials & Interfaces* 9: 26151–26160.
13. Moon I., Ki B., and Oh J. 2019. Three-dimensional porous stretchable supercapacitor with wavy structured PEDOT: PSS/graphene electrode. *Chemical Engineering Journal* 392:123794.
14. Dong Y., Zhang S., Du X., Hong S., Zhao S., Chen Y., Chen X., and Song H. 2019. Boosting the electrical double-layer capacitance of graphene by self-doped defects through ball-milling. *Advanced Function Materials* 29:1901127.

15. Zhang H., Li A., Yuan Y., Wei Y., Zheng D., Geng Z., Zhang H., Li G., and Zhang F. 2020. Preparation and characterization of colorful graphene oxide papers and flexible N-doping graphene papers for supercapacitor and capacitive deionization. *Carbon Energy* 2:656–674.
16. Wei D., Haque S., Andrew P., Kivioja J., Ryhanen T., Pesquera A., Centeno A., Alonso B., Chuvilin A., and Zurutuza A. 2013. Ultrathin rechargeable all-solid-state batteries based on monolayer graphene. *Journal of Chemistry A* 1:3177–3181.
17. Yin S., Zhang Y., Kong J., Zou C., Li C., Lu X., Ma J., Boey F., and Chen X. 2011. Assembly of graphene sheets into hierarchical structures for high-performance energy storage. *ACS Nano* 5:3831–3838.
18. Liu F., Song S., Xue D., and Zhang H. 2012. Folded structured graphene paper for high performance electrode materials. *Advanced Materials* 24:1089–1094.
19. Mukherjee R., Thomas A. V., Krishnamurthy A., and Koratkar N. 2012. Photothermally reduced graphene as high-power anodes for lithium-ion batteries. *ACS Nano* 6:7867–7878.
20. Liu X., Zhang J., Guo S., and Pinna N. 2016. Graphene/N-doped carbon sandwiched nanosheets with ultrahigh nitrogen doping for boosting lithium-ion batteries. *Journal of Materials Chemistry A* 4:1423–1431.
21. Mo R., Li F., Tan X., Xu P., Tao R., Shen G., Lu X., Liu F., Shen L., Xu B., Xiao Q., Wang X., Wang C., Li J., Wang G., and Lu Y. 2019. High-quality mesoporous graphene particles as high-cnergy and fast-charging anodes for lithium-ion batteries. *Nature Communications* 10:1474.
22. Lee B., Kim M., Kim S., Nanda J., Kwon S., Jang H., Mitlin D., and Lee S. 2020. High capacity adsorption-dominated potassium and sodium ion storage in activated crumpled graphene. *Advanced Energy Materials* 10:1903280.
23. Wang Y., Wang C., Wang Y., Liu H., and Huang Z. 2016. Boric acid assisted reduction of graphene oxide: A promising material for sodium-ion batteries. *ACS Applied Materials and Interfaces* 8:18860–18866.
24. Dan R., Chen W., Xiao Z., Li P., Liu M., Chen Z., and Yu F. 2020. N-doped biomass Carbon/RGO as a high performance anode for sodium-ion batteries. *Energy & Fuels* 34:3923–3930.
25. Ju Z., Li P., Ma G., Xing Z., Zhuang Q., and Qian Y. 2017. Few layer nitrogen-doped graphene with highly reversible potassium storage. *Energy Storage Materials* 11:38–46.
26. Dong S., Song Y., Fang Y., Zhu K., Ye K., Gao Y., Yan J., Wang G., and Can D. 2021. Microwave-assisted synthesis of carbon dots modified graphene for full carbon-based potassium ion capacitors. *Carbon* 178:1–9.
27. Qiao X., Sun J., Hou C., Bian S., Sun L., and Liao D. 2021. Precise synthesis of pillared graphene nanosheets with superior potassium storage via an in situ growth strategy. *New Journal of Chemistry* 45:14451–14457.
28. Manzeli S., Ovchinnikov D., Pasquier D., Yazyev O. V., and Kis A. 2017. 2D transition metal dichalcogenides. *Nature Reviews Materials* 2:10733.
29. Chhowalla M., Voiry D., Yang J., Shin H. S., Ping K. L. 2015. Phase-engineered transition-metal dichalcogenides for energy and electronics. *MRS Bulletin* 40:585–591.
30. Chen B., Chao D., Liu E., Jaroniec M., Zhao N. and Qiao S. 2020. Transition metal dichalcogenides for alkali metal ion batteries: Engineering strategies at atomic level. *Energy Environmental Science* 13:1096–1131.
31. Zhu Y., Peng L., Fang Z., Yan C., Zhang X., and Yu G. 2018. Structural engineering of 2D nanomaterials for energy storage and catalysis. *Advanced Materials* 30:1706437.
32. Cao L., Yang S., Gao W., Liu Z., Gong Y., Ma L., Shi G., Lei S., Zhang Y., Zhang S., Vajtai R., and Ajayan P. 2013. Direct laser-patterned micro-supercapacitors from paintable MoS_2 films. *Small* 9:2905–2910.

33. Acerce M., Voiry D., and Chhowalla M. 2015. Metallic 1T phase MoS_2 nanosheets as supercapacitor electrode materials. *Nature Nanotechnology* 10:313–318.
34. Liu Q., Ning J., Guo H., Xia M., Wang B., Feng X., Wang D., Zhang J., and Hao Y. 2021. Tungsten-modulated molybdenum selenide/graphene heterostructure as an advanced electrode for all-solid-state supercapacitors. *Nanomaterials* 11:1477.
35. Dong G., Fan Y., Liao S., Zhu K., Yan J., Ye K., Wang G., and Cao D. 2021. 3D tremella-like nitrogen-doped carbon encapsulated few-layer MoS_2 for lithium-ion batteries. *Journal of Colloid and Interface Science* 601:594–603.
36. Chan Y., Vedhanarayanan B., Ji X., and Lin T. 2021. Doubling the cyclic stability of 3D hierarchically structured composites of 1T-MoS_2/polyaniline/graphene through the formation of LiF-rich solid electrolyte interphase. *Applied Surface Science* 565:150582.
37. David L., Bhandavat R., and Singh G. 2014. MoS_2/graphene composite paper for sodium-ion battery electrodes. *ACS Nano* 8:1759–1770.
38. Yousaf M., Wang Y., Chen Y., Wang Z., Firdous A., Ali Z., Mahmood N., Zou R., Guo S., and Han R. 2019. A 3D trilayered CNT/$MoSe_2$/C heterostructure with an expanded $MoSe_2$ interlayer spacing for an efficient sodium storage. *Advanced Energy Materials* 9:1900567.
39. Jiang G., Xu X., Han H., Qu C., Repich H., Xu F., and Wang H. 2020. Edge-enriched MoS_2 for kinetics-enhanced potassium storage. *Nano Research* 13:2763–2769.
40. Soares D. and Singh G. 2020. Superior electrochemical performance of layered WTe_2 as potassium-ion battery electrode. *Nanotechnology* 31:455406.
41. Huang W., Hu L., Tang Y., Xie Z., and Zhang H. 2020. Recent advances in functional 2D MXene-based nanostructures for next-generation devices. *Advanced Function Materials* 30:2005223.
42. Wang C., Chen S., and Song L. 2020. Tuning 2D MXenes by surface controlling and interlayer engineering: Methods, properties, and synchrotron radiation characterizations. *Advanced Functional Materials* 30:2000869.
43. Wu C., Unnikrishnan B., Chen I., Harroun S., Chang H., and Huang C. 2020. Excellent oxidation resistive MXene aqueous ink for micro-supercapacitor application. *Energy Storage Materials* 25:563–571.
44. Levitt A., Alhabeb M., Hatter C., Sarycheva A., Dion G., and Gogotsi Y. 2019. Electrospun MXene/carbon nanofibers as supercapacitor electrodes. *Journal of Chemistry A* 7:269–277.
45. Cao M., Wang F., Wang L., Wu W., Lv W., and Zhu J. 2017. Room temperature oxidation of Ti_3C_2 MXene for supercapacitor electrodes. *Journal of the Electrochemical Society* 164:3933–3942.
46. Naguib M., Come J., Dyatkin B., Presser V., Taberna P., Simon P., Barsoum M., and Gogotsi Y. 2012. MXene: A promising transition metal carbide anode for lithium-ion batteries. *Electrochemistry Communications* 16:61–64.
47. Liu R., Cao W., Han D., Mo Y., Zeng H., Yang H., and Li W. 2019. Nitrogen-doped Nb2CTx MXene as anode materials for lithium ion batteries. *Journal of Alloys and Compounds* 793:505–511.
48. Wei S., Wang C., Zhang P., Zhu K., Chen S., and Song L. 2020. Mn^{2+} intercalated V_2C MXene for enhanced sodium ion battery. *Journal of Inorganic Materials* 35:139–144.
49. Zhu J., Wang M., Lyu M., Jiao Y., Du A., Luo B., Gental I., and Wang L. 2018. Two-dimensional titanium carbonitride Mxene for high-performance sodium ion batteries. *ACS Applied Nano Materials* 1:6854–6863.
50. Zhao R., Di H., Hui X., Zhao D., Wang R., Wang C., and Yin L. 2020. Self-assembled Ti_3C_2 MXene and N-rich porous carbon hybrids as superior anodes for high-performance potassium-ion batteries. *Energy & Environmental Science* 13:246–257.

12 Electrolyte Membrane for 2D Nanomaterials

S. Mohanapriya
Periyar University

S. Vinod Selvaganesh
Indian Institute of Technology-Madras

P. Dhanasekaran
CSIR-Central Electrochemical Research Institute-Madras Unit

CONTENTS

12.1 INTRODUCTION

Two-dimensional (2D) materials have been reported widespread use in condensed matter science, chemistry, biotechnology, materials engineering, and nanotechnology, among other disciplines [1]. Graphene and its derivative 2D materials such as transition metal dichalcogenides (TMDs), carbon nitride (g-C_3N_4), boron nitride (h-BN), black phosphorus, MXenes, and silicene used as catalysts and additives for membranes have captivated the research world for decades [2]. Meanwhile, 2D structures were believed to be thermodynamically unstable, the innovation of graphene

DOI: 10.1201/9781003178453-12

and its special properties culminated in a technological uprising. TMD materials, which are layered compounds with substantial in-plane interaction and weak out-of-plane interactions, may be exfoliated into 2D layers with single unit-cell thickness. 2D thin-film sheets have been investigated for years, current progress in nanoscale materials categorization and product processing has brought about a paradigm shift [3,4]. 2D electron confinement, for example, gives 2D materials intriguing electrical characteristics, which has inspired the development of next-generation electrical gadgets.

2D metals, unlike 2D layered materials, are usually made up of non-directional metallic bonds and have non-layered structures. Besides, 2D materials have outstanding mechanical stability and optical clarity due to their atomic thickness and elevation in anisotropy, which come up with new possibilities for designing 2D material-based optoelectronic systems and wearable devices. Furthermore, heterostructures can be formed by assembling disparate 2D materials without regard to lattice matching or processing compatibility [5]. There is a revival of technical and industrial curiosity in 2D structures with an atom thickness, thanks to recent advancements in sample processing, optical identification, conversion, and manipulation of 2D materials. As graphene and TMDs are packed into a few layers, van der Waals bonds form a weak force of attraction between the layers. The stacking series of bilayer 2D materials changes the crystal symmetry and equilibrium size, affecting physical properties such as band difference, phonon vibration frequency, and superconductivity. The combination of their electronic and optical properties, which can be engineered by chemical functionalization, makes 2D materials appealing in diverse fields.

Non-covalent interactions may cause physical adsorption, or physisorption, of molecular units onto the basal planes of 2 sheets, or chemical adsorption, or chemisorption, of reactive species undergoing chemical reactions. The heavy intra-layer covalent bonding and comparatively weaker interlayer van der Waals interaction of the analogous bulk materials, which facilitates their exfoliation to 2D layers, result in a layered structure that assists their 2D layer exfoliation. In layered materials, electronic mobility and hence conductivity are often 3- to 4-fold more effective across the layers than between the layers. This is due to good charge carrier localization between individual layers and charge carrier transport across several layers simultaneously at the same time.

12.2 ADVANTAGES OF 2D MATERIALS IN ELECTROLYTIC MEMBRANE

A solid-state electrolyte is a solid ionic conducting electrolyte material, and it is mainly utilized in electrochemical devices like fuel cells and solid-state batteries, besides wide their utility for molecular separation [6,7]. The most commonly used methods for solid-state electrolyte formulation are (a) polymer doping with organic or inorganic salts, resulting in so-called polymer electrolytes like those based on and (b) the use of a polymer network as mechanical reinforcement for an ion-conducting matrix, referred to as a composite electrolyte.

2D materials like graphene, MoS_2, MXene, metal-organic frameworks, and covalent organic framework nanosheets are fast getting prominence in fabricating high and permeability membranes. 2D-based membranes can generate highly selective separations due to their balanced tunability and specific mechanism of interlayer lengths and openings at nanoporous surfaces. Due to their ultra-thin thickness and remarkable mechanical and chemical stability, recent discoveries have revealed 2D fillers in composite materials are desirable components for modifying and enhancing membrane properties [8,9]. Nanoscale 2D materials have intriguing possibilities for enhanced molecular separations and transport. MXenes are 2D materials that have been extensively researched and applied to high-performance molecular separations and ionic/molecular transport. Because of its remarkable flexibility, hydrophilic surface, strong mechanical strength, and good electrical conductivity, the MXene family of materials has grown in prominence. 2D materials that possess good transport characteristics are incorporated into membrane microstructure because their physical and chemical properties drastically fine-tune the membrane characteristics. The larger surface area and more chemically active external surface structure 2D materials integrated composite membranes exhibit facile proton transport rate due to the creation of additional conducting channels for ion mobility. The planar structure of 2D materials is a more appropriate nano-configuration for the fabrication of thin selective layers, as 2D allows the full exploitation of their features, even at low filler loadings.

12.2.1 How 2D Fillers Differ from Other Nano Filler Materials

The membrane is composed of semi-crystalline glassy or elastomeric polymers. These polymers' structural and dynamical behavior differs, leading to a wide variety of molecular permeability characteristics. Cussler et al. developed an analytical model for calculating gas permeability and selectivity in composite membranes, including selective flakes, to anticipate layered material's impact on composite membrane separation performance [10]. Membrane nanoscopically thin flakes, according to the authors, is best described as nanoscale composite structures that cannot be explained by analytical models that assume bulk permeation action in the flakes. Because of its stability and intrinsic ability to be functionalized, graphene-based 2D materials are preferred over nanomaterials for membrane modification. Due to the barrier feature of the nanoplatelets, they are increasing the filler loading. With the use of 2D materials, rational design of predefined interlayer channels, reasonable functionalization, membrane nanopores, and new design of polymer structure is possible because 2D materials are easily tunable with several functional groups. In addition, incorporating 2D materials, interface design could be controlled that greatly facilitates improving selectivity and permeability of the membranes for dynamic liquid–liquid/liquid–gas separation.

12.3 IMPORTANT PARAMETERS FOR IMPROVING MEMBRANE PROPERTIES

The mechanism of transport or separation through 2D material incorporated membranes greatly influenced by the factors, namely (a) Membrane interlayer channels, (b) Membrane interposes, (c) Surface functionalization, (d) Membrane microdomains, and (e) Balancing hydrophilic/hydrophobic domains. The redesigned membranes' increased separation performance may be achieved because of their 2D structure, high hydrophilicity, tunable interlayer spacing, and superior mechanical strength. Finally, the influence of substrate chemical functional groups on the binding characteristics of 2D stacked materials with their underlying support should always be addressed. To assemble 2D nanosheets into ordered laminates, several methods namely polymer induction, vacuum suction, or external-force-driven techniques can be utilized predominantly. The rate of penetration and selectivity of membranes can both be improved by intelligently selecting the aperture size and permeability of in-plane nanopores. 2D nanosheets may be readily aggregated into laminar membranes having intermediate pathways for selective molecular transport, in addition to the in-plane intrinsic holes. It was shown for the first time by Geim et al. that graphene oxide (GO) membranes could enable unlimited water penetration while being impervious to liquids, vapors, and gases [11]. They discovered that the practically zero friction surface of the non-oxidized section of GO nanosheets improves the fast movement of water based on theoretical simulations. To provide firm and discerning transport channels in 2D materials-based membranes, three major approaches for creating nanostructures and regulating the chemical composition during membrane construction were proposed: (a) building laminar structures, (b) modification of interlayer transport pathways, and (c) directing superficial morphology. Herein, we have focused on recent and advanced 2D nanomaterials (current literature, novel methodology, and characterization) like graphene, Boron nitride, MoS_2, and MXene nanosheet used as effective filler for base electrolyte membranes and its Physicochemical properties toward energy community application.

12.4 GRAPHENE/GRAPHENE OXIDE THEIR COMPOSITES FOR ELECTROLYTE MEMBRANE APPLICATIONS

The new technology using graphene-based or composite 2D nanostructured electrolyte membrane offers incremental development such as scaling the size of the product or improving the order of magnitude. The new universal technology of graphene-based materials is better to change the product efficiency and is also environmentally friendly. As far as the graphene properties are considered, it undoubtedly has higher potential. Nobel laureates in Physics for 2010, A. Geim, and K. Novoselov, have acknowledged the excellent/intriguing novelty of the physical properties that could be exhibited in graphene nanostructured. Especially graphene has a higher Brunauer, Emmett and Teller (BET) surface area, a thin layer or monolayer, transparent like plastic, high electrical conductivity, superior thermal conductivity water-resistant membrane, elastic film, chemically inert, and superior stability, which are more useful for improving electrolyte membrane properties.

Graphene alone is not suitable or recommended materials for, especially electrolyte membrane applications such as storage and conversion applications. However, an optimum level of nanostructured graphene, when introduced into electrolyte membrane or separator, can significantly improve membrane properties like tensile strength, proton conductivity water uptake, Young modulus, and water permeability resulting in the overall enhanced performance of the designated membrane/separator in conversion, storage, gas sensor, and separation applications. In addition, it is more important to have a basic understanding of synthesis methodology, forming a stacking layer or monolayer, and constructing better composition or composite and physical characterization, which may further expand the electrolyte membrane properties.

12.4.1 Synthesis, Characterization, and Properties of Graphene/Polymer Electrolyte Membranes

An optimum graphene/graphene oxide powder was used as an effective additive/filler in the SPEEK electrolyte membrane in current decades because it allows accessible water uptake properties, gas permeability, and proton transport due to its larger BET area. Dai et al. synthesized an innovative composite membrane-like graphene-SPEEK via the conventional solution casting technique. First, an optimum level of graphene (15 mg) was dispersed in 10 mL of DMF followed by sonication. Followed by it, a required amount of SPEEK (1.5 mg) was gradually added to the above admixture and then stirred for 24 hours. Finally, the resultant admixture was transferred into a flat glass plate followed by evaporation at two different temperatures (60°C and 100°C) overnight. While introducing graphene 2D nanostructure into SPEEK membrane, physicochemical properties are improved, the vanadium ions reduced cross-overrate toward redox flow battery. Especially, the swelling ratio (15.2%) and ion exchange capacity (1.98 mmol g^{-1}) have been ameliorated as compared to pristine SPEEK membrane (21.2% and 2.0 mmol g^{-1}) [12].

Self-humidified composite membrane fabrication via solution casting techniques uses carboxyl-functionalized graphene and terminal anionic oxygen from phosphotungstic acid (PWA) to turn the composition into the SPEEK framework. The optimum composition of the SPEEK-PWA-graphene improves self-humidify behavior at 60°C. PWA and graphene with different compositions were mixed into SPEEK matrix, and the admixture was stirred for 48 hours. PWAs terminal and anionic oxygen species are highly reactive sites and could interact with graphene nanostructure and SPEEK network. The combination composition improves the proton conductivity (5.19×10^{-2} S cm^{-1}) and methanol uptake (91.2%), resulting in improved self-humidifying behavior for fuel cell application [13]. Suhaiman et al. observed that few-layer enriched graphene oxide incorporated into the SPEEK matrix showed solid interfacial interaction between surface groups (-O, -OH, -COOH) of graphene oxide and sulfonic group SPEEK, which resulted in improved proton conduction than SPEEK matrix. This result indicates that the accessibility of graphene oxides fillers developed with different surface oxygenated reactive groups has intensely influenced the electrostatic interaction and hydrogen bonding with the SPEEK electrolyte membrane [14].

Considering direct methanol fuel cell (DMFC), methanol permeability, and proton conductivity are important parameters for improving DMFC application. At the same time, the incorporation of graphene oxide nanosheet fillers into the SPEEK electrolyte membrane leads to retaining hydrophilic property via the hydrogen bonding network. Similarly, proton conductivity has tremendously enhanced while introducing an optimum amount of GO into the SPEEK electrolyte membrane. Moreover, incorporating layer-by-layer assembly of graphene oxide nanosheet in the SPEEK electrolyte membrane gradually decreases methanol permeability. The bilayer or layer-by-layer membrane was formed that is more helpful in blocking the methanol molecules and enhanced more than 50% fuel activity than SPEEK.

According to Cai et al., Zeolitic imidazolate-graphene oxide 2D SPEEK nanohybrid nanocomposite show superior membrane behavior and improved cell activity. First, 250 mg of SPEEK matrix is dissolved in DMF, then an optimum composition of ZIF-L@GO 2D nanostructured materials added dropwise followed by sonication for 4 hours and stirred overnight. Next, the above admixture was cast on a flat glass plate and dried for 16 hours at 80°C in the vacuum oven. The cross-linked structure of the ZIF-L@GO/SPEEK framework is illustrated in Figure 12.1. The cross-linked ZIF-L@GO/SPEEK bond formation was studied through X-ray Photoelectron Spectrometer (XPS) measurement. The optimum composition of ZIF-L@GO/SPEEK was found to retain the physical properties equivalent to SPEEK and improved stability than the SPEEK matrix [15].

Thimmappa et al. cast a thin layer of 2D Graphene oxide solution alone on a glass slide and then air-dried, effectively used as a membrane for the low temperature polymer electrolyte fuel cells (LT-PEFC) application. The as-prepared 2D graphene oxide showed improved H^+ conductivity of 0.001 mS cm^{-1} at atmospheric temperature. However, a thin layer of graphene oxide membrane has lower electrolyte resistances than Nafion 211, especially the LT-PEFC configuration. When the graphene oxide membrane thickness is reduced, the ion transport properties are superior at room temperature [16].

Similarly, Kumar et al. observed that free-standing graphene oxide exhibited higher proton conductivity $4–8\times10^{-2}$ S cm^{-1} at different temperatures ranging between 25°C and 90°C, showed a fuel cell performance of 8 mW cm^{-2} when tested in DMFC configuration [17]. Besides, the oxygen-containing surface group and sulfonic acid on the edges and basal plane are ascribed to the in-plane proton transport.

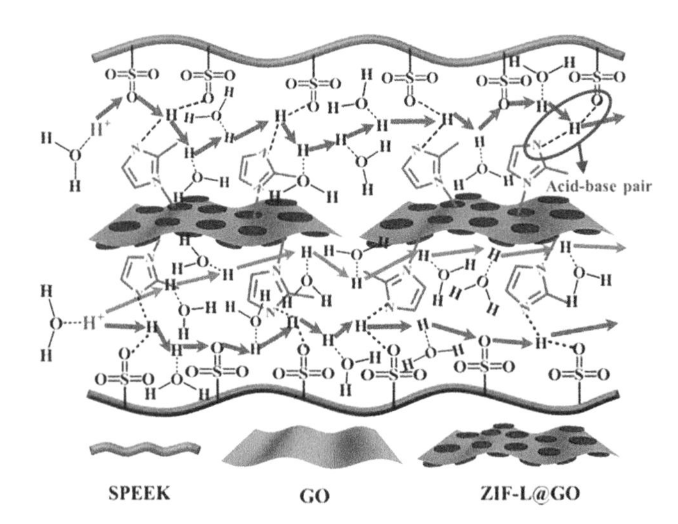

FIGURE 12.1 Schematic diagram of the formation of SPEEK/ZIF-L@GO. Adapted with permission from [15]. Copyright (2021) Elsevier publishing Ltd.

Kuila et al. explored an effective and simple technique for the fabrication of water-dispersible graphene-(SPEEK) electrolyte membrane used for energy storage application. According to Kuila et al., 0.1 g of graphite oxide was dispersed in 0.2 L of water, followed by sonication and centrifuged (remove unexfoliated graphite oxide). Separately, 0.3 g of SPEEK was dissolved in water and stirred at 60°C. The supernatant of SPEEK was slowly added to graphene oxide and stirred for 24 hours at 70°C. A required amount of hydrazine was added dropwise to the above admixture and refluxed at 100°C for 12 hours. Finally, SPEEK-Graphene (SPG) was filtered using a cellulose acetate membrane. The macromolecules nature of the SPEEK membrane is adsorbed on the graphene surface via π−π interactions. This modified nature of the SPEEK-graphene electrolyte was found to be highly stable even up to 6 months.

Figure 12.2a and b represents AFM micrographs image of graphene oxide and SPEEK-SPG. The AFM images indicated that graphene oxide has a thickness of 0.84 nm while SPG exhibited a higher thickness (1.34 nm). This is mainly due to the SPEEK membrane adsorption on a graphene surface and the development of the single-layer graphene sheet. SPG membrane shows larger rectangular areas representing its higher EDLC and superior capacitive behavior [18]. The Nafion-based perfluorinated membrane is generally used for LT-PEFC and HT-PEFC applications. However, Nafion suffers from deterioration even under standard polymer electrolyte fuel cells (PEFC) operating conditions, especially at higher temperatures. An optimum level of sulfonic acid-functionalized graphene is a promising filler for Nafion electrolyte membrane for improved proton-conducting in polymer electrolyte fuel cells. S-graphene-Nafion hybrid composite is achieved via embedding S-graphene in Nafion, which gives higher absorption water molecules and superior proton transport

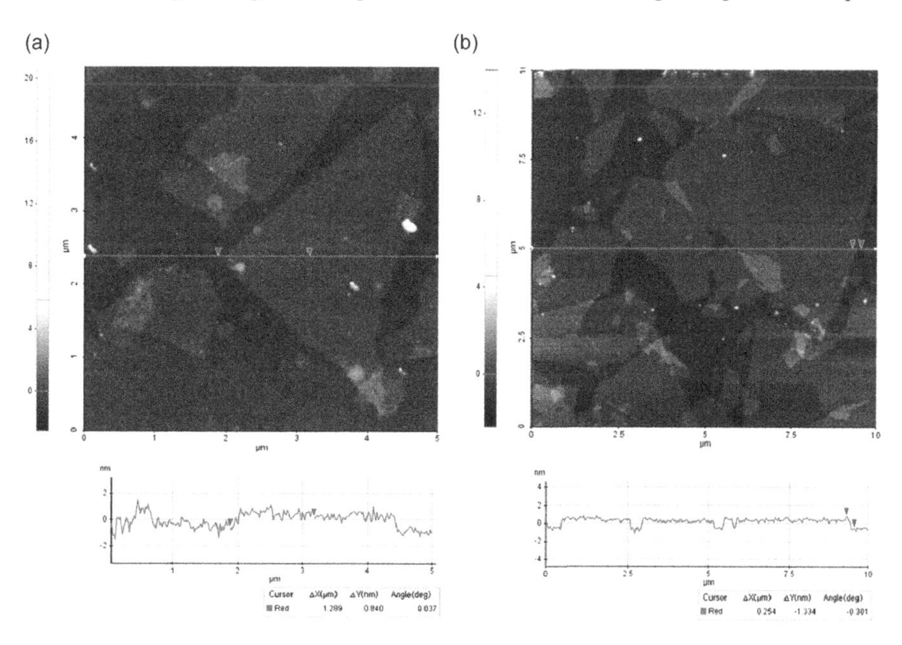

FIGURE 12.2 Atomic force microscopy of (a) graphene oxide and (b) SPEEK-Graphene composite (SPG). Adapted with permission from [18]. Copyright (2012) ACS publishing Ltd.

through the overall membrane under low relative humidity conditions. 1 wt.% of S-graphene-Nafion hybrid composite was observed to be 17 mS cm^{-1}, which is almost five times greater than recast Nafion membrane. This recast membrane shows a maximum of 300 mW cm^{-2} power density at 760 mA cm^{-2} under 20 RH conditions at 70°C [19]. Moreover, sulfonated graphene nanosheet was highly spread into Nafion structure resulting in hydrophilic domains and enhanced surface roughness, thus strengthening the proton conduction properties in a composite membrane. Similarly, 0.7 wt.% of graphene filler spread into the poly(vinyl alcohol) (PVA) matrix has a higher tensile strength (76%) and Young's modulus (62%) compared to PVA [20]. Sulfonated graphene nanoplates have naturally fascinated the ionic species of Nafion monomer to their region, which is rotated and polarized underneath the electric field, and rearranged proton channels in the solidified membrane. As a result, the trans-plane conductivity of the composite electrolytic membrane was observed to be 0.155 S cm^{-1} at 80°C, which is more than 48% higher than cast Nafion. Consequently, PEFC comprising composite electrolyte membrane delivered more than 15% higher fuel cell performance than Nafion at 80°C [21].

Yan et al. prepared a monolayer of thin graphene nanosheets sandwiched between Nafion membranes via chemical vapor deposition. Nafion ionomer spin-coated on graphene/copper surface, and then the Cu foil was an etching by $FeCl_3$-based materials. Finally, Graphene-Nafion film was transferred onto a Nafion electrolyte membrane and hot-pressed [22]. As a result, the graphene-Nafion formation reduced methanol permeability up to 68%, and the graphene monolayer was selectively allowed to protons than the Nafion membrane. Similarly, an optimum level of 5 wt.% reduced graphene oxide nanosheet in Nafion matrix has shown better proton conductivity. The optimum composition was synthesized via solution cast and hot press thermal reduction techniques. The composite active membrane showed increased proton conductivity more than 30 times from the recast Nafion membrane.

Su et al. fabricated GO-Nafion hybrid membranes via spin-coating techniques. The hybrid Nafion-graphene oxide composite shows inferior vanadium ion permeability than the Nafion. The ultrathin graphene oxide layer suppressed the vanadium cross-over and enhanced the proton conduction throughout the membrane. As a result, the composite membrane has a higher energy efficiency of 81%–88% and coulombic efficiency of 92%–98% than the Nafion membrane (energy efficiency 68%–79% and coulombic efficiency 73%–90%). Accordingly, as prepared composite membrane shows excellent battery stability even after 200 charge–discharge cycles, and the capacity decay rate is much lower (0.23%) than the Nafion membrane (0.44%) [23].

12.5 BORON NITRIDE NANOSHEET/POLYMER MEMBRANE SYNTHESIS AND PROPERTIES

In general, hydrocarbon-based or composite membranes have conjugated a more significant percentage of sulfonic acid. Sulfonic acid groups can prompt proton conductivity and hence enhancing ion exchange capacities based on the hopping mechanism. Through membrane stability test dry–wet cycles, hydrocarbon-based solid membranes significantly change volume shrinkage/expansion behavior. Thereby, the initial bond interfaces among the membrane and electrodes become weakened during stability tests (wet-dry cycles), damaging the membrane. Hence, reducing hydrocarbon-type membrane swelling and stability is a serious and challenging topic in the energy conversion application.

To overcome the above issues, 2D boron nitride (BN) has begun enhancing the mechanical strength when incorporated in a minimal amount within the composite. The additional advantages of BN nanosheet have low density, excellent chemical and thermal stability. Keun-Hwan et al. explored the intrinsic mechanical strength of boron nitride nanosheet and the marginal amount of (0.3 wt.%) boron nitrides nanoflakes (BNNF) significantly enhanced mechanical stability. The 1-pyrenesulfonic acid (PSA)-BNNF/SPEEK composites membrane fabricate via a solution casting method with a thickness of ~40 μm. The optimum composition of 0.3 wt.% of PSA-BNNF composites delivered a higher tensile strength of about 47.2 MPa, corresponding to 41% higher than bare SPEEK [24].

An optimum level of functionalized boron nitride is incorporated into the Nafion (N) membrane synthesis. Wu et al. via the vacuum filtration method. The optimum composition of boron nitride nanocomposites (NBNM) exhibits higher stability at fully humidifies conditions and a redox environment. The proton conductivity shown is 0.44 S cm^{-1} at 95% RH condition at 80°C than Nafion (0.13 S cm^{-1}). The higher proton conductivity might be attributed to the maximized availability of the microchannel and physicochemical environments in composite membranes. The composite membrane enhances proton conductivity rapidly at higher RH conditions because of the presence of water molecules as the H^+ carrier achieves the loosely arranged boron nitride sheet in the membrane and simplifies the state. In addition, the formation of long-range ionic channels and quasi-isotropic architecture simplifies the proton conduction [25].

Akel et al. explored the different amounts of 3, 5, 10, and 15 wt.% of nano-hexagonal boron nitride (NhBN) was added to Nafion polymer. The effect of NhBN was examined in terms of swelling property, thermal stability, methanol permeability, and proton conductivity. Generally, hydrocarbon-based membranes have a larger affinity to agglomerate, which is attributed to weakening mechanical and physicochemical behavior. While incorporating NhBN nanosheet improves water content throughout the membrane, it may considerably impact ionic channel or conductivity and thermal stability [26].

Hence, functionalized boron nitride-polymer composite membrane was found to improve the H^+ conductivity by order of magnitude with the increase in the temperature. Besides, nitrogen groups in the BN structure play a vital role in improving H^+ conduction and HSO_3^- group functionalization. In addition, the fBN-Nafion composite membrane exhibits lower weight than the Nafion membrane at the same temperature, which may lead to the discomfiture of breaking S-C bond due to the presence of hydrogen bond existing between the sulfonate group of Nafion and nitrogen atom of BN nanosheets. Moreover, the presence of functionalized BN limits the methanol permeation rate [27]. Similarly, hexagonal BN nanosheet combined with Sulfonated polyvinyl phosphonic acid and polysulfone (PVPA-SPSU) polymer, the resulting better thermomechanical stabilities. The uniform distribution of hexagonal BN nanosheet in polymer framework increases proton conduction. Incorporation of 5 wt.% h-BN into SPSU-PVPA results in a composite membrane that exhibits excellent proton conductivity of 9×10^{-3} (S cm^{-1}) at 150°C under anhydrous conditions [28].

According to Liu et al., reports wafer-scale hexagonal BN monolayer deposited on copper rolls via space confined CVD techniques (Figure 12.3). Further Nafion layers are functionalized on hexagonal BN nanosheets and then effectively transfer as grown Nafion-hexagonal BN films to SPEEK electrolyte membrane. During, self-discharge operating condition, Nafion/h-BN/SPEEK composite membrane exhibited higher long-term stability of ~78 hours than Nafion-SPEEK (51 hours) and SPEEK membrane (46 hours) [29]. The presence of monolayer h-BN nanosheet resulted in a positive effect on vanadium redox flow battery performance, and three times better selectivity is achieved than that of pure SPEEK membrane.

12.6 MODIFIED MOS_2/POLYMER MEMBRANES FOR ENERGY APPLICATIONS

Transition metal chalcogenides, predominantly molybdenum-based chalcogenides, have gained researchers' attention owing to their interesting physicochemical properties. MoS_2 has been the focus of various research groups, mainly attributed to its fascinating behavior like high anisotropy, distinct crystal structure, and low cost. The inter-layer distance calculated for MoS_2 is around 0.615 nm as compared to 0.335 nm for graphite. This is primarily attributed to the less strong physical affinity between the freely suspended S–Mo–S sheet [30]. This makes MoS_2 a promising alternative, eco-friendly nanomaterial to be implemented in diverse energy conversion and storage systems.

2D MoS_2 nanosheets in pristine form or composite have shown remarkable capacitance attributed to their high conductivity, larger surface area, and graphene-like morphology. Three-dimensional frameworks comprising 2D MoS_2 nanosheets exhibited enhanced stability that retains the integrity of the suspended ultrathin sheets having a large available surface area. This is considered a major advantage of using 2D MoS_2 ultrathin nanosheets as fillers. MoS_2 nanosheets polystyrene composite membrane was prepared through simply blending and casting technique. It was observed that the exfoliated sheets of approximately 3–4 layers remarkably enhanced Young's modulus up to 0.8 GPa as compared to 0.2 GPa observed for pristine polystyrene. In addition, the tensile strength also noticeably showed up to three-fold enhancement even with a very low-level loading of exfoliated MoS_2-nanosheets.

Hou et al. prepared a series of innovative quaternized poly(vinyl alcohol)/chitosan/molybdenum disulfide (QPVA/CS/MoS_2) membranes for alkaline DMFC. The optimum composition of QPVA/CS/MoS_2 was enhanced by the mechanical stability, which may be due to the incorporation of MoS_2. According to these authors, the addition of 1.0 wt.% of MoS_2 resulted in the lowest methanol permeability [31]. Wu et al. has adopted an innovative approach of in situ growing of MoS_2 to fabricate MoS_2-Nafion-based hybrid membranes showing superior selectivity for its application in DMFCs. He surmised that the judicious selection of the precursor for Mo, that is, $(NH_4)_2MoS_4$, and solvent media led to a strong interaction of Mo-precursor with sulfonic groups present in the Nafion. This resulted in the selective growth of MoS_2 flakes mostly surrounding the ionic clusters in the obtained MoS_2/Nafion hybrid membrane, which resulted in improved connectivity between the existent ionic clusters and consequently enhanced the ionic conductivity [32]. It is noteworthy

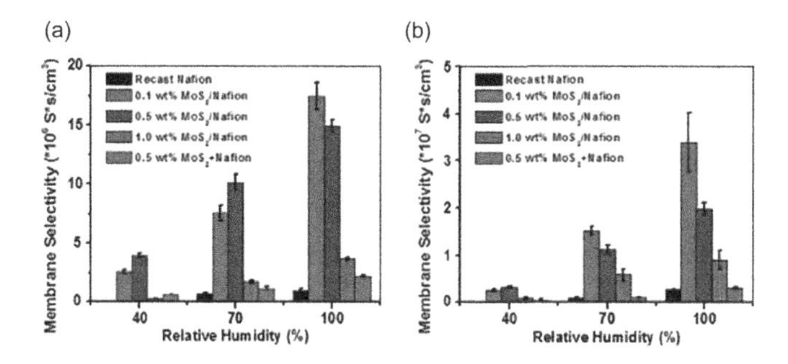

FIGURE 12.3 (a and b) Represents the membrane selectivity with respect to relative humidity for Nafion and 0.1–1.0 wt.% MoS_2/Nafion composite at 25°C and 50°C. Adapted with permission from [32]. Copyright (2013) ACS publishing Ltd.

that membrane selectivity has been enhanced by nearly two orders of magnitude for the designated MoS_2/Nafion composite membrane than with that of the Nafion when tested under stringent DMFC operation at 50°C and high methanol concentration. Figure 12.3 shows the comparative membrane selectivity for recast Nafion and various compositions of MoS_2/Nafion composite membranes at 25°C and 50°C.

Similarly, Wu et al. adopted Nafion in water as a medium to exfoliate MoS_2. According to his findings, strong non-covalent bonding interactions between 2D MoS_2 and Nafion were brought about by hydrophilic sulfonic acid groups and hydrophobic polytetrafluoroethylene backbone in stabilization and functionalization of Nafion–MoS_2 (N-MoS_2) nanocomposites. Figure 12.4 clearly illustrates the synchronized exfoliation of MoS_2 and functionalization of 2D MoS_2-nanosheets over Nafion. Interestingly, these interactions were found to be highly independent of the pH of the solvent/media. However, the extent of exfoliation and the size of the obtained MoS_2 nanosheets were severely influenced by the Nafion concentration [33].

Besides, an optimum composition of sulfonated poly (ether sulfone) (sPES) composited with sulfonated-MoS_2 was explored toward application in DMFC by

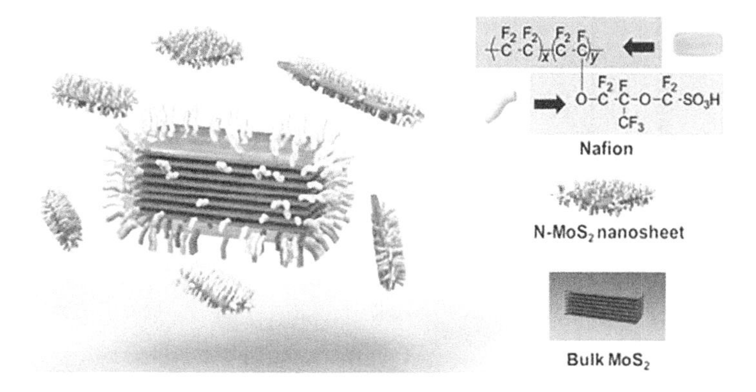

FIGURE 12.4 Illustration of the synchronized exfoliation of MoS_2 and functionalization of 2D MoS_2-nanosheets over Nafion. Adapted with permission from [33]. Copyright (2018) Springer publishing Ltd.

Kulshrestha and co-workers [34]. In this study, hydrothermally prepared 2D MoS_2 was synthesized by adopting a hydrothermal method and functionalized using 1, 3-propane sulfone as sulfonating media. Notably, the methanol cross-over from anode to cathode was drastically reduced due to the methanol permeation barrier created by s-MoS_2, which simultaneously offered a proper path for proton transport. 5 wt.% of s-MoS_2 into the composite membrane resulted in a reduction in methanol permeation up to 91% and four times higher electrochemical selectivity than sPES membrane, which further increased the cell performance as compared to sPES.

Exfoliated molybdenum disulfide (E-MoS_2) 2D nanosheets incorporated into chitosan membrane reveal superior IE capacity, water absorption, and proton conductivity than pre-chitosan-based electrolyte membrane. Composite membranes comprising 0.75% exfoliated MoS_2 showed better ionic conductivity and membrane selectivity. Besides, the presence of MoS_2 nanosheets in the membrane channels prevented methanol cross-over to a greater extent [35]. In a similar approach, Chen et al. have synthesized a chitosan-based composite membrane employing MoS_2-nanosheets with the primary intention of boosting membrane stability [36]. In this study, chemical conjugation of the exfoliated MoS_2-nanosheets was brought about by using thioglycolic acid to functionalize the exfoliated MoS_2 2D structure chemical conjugation with thioglycolic acid prior to the preparation of composite membrane. Blending with functionalized exfoliated MoS_2-nanosheets provided better thermal and mechanical stability to the chitosan membrane. The effect is primarily attributed to the strong interaction between the former and the latter. Thermogravimatric analysis (TGA) plot for pristine and composited chitosan membrane is shown in Figure 2.5.

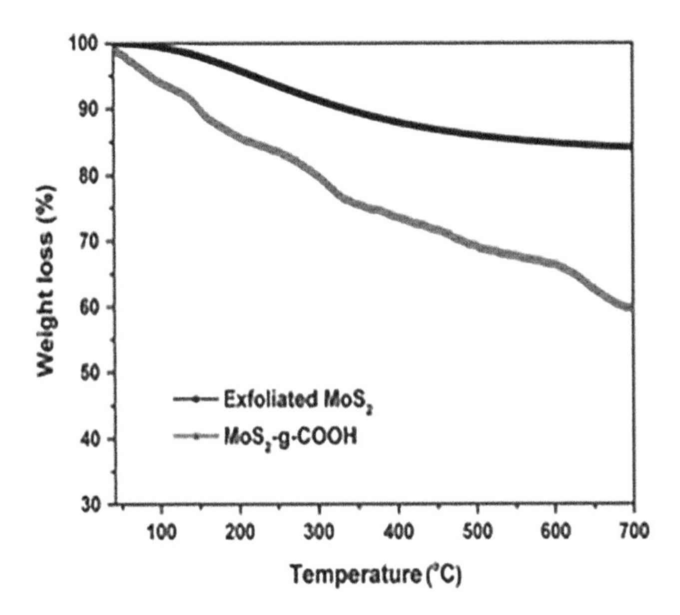

FIGURE 12.5 Comparative TGA plots for pristine and composited chitosan membrane with functionalized exfoliated MoS_2-nanosheets. Adapted with permission from [36]. Copyright (2016) Springer publishing Ltd.

12.7 2D MXene/POLYMER COMPOSITE MEMBRANES FOR ENERGY APPLICATIONS

MXenes are captivating materials comprising 2D transition metal nitrides or carbides obtained from the MAX phases via the selective chemical etching method. An optimum level of MXene-based fillers is recognized as the most promising filler material for enhancing the proton conduction behavior and thermo-mechanical stability of polymer matrix. MXene-based filler was used in polybenzimidazole (PBI) membrane for intermediated polymer electrolyte membrane fuel cell application. Interestingly, 3 wt.% of $Ti_3C_2T_x$-MXene 2D nanosheet incorporated into polybenzimidazole membrane that resulted in better physicochemical behavior, like Young's modulus and tensile strength for PBI membrane. According to the reports by Fei et al. about 3 wt.% of $Ti_3C_2T_x$-MXene incorporated in the PBI membrane when employed in high temperature-proton-exchange membrane fuel cells (PEMFC) resulted in improved performance of 200 mW cm^{-2} than to 130 mW cm^{-2} for pristine PBI membrane [37]. Besides, it was observed that the incorporation of $Ti_3C_2T_x$-MXene also resulted in the enhanced mechanical and thermal stability of PBI-based electrolyte membranes.

Cheng et al. reported an optimum level of $Ti_3C_2T_x$ nanosheet is used as an additive for quaternized polysulfone (QPSU) matrix and formed complex membrane for alkaline polymer electrolyte fuel cells (APEMFC). The designated composite membrane showed two-fold higher ionic conductivity and increased peak power density when tested in APEMFCs. This clearly shows that the quaternary ammonium group facilitates the OH^- conductivity [38]. Wang et al. adopted a quick solution casting technique to prepare a hybrid composite membrane by dispersing $Ti_3C_2T_x$ nanosheets into Nafion ionomer solution. The as-prepared hybrid membrane exhibited remarkable properties like thermal stability, mechanical strength, and water uptake [39]. In addition, 2D $Ti_3C_2T_x$-nanosheets were also investigated as an effective filler for the Nafion membrane. It is surmised that the incorporation of 2D MXene nanosheets as filler into the Nafion matrix resulted in a decrease in elongation properties. It is noteworthy that as-prepared hybrid membrane, when employed in PEMFCs, a power density of 200 mW cm^{-2} is achievable.

As proof of the thermal stability of 2D MXene-based materials, there are also reports wherein these materials are employed as fillers, even in solid oxide fuel cells (SOFCs). Xian et al. dispersed about 5 wt.% of $Ti_3C_2T_x$-MXene into $Sm_{0.2}Ce_{0.8}O_{1.9}$, and observed an improvement in the ionic conductivity. In addition, the performance of the SOFC cell was also observed to be doubled with the incorporation of $Ti_3C_2T_x$ into $Sm_{0.2}Ce_{0.8}O_{1.9}$-based membrane [40].

12.8 FUTURE SCOPE FOR 2D FRAMEWORK IN ELECTROLYTE MEMBRANE-BASED APPLICATION

2D material-based membranes have demonstrated remarkably high separation performance along with good transport properties that are desirable in sectors like nanofiltration, energy material device components, desalination, gas separation, and solvent dehydration. In current decades, several key concerns have been resolved.

Because of high-quality nanosheets and fine-tuning of the transport channel with the incorporation of single atomic layer thickness 2D materials, these membranes exhibit well-transport properties with good thermos-mechanical stabilities. Previous studies have demonstrated these membranes as building blocks for high-performance separation as well as better transportation. It is possible to highlight the use of 2D membrane materials for efficient separation, with an emphasis on two key issues: (a) membrane stability through extended times of operation and (b) the construction of large-area, effective uniform membranes. The structural stability of 2D membranes in water is one of the most complicated issues to resolve. Considering GO membranes as an example, water molecules will intercalate into interlayer channels due to the hydrophilic nature of GO nanosheets distributed with oxidized functional groups, breaking the previously formed barrier. As a result, it's essential to avoid 2D membrane swelling in order to improve durability. In an aqueous solution, the swelling of 2D material extends interlayer channels and reduces sieving property, resulting in 2D material disintegration and impeded long-term operation.

Constructing molecular bridges to enhance both the linkages between 2D materials and the interactions between the 2D material layer and the substrate has to be improved for constructing better-performing membrane. In addition to this, increasing the interfacial adhesion between the 2D material layer and the substrate is also desirable for fabricating stable 2D material-based membrane. Another issue with 2D material-based membranes in terms of practical use is membrane-scalable manufacturing. As a result, synthesis methodologies with a larger membrane area and their separation performance should be offered. 2D material-based membranes as a new family of superior membranes are capable of overcoming many challenges with their improved microdomain structure and phase well-separated composition with minimal thickness. However, the uneven distribution of pathways present in the membrane microdomains has to be regularized for better results along with exploring the fundamental mechanism behind nanochannel characterization and transport mechanism. Furthermore, there are no theoretical models for describing mass-transfer behavior happening within 2D membranes. The transport mechanism may be explored through theoretical calculations, and theoretical models must be constructed to envisage the impact of membrane assembly on the transport mechanism, such as pore size, molecular structure, electrical characteristic, and so on. Also, because the thickness of 2D material membranes is 1–2 orders of magnitude less than that of traditional membranes, specialized membrane modules are necessary to maintain structural integrity and, as a result, separation performance. In this line, the cost-efficient engineering of structurally robust, defect-free membranes for molecular transport/separation applications remains inspiring. Once integration methods with traditional porous supports have been refined, higher production scalability will be possible, leading to lower manufacturing costs. However, to attain higher industrial viability, cost reductions of at least 1–2 times higher are necessary, most likely influenced by economic forces. Such commercial forces often lead to further design optimization and process refinement in both research and industrial production. Materials development, essential knowledge acquisition, and the creation of analytical simulation and experimental techniques are all essential to adopt 2D material-based membranes for a diverse range of electrochemical technology.

ACKNOWLEDGMENT

S. Mohanapriya, P. Dhanasekaran, and S. Vinod Selvaganesh thank CSIR-for Senior Research Associateship (Scientist's Pool Scheme 9037-A, -9123-A, and 9178-A, respectively).

REFERENCES

1. H. W. Guo, Z. Hu, Z. B. Liu, and J. G. Tian, "Stacking of 2D materials," *Advanced Functional Materials*, 2021, 31, 1–32.
2. N. Glavin, R. Rao, V. Varshney, E. Bianco, A. Apte, A. Roy, E. Ringe, P. Ajayan, "Emerging applications of elemental 2D materials," *Advanced Materials*, 2020, 32, 1–22.
3. Z. Lin, C. Wang, Y. Chai, "Emerging group-VI elemental 2D materials: Preparations, properties, and device applications," *Small*, 2020, 16, 1–16.
4. X. P. Zhai, B. Ma, Q. Wang, and H. L. Zhang, "2D materials towards ultrafast photonic applications," *Physical Chemistry Chemical Physics*, 2020, 22, 22140–22156.
5. L. Huang, Z. Hu, H. Jin, J. Wu, K. Liu, Z. Xu, J. Wan, H. Zhou, J. Duan, B. Hu, J. Zhou, "Salt-assisted synthesis of 2D materials," *Advanced Functional Materials*, 2020, 30, 1–27.
6. C. Z. Liang, T. S. Chung, J. Y. Lai, "A review of polymeric composite membranes for gas separation and energy production," *Progress in Polymer Science*, 2019, 97, 101141.
7. C. Crivello, S. Sevim, O. Graniel, C. Franco, S. Pens, J. Puigmarti-Luuis, D. Munoz-Rojas, "Advanced technologies for the fabrication of MOF thin films," *Materials Horizons*, 2021, 8, 168–178.
8. A. M. Tandel, W. Guo, K. Bye, L. Huang, M. Galizia, H. Lin, "Designing organic solvent separation membranes: Polymers, porous structures, 2D materials, and their combinations," *Materials Advances*, 2021, 2, 4574.
9. L. Prozorovska, P. R. Kidambi, "State-of-the-art and future prospects for atomically thin membranes from 2D materials," *Advanced Materials*, 2018, 30, 1–24.
10. E. L. Cussler, D. F. Evans, "How to design liquid membrane separations," *Separation & Purification Reviews*, 1974, 3, 399–421.
11. R. R. Nair, H. A. Wu, P. N. Jayaram, I. V. Grigorieva, A. K. Geim, "Unimpeded permeation of water through helium-leak-tight graphene-based membranes," *Science*, 2012, 335, 442–444.
12. W. Dai, L. Yu, Z. Li, J. Yan, L. Liu, J. Xi, X. Qiu, "Sulfonated poly(Ether Ether Ketone)/graphene composite membrane for vanadium redox flow battery," *Electrochimica Acta*, 2014, 132, 200–207.
13. S. H. Lee, S. H. Choi, S. A. Gopalan, K. P. Lee, G. Anantha-Iyengar, "Preparation of new self-humidifying composite membrane by incorporating graphene and phosphotungstic acid into sulfonated poly(ether ether ketone) film," *International Journal of Hydrogen Energy*, 2014, 39, 17162–17177.
14. N. S. Suhaimin, J. Jaafar, M. Aziz, A. F. Ismail, M. H. D. Othman, M. A. Rahman, F. Aziz, N. Yusof, "Nanocomposite membrane by incorporating graphene oxide in sulfonated polyether ether ketone for direct methanol fuel cell," *Materials Today: Proceedings*, 2021, 46, 2084–2091.
15. Y. Y. Cai, Q. G. Zhang, A. M. Zhu, Q. L. Liu, "Two-dimensional metal-organic framework-graphene oxide hybrid nanocomposite proton exchange membranes with enhanced proton conduction," *Journal of Colloid and Interface Science*, 2021, 594, 593–603.

16. R. Thimmappa, M. Gautam, M. C. Devendrachari, A.R. Kottaichamy, Z. M. Bhat, A. Umar, M. O. Thotiyl, "Proton-conducting graphene membrane electrode assembly for high performance hydrogen fuel cells, *ACS Sustainable Chemistry and Engineering*, 2019, 7, 14189–14194.
17. R. Kumar, M. Mamlouk, K. Scott, "A graphite oxide paper polymer electrolyte for direct methanol fuel cells," *International Journal of Electrochemistry*, 2011, 2011, 1–7.
18. T. Kuila, A. K. Mishra, P. Khanra, N. H. Kim, M. E. Uddin, J. H. Lee, "Facile method for the preparation of water dispersible graphene using sulfonated poly(ether-ether-ketone) and its application as energy storage materials," *Langmuir*, 2012, 28, 9825–9833.
19. A. K. Sahu, K. Ketpang, S. Shanmugam, O. Kwon, S. Lee, H. Kim, "Sulfonated graphene-nafion composite membranes for polymer electrolyte fuel cells operating under reduced relative humidity," *Journal of Physical Chemistry C*, 2016, 120,15855–15866.
20. J. Liang, Y. Huang, L. Zhang, Y. Wang, Y. Ma, T. Guo, Y. Chen, "Molecular-level dispersion of graphene into poly(vinyl alcohol) and effective reinforcement of their nanocomposites," *Advanced Functional Materials*, 2009, 19, 2297–2302.
21. F. Fang, L. Liu, L. Min, L. Xu, W. Zhang, Y. Wang, "Enhanced proton conductivity of Nafion membrane with electrically aligned sulfonated graphene nanoplates," *International Journal of Hydrogen Energy*, 2021, 46, 17784–17792.
22. F. Fang, L. Liu, L. Min, L. Xu, W. Zhang, and Y. Wang, "Enhanced proton conductivity of Nafion membrane with electrically aligned sulfonated graphene nanoplates," *International Journal of Hydrogen Energy*, 2021, 46, 17784–17792.
23. L. Su, D. Zhang, S. Peng, X. Wu, Y. Luo, G. He, "Orientated graphene oxide/Nafion ultra-thin layer coated composite membranes for vanadium redox flow battery," *International Journal of Hydrogen Energy*, 2017, 42, 21806–21816.
24. K. H. Oh, D. Lee, M. H. Choo, K. H. Park, S. Jeon, S. H. Hong, J. K. Park, J. W. Choi, "Enhanced durability of polymer electrolyte membrane fuel cells by functionalized 2D boron nitride nanoflakes," *ACS Applied Materials and Interfaces*, 2014, 6, 7751–7758.
25. W. Jia, P. Wu, "Stable boron nitride nanocomposites based membranes for high-efficiency proton conduction," *Electrochimica Acta*, 2018, 273, 162–169.
26. N. M. Barkoula, B. Alcock, N. O. Cabrera, T. Peijs, "Flame-retardancy properties of intumescent ammonium poly(Phosphate) and mineral filler magnesium hydroxide in combination with graphene," *Polymers and Polymer Composites*, 2008, 16, 101–113.
27. W. Jia, B. Tang, P. Wu, "Novel composite proton exchange membrane with connected long-range ionic nanochannels constructed via exfoliated nafion-boron nitride nanocomposite," *ACS Applied Materials and Interfaces*, 2017, 9, 14791–14800.
28. M. S. Tutgun, D. Sinirlioglu, S. U. Celik, A. Bozkurt, "Preparation and characterization of hexagonal boron nitride and PAMPS-NMPA-based thin composite films and investigation of their membrane properties," *Ionics*, 2015, 21, 2871–2878.
29. J. Liu, L. Yu, X. Cai, U. Khan, Z. Cai, J. Xi, B. Liu, F. Kang, "Sandwiching h-BN monolayer films between sulfonated poly(ether ether ketone) and nafion for proton exchange membranes with improved ion selectivity," *ACS Nano*, 2019, 13, 2094–2102.
30. M. B. Khan, R. Jan, A. Habib, and A. N. Khan, "Evaluating mechanical properties of few layers MoS_2 nanosheets-polymer vomposites," *Advances Materials Science Engineering*, 2017, 2017, 1–7.
31. X. Jiang, Y. Sun, H. Zhang, L. Hou, "Preparation and characterization of quaternized poly(vinyl alcohol)/chitosan/MoS_2 composite anion exchange membranes with high selectivity," *Carbohydrate Polymers*, 2018, 180, 96–103.
32. K. Feng, B. Tang, P. Wu, "Selective growth of MoS_2 for proton exchange membranes with extremely high selectivity," *ACS Applied Materials and Interfaces*, 2013, 5, 13042–13049.

33. W. Jia, B. Tang, P. Wu, "Nafion-assisted exfoliation of MoS_2 in water phase and the application in quick-response NIR light controllable multi-shape memory membrane," *Nano Research*, 2018, 11, 542–553.
34. V. Yadav, N. Niluroutu, S. D. Bhat, V. Kulshrestha, "Sulfonated poly(ether sulfone) based sulfonated molybdenum sulfide composite membranes: Proton transport properties and direct methanol fuel cell performance," *Materials Advances*, 2020, 11, 820–829.
35. K. Divya, D. Rana, S. Alwarappan, M. S. S. Abirami Saraswathi, A. Nagendran, "Investigating the usefulness of chitosan based proton exchange membranes tailored with exfoliated molybdenum disulfide nanosheets for clean energy applications," *Carbohydrate Polymers*, 2019, 208, 504–512.
36. X. Yang, N. Meng, Y. Zhu, Y. Zhou, W. Nie, P. Chen, "Greatly improved mechanical and thermal properties of chitosan by carboxyl-functionalized MoS_2 nanosheets," *Journal of Materials Science*, 2016, 51, 1344–1353.
37. M. Fei, R. Lin, Y. Deng, H. Zian, "Polybenzimidazole/MXene composite membrnaes for intermediate temperature polymer electrolyte membrane fuel cell," *Nanotechnology*, 2018, 29, 035403.
38. X. Zhang, C. Fan, N. Yao, P. Zhang, T. Hong, C. Xu, J. Cheng, "Quaternary $Ti_3C_2T_x$ enhanced ionic conduction in quaternized polysulfone membrane for alkaline anion exchange membrane fuel cells," *Journal of Membrane Science*, 2018, 563, 882–887.
39. Y. Liu, J. Zhang, X. Zhang, Y. Li, J. Wang, "$Ti_3C_2T_x$ Filler effect on the proton conduction property of polymer electrolyte membrane," *ACS Applied Materials and Interfaces*, 2016, 8, 20352–20363.
40. H. Xian, C. Fan, P. Zhang, R. Wang, C. Xu, H. Zhai, T. Hong, J. Cheng, "Effect of MXene on oxygen ion conductivity of $Sm_{0.2}Ce_{0.8}O_{1.9}$ as electrolyte for low temperature SOFC," *International Journal of Electrochemical Science*, 2019, 14, 7729–7736.

13 Nanocomposites of 2D Materials for Enhanced Electrochemical Properties

Nitika Devi
Shoolini University

Alok Kumar Rai
University of Delhi (North Campus)

Rajesh Kumar Singh
Central University of Himachal Pradesh

CONTENTS

DOI: 10.1201/9781003178453-13

13.1 INTRODUCTION

Nanotechnology has provided a new world to material science, which not only helps in exploring the new applications in this area but also imparting the new proficiency in already existing approaches. Dimensions play a vital role in describing the particular type of nanomaterials and depending upon that they are classified as 0-dimensional, 1-dimensional, 2-dimensional, and 3-dimensional materials means at least one of their dimensions is in the nanometer range (1–100 nm). After the discovery of graphene, 2D materials found great attention because of their exceptional properties [1]. Later, it was found that beyond graphene there are other 2D materials that are also exceptionally good. These materials are MoS_2, WS_2, $MoSe_2$, and WSe_2 collectively known as transition metal dichalcogenides (TMDs), layered double hydroxides, LAPONITEs clay, hexagonal boron nitride (h-BN), black phosphorous (BP), a family of monoelemental compounds (Xenes), metal oxides, graphitic carbon nitride (g-C_3N_4), metal nitrides/carbides (MXenes), transition metal halides (e.g., PbI_2 and $MgBr_2$), transition metal oxides (e.g., MnO_2 and MoO_3), perovskite-type oxides (e.g., $K_2Ln_2Ti_3O_{10}$ and $RbLnTa_2O_7$ (Ln: lanthanide ion)), and 2D polymers. Although these materials are sufficient enough for particular use, many applications need to evolve their composite forms for overcoming some drawbacks associated with pristine form [2]. Usually, composite classification is done by making matrix material as a reference for classing the composites. Depending upon that we can have metal-matrix composites, organic-matrix composites, ceramic-matrix composites. Polymer composites and carbon matrix composites are two subclasses of organic-matrix composites. Another type of composite is heterostructure composites of 2D materials based on their structural features. Each composite type has its advantages as well drawbacks associated with them. For example, polymer composites have advantages like ease of formation, lightweight but for an application of high corrosive environment ceramic and metal-matrix composites are found to be more stable due to their nature of bonding that is either covalent or ionic. In addition to this, high strength and toughness of the metal-matrix composites provide a good mechanical strength to the composite [3–6].

Different approaches are employed for the synthesis of various types of 2D composites materials. But in general, it can be broadly divided into two types that are in situ approaches and ex situ approaches. In case of in situ approaches, examples are hydrothermal and solvothermal methods. In ex situ methods a specific composition and dimensions of 2D materials are fabricated then it is being combined with the other part of composite with covalent and non-covalent interactions [7]. Chen et al. [8] reported an in situ approach for the synthesis of graphene-reinforced Cu-matrix composites. In this work, ball milling is followed by the hot pressing sintering for obtaining the final product. Another report by Santalucia et al. [9] gave a comparison of in situ and ex situ approaches for synthesizing MoS_2/TiO_2 nanoparticles sheets composites. It was concluded that depending upon the particular synthesis approach, it is possible to incorporate a specific type of interactions, also in situ approach results in thinner and more defective MoS_2 slabs. The dimensionality of MoS_2 slabs is also lower as compared to the same product obtained by ex situ synthesis.

The abovementioned 2D-materials composites can be further utilized in different areas such as energy storage applications [10,11], sensing [12–14], and EMI shielding [15,16].

13.2 COMPOSITES OF 2D MATERIALS

Composite is a combination of two materials in specific composition so that resulting product has entirely different properties as compared to the individual components. Composites have been extensively used in material science for almost all applications so that either properties of the materials got enhanced or some drawbacks get improved. One component of the composite acts as a matrix and other component acts as reinforcement [17]. There are different types of possible 2D material composites based on the matrix.

13.2.1 Metal-Matrix Composites of 2D Materials

These types of composites metal and alloys are used as a matrix and 2D materials act as reinforcement. In order to achieve required properties of metal-matrix composites, the incorporation and dispersion of the reinforced material should be done in good manner so different techniques have been developed for the synthesis of this class of composites. In the case of graphene, techniques like metal alloying, spray forming, electrochemical deposition, and selective laser melting [18] are used. The development of these techniques is the result of the constant efforts for removing the agglomeration problem and increasing the interface binding between matrix and reinforcement material. All these factors significantly affect the mechanical strength of the material. Six main techniques that are used for forming the metal-matrix composites are ball milling, hot press sintering, stir casting, hot extrusion, wet chemistry, and compress shearing [4]. All the mentioned techniques have been successfully applied for the synthesis of various 2D material composites. Out of these techniques, milling, sintering, and casting are the most efficient ones for 2D materials composite synthesis. Nautiyal et al. [19] used ball milling for synthesizing copper matrix composites reinforced by reduced graphene oxide (rGO)-MoS_2. The advantage of using ball milling is uniform distribution but the associated drawback is that it can cause defects in the materials. In the hot press sintering process, high pressure and temperature are used for pressing and sintering. Usually graphite material die is used for compacting the materials as it is stable at high temperatures [18]. Liu et al. [20] used hot press sintering method for fabricating Ni-nanoparticles decorated graphene reinforced in Al matrix composites. Merit associated with this method is that no damage occurs to the material but the problem is contamination, which is mostly unavoidable in hot pressing sintering. Stir casting is also a metallurgical process like the sintering and ball milling; the difference is that in this process reaction occurs in liquid phase. A mechanical stirrer is used for mixing reinforcing and matrix material, which as a final product results in composite formation. In addition to already discussed processes stir casting offers some advantages related to the material size and yield and is also found to be cost-effective [18,20]. Sekar et al. [21] prepared MoS_2 and Al_2O_3 nanoparticles composites by using stir casting techniques. The method successfully synthesized the composite and was also capable of controlling the reinforcement percentage to a minimum of 0.5 wt.% for MoS_2. Other than the mentioned processes wet chemistry, hot extrusion, and compression shearing are also the possible routes for synthesis, etc. Microwave-assisted synthesis processes are also being used for composite formation as it provides a fast and efficient way of synthesis [22,23]. Many times two or more processes may be combined to get useful results. Such a synthesis of graphene reinforced in Cu matrix composite was reported by Nie et al. [24]. In this process, graphite oxide is synthesized by Hummer's method and then molecular level mixing of Cu matrix and RGO is done and in the last step hot press sintering is used for forming the compact mass of Cu-RGO composites. Schematic of the process is shown in Figure 13.1.

13.2.2 Ceramic-Matrix Composites

This class of composites is similar to the metal-matrix composites; the only difference is that the metal matrix is replaced by the ceramic matrix. Also, similar type of approaches has been used by the researchers for synthesizing the ceramic composites like mechanical mixing, colloidal dispersion, and solid-state sintering [25]. In the formation of ceramic-matrix composites the main problem is the high temperature of reaction, which sometimes result in diffusion and non-useful reactions. So, the need is to develop such methods, which can synthesize the 2D ceramic-matrix composites at low temperature or without forming any extra reaction by products. One such work is done by Guo et al. [26] in which formation of nanocomposites of 2D MXene with zinc oxide is done using a cold sintering process (schematic shown in Figure 13.2).

This developed process is capable of synthesizing the composite material with uniform distribution with relative density of 92%–98% without any oxidation and interdiffusion [26]. Colloidal dispersion mixing process has also been used for ceramic matrix-2D materials composite. In this process, both the phases are dispersed in the same type of colloidal solution and then both solutions are mixed for obtaining the final product. Surface treatment of material is important as oppositely charged surfaces will strongly bond as a result there will be strong binding in the component of composite [27]. Fan et al. [28] prepared ceramic matrix few layer graphene composite by using colloidal mixing. This method gave freedom for manipulating the doping and concentration level of the matrix and reinforced material. Liu et al. [29] prepared in situ Ni-nanoparticle decorated graphene reinforced in Al matrix, which results in improved mechanical properties and can be successfully used for different applications.

12.2.3 Metal-Organic Matrix/2D Polymer/Carbon Composites

These 2D material composites used polymer/carbon materials as a matrix. Such composites result in the enhancement of the properties of polymer materials with respect to strength, thermal stability, and conductivity. In some cases, mechanical properties of the polymer get enhanced by 200 times than the pristine polymer [6]. Synthesis of 2D-polymer composites involves two steps; the first one is exfoliation of the 2D material and then dispersion of this material into polymer matrix. Exfoliation

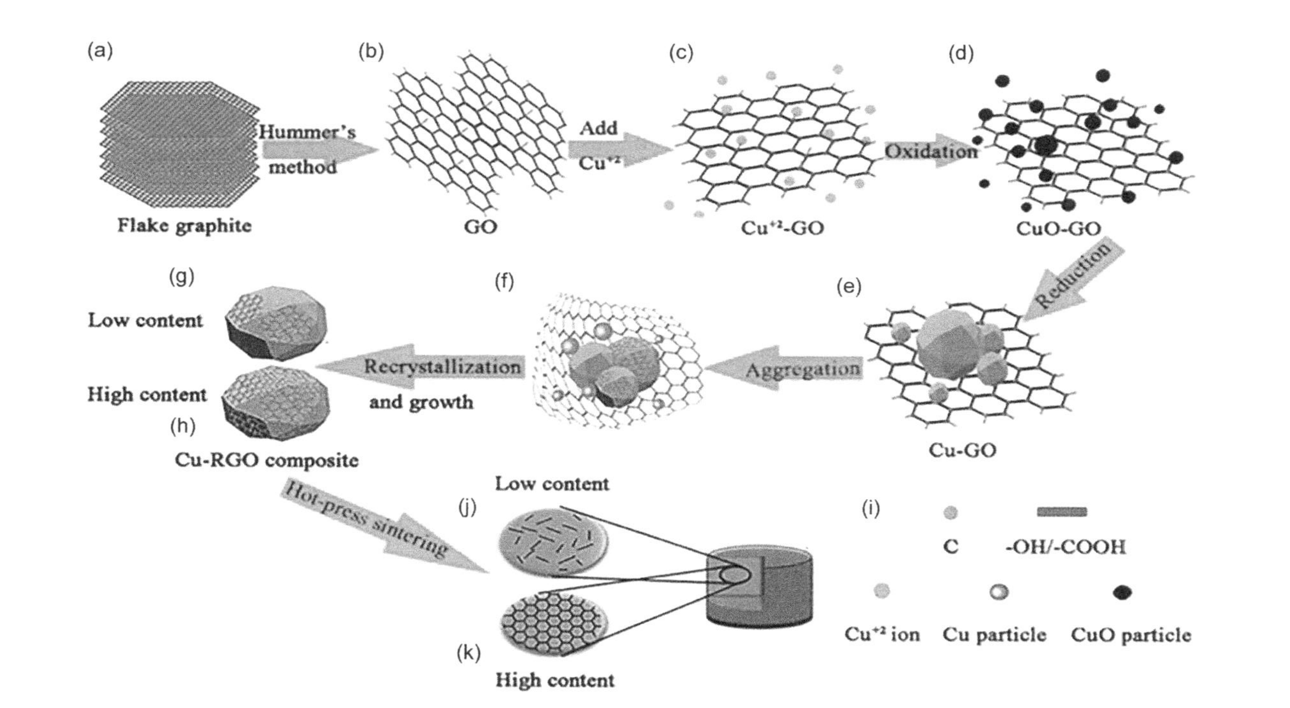

FIGURE 13.1 (a) Schematic representation of RGO-Cu matrix composites. (a) Graphite; (b) Hummer's method synthesis procedure for graphite oxide; (c) Cu^{2+} absorbed on the surface of GO; (d) oxidation of Cu^{2+}; (e) Cu-GO obtained by reducing Cu^{2+} with ascorbic acid; (f) small Cu particles aggregated; (g, h) Cu-RGO power with low and high content of RGO; (i) bulk Cu-RGO composite consolidated by hot-press sintering; (j, k) spatial distribution of RGO in copper matrix [24]. Adapted with permission from [24]. Copyright (2018) MDPI.

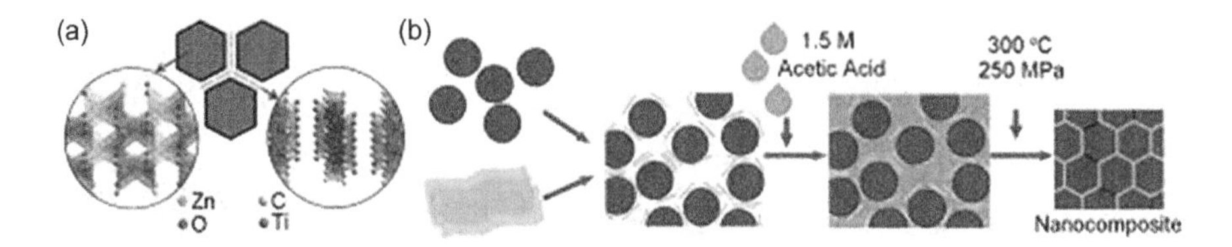

FIGURE 13.2 Schematic illustration showing: (a) the grain boundary of (1-*y*)ZnO -yTi3C2Tx nanocomposites and (b) the fabrication process via cold sintering. Adapted with permission from [26]. Copyright (2018) Wiley.

is a process used to synthesize 2D material, which was established after the discovery of graphene. In exfoliation, the interaction between the two layers of material is decreased either by functionalizing it or by ball milling. The exfoliation can be of several types depending upon the methods used such as electrochemical exfoliation and etching-assisted exfoliation [6]. Dispersion of 2D material in polymer matrix is done in some organic solvents. Interactions of matrix with fillers are very crucial as larger loading doesn't necessarily mean the enhanced properties. Many steps may be included in the synthesis method depending upon the product requirement. Song et al. [30] fabricated the composite of graphene with cellulose-derived carbon aerogels and polydimethylsiloxane (PDMS) polymer. Different steps involved in synthesis are shown in Figure 13.3. Resulting composites have been used for EMI shielding and found to have excellent EMI shielding capacity and thermal conductivity coefficient. Composite synthesis can be described in three main steps that is preparation

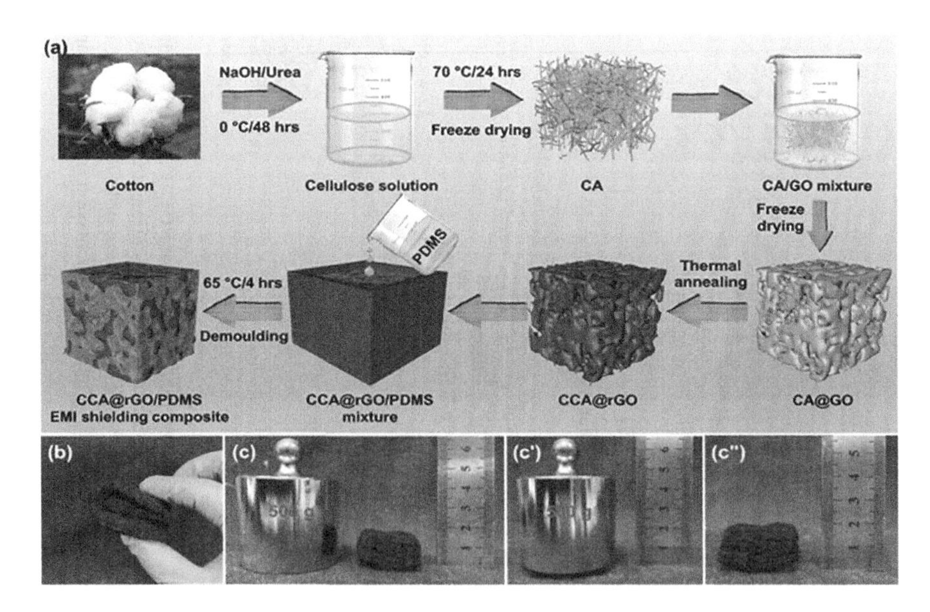

FIGURE 13.3 Schematic illustration of the fabrication procedure for CCA@rGO/PDMS EMI shielding composites (a) illustration of the flexibility (b) and resilience (c-c'') of CCA@ rGO aerogel. Adapted with permission from [30]. Copyright (2021) Springer.

of cellulose aerogel, formation of cellulose doped graphene, and in the last step the CCA@rGO/PDMS composite. Calculated composite mechanical strength is found to be 4.1 MPa and hardness was 42 HA.

Black phosphorous/polyaniline hybrid composites were developed by Moghaddam et al. [31] for supercapacitor applications. Composite showed a high specific capacitance of 354 F g^{-1} at 0.3 A g^{-1} current density. In situ polymerization of black phosphorous nanosheets and aniline is used for their composite formation. Fabrication of h-BN-rGO@PDA nanohybrids is also an example of polymer composites. Composite can be used as anti-corrosion filler for polyvinylbutyral (PVB) and showed a two-order higher performance than pristine PVB [32].

13.2.4 Heterostructure Composites of 2D Materials

2D materials have a layered structure so diverse 2D materials can be stacked in different orientations in order to give a heterostructure composite. Synthesis methods for developing the heterostructure composites are similar to those used for synthesizing the individual layer of composite like mechanical exfoliation, chemical vapor deposition (CVD), and molecular beam epitaxy (MBE). Depending upon the orientation of stacking layers, mainly two types of heterostructure are formed; vertical or lateral configured heterostructures. CVD synthesis is widely used for creating different heterostructure due to better control on orientation, no. of layers, and interface of different layers of 2D materials [3]. Schematic of vertically and laterally aligned heterostructure composites are shown in Figure 13.4. Mechanical exfoliation is capable of synthesizing 2D materials due to weak van der Waals forces interplay between the layers of 2D materials. Similar interactions are also responsible for the heterostructure formation. Heterostructure can significantly affect the properties of the individual sheets by combining the characteristics of two dissimilar materials. If we particularly discuss graphene, then it has a zero bandgap, which restricts its use in electronics applications. But TMDs are such 2D materials in which bandgap can be tailored, so heterostructures of graphene and TMDs provide efficient materials for electronics [33].

Cho et al. [35] fabricated a heterostructure device of 2D graphene/MoS_2 for NO_2 sensing. Two methods have been used for heterostructure device formation, i.e., mechanical exfoliation and CVD. Yamaguchi et al. [36] evaluated tunneling transport in a few monolayer-thick WS_2/graphene and MoS_2/graphene heterojunction. In this work, both TMDs and graphene were prepared by mechanical exfoliation and heterostructure was obtained by dry transfer process. Black phosphorous/ultrathin Ti_3C_2/ultrathin g-C_3N_4(BQ/TiC/UCN) composite fabrication and performance as a photocatalytic hydrogen production is reported by Song et al. [37]. Presence of MXene increases the charge transfer due to photoinduced charge carriers. In this report, UCN and TiC were prepared by the furnace heated melamine at 550°C for 4 hours and constant stirring of 72 hours Ti_3AlC_2 in HF solution, respectively. Dispersion of UCN and TiC is done in distilled water and then xBQ/TiC/UCN composite is formed by grinding and sonication in milli-Q water with 2D black phosphorous (BP) powder. Black phosphorus@laser-engraved graphene heterostructure is developed for the electronic skin application sensor. This heterostructure exhibits the high strain sensitivity (2,765), low detection (0.023%) limit, which is the result

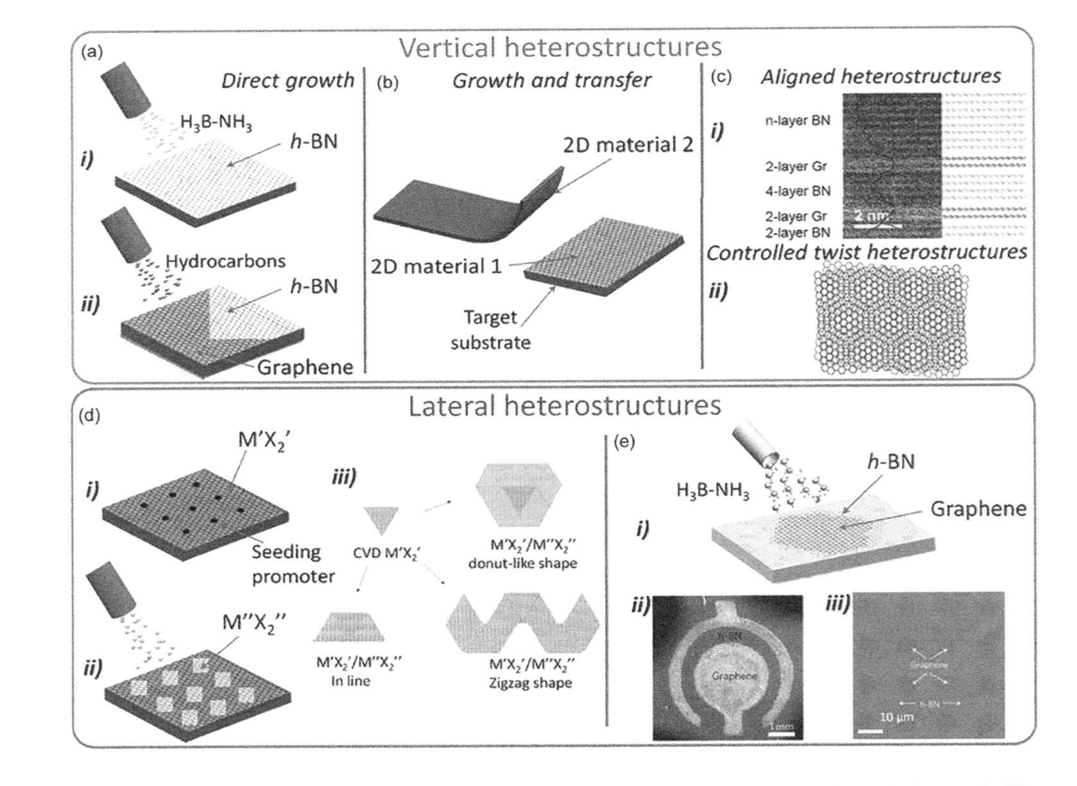

FIGURE 13.4 (a) CVD synthesis of Graphene and h-BN (i) Ammonia borane is used as a precursor for h-BN growth (ii) graphene is decorated on the h-BN grown surface. (b) Individual 2D sheets are prepared either by CVD or MBE then one layer is directly transferred to the other type of 2D material. (c) (i) Vertically aligned heterostructure (ii) Controlled twist heterostructure. (d) Different methods for synthesizing lateral heterostructures (i) After fabricating a 2D layer seeding is done on this layer (ii) Second layer is developed by using suitable precursor (iii) different shape like zig-zag, donut-like, linear can be grown by proper placement of seeds using processes like etching. (e)(i) h-BN/graphene heterostructure formation (ii) Scanning electron spectroscopy of h-BN/graphene heterostructure (iii) Optical image of h-BN/graphene heterostructure [34]. Adapted with permission from [34]. Copyright (2018) Nature.

of synergetic effect of different layered materials [38]. Scalable production of MoS_2/black phosphorous heterostructure is obtained by an in situ solvothermal synthesis in which black phosphorous is obtained by electrochemical exfoliation [39]. Composite showed a high catalytic activity with h low η_{10} overpotentials of 126, 237, 258 mV for hydrogen evolution reaction (HER) in acidic, alkaline, and neutral electrolytes, respectively [39].

13.3 SUPERCAPACITOR

Environment issues and increasing use of fossil fuel demand to develop alternative energy resources. Supercapacitors and batteries are such electrochemical alternatives, which can replace or reduce the use of fossil fuels. First, working of the supercapacitor will be described and then some of the works on 2D composites supercapacitors.

13.3.1 Mechanism

In conventional capacitors, energy is stored in the form of charges. Capacitors have a large value of power density but have very less energy density. Batteries on the other hand have a high value of energy density but low power density. Supercapacitors provided us an intermediate solution of balanced energy density and power density. The mechanism of working supercapacitors depends upon the type of supercapacitor. They are mainly of two types, i.e., electrochemical double-layer capacitor (EDLC) and pseudocapacitors. In EDLC, charging and discharging occurred because of absorption and desorption of the ions as in the case of conventional capacitors. But unlike the conventional capacitors in case of supercapacitors there is double layer charge deposition phenomenon, which results in high energy density and power density. A double layer of opposite charges is formed at the interface of electrode/electrolyte. In case of pseudocapacitors, chemical reactions are the main cause for the storage of energy. Because of abovementioned phenomenon, they are also named as Faradic and non-Faradic type supercapacitor. Faradic supercapacitors are pseudocapacitors and non-Faradic supercapacitors are EDLC [40,41]. A schematic of both types is shown below in Figure 13.5. Figure 13.5a shows a schematic of electrochemical double-layered supercapacitor. It can be seen from this schematic (a) that how the charges are accumulated as double layer and (b) electrochemical reaction are main cause for energy storage. Figure 13.5c giving a comparison of energy and power density of batteries and supercapacitors and EDLC supercapacitor are efficient enough in proving the good energy density and power density at the same time [41].

13.3.2 2D Material Composites as a Supercapacitor

Different materials like carbon nanotubes, transition metal oxide/hydroxides, 0D materials, and 2D materials are used as supercapacitor electrode materials. But the large surface area of 2D material makes them a suitable choice. In case of 2D materials graphene is used as an electrode material for EDLC supercapacitor as its high surface area (2,630 $m^2 g^{-1}$) results in large charge adsorption. Other than graphene, TMDs, MXene, and conducting polymers mainly work on the pseudocapacitance

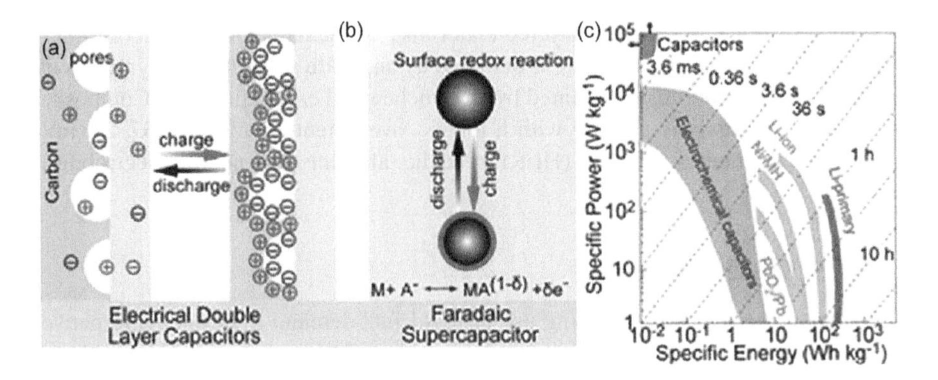

FIGURE 13.5 (a) Schematic representation of electrochemical double layer supercapacitor, (b) Pseudocapacitor, and (c) Ragone plot at different time constant [41]. Adapted with permission from [41]. Copyright (2019) Elsevier.

phenomenon. Phosphorene or black phosphorous is also being explored by researchers for its use as a supercapacitor electrode. Composite of abovementioned materials with metal oxide/hydroxide and ceramic will result in a hybrid-type supercapacitor in which energy and power densities are because of the contribution of both surface area and redox reactions. Such composite formation is important because both factors are important for its commercial applications [42]. Xiang et al. [43] reported in one of such effort in which reduced graphene oxide/Co_3O_4 composites are synthesized by in situ composite formation using hydrothermal synthesis. Morphological studies show that 20 nm Co_3O_4 nanoparticles are decorated on rGO sheets. Composites showed specific capacitance of 472 F g^{-1} at a scan rate of 2 mV s^{-1} with 82.6% retention of the capacitance at 100 mv^{-1} increased scanned rate with each step. Superior performance of the composite is due to the synergetic effect of redox reactions due to Co_3O_4 and conductivity of rGO matrix [43]. As discussed above, heterostructure composites formation can be done by stacking different 2D materials layer. Bissett et al. [44] synthesized such heterostructure composite, which is capable of giving 11 mF cm^{-2} at 5 mV s^{-1}. Composites were prepared by dispersing the mixture of exfoliating layers of MoS_2 and graphene in solution of *N*-methyl-2-pyrrolidone (NMP) dispersed isopropanol (IPA). Ramakrishnan et al. [45] also published a work on MoS_2-rGO composites for electrode material in Li-ion capacitors. Composite delivered a 52 Wh kg^{-1} and 60 W kg^{-1} energy density and power density, respectively. Polymer composites are also found to be used as an effective supercapacitor electrode. Wang et al. [46] synthesized MoS_2/polyaniline composites, which offered a high specific capacitance of 390 F g^{-1} with 86% of retention after 1,000 cycles. Composite is synthesized by in situ polymerization of exfoliated sheets of MoS_2 with aniline. 3D ternary composites of molybdenum disulfide/polyaniline/reduced graphene oxide aerogel are prepared by Sha et al. [47]. Composite formation of mass ratio 1:1 for MoS_2/rGO to PANI showed the highest specific capacitance of 618 F g^{-1} at 1.0 A g^{-1} with 96% of retention after 2,000 cycles [47]. MXene composite with 1T-MoS_2 also showed a high specific capacitance of 386.7 F g^{-1} at 1 A g^{-1}. 1T-MoS_2/Ti_3C_2 MXene heterostructures are synthesized using magneto-hydrothermal synthesis.

It is similar to the hydrothermal synthesis method, the only difference is the introduction of 9T magnetic field, which helps in the formation of the heterostructure composites. Also, morphology study showed that 1T MoS_2 nanosheets are successfully grown on Ti_3C_2 MXene. Electrochemical study cyclic voltammetry and galvanostatic charge–discharge results are shown in Figure 13.6. Figure 13.6 also explains the mechanism of storage in the heterostructure composites and the morphology image shows the stacked layers of the composite. Figure 13.6 given above showed the cyclic voltammetry and galvanostatic charge–discharge data for 1T-MoS_2, Ti_3C_2 MXene, 1T-MoS_2/Ti_3C_2 MXene heterostructures composites [48]. Figure 13.6a–d are the cyclic voltammetry and galvanostatic charge–discharge results of different samples and the large area under the curve indicates the high specific capacitance. Peaks in Figure 13.6a and c are direct consequences of the redox reactions. It is concluded by the electrochemical study that the high value of specific capacitance is because of the three factors that are 1T-MoS_2, Ti_3C_2 MXene, and extra H^+ storage formation in between the two composite layers as shown in Figure 13.6e. Figure 13.6f is the capacitance retention study that was found to be 96.8% after 20,000 cycles at 50 A g^{-1}, which is very significant for its electrode uses. Layer structure is preserved even after the 20,000 cycles as shown in SEM image of 1T-MoS_2/Ti_3C_2 MXene in Figure 13.6g [48].

Zhao et al. [49] described the development of metal-matrix composite of graphene, which explained the methods of synthesis and their application in a detailed manner. 2D material and modified 2D material like heteroatom doped grapheme [50], graphene, h-BN, MoS_2 [51], carbon-based 2D materials and their derivatives [52] have been reported for energy storage applications. VSe_2/N-doped carbon sphere composite gave an energy density of 85.41 Wh kg^{-1} at a power density of 701.99 W kg^{-1} [53]. Moghaddam et al. [54] review on TMDs and polymer composites describe all the aspects related to these composites. Black phosphorous composite is also getting attention due to its exceptional characteristics. Moghaddam et al. [31] synthesized the black phosphorous/polyaniline composite and analyzed its performance for supercapacitor application. Pseudocapacitor composite was capable of giving 354 F g^{-1} at a current density of 0.3 A g^{-1}. Another such composite thin film of polypyrrole/black phosphorus is fabricated by Luo et al. [55]. Composite films have an excellent stability and a high specific capacitance (497.5 F g^{-1}). So, this discussion can conclude that 2D material composite can give highly efficient supercapacitors with enhanced electrochemical properties.

13.4 BATTERIES

Batteries are an electrochemical device, which delivers electric energy at the expense of chemical energy.

13.4.1 MECHANISM

In battery, redox-reaction is a source for current generation that is at one electrode oxidation takes place and on another reduction. Electron loss at oxidized electrode is being forced to the pass through outer circuit which can be utilized for some useful work.

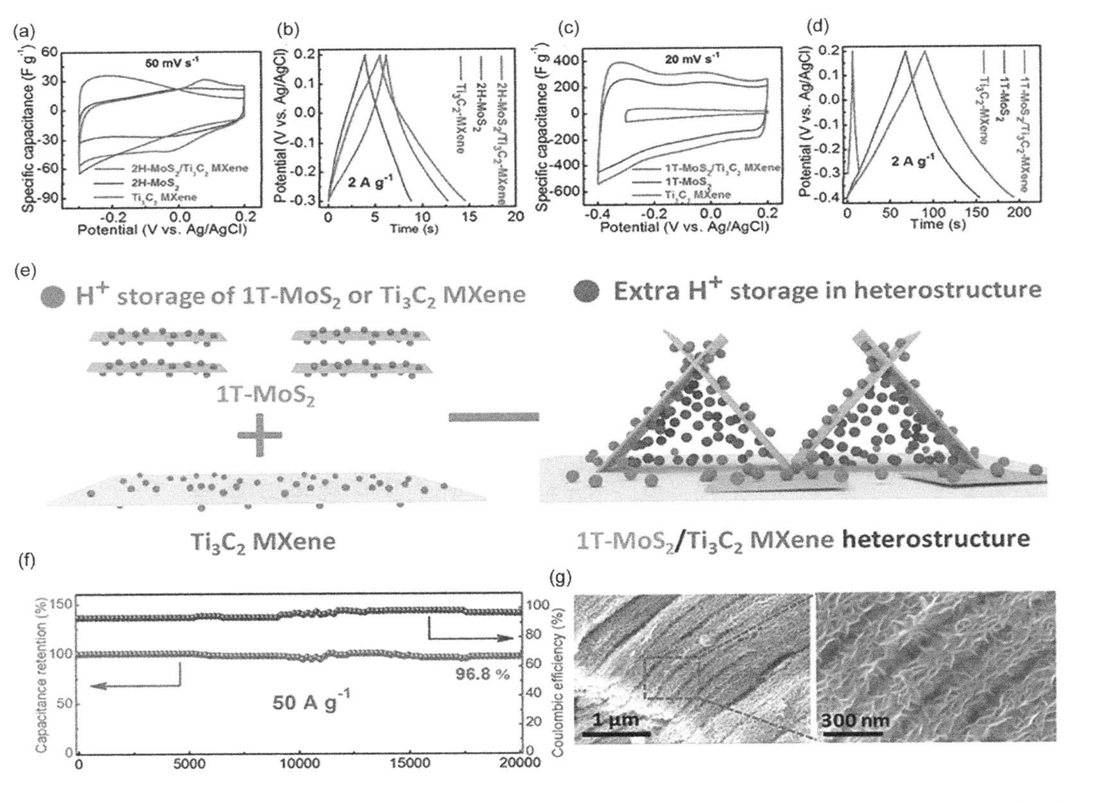

FIGURE 13.6 Electrochemical performance of 2H-MoS_2, 1T-MoS_2, Ti_3C_2 MXene, 2H-MoS_2/Ti_3C_2 MXene, and 1T-MoS_2/Ti_3C_2 MXene electrodes. (a) and (b) CV and GCD curves of 2H-MoS_2, Ti_3C_2 MXene, and 2H-MoS_2/Ti_3C_2 MXene. (c) and (d) CV and GCD curves 1T-MoS_2, Ti_3C_2 MXene, and 1T-MoS_2/Ti_3C_2 MXene. (e) Schematic diagram showing the H^+ ion storage of 1T-MoS_2 or Ti_3C_2 MXene and extra H^+ ion storage in 1T-MoS_2/Ti_3C_2 MXene heterostructure at charged–discharged state. (f) Capacitance retention after 20,000 cycles for 1T-MoS_2/Ti_3C_2 MXene electrode at 50 A g^{-1}. (g) SEM images of 1T-MoS_2/ Ti_3C_2 MXene electrode after long-term cycling [48]. Adapted with permission from [48]. Copyright (2020) Wiley.

This process is continuously repeated, which is called charging and discharging of the battery. For balancing this reaction an ion is also transported from one electrode to another [56].

13.4.2 2D Material Composites as a Battery

2D materials and their composites are also found to be suitable for battery electrodes. High conductivity and structural characteristics associated with these materials results in increased electrochemical performance. Graphene as well as other 2D materials composites are successfully being used for the electrodes in the battery and their synergistic effects result in highly efficient electrodes. Li et al. [57] synthesized graphene/BN thin film and used it as anode in the battery. Composite film was fabricated by vacuum filtration method by the dispersion of both GO and BN. Then vacuum-filtered thin film was annealed at 300°C for 2 hours in argon atmosphere. Resulting thin film was analyzed by cyclic voltammetry, galvanostatic charge–discharge, and electrochemical impedance spectroscopy. Composite thin film with 2 wt.% content showed excellent reversibility of 278 mAh g^{-1} and good retention on 200 cycles [57]. Metal oxide/graphene composites also offer good battery electrodes. Liang et al. [58] synthesized SnO_2/graphene composite and analyzed it as battery anode. Reported synthesis method is a facile green one-step approach, which also makes this cost-efficient and environmental friendly as well. Nanocomposite exhibited a reversible capacity of 690 mAh g^{-1} with 63% capacity retention after 20 cycles [58]. Heterostructure composite of g-C_3N_4/graphene has also been used for the cathode interlayer in Li-sulfur batteries [59]. As in Li-S batteries, performance is limited by the shuttle effect and the limited intrinsic electrical conductivity of sulfur. Shuttle effect is the main cause for limiting the performance of sulfur, which causes the dissolution of intermediate polysulfide, which leads to reversible loss of sulfur. This is because of the weak interaction between the cathode and dissolved lithium sulfide thus introducing an interlayer, which increases the interaction between dissolved lithium sulfide and interlayer. This type of arrangement can restrain the shuttle effect. Such an effort is done by Qu et al. [59]. They prepared an interlayer of g-C_3N_4/graphene by using a simple stirring and centrifugation of mixture of graphene and g-C_3N_4. Schematic Figure 13.7a, showing the arrangement of battery and how it is helping in restraining the shuttle effect. This interlayer is incorporated in between sulfur/carbon (S/KB) composites cathode and separator. Figure 13.7b image gave the morphology and how the S/KB layer is attached with g-C_3N_4/graphene. It is found that such arrangement results in strong stacking, which can also be approximated by overall performance. Figure 13.7c–e are the cyclic voltammetry, galvanostatic charge–discharge, and electrochemical impedance spectroscopy results. Resulting cell is capable of giving a high retention capacity and it remains good in the range from 0.1 to 2 C that is from 1,191.7 to 1,026.2 mAh g^{-1}, respectively, for 1,000 cycles. This study concludes that there is increase in Li^+ diffusion coefficient, which significantly increases the charge transfer at the electrolyte/electrode interface and thus reduces the shuttle effect [59]. Chen et al. [60] also developed Fe_3O_4 nanoparticle/MoS_2 nanosheet composites for battery application. Fe_3O_4 nanoparticle/MoS_2 nanosheet composite anode battery has highest delivering capacity of 1,033 and 224

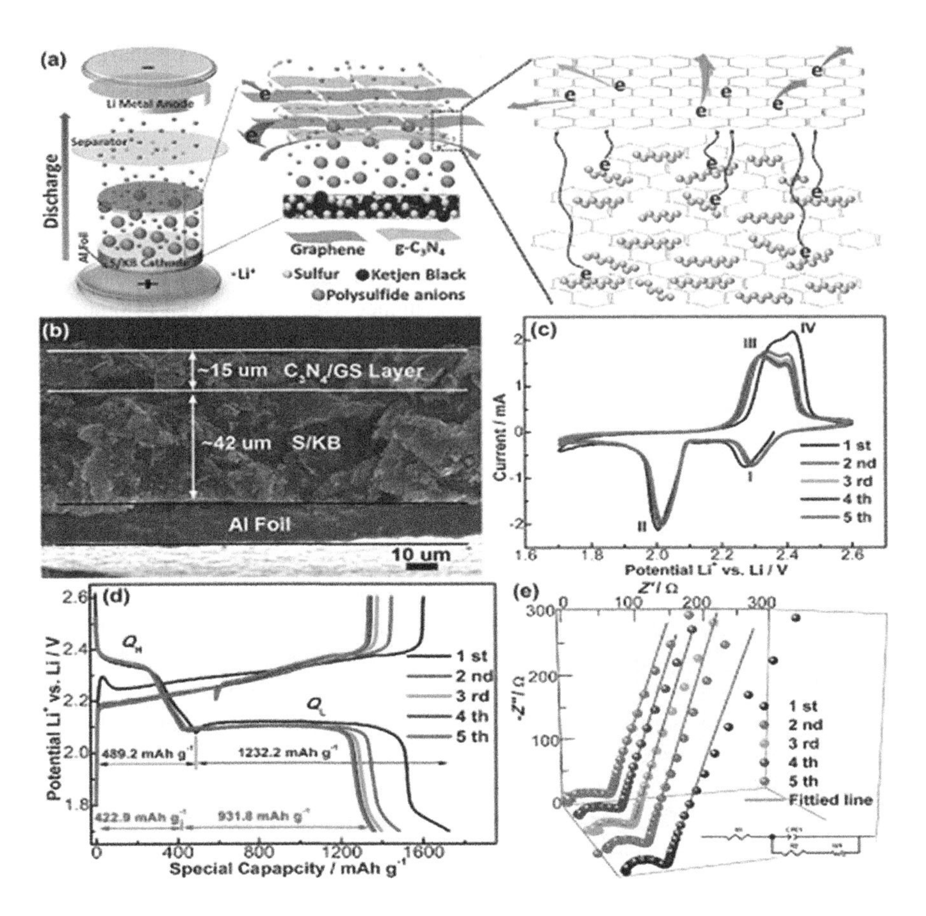

FIGURE 13.7 (a) Schematic of cell configuration, which is incorporated with g-C_3N_4/GS cathode interlayer. (b) SEM image of a section of the S/KB@C_3N_4/GS cathode. (c) CV curves at a scanning rate of 0.1 mVs^{-1}, (d) galvanostatic charge–discharge study at 0.1 C, and (e) EIS Nyquist plots for the S/KB@C_3N_4/GS cathode in the initial five cycles and the corresponding equivalent circuit model (inset) [59]. Adapted with permission from [59]. Copyright (2019) Wiley.

mAh g^{-1} at current densities of 2,000 and 10,000 mA g^{-1}. A sandwich-like silicon/$Ti_3C_2T_x$ MXene composite have a reversible capacity of 1,067.6 mAh g^{-1} at a current of 300 mA g^{-1} as reported by Zhang et al. [61].

Carbon and polymer composites are also being used for battery application like carbon/g-C_3N_4 composite is also used as an anode in battery [62]. 254 mAh g^{-1} is the reversible capacity offered by carbon/g-C_3N_4 composite, which have very high retention capacity that is 99.8% after 14,000 cycles. Kumar et al. [63] discussed the recent progress of graphene and its composites for Li-ion battery applications. Choi et al. [64] prepared anode material for battery that is crumpled graphene–molybdenum oxide composite. Composite showed a discharge capacity of 1,490 mAh g^{-1} at 2 A g^{-1} with a retention capacity of 47% after 100 cycles. Sb_2O_3/MXene (Ti_3C_2Tx) hybrid anode materials can give a reversible capacity of 295 mAh g^{-1} at 2A g^{-1} [65].

A ternary heterostructure of 2D transition metal chalcogenide-MXene-carbonaceous nanoribbon composite is used for boosting the sodium and potassium ion storage. Synergetic effect of all three components of this composite results in increased sodium (536.3 mAh g^{-1} @ 0.1 A g^{-1}) and potassium (305.6 mAh g^{-1} at 1.0 A g^{-1}) ion storage [66]. There are many reports that gave the evidences that nanocomposite of 2D materials can be used as electrode in the battery.

13.5 SENSOR

Sensor is a device, which responds as a result of some kind of input (mainly chemical compounds) and these responses are mostly in the form of electric signal.

13.5.1 Mechanism

2D material and its composite can be a good choice for sensing application because of their characteristics like high specific surface area and semiconducting nature. Sensors can be physical and chemical based on the type of parameters, which are recorded and analyzed. In physical sensing, any physical parameters like temperature and pressure are being recorded and in chemical sensing any chemical molecules interaction is being recorded. There are different parameters in terms of which we define the performance of the sensor. Sensor response, sensitivity, selectivity, response time, recovery time, and lowest limit are some of them. Sensor response is mostly taken in terms of resistance/current changes as a result of interaction of the sensor with external materials or parameters. Sensitivity is the response of the sensor per unit concentration and the selectivity is defined as the type of molecules, which can interact with sensor. Response time and recovery time are the time in which a sensor reaches a maximum value of current after being exposed to the sensing material and recovery time is the time in which the sensor reaches at 10% of the current value from its saturation value. All these parameters are crucial for defining the overall efficiency of a sensor [67]. Different types of sensing can be done by using 2D materials like H^+ ion sensing is important in the areas like food industries, monitoring waste, organic pollutants, and in chemical laboratories. pH meter works on the principle of measuring the electric potential difference, 2D material-based pH meters offer more sensitivity and compatible small-sized designs [68]. Glucose sensing is vital for monitoring the glucose level in the body. Metal ion sensing is the most important to monitor increased pollution level and other environmental activity.

13.5.2 2D Material Composites as a Sensor

Many researchers are devoted for developing the efficient use of 2D material composites for sensor applications. A recent report on TiO_2-C/g-C_3N_4 composites showed a successful use of this composite for the sensing of different gases [69]. In this work, heterostructure composites were fabricated by in situ growth method in which $Ti_3C_2T_x$ MXene mixed with melamine uniform suspension of this mixture was formed by ultrasonication using water as solvent. Suspended mixture is dried at 80°C for 12 hours and the powder formed is calcined at 550°C for 4 hours.

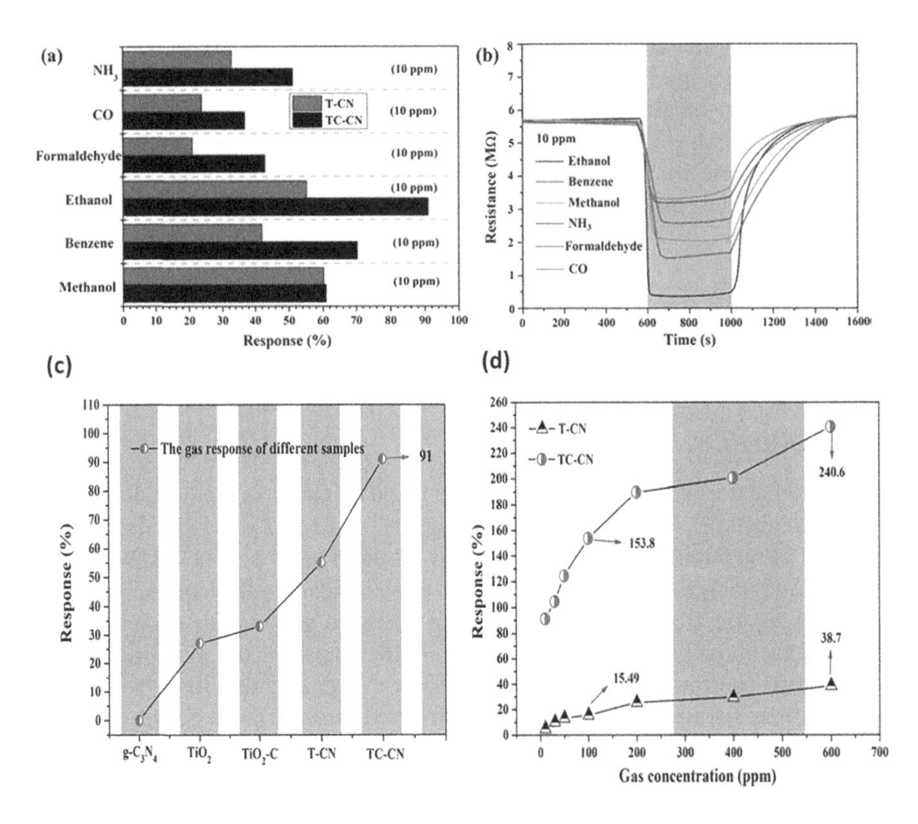

FIGURE 13.8 (a) Gas response of T-CN and TC-CN at UV light toward 10 ppm different VOCs at room temperature. (b) Dynamic response–recovery curve of TC-CN at UV light toward 10 ppm different VOCs at room temperature [69]. (c) Gas sensing response of different samples at UV light toward 10 ppm ethanol at room temperature. (d) Gas response of T-CN and TC-CN at UV light toward ethanol with different concentration at room temperature [69]. Adapted with permission from [69]. Copyright (2019) Elsevier.

Composite sensing response was found to be very good with high sensitivity and stability. Figure 13.8a and b show the response of the sensor toward different gases and TC-CN composite has high response and less recovery time as compared to the T-CN sample. TC-CN composite offers the highest sensitivity toward ethanol and little less sensitivity toward other gases. Figure 13.8c compares the response of different samples and in which TC-CN composite gives the highest response. Figure 13.8d shows the response of sensor at different concentrations of alcohol [69]. Therefore, in conclusion, out of all the samples, TC-CN composite has the best performance [69].

Metal nanoparticles decorated 2D materials are also being used for sensing application. Au nanoparticles decorated with a few layers of MoS_2 that are used for UV light-activated NO_2 sensing. Work involves a comparison of MoS_2 and Au-MOS_2 thin film for NO_2 sensing in which Au-MOS_2 thin film showed 10% more sensitivity toward 2.5 ppm [70]. g-C_3N_4-WO_3 composite materials are synthesized by hydrothermal synthesis and analyzed for acetone sensing by Chu et al. [71]. This composite sensor operating at 310°C temperature has responses of 58.2 and 1.6 for 1,000 ppm and 0.5 ppm, respectively. Hydrothermally grown GO-ZnO nanorods composite sheets are capable of sensing SO_2 and H_2. Sensors have a very good response toward both SO_2 and H_2 that is 5.82 and 5.45 at 100 ppm [72]. Zhou et al. [73] reported $Fe_2(MoO_4)_3$/MXene nanocomposites for gas sensing application and preparation method followed for composite formation is single-step hydrothermal method. Composites have very good stability and fast response and recovery time that were 18s and 24s. Glucose sensing by Graphene/AuNPs/chitosan composites is reported by Shan et al. [74]. Cathodic current variation is recorded with the variation of glucose concentration from 2.5 to 4.5 mM. Cathode current decreases with the increase in the glucose concentration. Heterostructure of 2D material has also been used for sensing, $MoSe_2$-MoO_3 heterostructure was used for the nitrite sensing [75]. Sensing of nitrite ion is important in water quality detection in remote areas so these composites can be successfully used in those areas. Black phosphorous composites can also be used for different sensing applications. Black phosphorous and dye composite thin films were used for chemical gas sensing and formation of thin films was carried out by using Langmuir–Blodgett (LB) technique [76]. Composite thin films have good sensitivity toward acid and alkali gases. Biosensing of exosomes by black phosphorus quantum dots (BPQDs) functionalized MXenes was done by Fang et al. [77]. BPQDs and MXene both show photothermal effect that makes them suitable as photothermal immunosensor for bioanalysis. Usually pH sensing is done by glass-based pH sensors, which consume space as well as also have drawback of fragile nature. These problems can be overcome by replacing these pH sensors with 2D-based composites materials. Amino-functionalized graphene and polyaniline composite film have been used for pH sensing by Xu et al. [78]. This composite works in a wide range of pH from 1 to 11 and have high sensitivity and fast response. Cysteic acid/graphene oxide nanocomposites are also reported for pH sensing uses [79]. So, 2D material composites started to play a significant role in sensing area as a result of researcher's efforts.

13.6 CONCLUSIONS

2D materials like graphene, TMD, black phosphorous, and few more are extensibly being used in various areas of material science. Although pristine 2D materials are efficient enough to be used in different applications, composite formation of these materials significantly increases the performance and also helps to overcome some limitations. Different types of composites are possible depending upon the matrix material. Metal-matrix, ceramic-matrix, and polymer composites are the popular ones among the 2D materials. There are different synthesis methods for composite formation and sometimes they can be formulated by simple single-step procedure. Energy storage is an area where 2D materials and their composites have shown excellent performance because of their properties like high surface area and high conductivity. Supercapacitor and battery electrodes made of these composites achieve significant results, which are also being used in commercial applications. 2D material composite-based sensors are also of much importance and researchers are continuously exploring these areas for better results. So, 2D material and their composites are capable of giving a brighter future to material science and technology.

REFERENCES

1. K.T. Ramesh, Nanomaterials. In: *Nanomaterials* (2009). Springer, Boston, MA, pp. 1–316. doi:10.1007/978-0-387-09783-1.
2. Z. Sayyar, Z. Jamshidi, The role of novel composite of 2D materials and their characterization, properties, and potential applications in different fields. In: M.A. Chowdhury, J.L.R. Armenta, M.M. Rahman, A. Asiri, and Inamuddin (eds.) *Composite Materials* (2021). IntechOpen, London, UK. doi:10.5772/intechopen.92707.
3. M.Y. Li, C.H. Chen, Y. Shi, L.J. Li, Heterostructures based on two-dimensional layered materials and their potential applications, *Mater. Today* 19 (2016) 322–335. doi:10.1016/j.mattod.2015.11.003.
4. X. Wen, R. Joshi, 2D materials-based metal matrix, *J. Phys. D Appl. Phys.* 53 (2020), 423001. doi:10.1088/1361-6463/ab9b5d.
5. Y. Xue, S. Zheng, H. Xue, H. Pang, Metal-organic framework composites and their electrochemical applications, *J. Mater. Chem. A* 7 (2019) 7301–7327. doi:10.1039/C8TA12178H.
6. W. Liu, B. Ullah, C.-C. Kuo, X. Cai, Two-dimensional nanomaterials-based polymer composites: Fabrication and energy storage applications, *Adv. Polym. Technol.* 2019 (2019) 1–15. doi:10.1155/2019/4294306.
7. B. Luo, G. Liu, L. Wang, Recent advances in 2D materials for photocatalysis, *Nanoscale* 8 (2016) 6904–6920. doi:10.1039/c6nr00546b.
8. Y. Chen, X. Zhang, E. Liu, C. He, C. Shi, J. Li, P. Nash, N. Zhao, Fabrication of in-situ grown graphene reinforced Cu matrix composites, *Sci. Rep.* 6 (2016) 1–9. doi:10.1038/srep19363.
9. R. Santalucia, T. Vacca, F. Cesano, G. Martra, F. Pellegrino, D. Scarano, Few-layered MoS_2 nanoparticles covering anatase TiO_2 nanosheets: Comparison between ex-situ and in-situ synthesis approaches, *Appl. Sci.* 11 (2021) 1–17. doi:10.3390/app11010143.
10. B. Wang, T. Ruan, Y. Chen, F. Jin, L. Peng, Y. Zhou, D. Wang, S. Dou, Graphene-based composites for electrochemical energy storage, *Energy Storage Mater.* 24 (2020) 22–51. doi:10.1016/j.ensm.2019.08.004.
11. R. Sahoo, A. Pal, T. Pal, 2D materials for renewable energy storage devices: Outlook and challenges, *Chem. Commun.* 52 (2016) 13528–13542. doi:10.1039/c6cc05357b.
12. J. Ping, Z. Fan, M. Sindoro, Y. Ying, H. Zhang, Recent advances in sensing applications of two-dimensional transition metal dichalcogenide nanosheets and their composites, *Adv. Funct. Mater.* 27 (2017) 1–18. doi:10.1002/adfm.201605817.
13. G. Neri, Thin 2D: The new dimensionality in gas sensing, *Chemosensors* 5 (2017). doi:10.3390/chemosensors5030021.
14. N. Rohaizad, C.C. Mayorga-Martinez, M. Fojtů, N.M. Latiff, M. Pumera, Two-dimensional materials in biomedical, biosensing and sensing applications, *Chem. Soc. Rev.* 50 (2021) 619–657. doi:10.1039/d0cs00150c.
15. F. Shahzad, M. Alhabeb, C.B. Hatter, B. Anasori, S.M. Hong, C.M. Koo, Y. Gogotsi, Electromagnetic interference shielding with 2D transition metal carbides (MXenes), *Science* 353 (2016) 1137–1140. doi:10.1126/science.aag2421.
16. J. Liang, Y. Wang, Y. Huang, Y. Ma, Z. Liu, J. Cai, C. Zhang, H. Gao, Y. Chen, Electromagnetic interference shielding of graphene/epoxy composites, *Carbon N. Y.* 47 (2009) 922–925. doi:10.1016/j.carbon.2008.12.038.
17. R.F. Gibson, *Principles of Composite Material Mechanics(Mechanical Engineering)* (2016), CRC Press, Taylor and Francis Group, U.S., Boca Raton, FL.
18. A. Naseer, F. Ahmad, M. Aslam, B.H. Guan, W.S.W. Harun, N. Muhamad, M.R. Raza, R.M. German, A review of processing techniques for graphene-reinforced metal matrix composites, *Mater. Manuf. Process.* 34 (2019) 957–985. doi:10.1080/10426914.2019.1615080.

19. H. Nautiyal, S. Kumari, O.P. Khatri, R. Tyagi, Copper matrix composites reinforced by rGO-MoS_2 Hybrid: Strengthening effect to enhancement of tribological properties, *Compos. Part B Eng.* 173 (2019) 106931. doi:10.1016/j.compositesb.2019.106931.
20. G. Liu, N. Zhao, C. Shi, E. Liu, F. He, L. Ma, Q. Li, J. Li, C. He, In-situ synthesis of graphene decorated with nickel nanoparticles for fabricating reinforced 6061Al Matrix Composites, *Mater. Sci. Eng. A* 699 (2017) 185–193. doi:10.1016/j.msea.2017.05.084.
21. A.K. Lakshminarayanan, Sridhar Idapalapati, M. Vasudevan, *Advances in Materials and Metallurgy* (2019) Springer, Singapore. doi:10.1007/978-981-13-1780-4.
22. N. Devi, R. Kumar, R.K. Singh. Microwave-assisted modification of graphene and its derivatives: Synthesis, reduction and exfoliation. In: A. Khan, M. Jawaid, B. Neppolian, A. Asiri (eds.) *Graphene Functionalization Strategies*. Carbon Nanostructures (2019), pp. 279–311. Springer, Singapore. doi:10.1007/978-981-32-9057-0_12.
23. N. Devi, S. Sahoo, R. Kumar, R.K. Singh, A review of the microwave-assisted synthesis of carbon nanomaterials, metal oxides/hydroxides and their composites for energy storage applications, *Nanoscale* 13 (2021) 11679–11711. doi:10.1039/d1nr01134k.
24. H. Nie, L. Fu, J. Zhu, W. Yang, D. Li, L. Zhou, Excellent tribological properties of lower reduced graphene oxide content copper composite by using a one-step reduction molecular-level mixing process, *Materials (Basel)* 11 (2018). doi:10.3390/ma11040600.
25. H. Porwal, S. Grasso, M.J. Reece, Review of graphene-ceramic matrix composites, *Adv. Appl. Ceram.* 112 (2013) 443–454. doi:10.1179/174367613X13764308970581.
26. J. Guo, B. Legum, B. Anasori, K. Wang, P. Lelyukh, Y. Gogotsi, C.A. Randall, Cold sintered ceramic nanocomposites of 2D MXene and zinc oxide, *Adv. Mater.* 30 (2018) 1–6. doi:10.1002/adma.201801846.
27. Y. Huang, C. Wan, Controllable fabrication and multifunctional applications of graphene/ceramic composites, *J. Adv. Ceram.* 9 (2020) 271–291. doi:10.1007/s40145-020-0376-7.
28. Y. Fan, L. Kang, W. Zhou, W. Jiang, L. Wang, A. Kawasaki, Control of doping by matrix in few-layer graphene/metal oxide composites with highly enhanced electrical conductivity, *Carbon N. Y.* 81 (2015) 83–90. doi:10.1016/j.carbon.2014.09.027.
29. J. Liu, H. Yan, K. Jiang, Mechanical properties of graphene platelet-reinforced alumina ceramic composites, *Ceram. Int.* 39 (2013) 6215–6221. doi:10.1016/j.ceramint.2013.01.041.
30. P. Song, B. Liu, C. Liang, K. Ruan, H. Qiu, Z. Ma, Lightweight, flexible cellulose - derived carbon aerogel @ reduced graphene oxide/PDMS composites with outstanding EMI shielding performances and excellent thermal conductivities, *Nano-Micro Lett.* (2021). doi:10.1007/s40820-021-00624-4.
31. A. Sajedi-moghaddam, C.C. Mayorga-martinez, D. Bous, E. Saievar-iranizad, M. Pumera, Black phosphorus nano flakes/polyaniline hybrid material for high-performance pseudocapacitors, *J. Phys. Chem. C* (2017). doi:10.1021/acs.jpcc.7b06958.
32. H. Huang, X. Huang, Y. Xie, Y. Tian, X. Jiang, X. Zhang, Fabrication of h-BN-rGO@PDA nanohybrids for composite coatings with enhanced anticorrosion performance, *Prog. Org. Coatings* 130 (2019) 124–131. doi:10.1016/j.porgcoat.2019.01.059.
33. Y. Gao, J. Ding, Synthesis of heterostructures based on two-dimensional materials. In: E.-H. Yang, D. Datta, J. Ding, G. Hader (eds.) *Synthesis, Modelling and Characterization of 2D Materials and their Heterostructures* (2020). Elsevier, Amsterdam. doi:10.1016/b978-0-12-818475-2.00013-1.
34. G. Iannaccone, F. Bonaccorso, L. Colombo, G. Fiori, Quantum engineering of transistors based on 2D materials heterostructures, *Nat. Nanotechnol.* 13 (2018) 183–191. doi:10.1038/s41565-018-0082-6.
35. B. Cho, J. Yoon, S.K. Lim, A.R. Kim, D.H. Kim, S.G. Park, J.D. Kwon, Y.J. Lee, K.H. Lee, B.H. Lee, H.C. Ko, M.G. Hahm, Chemical sensing of 2D graphene/MoS_2 heterostructure device, *ACS Appl. Mater. Interfaces* 7 (2015) 16775–16780. doi:10.1021/acsami.5b04541.

36. T. Yamaguchi, R. Moriya, Y. Inoue, S. Morikawa, S. Masubuchi, K. Watanabe, T. Taniguchi, T. Machida, Tunneling transport in a few monolayer-thick WS_2/graphene heterojunction, *Appl. Phys. Lett.* 105 (2014). doi:10.1063/1.4903190.
37. T. Song, L. Hou, B. Long, A. Ali, G.J. Deng, Ultrathin MXene "Bridge" to accelerate charge transfer in ultrathin metal-free 0D/2D Black Phosphorus/g-C_3N_4 heterojunction toward photocatalytic hydrogen production, *J. Colloid Interface Sci.* 584 (2021) 474–483. doi:10.1016/j.jcis.2020.09.103.
38. A. Chhetry, S. Sharma, S.C. Barman, H. Yoon, S. Ko, C. Park, S. Yoon, H. Kim, J.Y. Park, Black Phosphorus@Laser-engraved graphene heterostructure-based temperature–strain hybridized sensor for electronic-skin applications, *Adv. Funct. Mater.* 31 (2021) 1–14. doi:10.1002/adfm.202007661.
39. T. Liang, Y. Liu, Y. Cheng, F. Ma, Z. Dai. Scalable MoS_2/black phosphorus heterostructure for pH-universal hydrogen evolution catalysis. *ChemCatChem* 12 (2020) 2840–2848. doi:10.1002/cctc.202000139.
40. L. Zhang, X.S. Zhao, Carbon-based materials as supercapacitor electrodes, *Chem. Soc. Rev.* 38 (2009) 2520–2531. doi:10.1039/b813846j.
41. L. Lin, W. Lei, S. Zhang, Y. Liu, G.G. Wallace, J. Chen, Two-dimensional transition metal dichalcogenides in supercapacitors and secondary batteries, *Energy Storage Mater.* 19 (2019) 408–423. doi:10.1016/j.ensm.2019.02.023.
42. Y. Han, Y. Ge, Y. Chao, C. Wang, G.G. Wallace, Recent progress in 2D materials for flexible supercapacitors, *J. Energy Chem.* 27 (2018) 57–72. doi:10.1016/j.jechem.2017.10.033.
43. C. Xiang, M. Li, M. Zhi, A. Manivannan, N. Wu, A reduced graphene oxide/Co_3O_4 composite for supercapacitor electrode, *J. Power Sources* 226 (2013) 65–70. doi:10.1016/j.jpowsour.2012.10.064.
44. M.A. Bissett, I.A. Kinloch, R.A.W. Dryfe, Characterization of MoS_2-graphene composites for high-performance coin cell supercapacitors, *ACS Appl. Mater. Interfaces* 7 (2015) 17388–17398. doi:10.1021/acsami.5b04672.
45. K. Ramakrishnan, C. Nithya, R. Karvembu, Heterostructure of two different 2D materials based on MoS_2 Nanoflowers@ rGO: An electrode material for sodium-ion capacitors. *Nanoscale Adv.* 2019, 1(1), 334–341.doi:10.1039/C8NA00104A.
46. J. Wang, Z. Wu, K. Hu, X. Chen, H. Yin, High conductivity graphene-like MoS_2/ polyaniline nanocomposites and its application in supercapacitor, *J. Alloys Compd.* 619 (2015) 38–43. doi:10.1016/j.jallcom.2014.09.008.
47. C. Sha, B. Lu, H. Mao, J. Cheng, X. Pan, J. Lu, 3D ternary nanocomposites of molybdenum disulfide/polyaniline/reduced graphene oxide aerogel for high performance supercapacitors, *Carbon N. Y.* 99 (2016) 26–34. doi:10.1016/j.carbon.2015.11.066.
48. X. Wang, H. Li, H. Li, S. Lin, W. Ding, X. Zhu, Z. Sheng, H. Wang, X. Zhu, Y. Sun. 2D/2D 1T-MoS_2/Ti_3C_2 MXene heterostructure with excellent supercapacitor performance. *Adv. Funct. Mater.* 15 (2020), 0190302. doi:10.1002/adfm.201910302.
49. Z. Zhao, P. Bai, W. Du, B. Liu, D. Pan, R. Das, C. Liu, Z. Guo, An overview of graphene and its derivatives reinforced metal matrix composites: Preparation, properties and applications, *Carbon N. Y.* 170 (2020) 302–326. doi:10.1016/j.carbon.2020.08.040.
50. R. Kumar, S. Sahoo, E. Joanni, R.K. Singh, K. Maegawa, W.K. Tan, G. Kawamura, K.K. Kar, A. Matsuda, Heteroatom doped graphene engineering for energy storage and conversion, *Mater. Today* 39 (2020) 47–65. doi:10.1016/j.mattod.2020.04.010.
51. R. Kumar, S. Sahoo, E. Joanni, R.K. Singh, R.M. Yadav, R. Kumar, D.P. Singh, W.K. Tan, A. Pérez, S.A. Moshkalev, A review on synthesis of graphene, h-BN and MoS_2 for energy storage applications: Recent progress and perspectives, *Nano Res.* 12 (2019) 35–37. doi: 10.1007/s12274-019-2467-8.
52. R. Kumar, E. Joanni, R.K. Singh, D.P. Singh, S.A. Moshkalev, Recent advances in the synthesis and modification of carbon-based 2D materials for application in energy conversion and storage, *Prog. Energy Combust. Sci.* 67 (2018). doi:10.1016/j.pecs.2018.03.001.

53. J. Xu, S. Zhang, Z. Wei, W. Yan, X. Wei, K. Huang, Orientated VSe_2 nanoparticles anchored on N-doped hollow carbon sphere for high-stable aqueous energy application, *J. Colloid Interface Sci.* 585 (2021) 12–19. doi:10.1016/j.jcis.2020.11.065.
54. A. Sajedi-Moghaddam, E. Saievar-Iranizad, M. Pumera, Two-dimensional transition metal dichalcogenide/conducting polymer composites: Synthesis and applications, *Nanoscale* 9 (2017) 8052–8065. doi:10.1039/c7nr02022h.
55. S. Luo, J. Zhao, J. Zou, Z. He, C. Xu, F. Liu, Y. Huang, L. Dong, L. Wang, H. Zhang, Self-standing polypyrrole/black phosphorus laminated film: Promising electrode for flexible supercapacitor with enhanced capacitance and cycling stability, *ACS Appl. Mater. Interfaces* 10 (2018) 3538–3548. doi:10.1021/acsami.7b15458.
56. J. Zhang, L. Zhang, F. Sun, Z. Wang, An overview on thermal safety issues of lithium-ion batteries for electric vehicle application, *IEEE Access* 6 (2018) 23848–23863. doi:10.1109/ACCESS.2018.2824838.
57. H. Li, R.Y. Tay, S.H. Tsang, W. Liu, E.H.T. Teo, Reduced graphene oxide/boron nitride composite film as a novel binder-free anode for lithium ion batteries with enhanced performances, *Electrochim. Acta* 166 (2015) 197–205. doi:10.1016/j.electacta.2015.03.109.
58. J. Liang, W. Wei, D. Zhong, Q. Yang, L. Li, L. Guo, One-step in situ synthesis of SnO_2/graphene nanocomposites and its application as an anode material for li-ion batteries, *ACS Appl. Mater. Interfaces* 4 (2012) 454–459. doi:10.1021/am201541s.
59. L. Qu, P. Liu, Y. Yi, T. Wang, P. Yang, X. Tian, M. Li, B. Yang, S. Dai, Enhanced cycling performance for lithium–sulfur batteries by a laminated 2D g-C_3N_4/graphene cathode interlayer, *ChemSusChem* 12 (2019) 213–223. doi:10.1002/cssc.201802449.
60. Y. Chen, B. Song, X. Tang, L. Lu, J. Xue, Ultrasmall Fe_3O_4 nanoparticle/MoS_2 nanosheet composites with superior performances for lithium ion batteries, *Small* 10 (2014) 1536–1543. doi:10.1002/smll.201302879.
61. F. Zhang, Z. Jia, C. Wang, A. Feng, K. Wang, T. Hou, J. Liu, Y. Zhang, G. Wu, sandwich-like silicon/$Ti_3C_2T_x$ MXene composite by electrostatic self-assembly for high performance lithium ion battery, *Energy* 195 (2020) 117047. doi:10.1016/j.energy.2020.117047.
62. G. Weng, Y. Xie, H. Wang, C. Karpovich, J. Lipton, J. Zhu, J. Kong, L.D. Pfefferle, A.D. Taylor, A promising carbon/g-C_3N_4 composite negative electrode for a long-life sodium-ion battery, *Angew. Chemie* 131 (2019) 13865–13871. doi:10.1002/ange.201905803.
63. R. Kumar, S. Sahoo, E. Joanni, R. Kumar, W. Kian, K. Krishna, A. Matsuda, Recent progress in the synthesis of graphene and derived materials for next generation electrodes of high performance lithium ion batteries, *Prog. Energy Combust. Sci.* 75 (2019) 100786. doi:10.1016/j.pecs.2019.100786.
64. S.H. Choi, Y.C. Kang, Crumpled graphene-molybdenum oxide composite powders: Preparation and application in lithium-ion batteries, *ChemSusChem* 7 (2014) 523–528. doi:10.1002/cssc.201300838.
65. X. Guo, X. Xie, S. Choi, Y. Zhao, H. Liu, C. Wang, S. Chang, G. Wang, Sb_2O_3/MXene($Ti_3C_2T_x$) hybrid anode materials with enhanced performance for sodium-ion batteries, *J. Mater. Chem. A* 5 (2017) 12445–12452. doi:10.1039/c7ta02689g.
66. J. Cao, J. Li, D. Li, Z. Yuan, Y. Zhang, V. Shulga, Z. Sun, W. Han, Strongly coupled 2D transition metal chalcogenide-mxene-carbonaceous nanoribbon heterostructures with ultrafast ion transport for boosting sodium/potassium ions storage, *Nano-Micro Lett.* 13 (2021). doi:10.1007/s40820-021-00623-5.
67. D.J. Late, A. Bhat, C.S. Rout, Fundamentals and properties of 2D materials in general and sensing applications. In D.J. Late, A. Bhat, C.S. Rout (eds.), *Fundamentals and Sensing Applications of 2D Materials* (2019). Elsevier Ltd, Amsterdam. doi:10.1016/B978-0-08-102577-2.00002-6.
68. C.W. Lee, J.M. Suh, H.W. Jang, Chemical sensors based on two-dimensional (2D) materials for selective detection of ions and molecules in liquid, *Front. Chem.* 7 (2019) 1–21. doi:10.3389/fchem.2019.00708.

69. M. Hou, J. Gao, L. Yang, S. Guo, T. Hu, Y. Li, Room temperature gas sensing under UV light irradiation for $Ti_3C_2T_x$ MXene derived lamellar TiO_2-C/g-C_3N_4 composites, *Appl. Surf. Sci.* 535 (2021). doi:10.1016/j.apsusc.2020.147666.
70. Y. Zhou, C. Zou, X. Lin, Y. Guo, UV light activated NO_2 gas sensing based on Au nanoparticles decorated few-layer MoS_2 thin film at room temperature, *Appl. Phys. Lett.* 113 (2018) 2–7. doi:10.1063/1.5042061.
71. X. Chu, J. Liu, S. Liang, L. Bai, Y. Dong, M. Epifani, Facile preparation of g-C_3N_4-WO_3 composite gas sensing materials with enhanced gas sensing selectivity to acetone, *J. Sensors* 2019 (2019). doi:10.1155/2019/6074046.
72. V. Dhingra, S. Kumar, R. Kumar, A. Garg, A. Chowdhuri, Room temperature SO_2 and H_2 gas sensing using hydrothermally grown GO-ZnO nanorod composite films, *Mater. Res. Express* 7 (2020). doi:10.1088/2053-1591/ab9ae7.
73. S. Zou, J. Gao, L. Liu, Z. Lin, P. Fu, S. Wang, Z. Chen, Enhanced gas sensing properties at low working temperature of iron molybdate/MXene composite, *J. Alloys Compd.* 817 (2020) 152785. doi:10.1016/j.jallcom.2019.152785.
74. C. Shan, H. Yang, D. Han, Q. Zhang, A. Ivaska, L. Niu, Graphene/AuNPs/Chitosan nanocomposites film for glucose biosensing, *Biosens. Bioelectron.* 25 (2010) 1070–1074. doi:10.1016/j.bios.2009.09.024.
75. N. Vishnu, S. Badhulika, Single step synthesis of $MoSe_2$–MoO_3 heterostructure for highly sensitive amperometric detection of nitrite in water samples of industrial areas, *Electroanalysis* 31 (2019) 2410–2416. doi:10.1002/elan.201900310.
76. R. Wang, X. Yan, B. Ge, J. Zhou, M. Wang, L. Zhang, T. Jiao, Facile preparation of self-assembled black phosphorus-dye composite films for chemical gas sensors and surface-enhanced raman scattering performances, *ACS Sustain. Chem. Eng.* 8 (2020) 4521–4536. doi:10.1021/acssuschemeng.9b07840.
77. D. Fang, D. Zhao, S. Zhang, Y. Huang, H. Dai, Y. Lin, Black phosphorus quantum dots functionalized MXenes as the enhanced dual-mode probe for exosomes sensing, *Sensors Actuators, B Chem.* 305 (2020) 127544. doi:10.1016/j.snb.2019.127544.
78. W. Su, J. Xu, X. Ding, An electrochemical pH sensor based on the amino-functionalized graphene and polyaniline composite film, *IEEE Trans. Nanobiosci.* 15 (2016) 812–819. doi:10.1109/TNB.2016.2625842.
79. L. Liu, Y. Huang, J. Zhao, Z. Zhu, H. Zhang, C. Wang, All-solid-sate pH sensing material based on cysteic acid/graphene oxide nanocomposites (2015) 144–148. doi:10.12792/iciae2015.028.

14 Recent Developments in Group II-VI Based Chalcogenides and Their Potential Application in Solar Cells

Saif Ali, Faheem K. Butt, and Junaid Ahmad
University of Education Lahore

Zia Ur Rehman
Yangzhou University

Sami Ullah and Mashal Firdous
University of Education Lahore

Sajid Ur Rehman and Zeeshan Tariq
Minzu University of China

CONTENTS

DOI: 10.1201/9781003178453-14

14.1 INTRODUCTION

The II-VI-based compounds have gained much interest from researchers due to their distinctive properties and applications in numerous fields such as energy storage and conversion. These compounds show a large limit of bandgaps such that they can cover the whole solar spectrum. The wide bandgap elements majorly comprise the chalcogen compounds of Cd and Zn like ZnS, CdTe, and ZnTe. The ZnS has the widest bandgap value which is 3.4 eV. The elements with smaller bandgap comprise mercury chalcogenides like HgTe and HgSe. HgTe exhibits the negative bandgap value which is −3.0 eV. These compounds exhibit such types of applications that are exhibited by the developed semiconductors like Ge, Si, and some of the elements from group III-V. Moreover, the range of bandgap becomes continuous with the addition of ternary II-VI elements like CdHgTe. These types of materials are potential candidates for various applications such as solar cells, switches, electroluminescent diodes, radiation detectors, lasers, a passivation layer, phosphors, and infrared detectors [1,2].

Most of the II-VI elements are crystallized both in zincblend and wurtzite morphologies. The repeated and famous characteristic of these frameworks is that every single atom is surrounded or shared by the four atoms of other material. In zincblend, these frameworks are arranged in cubic assembly while in wurtzite, they are aligned in the hexagonal assembly. Generally, the elements from group II contain two electrons in the outermost shell of the atom with s^2, and elements from group VI contain six electrons in the outermost shell with s^2p^4 configuration. After combining the elements of group IIA with group VI, electronegativity remains strong due to which element from group IIA donates its two electrons to the element of group VI. As a result of this combination, octahedral type morphology obtained similar to NaCl. This type of bonding is prominent in IIA chalcogenides. Furthermore, when the elements from group IIB such as Zn, Hg, and Cd are combined with the elements of group VI, then, the ionization potential exhibits enough high value due to which these metals do not provide their whole electrons and only donate their free electrons with their neighbors, which belong to non-metal chalcogenides. As a result of this combination, tetrahedrally arrange covalent bonds developed due to which tetrahedrally based wurtzite and zincblende frameworks formed. Data related to II-VI materials are continuously being produced as a result of research work performed in this field. Consequently, these data are utilized in the development of new applications [1–3].

14.2 TWO-DIMENSIONAL (2D) NANOMATERIALS

Dimensional division of the nanomaterials is one of the approaches to categorize the nanomaterials because the same material can show remarkably different features by confining it into zero-dimensional (0D), one-dimensional (1D), two-dimensional (2D), and three-dimensional (3D) crystal framework [4]. Here, we are curious about 2D nanostructured materials (NSMs), therefore, 2D NSMs with their applications are illustrated in Figure 14.1. 2D NSMs are considered as the slimmest NSMs owing to their opacity and size on nanoscale or microscale. These NSMs contain a layered framework having vigorous bonds among the planes and feeble van der Waals among the layers. In the last few years, 2D NSMs like graphene, metal dichalcogenides, metal carbides/nitrides (MXenes) [5,6], metal oxides, hexagonal boron nitride (h-BN) [7], and metal halides have grasped much attention owing to their excellent features and extensive utilization in the field of electronics, energy storage and conversion, optoelectronics, sensors, lithium-ion batteries, catalysts, and solar cells [8,9].

The illustrative framework of graphene, h-BN nanosheets, and WSe_2 as a dichalcogenide has been shown in Figure 14.2. These compounds are assembled in a honeycomb-like morphology, but the ordering of the nearby atoms in the top and bottom layers of 2D NSMs is unlike. Every carbon atom in the structure of graphene is present next to other carbon atom in the top and bottom layers, but in the framework of h-BN every atom is present in the middle of the benzene ring on the top and bottom layers. Moreover, every atomic sheet of metal is sandwiched among the two sheets of X [9].

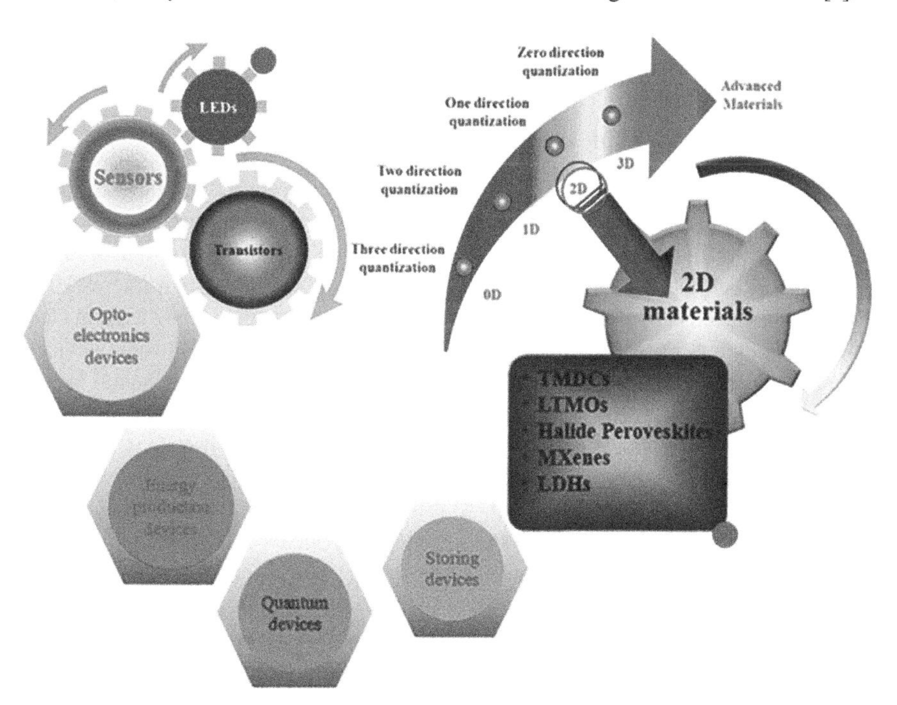

FIGURE 14.1 Different NSMs and applications of 2D NSMs. Adapted with permission from [8] Copyright (2020) Journal of Materials Chemistry C.

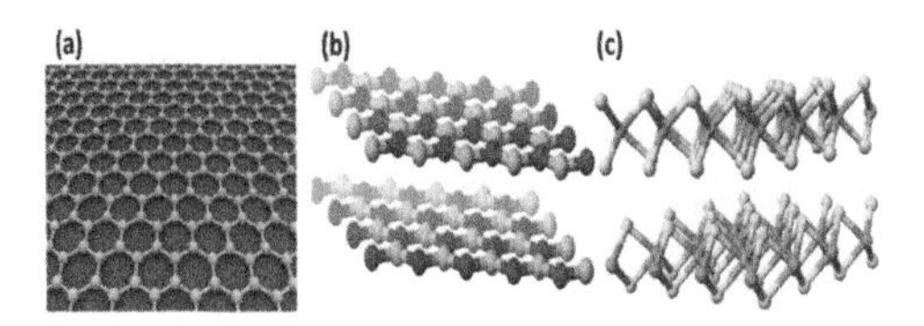

FIGURE 14.2 The framework of (a) graphene, (b) nanosheets of boron nitride (B: blue, N: pink), and (c) WSe_2 (W: blue, Se: yellow). Adapted with permission from [9] Copyright (2019) IntechOpen.

Transition metal dichalcogenides (TMDCs) have gained the attention of researchers in 2010 when the monolayer MoS_2 having direct bandgap was discovered and in 2011 fascinating electrical features of multilayered MoS_2 were observed [10]. However, the reliability of the mobility obtained from transistors based on MoS_2 was uncertain. At that time, W and Mo-based dichalcogenides such as WS_2, $MoTe_2$, MoS_2, WSe_2, $MoSe_2$, and WTe_2 were popular. Generally, TMDCs are written by MX_2 where M is a metal or transition metal and X represents chalcogen. More than 30 TMDCs have been fabricated by utilizing various fabrication techniques [11,12]. The band framework in 2D TMDCs can be tuned by varying the radioactive decay time of excitons or by ordering the heterolayers in bilayers or multi-layers, but in vertical ordering, the band arrangement is restricted to bilayer or tri-layer. At room temperature, 2D TMDCs exhibit electron mobilities of about $200\,cm^2/Vs$ having bandgap of 1–2 eV in the visible spectrum. In electronic gadgets, such type of materials is utilized in the operation of logic gates having large on/off fraction and small power [8].

14.3 IMPORTANCE OF II-VI BASED CHALCOGENIDES

Chalcogenides nanomaterials are attaining great importance due to their unusual quantum phenomena and important applications. Chalcogenide glasses are already being used as thin films in optical data storage. Because of the transmission of light at longer wavelengths than is possible with silica, chalcogenide fibers have great interest. The two-dimensional chalcogenides are said to be a helical spin state. Such a different electronic arrangement provides an ideal platform for the invention of spintronic devices. There is a wide bandgap in Chalcogenides semiconductors that is fundamental for the present electronic contraptions and energy applications because of their incredible optical transparency, controllable carrier fixation, and capable conductivity [13].

Chalcogenides are also beneficial for medical treatments such as laser surgery, which needs around a 3 μm wavelength. At this wavelength, Lanthanum sulfide fibers from gallium transmit well, and their non-toxic components and great melting temperature are suitable for minimally invasive surgery. In spectroscopy and sensing at similar wavelengths, chalcogenide fibers have various atmospheric transmission windows for important utilization [14]. Furthermore, with the advantage of inorganic nanoparticles' quantum confinement effect, they can be easily formed into various shapes, like prisms, wires, spheres, strands, rods, and even longer and more complex structures, like a tetrapod or hyper-branched. The involvement of light-absorbing lysine chalcogenides through inorganic receptors can also produce images of the carrier in charge, which may exceed the absorption contribution of fluorine derivatives. They offer the possibility to select the spectral window of the corresponding absorption profile with respect to the polymer since it is possible to make absorption quantum dots (QDs) to cover a wide spectral range [15].

Chalcogenide glasses are technically very important for read-write storage devices, as they can be quickly (in nanoseconds) interchanged within crystalline and amorphous phases by applying suitable thermal pulses. Chalcogenide glasses are attracting great interest with their mid-infrared transparency and highly nonlinear qualities. Due to their suitable amorphous semiconductor characteristics, chalcogens glass materials are applied in various technologies. Surface plasmon resonance (SPR) is a very versatile and precise technique for quantifying small changes in optoelectronics parameters. Therefore, optical materials based on chalcogenide glass (or anodized glass) may be a potential candidate for the manufacturing of SPR sensors to operate in the near-IR region [16].

In the mid-infrared region, the large optical window of chalcogenide glassy material allows scanning the whole spectrum area, including biomolecular vibrational patterns. Amorphous chalcogenides are resistant to chemical corrosion, which is better for their biological compatibility with biological components. Glass chalcogenides also have a hydrophobic surface quality; they attract nonpolar organic species because they return to the water. As a result, the optical signal of Devices based on chalcogenide glass has also demonstrated the possibility of producing low loss waveguides due to the huge refractive index between the glasses. Chalcogenide glass alloys can also be developed by adding metallic materials (such as Ag, Cu, Fe, Cd, and Zn) to get an order of magnitude greater thermal stability [17]. The porosity level of the tilt-controlled films, when the substrate is away from the deposition source, produces a controlled difference in the effective optical constants, making chalcogenide glass a promising material for chemical sensing [16].

In modern ages, Zinc sulfide has been achieving great attraction due to its potential importance in various fields such as biological labeling, solar cells, photocatalysts, photoconductors field-effect transistors, phosphors, sensors, optical, electroluminescent materials, and other light-emitting materials. ZnS is an important inorganic chalcogenide semiconductor with a broad bandgap extending from 3.5 to 3.7 eV [18].

Cadmium Telluride CdTe is a charming competition and produces a commercial thin-film unit of PV solar cells around the world. Thin-film solar cells of Champion CdTe have gained conversion capabilities, which is exceeding 16%. A thin film of CdTe is about a 10th of the diameter of human hair. The CdTe thin film absorbs visible light inside the micron of the material that works as the fundamental photoconversion layer. CdTe is a direct bandgap composite semiconductor, so it is used as photovoltaic (PV) energy conversion worldly having almost best match to the solar spectrum. The CdTe, central and transparent conducting oxide films (e.g., SnO_2 or Cd_2SnO_4) generate an electrical field, which is going to convert light absorbed by CdTe (intrinsic) layer into current and voltage. All of these layers with metal are deposited on incoming glass and form complete solar panels within a few hours [19] (Figure 14.3).

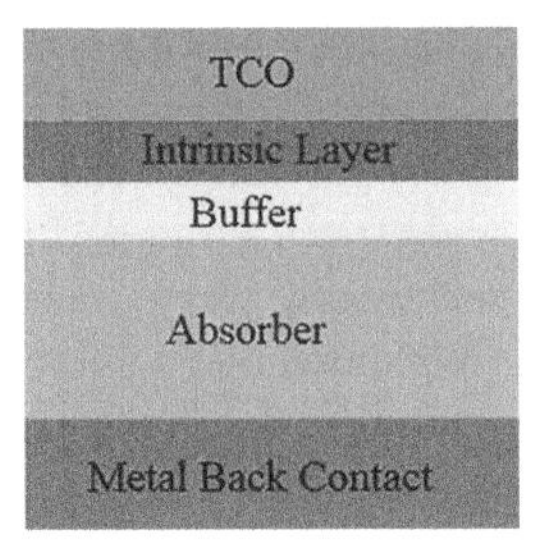

FIGURE 14.3 Physical structure of chalcogenide solar cells.

14.4 SYNTHESIS TECHNIQUES

PV solar cells are a hot topic in the energy sector. However, there are still many challenges to overcome before they can be implemented on a large scale. The efficiency of PV cells is limited by the quality and purity of their materials, which have been developed over decades through expensive research and development processes. To make this technology accessible for everyone, we need to find cheaper ways to produce high-quality solar cell materials while maintaining their performance characteristics. Several synthesis techniques including Microwave-assisted method [20,21], hydrothermal [22], sol–gel [23], thermal decomposition [24], electrochemical [25], and chemical bath deposition [21,26] are employed among other synthesis techniques. The abovementioned techniques are reported by most of the researchers to synthesize II-VI chalcogenides in a large quantity and they are non-toxic to the environment. These synthesis methods allow us to quickly manufacture these materials at a low cost with high yields and good quality control parameters. Here, in this section, we have discussed some important synthesis techniques for II-VI chalcogenides that are Microwave-assisted method, hydrothermal, sol–gel, and electrochemical synthesis. Moreover, we have listed a comparison table for various synthesis methods to fabricate II-VI chalcogenides.

14.4.1 Microwave-Assisted Synthesis

The microwave-assisted method is a methodology increasingly used for the synthesis of 2D nanomaterials. This method is widely used for the preparation of II-VI-based chalcogenides as it is a very easy, cost-effective, and green synthesis technique [27]. Guang Zhu et al. reported the microwave-assisted chemical bath decomposition synthesis technique to produce the CdS/CdSe QDs cosensitized TiO_2 films. They reported that the performance of the prepared cell was good and achieved the power conversion efficiency (PCE) of 3.06% and current density of 16.1 mA cm^{-2}. They further described that this technique allows rapid and direct deposition of QDs [21]. Similarly, Guang Zhu et al. in their other work reported the fabrication of CdSe QDs-sensitized TiO_2 films by using a one-step microwave-assisted chemical bath deposition method. They used the prepared films as photoanode for QDs-sensitized solar cells. As a result, they obtained 1.75% PCE and 12.1 mA cm^{-2} maximum circuit current density.

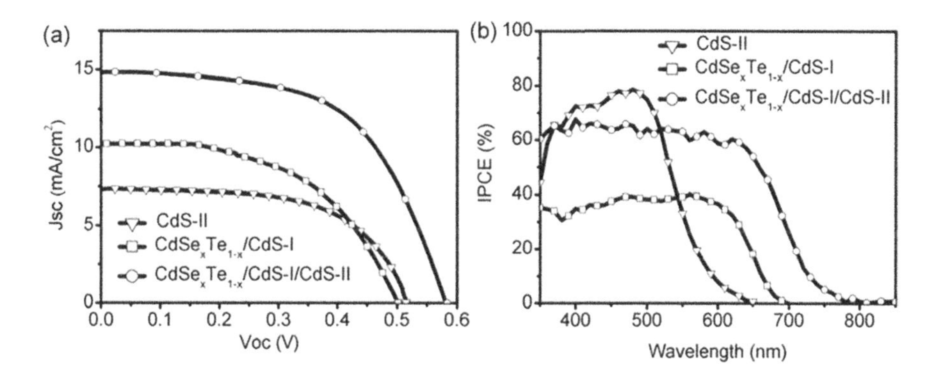

FIGURE 14.4 (a) J–V characteristics comparison of QDs-sensitized solar cells. (b) Spectra of incident photon to current conversion efficiency. Adapted with permission from [20] Copyright (2014) Chemical Communications.

Jianheng Luo et al. prepared CdSe$_x$Te$_{1-x}$–CdS QDs and core–shell type II via facile microwave aqueous method. They described that this is the rapid method to prepare the QDs. The core–shell of CdSe$_x$Te$_{1-x}$–CdS type II were used in QD-sensitized solar cells that showed a higher PCE of 5.04% [20] (Figure 14.4).

14.4.2 Hydrothermal Synthesis

Hydrothermal synthesis for the preparation of II-VI-based chalcogenides offers several advantages over other synthesis techniques such as easy experimental setup, high yield, low cost, and higher efficiency. This technique involved the mixing of proper agents and precursors into a solvent. This enables a crystalline nanostructure with an appropriate ratio. Li, Haiyan et al. fabricated the CdS nanoparticles decorated TiO_2 nanobelts via hydrothermal synthesis route. They reported that solar cells of CdS nanoparticles-decorated with TiO_2 nanobelts have 2.52% of PCE. While they further reported an efficiency of 2.84% when they deposited the nanobelts onto ZnO nanowires [22]. GuangchaoYin et al. successfully embedded CdS nanorods in CdTe absorbing layers by low cost and electrodeposition method and facile hydrothermal method. They obtained high-quality 3D heterostructures and examined their optical, structural, and electrical properties. They reported that the prepared 3D heterostructures have superior optical absorption properties. In addition, their studies indicated an 18.5% increase in the heterojunction solar cell energy conversion efficiency [28].

Mehdi Mousavi et al. prepared the $CdIn_2S_4$ nanostructures for solar cell application by using a facile hydrothermal method. They studied the optical, morphology, structure, and compositional properties by different characterization techniques. The prepared nanostructures of $CdIn_2S_4$ were used as a barrier layer in dye-sensitized solar cells; they showed notable enhancement in the efficiency of solar cells that was ~19%. Further, they described that the PCE of the solar cell was affected by the morphology and structure size [29].

14.4.3 Electrochemical Synthesis

Electrochemical synthesis is a simple and economical method to synthesize the II-VI-based chalcogenides for large-scale production at low temperatures. The architecture of nanomaterials can be tailored by this synthesis technique. This is the most suitable synthesis technique to prepare high-quality heterojunction solar cells. Chen-Zhong Yao et al. studied the vertical and high-density core–shell ZnO/CdS nanorods that were prepared by a two-step electrochemical deposition process. The prepared sample was used in QD-sensitized solar cells and showed improved performance. They reported 1.07% of PCE with J_{sc} of 5.43 mA cm^{-2} [30].

Shinya Higashimoto et al. reported a two-step route of template-free electrochemical deposition to prepare CdS nanotubes. They used the combination of polysulfide (S_x^{2-}/xS^{2-}), CdS NT, and carbon electrode for photo-electrochemical solar cells that exhibited the high efficiency of 80% at 510 nm due to effective light-harvesting performance. IPCE curve for the prepared CdS films fabricated for solar cells is shown in Figure 14.5b [25].

Dongjuan Xi et al. reported the hybrid films containing CdS nanorods arrays that were prepared by the electrochemical deposition technique. Due to the cathodic process, these nanorods were self-assembled on the gold-coated glass substrates. They further investigated their properties for solar cells and measured the open-circuit voltage of 0.84 V and 0.38% of overall PCE [31] (Table 14.1).

14.5 PROPERTIES OF II-VI BASED CHALCOGENIDES

In this section, we will briefly discuss the general properties of II-VI based Chalcogenides. Crystal structures of these materials are rather complex. Most II-VI based Chalcogenides crystallize in the wurtzite Zinc Blende and Wurtzite structures. Normally, these materials are found in Zinc-Blende structure at room temperature while at higher temperatures these are found in Wurtzite structure. One of the fascinating properties of these materials at atmospheric pressure is that they undergo a solid–solid phase transition below the melting point. This property enables these materials to apply in cooling systems now a day. Normally, CdSe and CdS crystallize in the HCP Wurtzite structure, CdTe and ZnTe in Zinc-Blende structure, whereas ZnS and ZnSe can exist in both of these structures. Bonding in II-VI-based Chalcogenides is a mixture of covalent and ionic types because the average number of electrons per atom is still four. As group-VI atoms are more electronegative than group-II atoms, so we can see ionicity in these materials. This ionic character varies considerably over the whole range and increases as the atomic size decreases. Therefore, ZnS is predominately ionic while the bonding in HgTe is nearly covalent. As we know that the ionic character has the effect of binding the valence electron rather tightly to the lattice atoms. Therefore, these materials have a wide direct bandgap. II-VI-based Chalcogenides with a wide bandgap have higher melting points. When they are heated to their melting point, the overheating effect occurs due to their high ionicity. It is difficult to produce bulk crystals from melt because of the greater vapor pressures at their melting points. The vapor phase, on the other hand, makes it simple

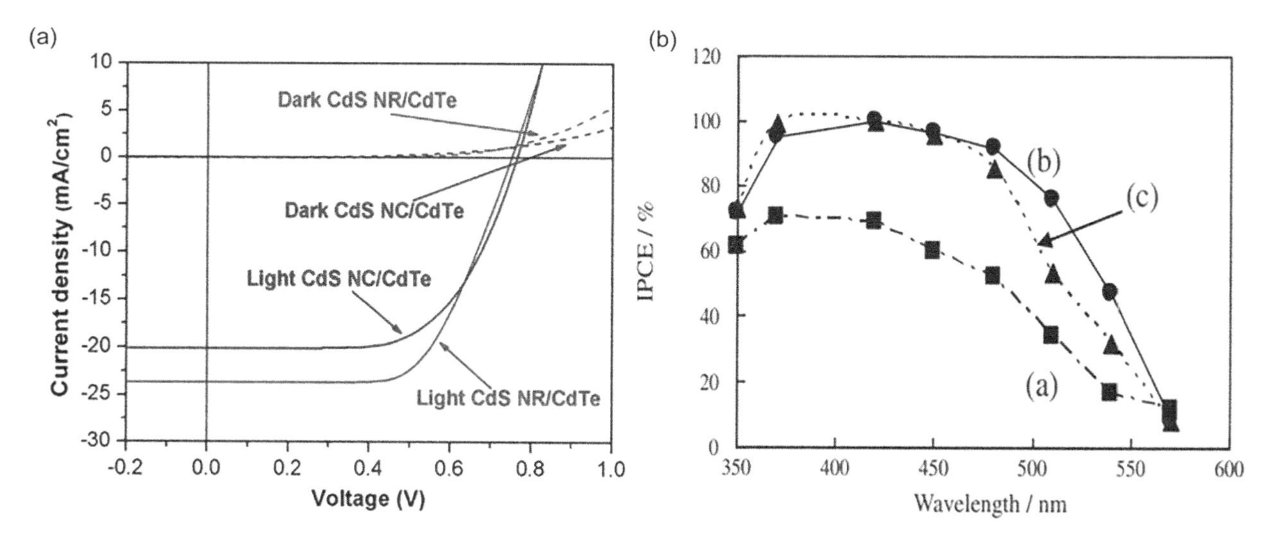

FIGURE 14.5 (a) Characteristics curve of photocurrent density and voltage for heterostructure solar cells. Based on 3D CdS nanorods/CdTe photo-electrode. (Adapted with permission from [28] Copyright (2017) Journal of Molecular Liquids.) (b) IPCE curve for the prepared CdS films fabricated for solar cells. Adapted with permission from [25] Copyright (2012) Electrochemistry communications.

TABLE 14.1
Comparison of Recent Synthesis Techniques and Solar Cell Efficiency

II-VI Chalcogenides	Synthesis Methods	Power Conversion Efficiency (%)	Application	Ref
CdS/CdSe quantum-dot (QD)-cosensitized TiO_2	Microwave-assisted chemical bath deposition (CBD)	3.06	QD-sensitized solar cells	[21]
TNARs/CdS/CdS	Hydrothermal, chemical bath deposition (CBD)	2.66	QD-sensitized solar cells	[32]
CdSe	CBD	6.8	QD-sensitized solar cells	[33]
CdS/CdSe QDs	Organometallic high-temperature injection method	5.32	CdS/CdSe-sensitized solar cells	[34]
CdS/CdSe QDs	Successive ion layer absorption and reaction (SILAR) method and CBD	4.62	CdS/CdSe-sensitized solar cells	[34]
CdS/CdSe QDs	SILAR	3.11	QD-sensitized solar cells	[35]
CdTe-CdS NCs	Hot-injection chemical precipitation method	4.2	QD-sensitized solar cells	[36]
CdTe nanocrystal	Solution processing	12.3	CdTe nanocrystal solar cells	[37]
ZnO/ZnSe core/shell nanorods	High temperature arrested precipitation	2.2	Liquid junction nanowire solar cells	[38]
Cu-doped ZnTe film	Electrochemical deposition	3.4–7.1	CdTe solar cells	[39]

to build bulk crystals and their films. As a result, it is vital to analyze the physical and chemical features of film and bulk crystal formation before introducing them. The parameters of some of the most common II-VI based Chalcogenides are shown in Table 14.2 [40–44].

14.6 APPLICATIONS IN SOLAR CELLS

In the era of science and technology, energy demand has been escalating day by day. The non-renewable resources are decreasing very fast with time. In this aspect, renewable resources such as hydropower, geothermal power, wind energy, and solar energy are better options to fulfill the energy requirement. Solar energy is the green and environment-friendly renewable energy resource available in large amounts in the universe. Solar cells or PV cells convert sunlight into electrical energy. This technology is sustainable and increased tremendously in recent years [46]. This remarkable increment in the usage of solar cells is due to advanced technologies, better construction materials, and public awareness about the advantages of solar energy. In this process when sunlight falls on the solar cell material the electrons in the material come in the conduction band and produce electricity. This phenomenon is called the

TABLE 14.2
Properties of Some II-VI Based Chalcogenides

Material Property	ZnS	ZnO	ZnSe	ZnTe	CdS	CdSe	CdTe
Melting point (K)	2038 (WZ, 150 atm)	2248	1797	1513	2023 (WZ, 100 atm)	1623	1370 [45]
Energy gap E_g at 300 K (eV) (ZB*/WZ*)	3.68/3.911	-/3.4	2.71/-	2.394	2.50/2.50	-/1.751	1.475
$dE_g/dT(\times10^{-4}eV/K)$ ZB/WZ	4.6/8.5	-/9.5	4.0/-	5.5/-	-/5.2	-/4.6	5.4/-
Structure	ZB/WZ	WZ	ZB/WZ	ZB	WZ	WZ	ZB
Bond length (μm)	2.342 (WZ)	1.977 (WZ)	2.454 [45]	2.636 [45]	2.530 [45]	2.630 [45]	2.806 [45]
Lattice constant [45] a_0 at 300 K (nm)	0.541	–	0.567	0.610	0.582	0.608	0.648
ZB nearest-neighbor dist	0.234	–	0.246	0.264	0.252	0.263	0.281
ZB density at 300 K (g/cm^3)	4.11	–	5.26	5.65	4.87	5.655	5.86
Heat capacity Cp (cal/mol K)	11.0	9.6	12.4	11.9	13.2	11.8	–
Ionicity (%)	62	62	63	61	69	70	72
Equilibrium pressure at c.m.p. (atm)	3.7	–	1.0	1.9	3.8	1.0	0.7
Minimum pressure at m.p. (atm)	2.8	7.82	0.53	0.64	2.2	0.4–0.5	0.23
Specific heat capacity (J/g K)	0.469	–	0.339	0.16	0.47	0.49	0.21
Thermal conductivity (W/(cm K))	0.27	0.6	0.19	0.18	0.2	0.09	0.01
Young/s modulus	10.8Mpsi	–	10.2Mpsi	–	45 GPa	5×10^{11} dyne cm^{-2}	3.7×10^{11} dyne cm^{-2}

m.p., melting point; c.m.p., congruent melting point; ZB, zinc blend; WZ, wurtzite.

PV effect. Many solar cells that are widely used are monocrystalline made with a single crystal of silicon have an efficiency of 26.5% reported but these monocrystalline solar cells are expensive [47]. Amorphous silicon solar cells can be fabricated at low temperatures and deposited on a substrate such as plastics and metal foils. They degrade when exposed to sunlight and consequently their performance decreases. Those solar cells made by a thin film of solar cell (SC) material have an advantage over the monocrystalline and amorphous silicon crystal. The reason is that polycrystalline solar cells have good efficiency and build a good electric field between the two different SC materials, which is known as a heterojunction. The semiconductor materials of II-VI group have great potential for the absorption and emission of light [48]. During the past few years, 2D layered materials have become an attractive field of research due to the promising and attractive properties of materials. Among these most emerging materials, II-VI chalcogenides have gained great potential in PV applications due to appropriate bandgap and other unique properties [49]. The detail about some chalcogenides materials is given below.

14.6.1 Cadmium-Related Chalcogenides

14.6.1.1 Cadmium Telluride

Cadmium telluride (CdTe) is a promising and attractive candidate for PV application. There are many unique properties of this material such as an appropriate bandgap (1.45 eV) and good optical absorption ability. The thin film of cadmium telluride with 2 mm thickness can absorb maximum (100%) incident light [4]. There is another benefit to use CdTe is that it has many techniques for deposition such as chemical vapor deposition (CVD), close-spaced sublimation (CSS), and electrode deposition. The efficiency of CdTe solar cell is 16.5%, which is prepared by the closed space sublimation. Other strategies are also helpful to attain the high-efficiency level, so this material is very attractive in the field of solar cells [50].

14.6.1.2 Cadmium Sulfide

The II-VI compounds such as CdS and CdSe have good ability for solar cell applications. CdS is a promising candidate in wide bandgap materials for PV applications. The composite of CdTe/CdS has good options to be tuned bandgap between 2.42 eV for CdS to 3.6 eV for ZnS. This combination enhances the wavelength spectral response for visible solar applications [51]. The red emission occurs due to the recombination of sulfur vacancy at 1.72 eV in CdS [52]. This red emission has been discussed by Lambe and Klick in the concept of phenomenological model [53] and in another complementary transition, an emission of 1.1 eV should be produced. The sum of these two emission energies is 2.82 eV. This value is very near and comparable with the bandgap of CdS, which can be improved with different strategies [54].

14.6.1.3 CdTe and CdS Heterojunction

CdTe and CdS heterojunction is also used for PV cells and showed good performance for solar energy. This thin layer is helpful to reduce the interfacial defects which are responsible for better efficiency [55]. This layer can be formed with those techniques which are operating at high temperatures like CVD or CSS. Better performance is achieved through heat treatment of CsTe/CdS due to the creation of interfacial layer, recrystallization, and defects passivation [56]. Figure 14.6 about the solar cell of CdTe and n-CdS composite is given below.

14.6.1.4 Cadmium Selenide

Solar cells based on the CdSe QDs have been described in some publications for PV applications with organic semiconductor polymer materials. There are some facile and efficient methods to fabricate the Cadmium Selenide (CdSe) nanomaterials using a single-source precursor [58]. The mechanism of CdSe to harvest the solar energy in the visible light

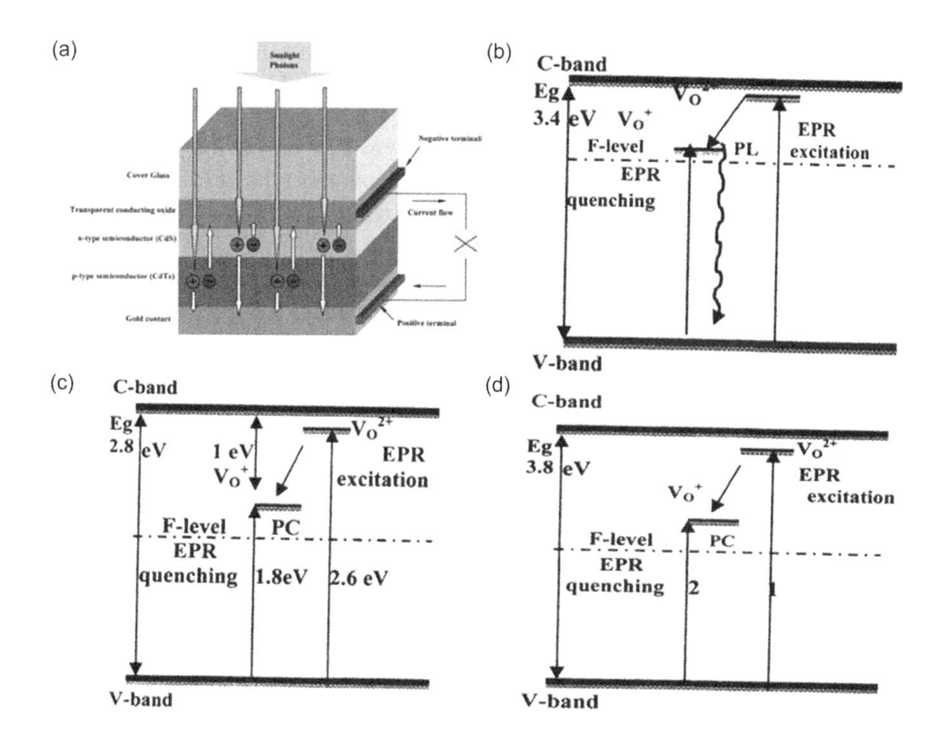

FIGURE 14.6 (a) A Schematic layout of a conventional p–n junction CdTe and CdS thin-film solar cell. (Adapted with permission from [48] Copyright (2006) Journal of Materials Chemistry.) (b) Scheme for recombination and electronic transitions-related oxygen vacancy presented to the bandgap of ZnO. (c) Recombination and electronic-transition scheme for the sulfur vacancy in the bandgap of ZnS. (d) Recombination and electronic-transition scheme on the selenium vacancy in the bandgap of ZnSe. Adapted with permission from [57] Copyright (2013) Journal of crystal growth.

spectrum resembles the CdS and this material is also useful for PV applications. The PL centered nearly at 1.0eV (1.2mm) is due to the recombination of charge carriers on Se vacancy [59]. The energy level of Se vacancy can be located at 1.82–1.0 eV$=E_C-0.82$eV which is very close to CdS. So this material has good ability for solar cells.

14.6.2 Zinc-Related Chalcogenides

14.6.2.1 Zinc Oxide

In PV applications, zinc oxide (ZnO) is an attractive and suitable material that can be used as nanoparticles and as thin films. Zinc oxide thin films are useful in solar cells because it has good conductivity and better transmittance for visible light radiation [60]. Zinc Oxide nanoparticles are also useful to build heterojunction solar cells. There is another advantage to use ZnO in solar cells is the facile fabrication of zinc oxide nanoparticles not needed for any further surface modification as compared to CdSe to produce solubility in organic solvents [61]. The information-related electron paramagnetic resonance (EPR) and photoluminescence (PL) quenching by light as a scheme about recombination are in the bandgap of ZnO via oxygen vacancy. In this scheme, PL emission arises due to the recombination of a hole from the valance band and electron caught on the energy level E (V_A $^+$). The EPR signal excitation is produced due to the conversion of V_A^{+2} to V_A $^{+1}$ during the process of an electron transition. Due to the fast hole-trapping occurs this type of transition starts p-type photoconductivity for the low intensity of light in all II-VI compounds [57]. This scheme is described in Figure 14.6b.

14.6.2.2 Zinc Sulfide

ZnS is another attractive chalcogenide material for solar cells application. The composite of the wide bandgap of ZnS 3.6 eV with CdS (2.42 eV) is also helpful for harvesting the maximum visible light radiation for solar energy. So this composite is also a suitable and attractive material for PV applications [51]. In Figure 14.6 during the reaction $V_S^{2+} + e \rightarrow V_S$ + transition 1 occurs and transfer V_S^{+2} to the EPR center, V_S $^+$. In the second transition come back the EPR center to the natural state following the reaction $V_S^+ + e \rightarrow V_S^0$. Due to this transition p-type photoconductivity was produced for the weak below bandgap [62]. The scheme of this process is shown in Figure 14.6c.

14.6.2.3 Zinc Selenium

ZnSe is another material for PV applications for solar cell applications. Its appropriate bandgap is suitable and helpful to increase its efficiency. In Figure 14.6d it is observed that after the irradiation mid-bandgap was much closer to the Fermi level between valance and conduction band. Due to this part of Se vacancies were caught electrons and transformed into the EPR center, V_{Se}^+. Similarly in the ZnS, ZnO, the photo-EPR center measurements play an important role and released quenching of EPR light. Shirakawa and Kukimoto and by Kishida et al. in Japan took measurements about the photosensitive band on ZnSe and this quenching spectra resembled their measurements [63,64]. In this scheme, the electronic transition coincided with anion vacancy and was constructed with permission as shown in Figure 14.6d.

14.6.2.4 Zinc Tellurium

Chemically Tellurium resembles selenium and sulfur and all materials belong to the class of chalcogens elements. The data about the absorption of light, recombination of photogenerated charge carriers, and EPR are very less related to the Te vacancies in the ZnTe composite. The energy level in this material associated with vacancy about Te, $V_{Te}^{2+/+}$ was measured at the position level of E_V +0.7 eV [65]. Due to the short data on an energy level, it is very difficult to find out the centers which are the reasons for PL peaks [66]. So we supposed the recombination scheme for ZnTe vacancy is similar to ZnO, ZnS, and ZnSe.

14.7 FUTURE PERSPECTIVES

Research studies are more and more directed to the rapid applications of II-VI-based chalcogenides. These materials are achieving much attraction because of their amazing properties. 2D-layered semiconducting materials such as ZnX and CdX materials have long been of enormous interest for a broad range of uses, such as transistor,

PV devices, heterojunction diodes, and photoconductors. 2D ZnX, CdX due to their bandgap, and high visual absorption coefficient are widely using solar cells that can store 100% of solar radiation. The efficiency of these materials is very high so low-cost, scalable solar cells energy can be produced. These trends will accelerate the development of the department, which is still rich in remarkable discoveries.

14.8 CONCLUSION

2D nanomaterials have gained significant attention from the scientist due to their high stability and reliability, applicability onto flexible substrates, aesthetical appearance, monolithic deposition, and superior temperature coefficient. Similarly, 2D chalcogenides nanomaterials are now widely used in solar cells due to their superior properties. They have enhanced the efficiency of solar cells. The II-VI-based chalcogenides nanomaterials are commonly used to enhance the efficiency, current density, and performance of solar cells. In this chapter, we have reviewed II-IV-based chalcogenides and their applications in different types of solar cells. Moreover, in-depth comparisons of various low-cost, scalable, economical, and easy synthesis techniques and types of various II-VI chalcogenides-based nanomaterials have been described in this chapter.

ACKNOWLEDGMENT

F. K. Butt acknowledges the funding support from HEC through grant number 7435/Punjab/NRPU/R&D/HEC/2017.

REFERENCES

1. Jain, M., ed. *II-VI Semiconductor Compounds*. World Scientific, Singapore, 1993.
2. Tamargo, M.C. *II-VI Semiconductor Materials and Their Applications*, Vol. 12. CRC Press, Boca Raton, FL, 2002.
3. Bhargava, R.N. 1997. 6.5 Light emitting diodes and electroluminescent devices based on widegap II-VIs. Properties of Wide Bandgap II-VI Semiconductors, (17).
4. Novoselov, K.S., D. Jiang, F. Schedin, T.J. Booth, V.V. Khotkevich, S.V. Morozov, and A.K. Geim. Two-dimensional atomic crystals. *Proceedings of the National Academy of Sciences* 102, no. 30 (2005): 10451–10453.
5. Khan, K., A.K. Tareen, M. Aslam, K. Hussain Thebo, U. Khan, R. Wang, S. Saqib Shams, Z. Han, and Z. Ouyang. A comprehensive review on synthesis of pristine and doped inorganic room temperature stable mayenite electride, [Ca24Al28O64] 4+(e–) 4 and its applications as a catalyst. *Progress in Solid State Chemistry* 54 (2019): 1–19.
6. Li, J., H. Luo, B. Zhai, R. Lu, Z. Guo, H. Zhang, and Y. Liu. Black phosphorus: a two-dimension saturable absorption material for mid-infrared Q-switched and mode-locked fiber lasers. *Scientific Reports* 6, no. 1 (2016): 1–11.
7. Liu, L., J. Park, D.A. Siegel, K.F. McCarty, K.W. Clark, W. Deng, L. Basile, J.C. Idrobo, A.-P. Li, and G. Gu. Heteroepitaxial growth of two-dimensional hexagonal boron nitride templated by graphene edges. *Science* 343, no. 6167 (2014): 163–167.
8. Khan, K., A.K. Tareen, M. Aslam, R. Wang, Y. Zhang, A. Mahmood, Z. Ouyang, H. Zhang, and Z. Guo. Recent developments in emerging two-dimensional materials and their applications. *Journal of Materials Chemistry C* 8, no. 2 (2020): 387–440.

9. Rafiei-Sarmazdeh, Z., S.M. Zahedi-Dizaji, and A. Kafi Kang. "Two-dimensional nanomaterials." In: Sadia Ameen, M. Shaheer Akhtar and Hyung-Shik Shin (eds.) *Nanostructures*. IntechOpen, London, UK, 2019.
10. Frindt, R.F. Single crystals of MoS_2 several molecular layers thick. *Journal of Applied Physics* 37, no. 4 (1966): 1928–1929.
11. Butler, S.Z., S.M. Hollen, L. Cao, Y. Cui, J.A. Gupta, H.R. Gutiérrez, T.F. Heinz, et al. Progress, challenges, and opportunities in two-dimensional materials beyond graphene. *ACS Nano* 7, no. 4 (2013): 2898–2926.
12. Wang, Q.H., K. Kalantar-Zadeh, A. Kis, J.N. Coleman, and M.S. Strano. Electronics and optoelectronics of two-dimensional transition metal dichalcogenides. *Nature Nanotechnology* 7, no. 11 (2012): 699–712.
13. Furdyna, J.K., S.N. Dong, S. Lee, X. Liu, and M. Dobrowolska. The ubiquitous nature of chalcogenides in science and technology. In *Chalcogenide* (pp. 1–30). Woodhead Publishing, Sawston, 2020.
14. Anscombe, N. The promise of chalcogenides. *Nature Photonics* 5, no. 8 (2011): 474–474..
15. Freitas, J.N., A.S. Gonçalves, and A.F. Nogueira. A comprehensive review of the application of chalcogenide nanoparticles in polymer solar cells. *Nanoscale* 6, no. 12 (2014): 6371–6397.
16. Singh, A.K. A short over view on advantage of chalcogenide glassy alloys. *Journal of Non-Oxide Glasses* 3 (2012): 1–4.
17. Garrido, J.C., F. Macoretta, M.A. Urena, and B. Arcondo. Application of Ag–Ge–Se based chalcogenide glasses on ion-selective electrodes. *Journal of Non-Crystalline Solids* 355, no. 37–42 (2009): 2079–2082.
18. Rahimi-Nasarabadi, M. Electrochemical synthesis and characterization of zinc sulfide nanoparticles. *Journal of Nanostructures* 4, no.2 (2014): 211–216.
19. nrel. Cadmium Telluride solar cells. (2021). Available from: https://www.nrel.gov/pv/cadmium-telluride-solar-cells.html.
20. Luo, J., H. Wei, F. Li, Q. Huang, D. Li, Y. Luo, and Q. Meng. Microwave assisted aqueous synthesis of core–shell CdSe x Te 1– x–CdS quantum dots for high performance sensitized solar cells. *Chemical Communications* 50, no. 26 (2014): 3464–3466.
21. Zhu, G., L. Pan, T. Xu, and Z. Sun. CdS/CdSe-cosensitized TiO_2 photoanode for quantum-dot-sensitized solar cells by a microwave-assisted chemical bath deposition method. *ACS Applied Materials & Interfaces* 3, no. 8 (2011): 3146–3151.
22. Li, H., M. Eastman, R. Schaller, W. Hudson, and J. Jiao. Hydrothermal synthesis of CdS nanoparticle-decorated TiO_2 nanobelts for solar cell. *Journal of Nanoscience and Nanotechnology* 11, no. 10 (2011): 8517–8521.
23. Cheng, Z., F. Su, L. Pan, M. Cao, and Z. Sun. CdS quantum dot-embedded silica film as luminescent down-shifting layer for crystalline Si solar cells. *Journal of Alloys and Compounds* 494, no. 1–2 (2010): L7–L10.
24. Kandula, S., and P. Jeevanandam. Visible-light-induced photodegradation of methylene blue using ZnO/CdS heteronanostructures synthesized through a novel thermal decomposition approach. *Journal of Nanoparticle Research* 16, no. 6 (2014): 1–18.
25. Higashimoto, S., K. Kawamoto, H. Hirai, M. Azuma, A. Ebrahimi, M. Matsuoka, and M. Takahashi. Fabrication of CdS nanotubes assisted by the template-free electrochemical synthesis method and their photo-electrochemical application. *Electrochemistry Communications* 20 (2012): 36–39.
26. Chang, C.-H., and Y.-L. Lee. Chemical bath deposition of CdS quantum dots onto mesoscopic TiO_2 films for application in quantum-dot-sensitized solar cells. *Applied Physics Letters* 91, no. 5 (2007): 053503.
27. Faraji, S., and F.N. Ani. Microwave-assisted synthesis of metal oxide/hydroxide composite electrodes for high power supercapacitors–a review. *Journal of Power Sources* 263 (2014): 338–360.

28. Yin, G., M. Sun, Y. Liu, Y. Sun, T. Zhou, and B. Liu. Performance improvement in three-dimensional heterojunction solar cells by embedding CdS nanorod arrays in CdTe absorbing layers. *Solar Energy Materials and Solar Cells* 159 (2017): 418–426.
29. Mousavi-Kamazani, M., M. Salavati-Niasari, M. Goudarzi, and Z. Zarghami. Hydrothermal synthesis of $CdIn_2S_4$ nanostructures using new starting reagent for elevating solar cells efficiency. *Journal of Molecular Liquids* 242 (2017): 653–661.
30. Yao, C.-Z., B.-H. Wei, L.-X. Meng, H. Li, Q.-J. Gong, H. Sun, H.-X. Ma, and X.-H. Hu. Controllable electrochemical synthesis and photovoltaic performance of ZnO/CdS core–shell nanorod arrays on fluorine-doped tin oxide. *Journal of Power Sources* 207 (2012): 222–228.
31. Xi, D., H. Zhang, S. Furst, B. Chen, and Q. Pei. Electrochemical synthesis and photovoltaic property of cadmium sulfide– polybithiophene interdigitated nanohybrid thin films. *The Journal of Physical Chemistry C* 112, no. 49 (2008): 19765–19769.
32. Chen, C., M. Ye, M. Lv, C. Gong, W. Guo, and C. Lin. Ultralong rutile TiO_2 nanorod arrays with large surface area for CdS/CdSe quantum dot-sensitized solar cells. *Electrochimica Acta* 121 (2014): 175–182.
33. Marandi, M., N. Torabi, and F. Ahangarani Farahani. Facile fabrication of well-performing CdS/CdSe quantum dot sensitized solar cells through a fast and effective formation of the CdSe nanocrystalline layer. *Solar Energy* 207 (2020): 32–39.
34. Pan, Z., H. Zhang, K. Cheng, Y. Hou, J. Hua, and X. Zhong. Highly efficient inverted type-I CdS/CdSe core/shell structure QD-sensitized solar cells. *ACS Nano* 6, no. 5 (2012): 3982–3991.
35. Tubtimtae, A., and M.-W. Lee. Effects of passivation treatment on performance of CdS/CdSe quantum-dot co-sensitized solar cells. *Thin Solid Films* 526 (2012): 225–230.
36. Marandi, M., and F. S. Mirahmadi. Aqueous synthesis of CdTe-CdS core shell nanocrystals and effect of shell-formation process on the efficiency of quantum dot sensitized solar cells. *Solar Energy* 188 (2019): 35–44.
37. Panthani, M.G., J. Matthew Kurley, R.W. Crisp, T.C. Dietz, T. Ezzyat, J.M. Luther, and D.V. Talapin. High efficiency solution processed sintered CdTe nanocrystal solar cells: The role of interfaces. *Nano Letters* 14, no. 2 (2014): 670–675.
38. Akram, M.A., S. Javed, M. Islam, M. Mujahid, and A. Safdar. Arrays of CZTS sensitized ZnO/ZnS and ZnO/ZnSe core/shell nanorods for liquid junction nanowire solar cells. *Solar Energy Materials and Solar Cells* 146 (2016): 121–128.
39. Jun, Y., K.-J. Kim, and D. Kim. Electrochemical synthesis of cu-doped znte films as back contacts to CdTe solar cells. *Metals and Materials* 5, no. 3 (1999): 279–285.
40. Isshiki, M., and J. Wang. Wide-bandgap II-VI semiconductors: Growth and properties. In: Kasap S., Capper P. (eds.) *Springer Handbook of Electronic and Photonic Materials*, p. 1. Springer, Cham, 2017.
41. Rudolph, P., N. Schäfer, and T. Fukuda. Crystal growth of ZnSe from the melt. *Materials Science and Engineering: R: Reports* 15, no. 3 (1995): 85–133.
42. Shetty, R., R. Balasubramanian, and W.R. Wilcox. Surface tension and contact angle of molten semiconductor compounds: I. Cadmium telluride. *Journal of Crystal Growth* 100, no. 1–2 (1990): 51–57.
43. Böer, K.W. *Survey of Semiconductor Physics-Electrons and Other Particles in Bulk Semiconductors*. Springer, New York, 1990.
44. Adachi, S.. *Properties of Semiconductor Alloys: Group-IV, III-V and II-VI Semiconductors*. Vol. 28. John Wiley & Sons, Hoboken, NJ, 2009.
45. Frackowiak, E., and F. Beguin. Carbon materials for the electrochemical storage of energy in capacitors. *Carbon* 39, no. 6 (2001): 937–950.
46. Jäger-Waldau, A. Status of thin film solar cells in research, production and the market. *Solar Energy* 77, no. 6 (2004): 667–678.

47. *Graphitic Carbon Nitride*. Available from: https://www.americanelements.com/graphitic-carbon-nitride#section-properties.
48. Afzaal, M., and P. O'Brien. Recent developments in II–VI and III–VI semiconductors and their applications in solar cells. *Journal of Materials Chemistry* 16, no. 17 (2006): 1597–1602.
49. Cai, H., Y. Gu, Y.-C. Lin, Y. Yu, D.B. Geohegan, and K. Xiao. Synthesis and emerging properties of 2D layered III–VI metal chalcogenides. *Applied Physics Reviews* 6, no. 4 (2019): 041312.
50. Wu, X., J. C. Keane, R. G. Dhere, C. Dehert, D. S. Albin, A. Dude, T. A. Gessert, S. Asher, D. H. Levi, and P. Sheldon. *Proceedings of the 17th European Photovoltaic Solar Energy Conference*. Munich, Germany 2, p. 995, 2001.
51. Zhou, J., Wu, X., Teeter, G., To, B., Yan, Y., Dhere, R.G. and Gessert, T.A. CBD-Cd1–xZnxS thin films and their application in CdTe solar cells. *Physica Status Solidi (b)* 241, no. 3 (2004): 775–778.
52. Vuylsteke, A. A., and Y. T. Sihvonen. Sulfur vacancy mechanism in pure CdS. *Physical Review* 113, no. 1 (1959): 40.
53. Lambe, J., and C.C. Klick. Model for luminescence and photoconductivity in the sulfides. *Physical Review* 98, no. 4 (1955): 909.
54. Sheinkman, M. K., I. V. Markevich, and T. V. Torchinskaya. The recharge-enhanced transformations of donor-acceptor pairs and clusters in CdS. *Journal of Physics and Chemistry of Solids* 43, no. 5 (1982): 475–479.
55. Ferekides, C.S., D. Marinskiy, V. Viswanathan, B. Tetali, V. Palekis, P. Selvaraj, and D. L. Morel. High efficiency CSS CdTe solar cells. *Thin Solid Films* 361 (2000): 520–526.
56. Burgelman, M., et al., Analysis of CdTe solar cells in relation to materials issues. *Thin Solid Films* 480 (2005): 392–398.
57. Babentsov, V., and R. B. James. Anion vacancies in II–VI chalcogenides: Review and critical analysis. *Journal of Crystal Growth* 379 (2013): 21–27.
58. Malik, M.A., N. Revaprasadu, and P. O'Brien. Air-stable single-source precursors for the synthesis of chalcogenide semiconductor nanoparticles. *Chemistry of Materials* 13, no. 3 (2001): 913–920.
59. Sheinkman, M. K., I. B. Ermolovich, and G. L. Belenki. Nature of infrared luminescence in CdSe single crystals and its correlation with photoconductivity. *Soviet Physics, Solid State* 10, no. 6 (1968): 1769–1772.
60. Alagappan, S.A., and S. Mitra. Optimizing the design of CIGS-based solar cells: A computational approach. *Materials Science and Engineering: B* 116, no. 3 (2005): 293–296.
61. Beek, W.J.E., M.M. Wienk, and R.A.J. Janssen. Efficient hybrid solar cells from zinc oxide nanoparticles and a conjugated polymer. *Advanced Materials* 16, no. 12 (2004): 1009–1013.
62. Leutwein, K., A. Räuber, and J. Schneider. Optical and photoelectric properties of the F-centre in ZnS. *Solid State Communications* 5, no. 9 (1967): 783–786.
63. Shirakawa, Y., and H. Kukimoto. The electron trap associated with an anion vacancy in ZnSe and ZnSxSe1– x. *Solid State Communications* 34, no. 5 (1980): 359–361.
64. Kishida, S., K. Matsuura, H. Nagase, H. Mori, F. Takeda, and I. Tsurumi. The photosensitive optical absorption bands in zn-treated and neutron-irradiated znse single crystals. *Physica Status Solidi (a)* 95, no. 1 (1986): 155–164.
65. Jansen, R.W., and O.F. Sankey. Theory of relative native-and impurity-defect abundances in compound semiconductors and the factors that influence them. *Physical Review B* 39, no. 5 (1989): 3192.
66. Garcia, J. A., A. Remon, V. Munoz, and R. Triboulet. Photoluminescence study of radiative transitions in ZnTe bulk crystals. *Journal of Crystal Growth* 191, no. 4 (1998): 685–691.

15 Photovoltaic Application of Graphene Oxide and Reduced Graphene Oxide

Perspectives on Material Characteristics and Device Performance

Tabitha A. Amollo
Egerton University

Vincent O. Nyamori
University of KwaZulu-Natal

CONTENTS

15.1 INTRODUCTION

The 21st century has seen exponential growth in demand for energy owing to the growing world population and day-to-day technological advancements. Thus, there is a need for a reliable and sustainable energy source for domestic and industrial use. Also, the adverse climate change experienced in the world today is largely occasioned by the emission of greenhouse gases. The burning of fossil fuels, as in automobiles

DOI: 10.1201/9781003178453-15

and industries, is a major contributor to the emission of such gases. Hence, there is a need not only for a reliable and sustainable but also clean energy resource. Solar energy comes in handy as it is a clean, reliable and sustainable source that is abundantly available worldwide. The effective conversion of solar energy to electricity requires high-performance solar cells devices.

Photovoltaic technology has evolved over the past decades giving rise to what's commonly referred to as the 'generations' of solar cells. To date, there have been four phases of photovoltaic technological evolution. The silicon-based solar cells comprise the first-generation solar cells. These have been in the market for over the past half-century. Even though silicon-based solar cells exhibit high power conversion efficiencies (PCEs), their fabrication is complex and costly, making them non-cost-effective to the global population. Besides, their rigidity limits their applications. The second-generation solar cells consist of thin-film technology that makes use of inorganic compounds such as amorphous silicon (a-Si:H), gallium arsenide (GaAs) and cadmium telluride (CdTe). These approaches are associated with high cost of production; for example, the thin film preparation necessitates the use of vacuum vapor deposition. The solution-processable thin-film solar cells, including polymer solar cells (PSCs), dye-sensitized solar cells (DSSCs), quantum dot solar cells and multijunction solar cells, form the third-generation solar cells. These are advantageous in terms of production and materials cost, environmental safety, flexibility and lightweight property. Perovskite solar cells (PVSCs), which form the fourth-generation solar cells, were introduced to overcome the challenges experienced in the synthesis of dyes for DSSCs. The fourth-generation solar cells also include types of PSCs in which the active layer comprises a polymer/nanoparticle composite. This chapter focuses on organic photovoltaics (OPV), that is, the PSCs and DSSCs.

Functional materials are a prerequisite to high-performance solar cells. As the photovoltaic technologies evolved with time, equally have the materials developed for the solar cells advanced, from crystalline silicon through the inorganic compounds to the polymers, small molecule compounds, dyes and perovskites. Thus, the scientific research community is focused on developing cutting-edge energy materials for photovoltaic application. Among the functional nanomaterials under intense research is graphene. Graphene is the first truly 2D nanomaterial composed of a monolayer of sp^2-hybridized carbon atoms. It is characterized by outstanding properties such as high carrier mobility, optical transparency, mechanical strength, flexibility and thermal conductivity. Because of these unique properties, graphene has been utilized in various categories of solar cells, mainly PSCs, DSSCs and PVSCs. Good results have been reported for such studies, and yet more studies on the same are ongoing.

Nevertheless, research reports have indicated challenges in using pristine graphene in the solution-based processing of solar cells owing to its hydrophobic nature. For this reason, graphene derivatives have been/are being developed for application in solar cells. This chapter focuses on the application of graphene derivatives in PSCs and DSSCs. The nanomaterials under focus are graphene oxide (GO), reduced graphene oxide (rGO) and their nanocomposites.

15.2 OVERVIEW OF ORGANIC PHOTOVOLTAICS OPERATION

OPV is a class of solar cells comprising the PSCs, DSSCs and PVSCs. These have the advantage of solution-based-processability, implying the capability of large-scale production. Under illumination, the absorber material in OPV generates an exciton that diffuses and dissociates to holes and electrons at the interface. This is followed by charge transfer to the electrodes (outward circuitry/load). Thus, the generation of charge carriers, diffusion and separation are the determinant factors of the PCE. These processes are mainly governed by the optoelectronic properties of the materials of the solar cell. Device engineering, morphological characteristics of the formed thin films and fabrication conditions are other factors that influence OPV performance. Generally, the photovoltaic performance of solar cells is characterized by the open-circuit voltage (V_{oc}), short-circuit current density (J_{sc}), fill factor (FF) and the ultimate PCE (η).

15.2.1 Polymer Solar Cells

PSCs can be fabricated in the planar or bulk heterojunction (BHJ) device structures. The latter structure is more commonly used as it enables longer exciton diffusion length. The planar structure has the donor and acceptor material intimately contacted together, while in the BHJ structure, the donor and acceptor materials are intermixed to form a blend. The typical BHJ structure is formed by the anode (transparent conductive electrode (TCO))/hole transport layer (HTL)/active layer/electron transport layer (ETL)/cathode, as shown in Figure 15.1a. The photoactive medium of the prototype PSC consists of poly-3-hexylthiophene (P3HT) and (6-6) phenyl-C_{61}-butyric acid methyl ester (PCBM) as the donor polymer and the fullerene acceptor, respectively. Poly(3,4-ethylenedioxythiophene):poly(styrenesulfonate) (PEDOT:PSS) and LiF/Ca forms the HTL and the ETL, respectively, while the anode is usually of indium tin oxide (ITO) and the cathode is of Al. The operation of the PSCs is explained in Section 15.2. The active layer may consist of a polymer donor and fullerene acceptor materials, a polymer donor and small molecule electron acceptor (SMA) materials or a polymer donor and polymer acceptor materials. The interfacial layers, HTL and ETL, create an ohmic contact between the active layer and the electrodes and promote the injection of charge carriers.

Though PEDOT:PSS ensures effective transfer of holes to the anode, its acidic and hydrophilic nature causes interface instability thus, leads to device degradation. Similarly, the low work function metals used in the ETL like Ca and Li are easily oxidized, causing an increase in series resistance at the interface thus, leads to device degradation [1]. Though characterized by high carrier mobility, the fullerene acceptors suffer from morphological instability, low light-harvesting capability, and difficulty in bandgap engineering [2]. Also associated with the fullerenes is photon energy loss through recombination processes occasioned by the electron disorder and low dielectric constant of the materials [3]. Because of these factors, fullerene-based PSCs manifest low V_{oc} and J_{sc}. Non-fullerenes, which include polymeric and SMAs, have been developed to overcome the challenges of fullerenes. These are designed

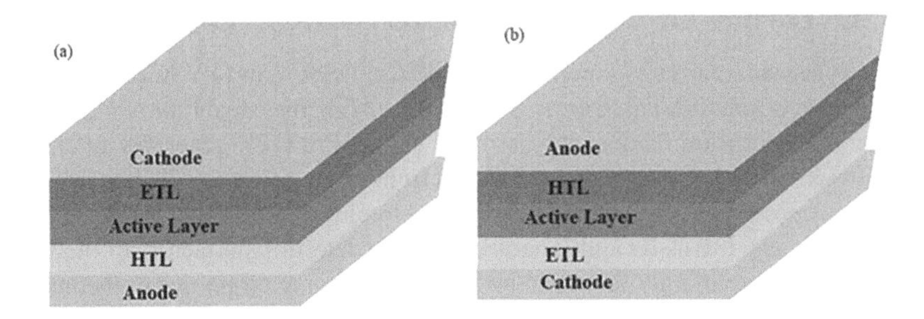

FIGURE 15.1 PSCs of (a) conventional and (b) inverted BHJ device structure.

in such a way as to offer complementary optical absorption to the donor materials. Also, the non-fullerene acceptors have the advantage of tunable energy levels [4] and low energy loss [5]; thus, high V_{oc} values are achievable with this class of materials. Nonetheless, the non-fullerenes are associated with lower carrier mobilities compared to their fullerene counterparts.

PSCs can also be fabricated in the inverted structure of TCO/ETL/active layer/HTL/cathode, as shown in Figure 15.1b. The inverted structure was enabled by the tunability of the work function of ITO to form a low work function cathode [6]. In the inverted structure, a high work function transition metal oxide like molybdenum oxide and vanadium oxide or PEDOT:PSS form the HTL, while a low work function compound forms the ETL. The anode comprises a high work function metal like Au or Ag. The inverted configuration is advantageous in terms of device stability [7], vertical phase separation in the active layer [8] and provides provision for design flexibility for tandem solar cells. Generally, good photovoltaic performance has been recorded for the PSCs, with those devices utilizing non-fullerene acceptors outperforming their fullerene-based counterparts [9]. Nevertheless, the PSCs have limitations of low carrier mobility and non-optimal optical absorption. Research on material development for PSCs is still ongoing with a view of optimizing not only the optical and electronic properties of the donor and acceptor materials but also interfacial layer materials for effective charge transfer. To this end, graphene nanomaterials are under intense investigation.

15.2.2 Dye-sensitized Solar Cells

DSSCs are fabricated in the TCO/counter electrode (CE)/electrolyte/dye/photoanode/TCO structure, as shown in Figure 15.2. The operation of DSSCs begins with the absorption of incident light by the dye promoting an electron from its highest occupied molecular orbital (HOMO) to its lowest unoccupied molecular orbital. This is followed by the injection of the electrons to the photoanode by the dye for further diffusion to the cathode (external circuitry). The electrons then return to the CE and regenerate the electrolyte which had regenerated the oxidized dye. In typical DSSCs, fluorine doped-tin oxide (FTO) or ITO forms the TCO, titania (TiO_2) forms the photoanode, platinum nanoparticles forms the CE. The dyes used in DSSCs are based on ruthenium, while the electrolytes are based on redox pairs like those of cobalt or

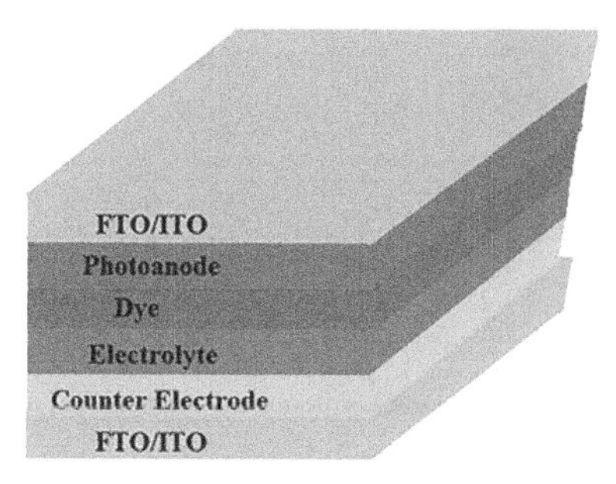

FIGURE 15.2 DSSCs device structure.

iodide/triiodide. The photoanode facilitates dye adsorption and charge transfer while the CE serves as a catalyst for the reduction reaction of the redox pairs in the electrolyte. Titania is mostly used as the photoanode because of its favorable optoelectronic properties viz. wide energy bandgap of 3.2 eV and optical absorption at wavelengths only below 388 nm, thermal stability, non-toxicity and chemical inertness [10]. The major challenge with the DSSCs is the poor stability of the dyes and the electrolytes.

15.3 GRAPHENE OXIDE AND REDUCED GRAPHENE OXIDE

GO is an oxygen-functionalized derivative of graphene. It consists of oxygen functional groups on the graphene backbone viz. the epoxy and hydroxyl groups on the basal planes and the carboxylic groups at the edges. The hybridization system of these oxygen functional groups differs in that the epoxy and hydroxylic groups' atoms are sp^3-hybridized while those of the carboxylic groups are sp^2-hybridized. The result of this mixed hybridization system is a disruption of the sp^2 conjugation network of the graphene lattice. For this reason, GO exhibits insulating behavior. Nevertheless, charge transfer does occur in GO via the hopping mechanism within the localized sp^2 sites. The polar oxygen functional groups on GO make it hydrophilic; thus, it's compatible with solution-based device processing. Homogenous solutions of GO in water or organic solvents can be obtained by stirring or sonication. The common synthesis route for GO involves the exfoliation of graphite flakes/powder using strong oxidizing agents such as potassium permanganate. This was first demonstrated in 1859 by Brodie [11], but the synthesis protocol has developed over time. Today, the commonly used method is the modified Hummers method, which was first demonstrated in 1958 [12].

rGO is obtained from GO via chemical or thermal reduction. The reduction treatment is aimed at reducing the oxygen functional groups. This is a means of restitution of the disrupted sp^2 conjugation network of graphene thus, restoring the electrical conductivity behavior. Chemical reduction of graphene involves reducing agents such as hydrazine, sodium borohydride and hydroquinone [13]. On the other hand, thermal reduction involves thermal annealing of the GO samples under the flow of a reducing gas like hydrogen or nitrogen. Typically, the GO sample, in a crucible, is placed in a quartz furnace and heated at temperatures as low as 150°C to as high temperatures as 2,000°C. In this procedure, the quality of the rGO samples

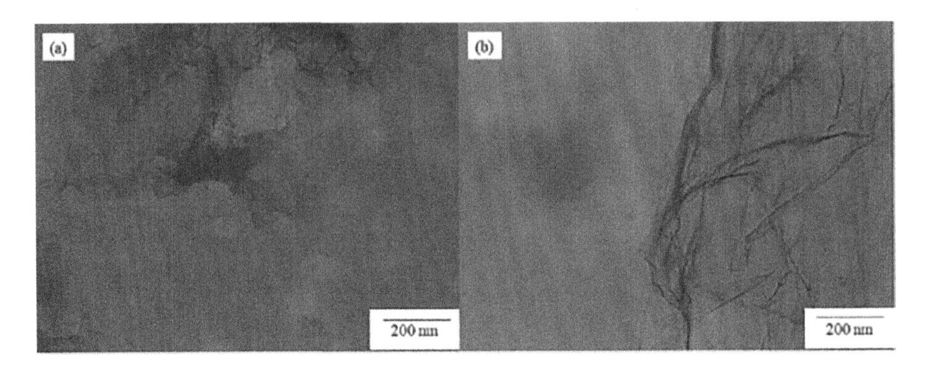

FIGURE 15.3 TEM images of (a) Hydrazine hydrate reduced rGO and (b) thermally reduced rGO at 500°C.

produced is dependent on the annealing temperature. For example, thermal reduction at lower temperatures yields rGO with lots of defects, but as the annealing temperature increases, a reduction in the number of defects is observed [14]. Also, although both the chemical and thermal reduction routes yield rGO, the quality of the samples produced differ. As exemplified in Figure 15.3, the hydrazine hydrate reduced rGO manifests a more wrinkled structure (with more folds) than the thermally reduced rGO. The sample quality is key to the performance of devices based on these materials; thus, one would need to optimize not only the synthesis methodology but also their corresponding conditions.

15.3.1 Application of Graphene Oxide, Reduced Graphene Oxide and Their Nanocomposite in Polymer Solar Cells

Both GO, rGO and their nanocomposites have been utilized in the interfacial and active layers of PSCs. In both cases, the graphene materials influence the performance of the devices, albeit via different mechanisms. Recently, we reported the use of graphene nanomaterials, namely GO, rGO and reduced graphene oxide germanium dioxide (rGO-GeO_2) in the photoactive medium of PSCs. Similarly, we have employed GO and reduced graphene oxide germanium quantum dots (rGO-Ge) in the HTL of PSCs of the structure glass/ITO/PEDOT:PSS/active layer/LiF/Al. The active layer of the PSCs consisted of P3HT:PCBM. The optimum concentrations for GO, rGO and rGO-GeO_2 in the active layer were 10, 3 and 3 wt.%, respectively, while for the GO and rGO-Ge in the HTL, the optimum concentration was 0.125 wt.%. Reference devices were also fabricated for all the experiment sets. For all the fabricated devices, the use of the graphene nanomaterials improved their photovoltaic performance, albeit via different mechanisms. The electrical properties of these devices are listed in Table 15.1. GO in the active layer improved the device's PCE by 120% [15]. This resulted from improved J_{sc} occasioned by the high optical absorption of the GO modified active layer device. The high J_{sc} of 18.2 mA cm^{-2} indicated high photogeneration of charge carriers and effective charge transport in the device, which corroborated the enhanced carrier mobility by two orders of magnitude.

TABLE 15.1
Electrical Parameters of BHJ Solar Cells Modified with Graphene Nanomaterials in the Active Layer and HTL

HTL	Active Layer	V_{oc} (volts)	J_{sc} (mA cm^{-2})	FF (%)	PCE (%)	Improved PCE index (%)	Ref
PEDOT:PSS	P3HT:PCBM:GO	0.57	18.2	43	4.4	120	[15]
PEDOT:PSS	P3HT:PCBM :rGO	0.52	10.0	38	2.0	25	[17]
PEDOT:PSS	P3HT:PCBM:rGO-GeO_2	0.44	17.0	31	2.3	53	[17]
PEDOT:PSS-GO	P3HT:PCBM	0.53	14.0	38	2.8	40	[15]
PEDOT:PSS/rGO-Ge	P3HT:PCBM	0.51	10.3	46	2.4	50	[16]

The FF of the device was comparable to the reference device, while there was a slight improvement in the V_{oc}. The active layer film was of smooth morphology following the high crystallinity of the GO sample. This aided the charge carriers separation process. An improved PCE of 40% was achieved in devices with GO in the HTL. This resulted from improved J_{sc}. Also, the carrier mobilities improved by two orders of magnitude. The GO material served to lower the injection barrier at the interface thus, favoring extraction of holes. Also, given the insulating behavior of GO, it aids in selective blocking of electrons at the interface, thereby reducing recombination processes. GO inclusion in both the active layer and HTL resulted in the formation of interconnected pathways for charge transport. The device with rGO-Ge in the HTL manifested improved PCE by 50%, arising from improved J_{sc} and FF [16]. This was achieved even though the HTL film was characterized by agglomerates, which are potential sites for current leakages. The rGO-Ge nanocomposite used was of good electrical conductivity thus, provided more percolation pathways in the HTL. This served to reduce non-geminate charge recombination. Also, the Ge quantum dots of high electron affinity served to aid selective electron blocking. Thus, the inherent properties of the HTL materials served to improve the photovoltaic performance of the devices outdoing the detrimental effects of unfavorable morphology. rGO-GeO_2 nanocomposite and rGO employed in the photoactive medium of PSCs resulted in improved PCE by 53% and 25%, respectively [17]. The rGO-GeO_2 nanocomposite in the active layer resulted in remarkable improvement in J_{sc} to 17 mA cm^{-2}, though with lower V_{oc} and FF values. The charge carrier mobility in these devices improved by order of magnitude. The high electron affinity of Ge resulted in a high built-in potential difference between the donor and acceptor phases causing effective exciton dissociation. This served to limit geminate charge carriers recombination. The lower V_{oc} and FF were due to increased non-geminate charge recombination occasioned by the rough morphology of the active layer film. Nevertheless, the effective charge separation in the devices yielded a remarkable improvement in the J_{sc}. An increase in PCE by 25% was achieved by employing rGO in the active layer of PSCs. In terms of V_{oc} and FF, the rGO modified devices performed comparably to the reference device. But there was a slight improvement in the J_{sc} which resulted in the improvement of the PCE. Also, the charge carrier mobility was improved by two orders of magnitude, implying more percolation pathways upon inclusion of rGO. For all the devices fabricated with the graphene nanomaterials either in the HTL or active layer, there was observed improvement in charge carrier mobility which yielded an improved J_{sc}.

GO was utilized as the HTL material for poly [N-9′-heptadecanyl-2, 7-carbazole-alt-5,5-(4′, 7′-di-2-thienyl-2′, 1′, 3′benzothiadiazole)] PCDTBT:PC_{71}BM based solar cells [18]. At an optimum GO concentration of 1 mg/ml, a V_{oc}, J_{sc}, FF and PCE of 0.8 V, 8.14 mA cm^{-2}, 42% and 2.73%, respectively, were achieved. The work demonstrated the effect of material concentration and thickness of the HTL. At a higher concentration of GO and consequently the thickness, the device performance deteriorated with lower J_{sc} and high series resistance being manifested.

The effect of oxidation level of GO for application as HTL in PSCs was investigated by Wu et al. [19]. Table 15.2 shows the electrical parameters of P3HT:PCBM-based PSCs with GO at different oxidation levels as the HTL. The V_{oc}, FF and PCE increased with an increase in the oxidation level. The increase in V_{oc} (0.34–0.56 V) with increasing oxidation level (50%–400%) was attributed to an increased work function. At the optimum oxidation level (400%), better dispersion and smooth morphology of GO were observed. The attained work function (5.1 eV) at the optimum oxidation level matches that of P3HT (5.2 eV), implying a lowering of the Schottky barrier at the interface. Thus, it yielded an ohmic contact for charge extraction. However, the photovoltaic performance of the device with GO oxidation level beyond the optimum dropped due to poor electrical conductivity. Nevertheless, P3HT:PCBM-based devices with GO HTLs at optimum oxidation level performed better due to increased dispersion (yielding better film morphology) and work function (yielding an ohmic contact at the interface). This observation echoes the interplay of the various factors that affect the performance of PSCs like material properties, energy level alignment at the interface and film morphology. This study showed that the optoelectronic properties and dispersibility of GO could be optimized by reasonable control of the oxidation level.

Singly applied (spin-coated) GO/PEDOT:PSS HTL yielded an improved performance in P3HT:PCBM solar cells than either GO or PEDOT:PSS only HTL devices [20]. While the V_{oc} and FF of the devices were comparable, the double-layered HTL device attained a PCE of 4.83% with J_{sc} of 15.42 mA cm^{-2} while the GO and PEDOT:PSS devices attained PCEs of 3.16% with J_{sc} of 11.18 mA cm^{-2}

TABLE 15.2
Electrical Parameters of P3HT:PCBM-Based PSCs with GO at Different Oxidation Levels as the HTL

HTL Material	V_{oc} (volts)	J_{sc} (mA cm^{-2})	FF (%)	PCE (%)
PEDOT:PSS	0.60	8.22	60	3.2
GO(50 wt.% $KMnO_4$)	0.34	7.93	33	1.1
GO(100 wt.%$KMnO_4$)	0.41	7.69	52	2.0
GO(400 wt.%$KMnO_4$)	0.56	7.71	64	3.0
GO(600 wt.%$KMnO_4$)	0.54	7.67	62	2.7

Adapted with permission from [19]. Copyright (2015) Royal Society of Chemistry.

and 4.00% with J_{sc} of 13.38 mA cm^{-2}, respectively. Also, the GO/PEDOT:PSS HTL device exhibited superior environmental stability. The performance gains in these devices were attributed to the occurrence of molecular-level interaction at the GO–PEDOT:PSS interface. Such an interaction would yield a linking bridge for charge transport, enabling the transport of holes from the donor phase of the active layer to the anode, leading to superior photovoltaic performance. Also, the strong molecular interaction acted as a barrier to prevent absorption of water molecules and impede the diffusion of molecules between the active layer and the anode thus, improving the device's stability.

Sulfonated reduced graphene oxide (S-rGO) obtained by treating GO with concentrated sulfuric acid was used as HTL in inverted P3HT:PCBM solar cells [21]. The S-rGO showed tunability of optical bandgap and electrical conductivity with the level of sulfonation and reduction of GO. The sample, S-rGO$_1$, which was the most reduced and least sulfonated, exhibited a C/O ratio of 2.59 and electrical conductivity of 1.4 S cm^{-1}. The other samples, S-rGO$_2$ and S-rGO$_3$, manifested C/O ratios of 1.10 and 0.73 and electrical conductivities of 1.1 and 0.73 S cm^{-1}, respectively. The optimized S-rGO HTL device exhibited comparable performance (a PCE of 2.80%) to the reference PEDOT:PSS HTL device (PCE of 2.75%). This was fabricated with S-rGO$_2$ as the HTL, while S-rGO$_1$, the most reduced sample, yielded an inferior performance with a PCE of 1.54%. It should be noted here that the S-rGO$_1$ sample, which was the most reduced and exhibited the highest electrical conductivity yielded an inferior photovoltaic performance; this underscores the fact that various factors besides the material properties influence the performance level of PSCs.

A nanocomposite of rGO/Pt nanoparticles (Pt NPs) synthesized through in situ crystallization was employed in the HTL of a P3HT:PC$_{71}$BM PSC [22]. They used ethylene glycol (EG), sodium citrate (SC) and ascorbic acid (AA) as the reducing agents in the crystallization process. With EG as the reducing agent, rGO/Pt NPs composite with the most reduced GO and Pt ions and the highest content of Pt NPs uniformly distributed on the rGO nanosheets were produced. Consequently, this nanocomposite exhibited the highest electrical conductivity and the lowest HOMO level to match that of P3HT. PSCs with EG rGO/Pt NPs as the HTL exhibited the best photovoltaic performance with a PCE of 3.6%. This was a slight improvement compared to PEDOT:PSS HTL device with a PCE of 3.68%. Whereas, V_{oc} and FF were comparable, the EG rGO/Pt NPs and PEDOT:PSS HTL devices yielded a Jsc of 11.01 and 10.02 mA cm^{-2}, respectively. The improved performance was attributed to better energy level alignment at the HTL active layer interface.

Another study investigated the effect of the size of GO nanosheets on the photovoltaic performance of BDT-thieno[3,4-b]thiophene (PTB7):PC$_{71}$BM-based PSCs [23]. GO nanosheets of small size (SGO), medium size (MGO) and large size (LGO) prepared by varying the sonication time from 3, 6 and 10 minutes, respectively, were employed in the active layer. The GO nanosheets' sizes ranged from hundreds of nanometers to micrometers. MGOs sized at hundreds of nanometers yielded the best performance with a PCE of 9.21% at FF of 69.4% and a PCE of 8.27% at FF of 70.5% for inverted and conventional device structures, respectively. These devices' photovoltaic performance was superior to the reference devices without GO, with an increase of 16% and 23% in the inverted and conventional device structures, respectively. The LGO manifested poor dispersion capability leading to aggregated morphology of the active layer film. The space charge limited current mobility was highest in the MGO-modified devices. Consequently, the high charge transport in these devices reduced non-geminate recombination leading to the high FF values.

A solution-processed rGO (FGr) functionalized by introducing the vinyl group and Si-O-Si groups through vinyl triethoxysilane on graphene surface was used as ETL in PTB7:$PC_{71}BM$ and P3HT:$PC_{71}BM$-based PSCs [24]. The *J–V* characteristics of the devices are shown in Figure 15.4. A PCE of 9.47% and 4.05% was achieved in the PTB7:$PC_{71}BM$ and P3HT:$PC_{71}BM$ FGr ETL devices, respectively. This was an improvement from 8.94% (PTB7:$PC_{71}BM$) and 3.52% (P3HT:$PC_{71}BM$) in the PFN ETL devices. The FGr ETL devices exhibited remarkable stability, with the PCE dropping only by 7.4% after 61 days of storage in a nitrogen glove box. The same group tested rGO ETL devices which yielded a PCE of 8.96% and 3.74% for PTB7:$PC_{71}BM$ and P3HT:$PC_{71}BM$, respectively. The better electron extraction capabilities of the FGr ETL were attributed to the formation of a homogenous ETL layer resulting from the high dispersion of FGr in the organic solvent. Thus, this work echoes the influence of morphological characteristics of not only the active layer but also the interfacial layers on the photovoltaic performance of PSCs.

TiOx/rGO was employed as ETL in inverted P3HT:$PC_{61}BM$ PSCs [25]. The devices registered a V_{oc}, J_{sc}, FF and PCE of 0.66 V, 8.07 mA cm^{-2}, 51% and 2.7%, respectively. This was an improved performance in comparison to the TiOx device, which recorded a V_{oc}, J_{sc}, FF and PCE of 0.62 V, 8.42 mA cm^{-2}, 48% and 2.5%, respectively. The improved V_{oc} and PCE indicated that rGO in the ETL promotes electron extraction from the active layer to the cathode. Also, the devices with TiOx/rGO as ETL exhibited good stability, with the PCE reducing by only 27% after 90 days of storage in ambient conditions. Another study reported using rGO-anatase titania nanocomposite in the HTL and the active layer of P3HT:$PC_{61}BM$ PSCs [26]. The device with the nanocomposite in the HTL yielded a V_{oc}, J_{sc}, FF and PCE of 0.51 V, 6.56 mA cm^{-2}, 28% and 0.93%, respectively, while the nanocomposite-modified active layer device realized a V_{oc}, J_{sc}, FF and PCE of 0.55 V, 13.59 mA cm^{-2}, 43% and 3.22%, respectively.

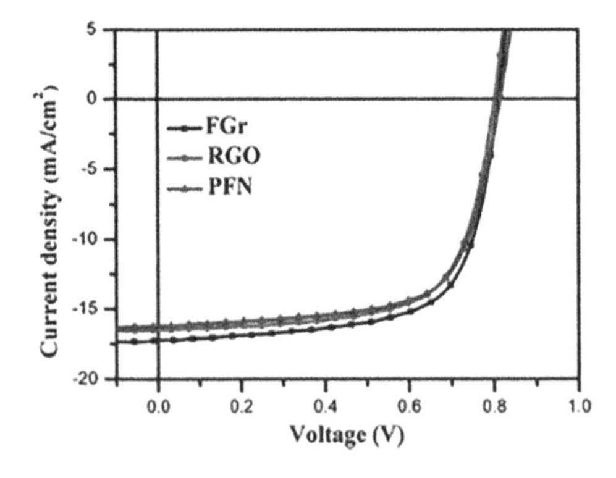

FIGURE 15.4 Current–voltage characteristics of PSCs with FGr, rGO and PFN as the ETL materials. Adapted with permission from [24]. Copyright (2018) Elsevier.

The reference devices with PEDOT:PSS as HTL and P3HT:$PC_{61}BM$ as active layer registered a V_{oc}, J_{sc}, FF and PCE of 0.51 V, 6.04 mA cm^{-2}, 37% and 1.1%, respectively. The improved photovoltaic performance of the nanocomposite active layer modified devices resulted from improved photon harvesting occasioned by the presence of the nanocomposite. This yielded an enhanced J_{sc} and FF.

15.3.2 Application of Graphene Oxide, Reduced Graphene Oxide and Their Nanocomposite in Dye-sensitized Solar Cells

The common application of GO, rGO and their nanocomposite in DSSCs include a transparent electrode, CE material and a photoanode additive. Graphene is a suitable replacement for the Pt CE material due to its high catalytic activity, specific surface area and chemical stability. The catalytic activity is further improved in GO because of the presence of the oxygen functional groups. Graphene comes in handy as a photoanode additive, as it facilitates electron transfer from the conduction band of TiO_2 to the cathode via the formation of charge percolation pathways and favorable energy level alignment. For effective electron transfer, the TiO_2 species should be uniformly and densely populated on the graphene nanosheets. Zhao et al. [27] reported the photovoltaic performance of DSSCs fabricated with rGO produced by in situ thermal reduction of rGO as CE. GO was first adsorbed on the FTO substrate to form a film. rGO film was then formed by thermal annealing of the GO film in a nitrogen atmosphere. For comparison, rGO film was also directly adsorbed on the substrate. The rGO CE formed by in situ annealing was observed to be more uniform and adhered strongly to the substrate compared to the directly adsorbed rGO CE. Consequently, the devices with the in situ annealed rGO CE yielded a better performance with a V_{oc}, J_{sc}, FF and PCE of 0.68 V, 12.93 mA cm^{-2}, 73% and 6.35%, while directly adsorbed rGO CE device recorded a V_{oc}, J_{sc}, FF and PCE of 0.60 V, 10.34 mA cm^{-2}, 19% and 1.20%. The poor performance of the directly adsorbed rGO CE device was due to the aggregation of the rGO sheets and the extremely poor contact with the FTO substrate. On the other hand, the good performance of the in situ rGO CE device was attributed to a smooth film, evidenced by a high FF and strong adhesion to the substrate. The reference Pt CE device yielded a V_{oc}, J_{sc}, FF and PCE of 0.74 V, 13.83 mA cm^{-2}, 74% and 7.53%, respectively. This study underscores the influence of film preparation techniques on the photovoltaic performance of solar cells.

rGO-coated polyaniline (PANi/rGO) nanocomposite prepared via in situ polymerization was used as the CE in DSSCs [28]. Figure 15.5 shows the current–voltage characteristics of the devices fabricated with the nanocomposite as CE. The nanocomposite CE device yielded a PCE of 3.98%, which was a better performance in comparison to PANi and rGO CE devices with a PCE of 1.83% and 1.07%, respectively. The better PCE achieved by the nanocomposite CE device resulted from better J_{sc} and FF being 12.58 mA cm^{-2} and 55%, respectively. The J_{sc} and FF of the PANi CE device were 11.03 mA cm^{-2} and 47%, while those of the rGO CE device were 9.41 mA cm^{-2} and 36%, respectively. The nanocomposite offered a larger surface area for better catalytic activity, resulting in the high J_{sc} and FF. The rGO CE performed poorly because of surface defects, which adversely affected the electrical conductivity and catalytic activity. Nonetheless, the Pt CE device performed better

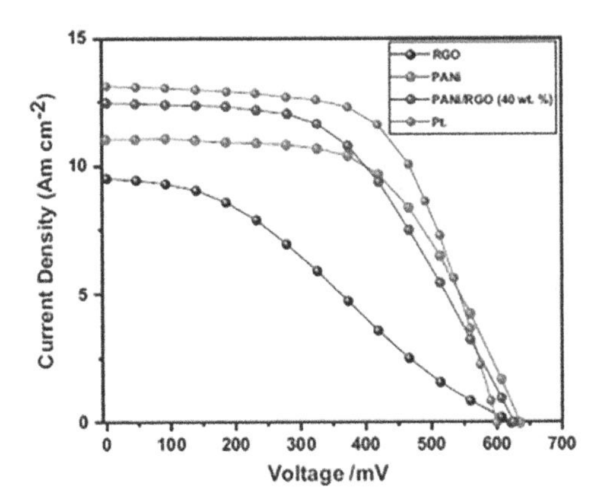

FIGURE 15.5 Current–voltage characteristics of DSSCs with rGO, PANi, PANi/rGO and reference Pt CEs. Adapted with permission from [28]. Copyright (2020) Copyright The Authors, some rights reserved; exclusive licensee [Elsevier]. Distributed under a Creative Commons Attribution License 4.0 (CC BY) https://creativecommons.org/licenses/by/4.0/.

than the nanocomposite one with a $J_{sc,}$ FF and PCE of 13.11 mA cm^{-2}, 63% and 4.75%, respectively. This was attributed to the superior electrical conductivity of Pt CE, yielding the highest J_{sc} and FF.

Sudhakar et al. [29] utilized PEDOT:PSS/rGO nanocomposite prepared by simple sonication as the CE in DSSCs. The device's electrical performance is shown in Table 15.3. DSSCs with PEDOT:PSS/rGO as CE recorded a comparable performance to that of Pt. The good performance was attributed to good catalytic activity and electrical conductivity of the nanocomposite CE, as well as good adherence of the nanocomposite to the substrate. Additionally, the nanocomposite improved the stability of the device, with insignificant changes being observed upon irradiation of the device with the light of different intensities.

rGO-TiO_2 composite nanofibers with rGO at various concentrations were utilized as a photoanode additive in DSSCs [30]. Table 15.4 shows the electrical parameters of the fabricated devices. The enhanced performance at 2 mg rGO-TiO_2 loading was

TABLE 15.3
Electrical Parameters of DSSCs with of PEDOT:PSS/rGO Nanocomposite as the CE

HTL Material	V_{OC} (volts)	J_{sc} (mA cm^{-2})	FF (%)	PCE (%)
Pt	0.75	17.3	74	9.64
rGO	0.73	12.9	71	6.70
PEDOT:PSS	0.76	11.2	72	6.10
PEDOT:PSS/rGO	0.78	16.1	76	9.57

Adapted with permission from [29]. Copyright (2020) ACS Publications.

TABLE 15.4
Electrical Parameters of DSSCs with Photoanode of rGO-TiO_2 Composite Nanofibers

HTL Material	V_{oc} (volts)	J_{sc} (mA cm^{-2})	FF (%)	PCE (%)
TiO_2	0.67	9.73	59	3.83
2 mg rGO-TiO_2	0.65	10.82	59	4.10
4 mg rGO-TiO_2	0.65	10.92	62	4.43
6 mg rGO-TiO_2	0.66	9.64	62	3.93

Adapted with permission from [30]. Copyright (2019) Elsevier.

attributed mainly to effective electron transfer in the devices. The reduced diameter of the nanofibers in this sample resulted in increased surface area. The best performance of the 4 mg rGO-TiO_2 photoanode was attributed to improved electron transport and dye adsorption. Further increase in the amount of GO in the composite resulted in a drop in performance. This was attributed to a decreased dye adsorption with increased diameter TiO_2. Also, the GO would compete for solar absorption with the dye. Overall, the devices with rGO-TiO_2 photoanode performed better than the pristine TiO_2 photoanode device.

rGO prepared by microwave exfoliation method was utilized as a photoanode additive and as a transparent layer deposited prior to TiO_2 photoanode in DSSCs with quasi-solid polymer electrolyte [31]. The devices with rGO as a photoanode additive performed poorly, recording a lower PCE c. 0.2% than that of the reference device c. 0.22%. The rGO sample was of high crystallinity, which hampered interaction with TiO_2 in the photoanode. This poor interaction impeded electron transfer, which led to poor results. The device with rGO as a transparent electrode yielded an improvement in PCE c. 0.58%. The highly crystalline rGO as a working electrode enabled better electron transfer leading to improved performance. Also, the better performance was attributed to a large surface area for dye adsorption and consequently an increased light harvesting. This corroborated with the better J_{sc} in this device. It is usually the case that highly crystalline samples are desirable for application in functional devices like solar cells. This study, however, shows that molecular-level interaction is not favored by highly crystalline samples.

15.4 CONCLUSION

In summary, this chapter examined the application of GO, rGO and their nanocomposite (graphene nanomaterials) in the interfacial and photoactive layers of PSCs, as well as the CE and photoanode additive of DSSCs. In all cases, the graphene nanomaterials improved the devices' performance, albeit via different mechanisms. As the interfacial layer material, the graphene nanomaterials promote charge carrier extraction from the active layer to the respective electrodes. However, high concentrations of the graphene nanomaterial in the ETL impair optical transmittance leading to decreased light absorption in the device. In the active layer, the graphene nanomaterials improve light harvesting and charge transport. In both cases, the inclusion of the graphene nanomaterials results in improved environmental stability of the devices. Besides the good catalytic activity of the graphene nanomaterials, their use as a CE in DSSCs affects a larger surface area for increased catalytic activity. As a photoanode additive, the graphene nanomaterials improved the electron transport and affect a large surface area for dye adsorption. Nonetheless, a high concentration of graphene nanomaterials competes with the dye in light harvesting thus, impairs device performance. Generally, the graphene nanomaterials create percolation pathways for charge transfer thus, improves charge carrier transport in the devices; this, in turn, reduces charge recombination effects.

Material properties and preparation process/technique are determinants to the performance of OSCs. Both GO and rGO are hydrophilic, which makes them compatible with solution-based device processing. Even though the oxygen functional groups in GO effects insulating behavior, charge transport occurs *via* the hopping mechanism. Thus, GO facilitates effective charge transport in OSCs. The oxidation level of GO influences the performance of the device. A higher oxidation level is desirable for the formation of smooth films and better interfacial energy alignment because of increased dispersibility and work function, respectively. The dispersibility of GO and rGO-based nanomaterials is affected by the sample sizes. Small-sized samples exhibit higher dispersibility and hence, form better film morphology. The patterning of the devices and the thin film preparation technique also influence the photovoltaic performance. For example, a HTL of GO layer applied prior to PEDOT:PSS performs better than either GO or PEDOT:PSS HTLs because of better molecular-level interaction, improving charge transport. Similarly, a thin film of rGO formed by in situ thermal annealing on the substrates performs better than rGO that's directly adsorbed on the substrate because of better film morphology. The level of reduction of rGO influences the devices' performance. The reduction level is dependent not only on the control of the reduction process but also on the reducing agent used. rGO sample with a low C/O ratio exhibits better electrical conductivity, which translates to better photovoltaic performance. However, at a high level of reduction of the oxygen moieties, the sample becomes hydrophobic, which leads to the formation of agglomerates on the films. In all cases, the optimum concentration of the graphene nanomaterials is key to high-performance devices. High concentration leads to the formation of thicker films, which yield poor performance.

REFERENCES

1. Chen, L.M., Hong, Z., Li, G., Yang, Y. 2009. Recent progress in polymer solar cells: Manipulation of polymer: fullerene morphology and the formation of efficient inverted polymer solar cells. *Advanced Materials*, 21: 1434–49.
2. Kim, T., Kim, J.-H., Kang, T.E., Lee, C., Kang, H., Shin, M., Wang, C., Ma, B., Jeong U., Kim, T., Kim, B. 2015. Flexible, highly efficient all-polymer solar cells. *Nature Communications*, 6: 1–7.
3. Li, W., Yao, H., Zhang, H., Li, S., Hou, J. 2017. Potential of nonfullerene small molecules with high photovoltaic performance. *Chemistry–An Asian Journal*, 12: 2160–71.

4. Zhao, F., Dai, S., Wu, Y., Zhang, Q., Wang, J., Jiang, L., Ling, Q., Wei, Z., Ma, W., You, W., Wang, C., Zhan X. 2017. Single-junction binary-blend nonfullerene polymer solar cells with 12.1% efficiency. *Advanced Materials*, 29: 1700144.
5. Li, Y., Liu, X., Wu, F.-P., Zhou, Y., Jiang, Z.-Q., Song, B., Xia, Y., Zhang, Z-G., Gao, F., Inganäs O., Li, Y., Liao, L.-S. 2016. Non-fullerene acceptor with low energy loss and high external quantum efficiency: Towards high performance polymer solar cells. *Journal of Materials Chemistry A*, 4: 5890–7.
6. Li, G., Chu, C.-W., Shrotriya, V., Huang, J., Yang, Y. 2006. Efficient inverted polymer solar cells. *Applied Physics Letters*, 88: 253503.
7. Hsieh, C.-H., Cheng, Y.-J., Li, P.-J., Chen, C.-H., Dubosc, M., Liang, R.-M., Hsu, C.-S. 2010. Highly efficient and stable inverted polymer solar cells integrated with a cross-linked fullerene material as an interlayer. *Journal of the American Chemical Society*, 132: 4887–93.
8. Xu, Z., Chen, L.M., Yang, G., Huang, C.H., Hou, J., Wu, Y., Li, G., Hsu, C.-S.,Yang, Y. 2009. Vertical phase separation in Poly (3-hexylthiophene): Fullerene derivative blends and its advantage for inverted structure solar cells. *Advanced Functional Materials*, 19: 1227–34.
9. Amollo, T.A., Mola, G.T., Nyamori, V.O. 2020. Organic solar cells: Materials and prospects of graphene for active and interfacial layers. *Critical Reviews in Solid State and Materials Sciences*, 45: 261–88.
10. Pagliaro, M., Palmisano, G., Ciriminna, R., Loddo, V. 2009. Nanochemistry aspects of titania in dye-sensitized solar cells. *Energy & Environmental Science*, 2: 838–44.
11. Brodie, B.C. 1859. XIII. On the atomic weight of graphite. *Philosophical transactions of the Royal Society of London*, 149: 249–59.
12. Hummers Jr., W.S., Offeman, R.E. 1958. Preparation of graphitic oxide. *Journal of the American Chemical Society*, 80: 1339.
13. Park, S., An, J., Potts, J.R., Velamakanni, A., Murali, S., Ruoff, R.S. 2011. Hydrazine-reduction of graphite-and graphene oxide. *Carbon*, 49: 3019–23.
14. Huh, S.H. 2011. Thermal reduction of graphene oxide. *Physics and Applications of Graphene-Experiments*: Edited by Mikhailov, S. Intech, Croatia: 73–90.
15. Amollo, T.A., Mola, G.T., Nyamori, V.O. 2018. High-performance organic solar cells utilizing graphene oxide in the active and hole transport layers. *Solar Energy*, 171: 83–91.
16. Amollo, T.A., Mola, G.T., Nyamori, V.O. 2018. Polymer solar cells with reduced graphene oxide–germanium quantum dots nanocomposite in the hole transport layer. *Journal of Materials Science: Materials in Electronics*, 29: 7820–31.
17. Amollo, T.A., Mola, G.T., Nyamori, V.O. 2020. Improved short-circuit current density in bulk heterojunction solar cells with reduced graphene oxide-germanium dioxide nanocomposite in the photoactive layer. *Materials Chemistry and Physics*, 254: 123448.
18. Rafique, S., Abdullah, S.M., Alhummiany, H., Abdel-Wahab, M.S., Iqbal, J., Sulaiman, K. 2017. Bulk heterojunction organic solar cells with graphene oxide hole transport layer: Effect of varied concentration on photovoltaic performance. *The Journal of Physical Chemistry C*, 121: 140–6.
19. Wu, R., Wang, Y., Chen, L., Huang, L., Chen, Y. 2015. Control of the oxidation level of graphene oxide for high efficiency polymer solar cells. *RSC Advances*, 5: 49182–7.
20. Hilal, M., Han, J.I. 2018. Significant improvement in the photovoltaic stability of bulk heterojunction organic solar cells by the molecular level interaction of graphene oxide with a PEDOT: PSS composite hole transport layer. *Solar Energy*, 167: 24–34.
21. Ali, A., Khan, Z.S., Jamil, M., Khan, Y., Ahmad, N., Ahmed, S. 2018. Simultaneous reduction and sulfonation of graphene oxide for efficient hole selectivity in polymer solar cells. *Current Applied Physics*, 18: 599–610.

22. Hosseini, M., Naji, L., Fakharan, Z. 2020. Influences of synthesis parameters on the physicochemical and electrochemical characteristics of reduced graphene oxide/Pt nanoparticles as hole transporting layer in polymer solar cells. *Synthetic Metals*, 263: 116366.
23. Kim, J.-H., Sin, D.H., Kim, H., Jo, S.B., Lee, H., Han, J.T., Cho, K. 2019. Improved charge transport and reduced non-geminate recombination in organic solar cells by adding size-selected graphene oxide nanosheets. *ACS Applied Materials & Interfaces*, 11: 20183–91.
24. Wang, Y., Bao, X., Pang, B., Zhu, Q., Wang, J., Zhu, D., Yu, L., Yang, R., Dong, L. 2018. Solution-processed functionalized reduced graphene oxide-an efficient stable electron buffer layer for high-performance solar cells. *Carbon*, 131: 31–7.
25. Zhang, Y., Yuan, S., Li, Y., Zhang, W. 2014. Enhanced electron collection in inverted organic solar cells using titanium oxide/reduced graphene oxide composite films as electron collecting layers. *Electrochimica Acta*, 117: 438–42.
26. Shoyiga, H.O., Martincigh, B.S., Nyamori, V.O. 2021. Hydrothermal synthesis of reduced graphene oxide-anatase titania nanocomposites for dual application in organic solar cells. *International Journal of Energy Research*, 45: 7293–314.
27. Zhao, G., Feng, C., Cheng, H., Li, Y., Wang, Z.-S. 2019. In situ thermal conversion of graphene oxide films to reduced graphene oxide films for efficient dye-sensitized solar cells. *Materials Research Bulletin*, 120: 110609.
28. Seema, H., Zafar, Z., Samreen, A. 2020. Evaluation of solution processable polymer reduced graphene oxide transparent films as counter electrodes for dye-sensitized solar cells. *Arabian Journal of Chemistry*, 13: 4978–86.
29. Sudhakar, V., Singh, A.K., Chini, M.K. 2020. Nanoporous reduced graphene oxide and polymer composites as efficient counter electrodes in dye-sensitized solar cells. *ACS Applied Electronic Materials*, 2: 626–34.
30. Patil, J.V., Mali, S.S., Shaikh, J.S., Patil, A.P., Patil, P.S., Hong, C.K. 2019. Influence of reduced graphene oxide-TiO_2 composite nanofibers in organic indoline DN350 based dye sensitized solar cells. *Synthetic Metals*, 256: 116146.
31. Surana, K., Konwar, S., Singh, P.K., Bhattacharya, B. 2019. Utilizing reduced graphene oxide for achieving better efficient dye sensitized solar cells. *Journal of Alloys and Compounds*, 788: 672–6.

16 Revolutionizing the Field of Solar Cells by Utilization of Nanoscale Metal Oxide/Hydroxide Based 2D Materials

Shaan Bibi Jaffri and Khuram Shahzad Ahmad
Fatima Jinnah Women University

CONTENTS

16.1 INTRODUCTION

For the sustainable future of humanity, the quest for pursuit of inexhaustible energy needs to be addressed fully in addition to the resolution of other environmental issues, i.e., climatic change, negative impacts associated with the mining of fossil fuels, impoverishment of resources, global scale energy scarcity. Consumption of energy at the global scale is anticipated to augment in a twice manner by 2050, in comparison to the global energy consumption rate of 13.5 terawatt (TW) in 2001, due to swiftly cumulative ecospheric populace and monetary expansions. Amplified energy demand at global scale points the need for the exploration of substitution of the fossil fuels with carbon-free and inexhaustible energy sources. Solar energy accounts for the most plenteous renewable resource among all energy resources. Generation of the solar energy can be done by conversion of solar aurora into thermal or electrical energy by employment of different photovoltaic (PV) contraptions. There is a global scale installation of a huge number of gigawatts of the renewable energy technologies in form of wind turbines, hydropower production tech savvy, and solar PV segments in addition to the

DOI: 10.1201/9781003178453-16

geothermal and tidal energy [1]. Remarkable features associated with the renewable energy technologies signifies the domination of future by the one-thirds of the energy generation from PV technologies based on the solar energy [2]. Nevertheless, the limitation of different kinds of renewable energy sources by means of the geographical and ecological factors makes them difficult for energy harvesting on a global scale. Hence, the suitability of the solar energy due to its green nature is expressed by lesser limitation of solar energy by the terrain and environmental factors [3].

Exploitation of solar energy is preferable over other renewable energy resources since it is clean, unobtrusive, infinite, sustained, and unswerving. Such advantageous features marks the successful candidacy of the solar energy for fulfilling the ever growing electrical energy demand at global scale due to the population boom and enlargement of infrastructure. The attractiveness associated with the solar energy for fulfilling the global energy demands is due to its sustainability and affordability. Therefore, solar PV technologies are the most dominant technologies and growing at an annual rate of 60% as per estimations done between the years 2000 and 2016. In the present era, despite heavy pressure on the fossil fuels, the production of total electricity is insufficient to satisfy the global energy requirements. In the current age, the derivation of energy from fossil fuels is known for its finiteness and thus, becoming deficient. Even the utilization of the presently available energy sources obtained from fossil fuels is giving rise to a wide range of environmentally deteriorative phenomenon, e.g., global warming. Such events are increasing with the passage of time [4]. Therefore, human intellect and logic must be used for achieving the goals of sustainable development in presence of such challenges. In this regard, solar energy, among all other renewable energy resources has a very good future prospect in derivation of the electrical energy in a sustainable mode [5].

Abundantly available solar energy can be scaled up for meeting world energy demands [6] since earth receives 173,000 terawatts of solar energy on continuous basis. By utilization of only 0.02% of this solar energy, solar cells carry out the direct conversion of the solar to electrical energy. The research in this regard is ever increasing and emerging with passage of the time and through rigorous efforts, an enhancement in the efficiency has been achieved [7–10]. Considering the augmenting efficiency of the perovskite solar cellc (PSCs), at present the silicon crystal cells and PSCs are comparable [11].

PSCs have been named so because they have a crystal structure specialized in light absorption, mimicking the structure of the mineral $CaTiO_3$. Presently, number of synthesized compounds with ABX_3 stoichiometric characteristics have been investigated for their light absorptive behavior, where A and B expressed the cationic species and X represents the anionic species [12]. The crystalline structural form of the ABX_3 perovskite usually consists of an angle sharing $[BX_6]$ octahedra with the cationic A type species and occupies 12-fold synchronization location, which is being formed at the central position of cube of eight in such octahedra. Such crystalline geometry is ideal one and is a rare one because the naturally existing mineral perovskite form is also known for slight distortion. Oxides-based perovskites are the most studied ones. Perovskite group of compounds is quite famous for their flexibility in the accommodation of a great number rather all elements existing in the periodic system.

Zhou and co-workers [13] have for the first time developed such colossal super-alkali perovskite, e.g., Li_3O, Li_2F, and H_5O_2. The developed super-alkali perovskites were studied via density functional theory (DFT) and molecular dynamics simulation. DFT and simulation results were indicative of the unstable dynamics functionality of the super-alkali perovskites inclusive of metallic atoms at ambient temperature and pressure. Conversely, considerably stabilized dynamics functionality was obtained for cubic $H_5O_2MBr_3$ (M = Ge, Sn, Pb) and $H_5O_2PbI_3$ having super-alkali perovskites-based composition. In addition to favorable dynamics, the modified super-alkali perovskites also exhibited remarkable tolerance factors, generation energies that were negative, bandgaps that can be easily tailored, smaller efficacious hole sizes and electronic masses. Considering the bandgap issue of $CH_3NH_3PbI_3$ perovskite material, i.e., 1.55 eV and it exceeds the most favorable range of 1.1–1.4 eV in terms of the single junction PSCs, another DFT-based study has been conducted by Zhou and co-workers [14]. Organized DFT-based investigation included the cubic $(Li_3O)M(BH_4)_{3-x}Br_x$ (M ¼ Ge, Sn and Pb; x ¼ 0–3) perovskites for bandgap suitability and electronic structures tailoring employing first principle calculations. An auspicious PCE of 23.12% was indicated by DFT studies for cubic $(Li_3O)Ge(BH_4)Br_2$ and $(Li_3O)Pb(BH_4)_2Br$ perovskites with other favorable features [14] as obtained in case of cubic $H_5O_2MBr_3$ (M = Ge, Sn, Pb) and $H_5O_2PbI_3$ super-alkali perovskites [13]. The functionality of perovskites for obtaining high-quality PSCs has also been altered in an efficient way through utilization of different additives, e.g., 2-aminoethanesulfonamide hydrochloride. Upon doping-based modification, the perovskite thin films expressed an overall performance improvement, which can be attributed to the better morphological and crystalline characteristics in addition to trap states considerably reduced in addition to reduced hysterical response [15].

For any PSC device to work in an efficient manner, there should be a perfect synchronicity between the band alignment of the constituents including hole transport layer (HTL), electron transport layer (ETL), and active absorber perovskite layer. The role of HTL in an efficacious extraction of holes and transferal cannot be overlooked. The most frequently used HTL is poly (3,4-ethylenedioxythiphene):poly(styrenesulfonate) (PEDOT:PSS) [16,17]. Among various strategies adopted for improvement of the PSCs, utilization of diverse ETLs has been highly emphasized. For instance, in a recent work by Yang et al. [18], a novel inorganic $CsPbIBr_2$-based PSC was fabricated employing In_2S_3 ETL, which was processed at a low temperature of 70°C. The modified device exceeded in gaining a PCE of 5.59% with a profound reduction in the hysteresis extent. This work signified the metallic sulfides of transition group to be future candidates for ETLs in PSCs. In addition to experimentation with different materials for use as ETLs in PSCs, ETLs have been modified by means of doping with effective materials to enhance their functionality and thus positively impact the overall efficiency of the PV device. PSCs have been especially emphasized considering the future candidacy of PSCs in competing with the silicon solar cells. Furthermore, this chapter also provides a detailed account of the compositional aspects of the presently employed ETLs. Thorough comprehension of the advancements done in the doping of the ETLs provides a layout for the better selection of doping materials for ETLs for obtaining PSCs with commendable efficiencies and stability so that they can be commercialized and used for benefit derivation in the real life [19–22].

16.2 ADVANCEMENTS IN THE DOPED ELECTRON TRANSPORT LAYERS (ETLs)

Commercial-scale functionality of the PSCs depends upon all the components either in mesoscopic or planar configuration. In this regard, PV performance is predominantly controlled by the electronic transferal characteristics of the perovskite active absorber layer. PSCs are based on $CH_3NH_3PbI_3$ function in a better manner if developed in the mesoscopic fashion, while PSCs having $CH_3NH_3PbI_{3-x}Cl_x$ express better functionality in the planar mode in terms of the various modes of the electronic diffusion in both cases [23–27]. However, the presence of ETL is mandatory in both types of configurations. Most frequently used ETLs must be specialized in overcoming different issues, e.g., oxygen vacant sites assuming the role of trap sites. For addressing the issues associated with the functionality of different ETLs, doping strategy has been adopted. Most recent advances done in doping of different ETLs has been comprehensively elucidated in the following section:

16.2.1 Doped TiO_2 ETL

Jeon et al. succeeded in achieving good quality PSC comprising of the mixed halide and cationic formulation of $(FAPbI_3)_{0.85}(MAPbBr_3)_{0.15}$ composition with a certified PCE of 17.9% [27]. Such PSC device consisted of the record solar cell configuration with mesoporous TiO_2 ETL, permeated by perovskite precursor solution in the liquid form leading to the formation of the perovskite film in solidified form following the annealing step. Most of the optoelectronic applications, e.g., electrodes having higher surface area make use of mesoporous TiO_2 [28]. Particularly, it has been used in the dye-sensitized solar cells (DSSCs) for collection and transportation of the electrons that are photo-injected from sensitizer with surface-adsorbed features. Substitutional dopants utilization has been done for the improvement of such cells making use of the mesoporous TiO_2 ETL. For achieving an alleviation in the conduction band edge of the mesoporous TiO_2 ETL, lithium intercalation has been profoundly done for the facilitation of the electronic injection and transportation inside TiO_2 ETL [29].

Lithium salts have been utilized for the n-doping of the mesoporous TiO_2 ETL by means of the easier treatment of the device films. Modification of mesoporous TiO_2 ETL by Li^+ doping exhibited auspicious electrical characteristics, by causing assuagement of the trap states and thus subsequently ensuring the rapid electronic transferal. Scrupulous comparison of the Li^+ doped ETL devices and undoped devices exhibited remarkably conspicuous performances for the modified TiO_2 ETL PSC comprising of the $CH_3NH_3PbI_3$ light harvester in addition to the improvement in the PCE from 17% to over 19% with insignificant hysterical performance < 0.3%. Results in this regard expressed a significant difference of the backward and the forward scans in case of the pristine device in comparison to the device having Li^+-doped TiO_2 ETL. Doping mechanism and chemistry present a commendable approach for the modification of the electronic band morphologies and surficial states of the substances. In comparison to the DSSCs based on the liquid electrolyte, the effect of ETL doping in PSC has not been as much explored till 2010, except fewer reports expressing the significant role of the doping with aliovalent substitutions using Y^{3+}, Nb^{5+}, or Ta^{5+} doping into TiO_2 ETL elucidating the effectiveness of these dopants in enhancing TiO_2 ETL electrical properties. Among these aliovalent substitutions, Nb^{5+} has been considered as one of the suitable n-type dopant for the enhancement in the charge assortment characteristics. In case of DSSCs, different PV researchers investigated their influence for generation of Nb:TiO_2 (NTO) compact layer and mesoporous layer comprising of the NTO nanoparticles (NPs). Nevertheless, previous investigations based on NTO NPs have not isolated the pristine effect of Nb doping on PV characteristics since NTO NPs expressed various sizes or diverse phases at different levels of Nb^{5+}.

Results for Nb:TiO_2 ETL expressed the lighter Nb doping, i.e., 0.5% and 1.0% enhancing the optical bandgap. Doping in greater amounts, i.e., 5.0% lead to the subsequent reduction in the optical bandgaps. There was a profound similarity between the fermi levels of the conduction band in case of both light Nb doping and undoped TiO_2. However, fermi level was assuaged to ~0.3 eV for heavier doping of the Nb into TiO_2. Furthermore appreciable PCEs were obtained for lightly doped TiO_2 with Nb with 10% PCE > undoped TiO_2 (12.2%–13.4%), and 52% > heavily doped TiO_2 (8.8%–13.4%). Such commendable PCEs of the modified PSC are attributable to the faster electronic injection/transferral and well-kept-up electronic lifespan [30]. Doping enhances the conductivity of the host ETL by not only transforming the electronic mobility pattern but also the amount of electronic concentration being supplied to TiO_2 photo-anode is increased. Furthermore, the locations of conduction bands are altered leading to the passivation of the surficial defects and impacting host morphologies. Apart from PSCs, Nb-doped TiO_2 finds employment in different applications. In case of Nb doping, often the simplified doping effects can be seen in case of the utilization of the TiO_2 films as scaffolds.

In a recent investigation by Liu et al. [31], facile route using solution processing at low temperature was adopted for doping Zn into TiO_2 crystal lattice. Zn doping into TiO_2 ETL expressed an alleviated trap states density and improved conductivity in comparison to the pristine films without any doping. Results were also indicative of the uplifting of the TiO_2's fermi level causing an improvement in the carrier detachment and conveyance. PSC device modified with Zn-doped TiO_2 ETL and having a perovskite composition of the $CH_3NH_3PbI_3$ attained PCE of 17.6%, which is almost 27.5% higher than the archetypal device without Zn doping, i.e., 13.8%. Ag doping into TiO_2 ETL is associated with the reduction in the bandgap in a correlative manner as the concentration of Ag is increased there will be further alleviation in the bandgap. Furthermore, it is also associated with the enlargement of the J_{sc} due to an enhanced electrical conductivity. Nevertheless, there is also an issue of the suffering of planar structured PSC from severe J–V hysterical behavior and thus, results in the misjudgment of the PV functionality. In another recent investigation, Ag doping was done in TiO_2 making up meso-Ag:TiO_2 ETL, results expressed that an enhanced PCE of 17.7% was achieved by optimization of the different parameters of the modified PSC in addition to the alignment of the energy bands between active absorber perovskite layer and meso-Ag:TiO_2 ETL.

TiO_2 ETL morphology, surficial trap sites, and lattice structure have often been modified via doping. Doping impacts the electrical characteristics of the host material. Recently, Al and Mg doping was done in TiO_2 with $CH_3NH_3PbI_3$ as an active perovskite absorber layer in mesoscopic configuration. Al- and Mg-doped TiO_2

ETL containing PSC exhibited 22% higher in comparison to the control device without Al- and Mg-modified device. Results were indicative of the superior nature of Al doping in enhancing the V_{oc}, J_{sc}, and FF in an effective manner, while Mg doping only enhanced the V_{oc} to some extent. The augmented generation of electron–hole pair is due to an alleviation of the bandgap of doped TiO_2. This phenomenon leads to the injection of even more electrons from the active absorber perovskite into the ETL. Mg has been used for modification of the physical and optoelectronic characteristics of TiO_2. Mg doping in TiO_2 ETL is associated with facilitation of the charge transportation and suppression of the charge carrier concurrence between the TiO_2 ETL and active perovskite absorber layer. Mg-doped TiO_2 ETL modified PSC device obtained a PCE 15.73%, which is commendable in terms of the efficiency obtained for carbon-based planar PSCs. The augmentation in PCE is due to rapid extraction of the charges, enhanced electrical conductivity, and kerbed charge concurrence of the Mg-doped TiO_2 ETL. Furthermore, the procedural steps were conducted maintaining temperature of 100°C, paving a way for the development of the cost-effective and remarkably stable PSCs possessing compatibility with supple substrates. Elemental doping is associated with the dual functionality of enhancing the conductivity, carrier transferral, and also tailors the energy band morphology subsequently giving rise to the improved quality of the ETL quality. Recently, Ni was used as a dopant for development of the Ni-doped rutile TiO_2 ETL having higher crystallinity index and HTL was composed of the copper phthalocyanine (CuPc) in a heterojunction PSC device having carbon composition in planar configuration. There is an upward shifting of the fermi level of ETL due to Ni doping and augmentation in the charge transferral of the ETL, subsequently enhancing the charge conveyance and extraction. PSC modified with Ni-doped rutile TiO_2 ETL achieved a PCE of 17.46% with 0.01 M Ni doping, which is so far highest efficiency reported for carbon-based PSCs. In addition to an enhanced PCE, Ni-doped rutile TiO_2 ETL containing PSC with CuPc HTL exhibited an outstanding stability with negligible deterioration in the PCE even after storage in ambient air for 1,200 hours.

Specifically, doping can perform specialized functions of carrier concentration increment, e.g., Nb^{5+}, or mobility improvement, e.g., Sn^{4+} or the reduction in the surficial defect trap states can also be achieved, e.g., Al^{3+} or Zr^{4+}. Otherwise, there is also a great room for the TiO_2 surficial modification by means of using functional molecules or ionic species in form of a monolayer for trap states reduction. Sn doping for TiO_2 ETL has been reported but it is marked by requirement of higher temperature for processing at 450°C, which is in turn associated with the prevention of the doped TiO_2 ETL employment in PSCs that are manufactured at a lower temperature. Hence development of an efficacious method for doping TiO_2 ETL below 100°C is required for commercialization of the PSCs at industrial scale. Recently, PSC was developed by the successful deposition of the Sn-doped TiO_2 films at lower temperature as an ETL. In comparison to the archetypal TiO_2 ETL, Sn-doped TiO_2 ETL exhibited higher efficiency for the transferal and extraction of the photo-generated electron, elucidating an alleviated trap-state density and augmented conductivity. Sn-doped TiO_2 ETL containing PSC yielded a PCE of 17.2% exceeding the pristine device up to 29.3%.

16.2.2 Doped $SrSnO_3$ ETL

Perovskite device functionality has been considerably improved by means of a myriad of approaches, e.g., meticulous choice of ETL and HTL, crystalline growth control of the perovskite film, interfacial engineering, and device configurational device. Nevertheless, the dependence of the higher functionality PSC is dependent upon the mesoscopic TiO_2 ETL [32], which needs an augmented sintering temperature >450°C, presenting a conspicuous drawback for the commercialization of the economically viable PSCs. Hence, PV community has been putting efforts on the elimination of the ETLs processed at higher temperature inside PSCs, for inducing a greater level of simplicity in the planar PSC. Generally, PSCs with the planar heterojunction configuration are developed with two types of structure, i.e., inverted (p-i-n) assemblage or conventional (n-i-p) assemblage, on the basis of the charge restrictive contact. Thus, for the purpose of charge carrier transportation and extraction, the development of planar PSC with higher efficiency with ETLs and HTLs having low-temperature processing is highly desirable. Recently, TiO_2 ETL has been replaced successfully with $SrSnO_3$ ETL, having good conductivity and low-temperature processing. $SrSnO_3$ possessing ABO_3 perovskite structure having orthorhombic geometry is employed in a myriad of applications, e.g., transparent conducting oxide (TCO) [33], batteries having Li^+ ion, or DSSCs. In spite of the favorable features of the $SrSnO_3$ in terms of wider bandgap, higher conductivity, and remarkable charge carrier maneuverability, PSC devices are yet far away from commercialization at the industrial scale. Furthermore, there is also a need for the efficient way of the synthesis inclusive of the traditional solid-state procedures since the conventional synthetic modes require higher temperature and very extensive reaction timings.

In a recent investigation by Guo et al. [34], $SrSnO_3$ perovskite nanoparticles were fabricated by a low-temperature process in colloidal form and are used as an ETL. Furthermore, the modification of PSC containing $SrSnO_3$ ETL was carried by doping with yttrium, which possessed remarkable electronic conductivity, rapid electronic transferal, perfect band alignment between ETL and active absorber layer in comparison to the controlled device having undoped $SrSnO_3$ ETL in seamless conformity with the theoretical calculations. As a result, such factors boosted the PV functionality of the overall champion device exhibiting an average efficiency of 17.8% and PCE of 19% with negligible alleviation in the J–V hysterical features and an extensive period stability.

16.2.3 Doped ZnO ETL

PSCs have also been developed with composite ETLs comprising ZnO or TiO_2 and polymeric substances by means of formation of mixture prior to deposition. However, some investigations report the utilization of composite ETLs as a replacement for bilayer ETLs in polymeric ETLs. For instance, polymeric solar cells comprise of the amalgamated ETL based up on polyethylene oxide (PEO), poly(ethylene glycol), and polyethylenimine (PEI), with ZnO films. Such ETLs are known for causing an alleviation in the bulk traps inside semiconducting oxide film, which is in return associated with the reduction in the probability of trap-propped interfacial concurrence of charges, improving overall device functionality [35]. Nevertheless, such composite

ETLs having polymer constituent have not been utilized in PSCs. Utilization of ZnO as an ETL is marked by processing at low temperature and higher electronic mobility. Furthermore, a considerable reduction in the trap states density has been achieved by Rehman et al. [36], aimed at obtaining stabilized PCE so that the scaling up of such doped ETLs containing PSCs can be made possible [36]. Further researches in this regard are also supportive of the B:ZnO to be an efficient photo-anode for possessing improved electronic transferal properties leading to higher efficiencies by the dint of comparatively larger surficial area and good potential for light harvesting in comparison to single layers comprising of only ZnO.

Indium (In) has been considered as a suitable dopant for ZnO ETL due to reduced reactivity and remarkably resistive in terms of oxidative environment in comparison to Al or Ga. Furthermore, In doping is also known for causing an augmentation in the carrier amount of the ZnO and thus resulting in a superior level conductivity. In-doped ZnO having nano-fibrous morphology has also been utilized as ETL material developed by means of electrospinning route in PSCs. In-doped ZnO modified device achieved PCE of 17.18% with non-existent hysterical behavior due to an outstanding porosity and crystallinity.

ZnO is used as an alternative ETL to TiO_2 in PSCs. In this case, ZnO has been used as an archetypal material in addition to its modified forms obtained by extrinsic doping with different elements, e.g., Al, In, Li, Ga, Mg, W, or Cu. Among different elements, gallium (Ga) has been preferred over other dopants due to its three valence electrons, furthermore, there is a close proximity between the ionic radius of Ga^{3+}(0.062 nm) and Zn^{2+} (0.074 nm), and also the estimated covalent bond lengths are 1.92 Å and 1.97 Å for Ga-O and Zn-O, respectively. Such characteristics are suggestive of the minimal distortion of the ZnO wurtzite structure even after incorporation of the highest concentration of Ga^{3+} in ZnO in comparison to In-doped ZnO or Al-doped ZnO. ETL materials utilized in PSC devices, the potential of electronic extraction is usually determined by conductivity and surface work function (WF). If there is a perfect matching between the WF of the used ETL and electronic affinity of the perovskite absorber layer, then there is a considerable reduction in the Schottky barrier and formation of the Ohmic contact for facilitation of the electronic extraction and assemblage. There is a profound adjustability associated with the surface WF of ZnO, which is estimated by the polarization extent, morphological features, doping concentration, and surficial defects. The majority of the ETLs processing relying upon elevated temperatures processability makes them commercially unattractive. Recently, excimer laser annealing (ELA) was employed for treating Ga-doped ZnO ETL at ambient temperature. ELA treatment gave rise to the enhanced optically transparent features and electrical conductivity, subsequently improving the light absorptive aspects, electronic injection, and depression of the charge concurrence. In a recent investigation by Chen et al. [37], PSC containing Ga-doped ZnO ETL succeeded in obtaining a PCE of 21.132% with good surficial WF of 3.9 eV, reducing the interlayer connexion barrier and optimization of energy levels alignment [37].

Utilization of the doped ZnO is associated with the shifting of the Fermi level toward conduction band, which signifies a greater favorability in conductivity enhancement and WF reduction. Al doping increasing charge mobility and band energy matching can be attributed to the reaction occurring between ZnO having basic nature and protonic specie present on $CH_3NH_3^+$. Using pristine ZnO as an ETL in PSC expressed lower cell efficiencies but ZnO nanorods doped with aluminum (Al) expressed profound alleviation in the charge recombination at the interface between active absorber perovskite layer and Al:ZnO ETL. Similar enhanced PCE of 8.5%–10.7% was reported by Dong et al. [38] for PSC having nanorod (NR)/Al:ZnO/$CH_3NH_3PbI_3$/SpiroOMeTAD/Au/ZnO architecture indicative of the role of Al doping in reducing the charge recombination. Such an auspiciousness of Al can be ascribed to its remarkable ionic radii and optical transmission functionality, which enhanced charge carriers and electronic movement in a quantitative manner, subsequently increasing conductivity of the device. Al doping in ZnO has also been tested for inverted PSCs over ambient conditions for preparation of a novel Al:ZnO ETL via radio-frequency magnetron sputtering. The deposition of this layer was done between organic PTB7 and PC71BM mixture active layer and the TCO based on indium tin oxide (ITO) electrode. Through this approach, Lee et al. [39] succeeded in obtaining higher PCEs of 7.87% in comparison to the pristine PSCs yielding up to 4.19% only [39]. Another interesting feature associated with Al: ZnO-based PSCs is the observation of the stability potential of such devices showing 75% of their original functionality being conserved even after an aging of 1,000 hours.

Although, utilization of solution processed ZnO ETLs signifies easier and economically facile mode. Nevertheless, the fact that ZnO possesses comparatively open structure with closely packed lattice hexagonal crystallites. In such geometry, native Zn atoms express occupancy of only half of the tetrahedral locations, while the tetrahedral sites remain vacant. These vacant locations behave as defect states, which can be lying very deep as 0.2 eV within ZnO bandgap Therefore, PSCs having pristine ZnO as ETL is often observed to cause the jumping of the photo-generated electronic species from the conduction band of the perovskite material to such trap sites, even before final extraction by the electrode material. Such jumping mechanism leads to the trap-facilitated charge carriers recombination at the interfacial region expressing a serious drawback for ZnO ETL. Thus, such mechanistic issues are overcome by using doping mechanism. Lithium (Li)-doped ZnO used in PSCs having triple cationic composition, i.e., $MA_{0.57}FA_{0.38}Rb_{0.05}PbI_3$. An interstitial doping-based phenomenon using Li:ZnO as an ETL having Li intercalated in the ZnO lattice caused the passivation of inherent ZnO defects by downshifting the fermi levels [40]. In addition to different dopants, iodine (I) has also been employed in PSCs fabricated via spin coating using ZnO:I nanopillar in planar architecture. By means of I doping, the crystalline growth of one-dimensional (1D) ZnO was detected along [0001] plane signifying hexagonal crystalline pattern being suppressed. Such a modified PSC succeeded in attaining 18.24% PCE for ZnO:I nanopillars having $CH_3NH_3PbI_3$ as active absorber layer [41]. Efficiency in increment in the PSCs was also obtained for nitrogen-doped ZnO and Erbium-doped ZnO ETLs.

16.3 CONCLUSIONS AND PROSPECTS

Sustainable development goal 7, i.e., clean and affordable energy can be realized via adoption of green chemistry for development of energy devices, which are not only efficient but also benefits a larger population consolidating the concept of environmental sustainability. With a modifying human life style being demanding of greater utilization of power, the finite fossilized resources are being exploited in an unregulated manner. This type of consumptive pattern is not only weakening the concept of sustainability where we are depriving our generations from benefits we enjoyed from nature, but we are also damaging our current resources. Switching to sustainable materials via engineering can prove to be a savior. Perovskite solar cells are deemed as next-generation PV technology by the dint of their solution processability, lower costs, and remarkable photo-responses. Today, a large number of investigations reporting different features, designing, functionality, stability, etc. of perovskites are being published and an effort is being done to resolve the obstacles faced by the PV researchers when it comes to the industrial modulation of these cells. Top ranking PV commercial companies have even tried to turn this dream into reality, e.g., Saule Technologies, Solaronix, Toshiba, Oxford PV, and Slliance. However, the lab-based work requires further upgradation for such commercial companies to make efficient contraptions and make them available to general public at affordable costs. The future researchers working on the performance enhancement of the perovskite solar cells in terms of doping must investigate the neoteric modes for the optimized development of the modified perovskite solar cell devices so that the dream of making perovskite solar cells as a replacement to costly silicon-based PV devices can be realized on a practical scale. The revolutionary characteristics of the perovskite solar cells-based devices are associated with the transfiguration of human lives on a global scale for production of cheap energy.

REFERENCES

1. D. Khojasteh, D. Khojasteh, R. Kamali, A. Beyene, G. Iglesias. Assessment of renewable energy resources in Iran; with a focus on wave and tidal energy, *Renew. Sust. Energy. Rev.* 81 (2018) 2992–3005.
2. S. Bilgen, K. Kaygusuz, A. Sari. Renewable energy for a clean and sustainable future, *Energy Sourc.* 26 (2004) 1119–1129.
3. K. Hansen, B.V. Mathiesen. Comprehensive assessment of the role and potential for solar thermal in future energy systems, *Solar Energy.* 169 (2018) 144–152.
4. L. Calió, C. Momblona, L. Gil-Escrig, S. Kazim, M. Sessolo, A. Sastre-Santos, H.J. Bolink, S. Ahmad. Vacuum deposited perovskite solar cells employing dopant-free triazatruxene as the hole transport material, *Solar Energy Mat. Solar Cells.* 163 (2017) 237–241.
5. P. Liu, B. Xu, Y. Hua, M. Cheng, K. Aitola, K. Sveinbjörnsson, J. Zhang, G. Boschloo, L. Sun, L. Kloo. Design, synthesis and application of a π-conjugated, non-spiro molecular alternative as hole-transport material for highly efficient dye-sensitized solar cells and perovskite solar cells, *J. Power Sourc.* 344 (2017) 11–14.
6. M. Grätzel. Photoelectrochemical cells, *Nature* 414 (2001) 338.
7. J.V. Milić, J.H. Im, D.J. Kubicki, A. Ummadisingu, J.Y. Seo, Y. Li, M.A. Ruiz-Preciado, M.I. Dar, S.M. Zakeeruddin, L. Emsley, M. Grätzel. Supramolecular engineering for formamidinium-based layered 2D perovskite solar cells: Structural complexity and dynamics revealed by solid-state NMR spectroscopy, *Adv. Energy Mat.* 1 (2019) 1900284.
8. K. Yao, H. Zhong, Z. Liu, M. Xiong, S. Leng, J. Zhang, Y.X. Xu, W. Wang, L. Zhou, H. Huang, A.K. Jen. Plasmonic metal nanoparticles with core-bishell structure for high-performance organic and perovskite solar cells, *ACS Nano.* 13 (2019) 1–10.
9. E.A. Alharbi, M.I. Dar, N. Arora, M.H. Alotaibi, Y.A. Alzhrani, P. Yadav, W. Tress, A. Alyamani, A. Albadri, S.M. Zakeeruddin, M. Grätzel. Perovskite Solar Cells Yielding Reproducible Photovoltage of 1.20 V, *Research.* 1 (2019) 8474698.

10. R. Singh, A. Giri, M. Pal, K. Thiyagarajan, J. Kwak, J.J. Lee, U. Jeong, K. Cho. Perovskite solar cell with MoS_2 electron transport layer, *J. Mat. Chem. A.* 7 (2019) 1–10.
11. J. Luo, W.G. Yang, B. Liao, H.B. Guo, W.M. Shi, Y.G. Chen. Improved photovoltaic performance of dye-sensitized solar cells by carbon-ion implantation of tri-layer titania film electrodes, *Rare Metals.* 34 (2015) 34–39.
12. Z. Yi, N.H. Ladi, X. Shai, H. Li, Y. Shen, M. Wang. Will organic–inorganic hybrid halide lead perovskites be eliminated from optoelectronic applications? *Nanoscale Adv.* 1 (2019) 1–10.
13. T. Zhou, M. Wang, Z. Zang, L. Fang, Stable dynamics performance and high efficiency of ABX3-type super-alkali perovskites first obtained by introducing H_5O_2 Cation, *Adv. Energy Mat.* 9 (2019) 1900664.
14. T. Zhou, Y. Zhang, M. Wang, Z. Zang, X. Tang, Tunable electronic structures and high efficiency obtained by introducing superalkali and superhalogen into AMX3-type perovskites, *J. Power Sourc.* 429 (2019) 120–126.
15. W. Chen, K. Sun, C. Ma, C. Leng, J. Fu, L. Hu, S. Lu, Eliminating J-V hysteresis in perovskite solar cells via defect controlling, *Org. Elect.* 58 (2018) 283–289.
16. L. Hu, M. Li, K. Yang, Z. Xiong, B. Yang, M. Wang, K. Sun, PEDOT:PSS monolayers to enhance the hole extraction and stability of perovskite solar cells. *J. Mat. Chem. A.* 6 (2018) 1–10.
17. L. Hu, J. Fu, K. Yang, Z. Xiong, M. Wang, B. Yang, J. Li, Inhibition of in-planc chargc transport in hole transfer layer to achieve high fill factor for inverted planar perovskite solar cells. *Solar RRL.* 3 (2019) 1900104.
18. B. Yang, M. Wang, X. Hu, T. Zhou, Z. Zang, Highly efficient semitransparent $CsPbIBr_2$ perovskite solar cells via low-temperature processed In_2S_3 as electron-transport-layer. *Nano Energy.* 57 (2019) 1–10.
19. B. Feng, J. Duan, L. Tao, J. Zhang, H. Wang. Enhanced performance in perovskite solar cells via bromide ion substitution and ethanol treatment, *Appl. Surf. Sci.* 430 (2018) 603–612.
20. J. Shi, X. Xu, D. Li, Q. Meng. Interfaces in perovskite solar cells, *Small.* 11 (2015) 2472–2486.
21. Z.H. Bakr, Q. Wali, A. Fakharuddin, L. Schmidt-Mende, T.M. Brown, R. Jose. Advances in hole transport materials engineering for stable and efficient perovskite solar cells, *Nano Energy.* 34 (2017) 271–305.
22. L.S. Oh, D.H. Kim, J.A. Lee, S.S. Shin, J.W. Lee, I.J. Park, M.J. Ko, N.G. Park, S.G. Pyo, K.S. Hong, J.Y. Kim. Zn_2SnO_4-based photoelectrodes for organolead halide perovskite solar cells, *J. Phy. Chem. C.* 118 (2014) 22991–22994.
23. P. Docampo, S. Guldin, U. Steiner, H.J. Snaith. Charge transport limitations in self-assembled TiO_2 photoanodes for dye-sensitized solar cells, *J. Phy. Chem. Lett.* 4 (2013) 698–703.
24. M. Li, Y. Huan, X. Yan, Z. Kang, Y. Guo, Y. Li, Y. Zhang, Efficient yttrium (III) chloride-treated TiO_2 electron transfer layers for performance-improved and hysteresis-less perovskite solar cells, *ChemSusChem.* 11 (2018) 171–177.
25. D. Yang, R. Yang, K. Wang, C. Wu, X. Zhu, J. Feng, S.F. Liu, High efficiency planar-type perovskite solar cells with negligible hysteresis using EDTA-complexed SnO_2, *Nat. Comm.* 9(2018) 1–11.
26. J.P. Correa Baena, L. Steier, W. Tress, M. Saliba, S. Neutzner, T. Matsui, F. Giordano, T.J. Jacobsson, A.R. Srimath Kandada, S.M. Zakeeruddin, A. Petrozza, A. Abate, M.K. Nazeeruddin, M. Grätzel, A. Hagfeldt, Highly efficient planar perovskite solar cells through band alignment engineering, *Energy Environ. Sci.* 8 (2015) 2928–2934.
27. N.J. Jeon, J.H. Noh, W.S. Yang, Y.C. Kim, S. Ryu, J. Seo, S.I. Seok. Compositional engineering of perovskite materials for high-performance solar cells, *Nature.* 517 (2015) 476.

28. E.J. Crossland, N. Noel, V. Sivaram, T. Leijtens, J.A. Alexander-Webber, H.J. Snaith. Mesoporous TiO_2 single crystals delivering enhanced mobility and optoelectronic device performance, *Nature*. 495 (2013) 215.
29. A. Abate, D.J. Hollman, J. Teuscher, S. Pathak, R. Avolio, G. D'Errico, G. Vitiello, S. Fantacci, H.J. Snaith. Protic ionic liquids as p-dopant for organic hole transporting materials and their application in high efficiency hybrid solar cells, *J. Americ. Chem. Soc.* 135 (2013) 13538–135348.
30. D.H. Kim, G.S. Han, W.M. Seong, J.W. Lee, B.J. Kim, N.G. Park, K.S. Hong, S. Lee, H.S. Jung. Niobium doping effects on TiO_2 mesoscopic electron transport layer-based perovskite solar cells, *ChemSusChem*. 8 (2015) 2392–2398.
31. X. Liu, Z. Wu, Y. Zhang, C. Tsamis. Low temperature Zn-doped TiO_2 as electron transport layer for 19% efficient planar perovskite solar cells, *Appl. Surf. Sci.* 471 (2019) 28–35.
32. H. Tan, A. Jain, O. Voznyy, X. Lan, F.P. De Arquer, J.Z. Fan, R. Quintero-Bermudez, M. Yuan, B. Zhang, Y. Zhao, F. Fan. Efficient and stable solution-processed planar perovskite solar cells via contact passivation, *Science*. 355 (2017) 722–726.
33. K.P. Ong, X. Fan, A. Subedi, M.B. Sullivan, D.J. Singh. Transparent conducting properties of $SrSnO_3$ and $ZnSnO_3$, *APL Mat.* 3 (2015) 062505.
34. H. Guo, H. Chen, H. Zhang, X. Huang, J. Yang, B. Wang, Y. Li, L. Wang, X. Niu, Z. Wang. Low-temperature processed yttrium-doped $SrSnO_3$ perovskite electron transport layer for planar heterojunction perovskite solar cells with high efficiency, *Nano Energy*, 59 (2019) 1–9.
35. X. Chen, S. Yang, Y.C. Zheng, Y. Chen, Y. Hou, X.H. Yang, H.G. Yang. Multifunctional inverse opal-like TiO_2 electron transport layer for efficient hybrid perovskite solar cells, *Adv. Sci.* 2 (2015) 1500105.
36. F. Rehman, K. Mahmood, A. Khalid, M.S. Zafar, M. Hameed. Solution-processed barium hydroxide modified boron-doped ZnO bilayer electron transporting materials: Toward stable perovskite solar cells with high efficiency of over 20.5%, *J. Colloid Inter. Sci.* 535 (2019) 353–362.
37. Y. Chen, Y. Hu, Q. Meng, H. Yan, W. Shuai, Z. Zhang. Natively textured surface of Ga-doped ZnO films electron transporting layer for perovskite solar cells: Further performance analysis from device simulation, *J. Mat. Sci. Elect.* 30 (2019) 4726–436.
38. J. Dong, Y. Zhao, J. Shi, H. Wei, J. Xiao, X. Xu, J. Luo, J. Xu, D. Li, Y. Luo, Q. Meng, Impressive enhancement in the cell performance of ZnO nanorod-based perovskite solar cells with Al-doped ZnO interfacial modification, *Chem. Commun.* 50 (2014) 13381–13384.
39. S.J. Lee, S. Kim, D.C. Lim, D.H. Kim, S. Nahm, S.H. Han, Inverted bulk-heterojunction polymer solar cells using a sputter-deposited Al-doped ZnO electron transport layer, *J. Alloys Comp.* 777 (2018) 1–10.
40. M.A. Mahmud, N.K. Elumalai, M.B. Upama, D. Wang, M. Wright, T. Sun, A. Uddin, Simultaneous enhancement in stability and efficiency of low-temperature processed perovskite solar cells, *RSC Adv.* 6 (2016) 86108–86125.
41. Y.Z. Zheng, E.F. Zhao, F.L. Meng, X.S. Lai, X.M. Dong, J.J. Wu, X. Tao, Iodine-doped ZnO nanopillar arrays for perovskite solar cells with high efficiency up to 18.24%, *J. Mat. Chem. A.* 5 (2017) 12416–12425.

17 2D Materials for Flexible Photo Detector Applications

Aruna Pattipati and Joseph Chennemkeril Mathew
Dayananda Sagar College of Engineering

CONTENTS

17.1 INTRODUCTION

Over the past few years, two-dimensional (2D) photodetectors have gained a lot of importance and are widely studied due to their features such as fast response, high responsivity and spectrum specificity [1]. Photodetectors (PDs) based on 2D materials have greater prospects due to their flexible tuning, lack of dangling bonds, high mobility and so on. These PDs can detect a wide range of wavelengths from visible light to terahertz band. 2D materials have excellent optoelectronic properties due to their ultrafast charge transport and tunable photon absorption. These materials include semimetal Graphene, semiconductor black phosphorus (BP) and transition metal dichalcogenides (TMDCs) and so on.

Organic materials have advantages like low fabrication cost, high flexibility and large area scalability. 2D materials have features like high carrier mobility and tunable optical absorption. These 2D materials-based PDs have unique advantages like fast and high responsivity and spectrum specificity [2,3]. To tune the photoelectric properties of 2D materials, organic materials can be attached to the surface either by solution or epitaxial growth.

In 2D materials, atoms are connected by covalent bonds and the layers are connected by Van der Waals bonds. Quantum confinement effect in the out-of-plane direction in 2D materials makes them exhibit excellent electronic and optoelectronic properties. High mobility and in-plane thermal conductivity of Graphene [4], strong interaction of TMDCs with light [5], anisotropy and direct bandgap of BP make them suitable candidates for wide band detectors [6]. 2D materials have applications

DOI: 10.1201/9781003178453-17

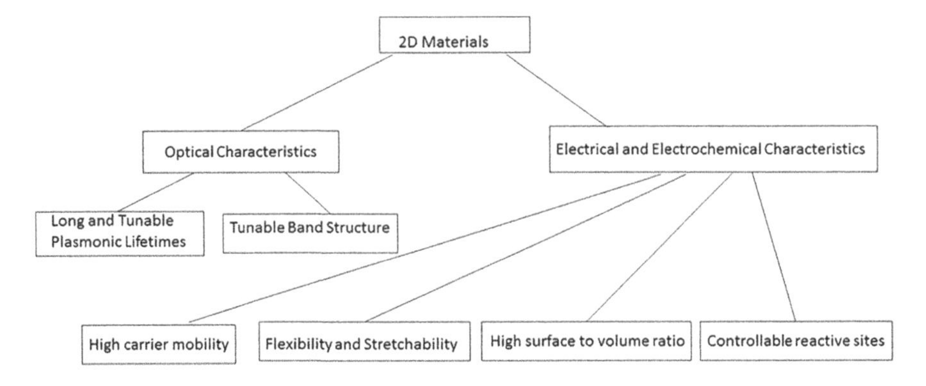

FIGURE 17.1 Characteristics of two-dimensional materials.

in emerging fields of optoelectronics such as sensing, motion detectors, digital imaging, spintronics, low power electronics, photonics, plasmonics and super capacitors. Characteristics of 2D materials are shown in Figure 17.1.

17.2 PHOTODETECTORS

PDs are devices, which can detect light and have a variety of applications such as optical communication, environmental monitoring, military and biomedical applications. Utilizing the property of absorption tunability and high carrier mobility of 2D materials and combining those with organic materials to form thin-film heterojunctions with high and fast responsivity with a wide spectra region PDs. Organic PDs have usage in biomedical science, education, environmental monitoring, optical communication, computer vision and sensory imaging [7]. Although silicon and other traditional materials-based PDs are being used, they have limitations in operational wavelengths as well as responsivity and speed. Ease of preparation and integration of 2D materials with other materials along with their high mobility and strong interaction with light makes them achieve high sensitivity and responsivity. Hence a lot of research on these types of materials is going on in this line.

Contrary to conventional optoelectronic devices, flexible optoelectronic devices are fabricated on flexible substrates, which revolutionized their applications. Flexible optoelectronic devices are compact and mechanically flexible as they can be bent, twisted or even rolled, which makes them find various applications in bioelectronics for wearable and implantable devices, which is not possible with conventional devices.

17.3 MATERIALS USED IN PHOTODETECTORS

In this chapter few 2D materials useful for PDs in the various wavelength regions are discussed. Graphene has extraordinary optical, electrical, magnetic and mechanical properties due to which it is a widely studied 2D material. It is an allotrope of carbon, which was discovered in 2004. It consists of a flat monolayer of sp^2 bonded carbon atoms. Graphene is a 2D sp^2 bonded carbon sheet with a high theoretical surface area of around $2{,}630\,m^2g^{-1}$ arranged in a hexagonal honeycomb lattice [8]. It has a high electrical conductivity with a carrier mobility of $15{,}000\,cm^2V^{-1}s^{-1}$ at room temperature. 2D lattice in Graphene suppresses the back scattering of charge carriers and hence the effective mass of Graphene is almost zero. Graphene has zero bandgap and hence it will be in ON state which makes it useful in analog devices like sensors rather than digital devices. It has a wide absorption band, short carrier lifetime and high carrier mobility as mentioned earlier. Graphene has an excellent mechanical flexibility, strong atomic bonding and outstanding young modulus and hence a promising candidate for flexible electronics. It also has good optical properties and it is being studied to be used as a conductor for transparent and flexible devices as it has a transmittance of around 97%, which can also replace Indium Tin Oxide (ITO).

Lack of bandgap didn't pose any limitation for Graphene to be used in applications like heat spreaders, transparent conductive films, acoustic speakers and mechanical actuators. It offers the fastest charge transport, stiffness and thermal conductivity [9]. Graphene PDs are used in ultra-wideband range (300 nm ~ 6 μm) although these devices have problems like low responsivity, low external quantum efficiency and absence of spectral selectivity due to weak light absorption and quick recombination of carriers.

Graphene is used in devices like wearable sensing devices and for the detection of oral bacteria-based sensors are examples for flexible biocompatibility and compactness [10]. Graphene is also used in devices like tattoo sensors, electrocardiograms (ECGs), electromyograms (EMGs) and electroencephalograms. Functionalization of Graphene with biomolecules enables targeted detection of specific cancer cells [11]. Graphene changes its electrical characteristics due to environmental stimuli and this characteristic is utilized in the development of Graphene-based flexible sensor systems. As Graphene is highly stable, this is used as electrodes/channels in sensors and energy storage devices such as super capacitors and batteries [12]. Graphene can be produced on a large area by a simple chemical vapor deposition technique. Unique advantages of Graphene make it useful in PD applications especially in the far-infrared and terahertz regions.

TMDCs are extensively studied for use in flexible devices after Graphene. TMDCs can be fabricated as a monolayer and this family can exhibit various electrical properties like metallic, half metallic and semiconducting making them useful in various types of devices. Semiconducting TMDCs have bandgaps of 1–2 eV and among TMDCs, Tungsten disulfide (WS_2) has a high optical absorption coefficient and a large exciton binding energies of 700–800 meV making it a better alternative to Graphene. These materials can be easily coated on flexible substrates for wearable devices. Utilizing semiconducting properties of TMDCs, Lim et al. fabricated a TMDC-based MoS_2 PD [13], which maintained their characteristics even with 10^5 bending cycles. Gas sensors were also reported with MoS_2 channel and reduced Graphene oxide electrodes [14]. TMDCs have a layered structure of either a single or a few layers. The general formula for TMDCs is MX_2 (M=Mo/W/Re and X= S/Se/Te) like Molybdenum disulfide (MoS_2), Tungsten disulfide (WS_2) and Molybdenum diselenide ($MoSe_2$).With various combinations of chalcogenides and metals, there are around 40 TMDCs and these have a variety of properties such as charge density wave, semimetal and superconductivity [15,16]. The direct bandgap of a monolayer MoS_2 changes from 1.8 eV to indirect bandgap of 1.2 eV when the number of relevant layers increases [17]. Due to its symmetry in the electronic state, topological semimetals like Platinum diselenide ($PtSe_2$), Palladium ditelluride ($PdTe_2$) and Platinum ditelluride ($PtTe_2$) have unique carrier properties and can remarkably perform in broadband PDs [18]. Due to these properties, they are used in tunable exciton devices and spin valley lasers [19] with a disadvantage of being indirect bandgap materials having low mobility [20].

Due to the limitations of Graphene and TMDCs, BP and other group V elements layered materials are also considered as potential 2D materials for their usage in optoelectronic applications due to their high mobility and adjustable bandgap [21]. The bandgap of BP changes with the number of atomic layers and can be directly coupled with light. It has a strong optical conductivity in the visible to mid-infrared range; it is a promising candidate for PD applications in this range [22]. However, its application is limited due to its poor stability when exposed to water, light and oxygen [23].

Black arsenic (BA) has a single layered puckered structure resembling BP making it to have anisotropy and is more pronounced than that of BP [24]. Preparation and application of large area ultra-thin film samples is possible as BA is more stable than BP.

In recent years, a new semiconducting 2D material, Antimonene is also used for flexible PDs with a hybrid structure showing a good response and on/off ratio. Compared to other 2D materials, Antimonene has a short layer distance and small binding energy suitable for exfoliation and surface modification methods. Though bulk antimony exhibits metal electrical transport characteristics, 2D layered antimony exhibits indirect bandgap semiconducting properties [25]. Owing to these characteristics, its usage is limited in the development of optoelectronic devices. These 2D layered semiconducting materials, arsenic and antimony, have high carrier concentration and mobility making them widely studied materials for their usage in transistors, quantum spin devices and other optoelectronic devices. At present, studies on these materials are limited and need to be developed.

Extensive research is going on organic–inorganic perovskite 2D nanoplatelet or nanosheets (NSs) due to their large lateral size, narrow band absorption, long diffusion length, excellent charge transport properties and long carrier lifetime making them excellent candidates for photovoltaic and optoelectronic applications [26,27]. Organic–inorganic halide perovskites have gained more attention than their inorganic counterparts as they have advantages like easy solution processing at room temperature, good stability and excellent optoelectronic properties [28]. By controlling the thickness of the 2D NSs, bandgap tuning can be achieved in this material. A white light converter was fabricated using the composition $CH_3NH_3PbBrI_2$ NSs on a blue LED chip. Parveen et al. [29] reported a 2D perovskite PD with a stable and very fast rise/fall time (24 μs/103 μs) along with high responsivity and detectivity of ~1.93 A/W and 1.04×10^{12} Jones, respectively. As reported, storage, operational and temperature-dependent stability studies reveal high stability of the 2D perovskite NSs under the ambient condition with high humidity.

Recently buckled Graphene analogues (Xenes) have created interest among researchers and are being studied theoretically for their usage in flexible nanoelectronics [30]. MXenes are a new type of 2D material composed of elements $M_{n+1}X_nT_x$ elements where M is a transition metal element, X is Carbides or Nitrides and T is a group or modification on the surface of the 2D material and n takes the value in the range 1–3. MXenes have the advantages like good optical transmittance, fast charge transfer, tunable bandgap, more active sites, ease of modification and low cost of fabrication [31]. Currently most of the MXenes exhibit metal-like properties with small bandgaps [32] and provide a platform for photon electron coupling on the surface. Hence they are widely used in biological, chemical and optical sensors [32]. Room temperature solid-state properties of selected 2D crystalline materials are mentioned in a previous report [9].

Polymer PDs have gained a lot of importance in recent years due to their ease of fabrication on flexible substrates at low cost. Many of the polymer PDs are based on bulk-heterojunction structure [33,34] with donor–acceptor (D-A) blended active layers. To overcome the weak absorption of fullerenes and hard-to-modify bandgap, non-fullerene acceptor-based PDs are being studied in recent times [35]. Organic semiconductors have excitation energy in the range 0.3–1 eV, which is higher than that of inorganic materials [36] and hence these have efficient exciton dissociation [37], which in turn improves the device efficiency of a PD. Planar heterojunction (PHJ) devices, due to their layered structure have several advantages like high carrier mobility, ease in fabrication and reproducibility [38]. PHJ structure refers to the contact surface of two materials in the D-A plane. To enhance the contact area of D-A, which increases the exciton dissociation, the PHJ concept was proposed in the polymer fullerene system [39].

The performance of heterojunctions of single-layer organic small molecules and polymers is inferior due to the weak interaction between the interfaces. To overcome this, 2D materials are incorporated into bulk/multilayer heterojunctions [40]. For obtaining the best charge separation efficiency, the thickness of the multilayer should be less than the exciton diffusion length. Weak electrons in 2D materials interact with photons and hence enhance photocurrent response [41]. 2D materials and organic interface play an important role in the performance of the device. To enhance the photo trapping effect, which increases the light absorption of 2D materials, plasmonic effect is also utilized [42].

Efficiency of the gold nanoparticles (AuNPs) doped P3HT device was enhanced due to the charge transfer between P3HT and AuNPs as reported earlier [43]. The layers on ITO-coated glass substrate consisted of thin films of PEDOT: PSS, P3HT with Au nanoparticles and Al metal contact on which the photosensing studies were carried out. Photoresponse curve of the device is shown in Figure 17.2.

Following are things to be considered to have a good responsivity of a PD: (a) building an effective electric field in the photoactive layer, (b) to select a material to trap the charge carriers and (c) to prolong the trapping time within the photoactive layer.

The effect of doping concentration of AuNPs on various properties of the device was reported and is shown in Table 17.1 as reported earlier [43]. The layers on ITO-coated glass substrate consist of thin films of PEDOT: PSS, P3HT with Au nanoparticles and Al metal contact on which the photosensing studies were carried out and the photosensing parameters were determined as shown in Table 17.1.

In recent times, researchers are using 2D materials to trap the carriers using a hybrid structure strategy [44]. According to a recent report, a variety of 'all 2D' heterostructure PDs were studied [45]. Plastic PDs utilizing the low bandgap polymers were also studied for fast response and high sensitivity [46]. Polymer organic PDs lead the way for 2D-based optoelectronic devices. As per previous reports of Aruna et al.

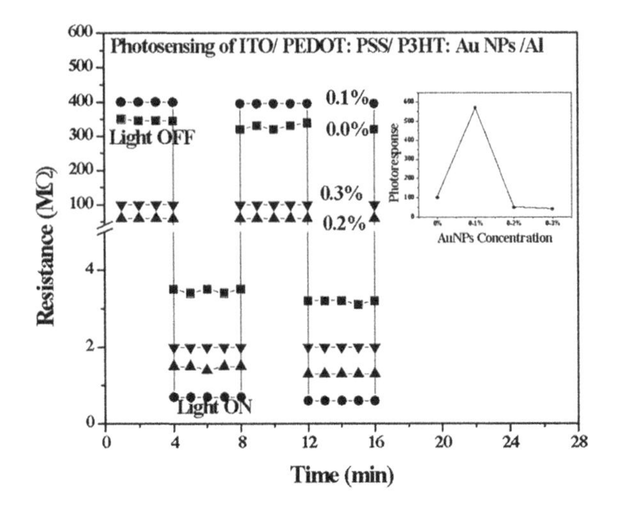

FIGURE 17.2 Photosensing of ITO/PEDOT: PSS/P3HT: AuNPs (170nm)/Al devices for different concentrations of AuNPs. Inset: Photoresponse Curve for different AuNPs concentrations. Adapted with permission from [43]. Copyright (2021) Elsevier Publishers.

TABLE 17.1

Optical and Photosensing Parameters of the ITO/PEDOT: PSS/P3HT: AuNPs (170nm)/Al Devices

Concentration of AuNPs in P3HT Thin Films (%)	Bandgap (eV)	On/off Ratio (Resistance)	Dark Current I_{dark} (nA)	Photo Current I_{light} (nA)	Responsivity (μA/W)	Photoresponse $\Delta I_{resp} = I_{light}/I_{dark}$
0	2.10	100	0.28	28.5	7.1	101
0.1	1.75	571	0.25	142.9	35.7	571
0.2	1.90	50	1	50	12.2	50
0.3	1.95	40	1.6	66	16.1	41

Adapted with permission from [43]. Copyright (2021) Elsevier Publishers.

[47], the photoresponse with respect to different wavelengths is shown in Figure 17.3 indicating the variation of the responsivity with respect to the wavelength used. The layers on ITO-coated glass substrate consist of thin films of C_{60}-doped P3HT and Al metal contact on which the photoresponse studies in light and dark were carried out.

17.4 PHOTODETECTORS FOR DIFFERENT WAVELENGTHS

PDs in the visible wavelength region (400–750nm), near-infrared region (750 nm–1.1 μm) and short wave infrared region (1–3 μm) have applications in the fields like biomedical imaging, night vision thermal imaging and communication [48,49]. Mueller et al. studied Graphene-based PD, which adopted a metal Graphene metal structure and achieved a responsivity of 6.1 mAW^{-1} at a wavelength of 1.55 μm [50].

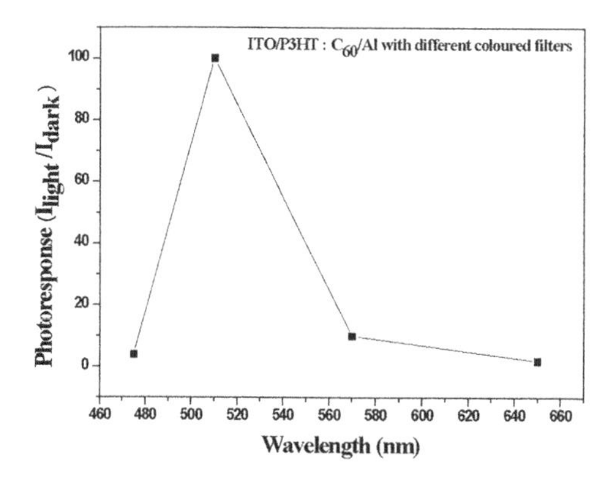

FIGURE 17.3 Photoresponse characteristics of P3HT:C60 device. Adapted with permission from [47]. Copyright (2018) Elsevier Publishers.

Yin et al. studied TMDCs-based single-layer MoS_2 phototransistors and it showed a better optical response compared to Graphene-based devices [51]. Huang et al. demonstrated a high-performance PD based on BP in the wavelength region 400–900 nm [52]. Tan et al. demonstrated a tunable black phosphor carbide infrared phototransistor with a wavelength of 2,004 nm and it showed a responsivity of 2,163 AW^{-1} [53]. PDs in the mid-infrared region (3–30 μm) have applications in the fields of free space communication, environmental monitoring and biomedicine [54,55]. Liu et al. reported a broadband PD with two Graphene layers sandwiching a thin tunnel barrier and the responsivity of this detector was more than 1 A W^{-1} at 3.2 μm [56]. BP has a moderate bandgap and has a high mobility and low dark current and can be integrated with traditional materials like Silicon. Guo et al. reported a PD based on BP with photoresponsivity up to 82 A W^{-1} at 3.39 μm [57]. BA phosphorus is an alloy of BP and arsenic atoms and is similar to BP and has a narrow bandgap and is semiconducting in nature [58]. Amani et al. reported a photoconductor based on BA phosphorus in the wavelength region 3.9–4.6 μm [59]. Utilizing the excellent optoelectronic properties of 2D materials, they can be used in PDs, which can detect light in the terahertz region (30 μm–3 mm). Zak et al. reported a THz PD based on split-bow-tie antenna integrated with Graphene field effect transistor [60].

17.5 CONCLUSION

2D materials have great potential for incorporating in devices while fabricating PDs for excellent efficiency. With continuous emergence of 2D materials, their excellent properties have been utilized in the area of PDs. Still there are a few challenges faced by the scientific community, one of which is the production of these 2D materials in a large area as most of the high-quality 2D materials are obtained by micro-mechanical exfoliation. Another challenge is that the contact barrier between 2D material and the electrode is relatively high because of which the stability of the device is poor. By overcoming these limitations, one can achieve PDs with high efficiency. Flexible PDs based on 2D materials will have remarkable progress and great prospects in near future.

REFERENCES

1. Kawasaki T., Sugawara K., Dobroiu A., Eto T., Kurita Y., Kojima K., Yabe Y., Sugiyama H., Watanabe T., Suemitsu T., Ryzhii V., Iwatsuki K., Fukada Y., Kani J., Terada J., Yoshimoto N., Kawahara K., Ago H., Otsuji T. (2015) Graphene-channel FETs for photonic frequency double-mixing conversion over the sub-THz band. *Solid-State Electron.* 103: 216–221.
2. Konstantatos G., Badioli M., Gaudreau L., Osmond J., Bernechea M., de Arquer F.P.G., Gatti F., Frank H.L.K. (2012) Hybrid Graphene-quantum dot phototransistors with ultrahigh gain. *Nat. Nanotechnol.* 7: 363–368.
3. Wang P., Liu S., Luo W., Fang H., Gong F., Guo N., Chen Z.G., Zou J., Huang Y., Zhou X., Wang J., Chen X., Lu W., Xiu F., Hu W. (2017) Arrayed van der Waals broadband detectors for dual-band detection. *Adv. Mater.* 29: 1–8.
4. Morozov S.V., Novoselov K.S., Katsnelson M.I., Schedin F., Elias D.C., Jaszczak J.A., Geim A.K. (2008) Giant intrinsic carrier mobilities in Graphene and its bilayer. *Phys. Rev. Lett.* 100: 1–4.
5. Mak K.F., Shan J. (2016) Photonics and optoelectronics of 2D semiconductor transition metal dichalcogenides. *Nat. Photonics* 10: 216–226.
6. Rodrigues M.J.L.F., de Matos C.J.S., Ho Y.W., Peixoto H., de Oliveira R.E.P., Wu H.Y., Neto A.H.C., Viana-Gomes J. (2016) Resonantly increased optical frequency conversion in atomically thin black phosphorus. *Adv. Mater.* 28: 10693–10700.
7. Chen L., Deng J.X., Kong L., Cui M., Chen R.G., Zhang Z.J. (2015) Optical properties of rubrene thin film prepared by thermal evaporation. *Chin. Phys. B.* 24: 1–5.
8. Zhu Y., Murali S., Stoller M.D., Ganesh K.J., Cai W., Ferreira P.J., Pirkle A., Wallace R.M., Cychosz K.A., Thommes M., Su D., Stach E.A., Ruoff R.S. (2011) Carbon based super capacitors produced by activation of Graphene. *Science.* 332: 1537–1541.
9. Akinwande D., Petrone N., Hone J. (2014) Two-dimensional flexible nanoelectronics. *Nat. Commun.* 5: 1–12.
10. Vellappally S., Al Kheraif A.A., Anil S., Wahba A.A. (2019) IoT medical tooth mounted sensor for monitoring teeth and food level using bacterial optimization along with adaptive deep learning neural network. *Measurement.* 135: 672–677.
11. Joe D.J., Hwang J., Johnson C., Cha H.Y., Lee J.W., Shen X., Spencer M.G., Tiwari S., Kim M. (2016) Surface functionalized Graphene biosensor on sapphire for cancer cell detection. *J. Nanosci. Nanotechnol.* 16: 144–151.
12. Chee W.K., Lim H.N., Zainal Z., Huang N.M., Harrison I., Andou Y. (2016) Flexible Graphene-based super capacitors: A review. *J. Phys. Chem. C.* 120: 4153–4172.
13. Lim Y.R., Song W., Han J.K., Lee Y.B., Kim S.J., Myung S., Lee S.S., An K.S., Choi C.J., Lim J. (2016) Wafer-scale, homogeneous MoS_2 layers on plastic substrates for flexible visible-light photodetectors. *Adv Mater.* 28: 5025–5030.
14. He Q., Zeng Z., Yin Z., Li H., Wu S., Huang X., Zhang H. (2012) Fabrication of flexible MoS_2 thin film transistor arrays for practical gas sensing applications. *Small.* 8: 2994–2999.
15. Shishidou T., Freeman A.J., Asahi R. (2001) Effect of GGA in the half-metallicity of the itinerant ferromagnet CoS_2. *Phys. Rev. B.* 64: 180401 (1–4).
16. Xi X., Wang Z., Zhao W., Park J.-H., Law K.T., Berger H., Forró L., Shan J., Mak K.F. (2015) Ising pairing in superconducting $NbSe_2$ atomic layers. *Nat. Phys.* 12: 139–143.
17. Splendiani A., Sun L., Zhang Y., Li T., Kim J., Chim C.Y., Galli G., Wang F. (2010) Emerging photoluminescence in monolayer MoS_2. *Nano Lett.* 10: 1271–1275.
18. Zyuzin A.A., Silaev M., Zyuzin V.A. (2018) Non linear chiral transport in Dirac semimetals. *Phys. Rev. B.* 98: 205149 (1–6).
19. Ye Y., Xiao J., Wang H., Ye Z., Zhu H., Zhao M., Wang Y., Zhao J., Yin X., Zhang X. (2016) Electrical generation and control of the valley carriers in a monolayer transition metal dichalcogenide. *Nat. Nanotechnol.* 11: 598–602.

20. Britnell L., Ribeiro R.M., Eckmann A., Jalil R., Belle B.D., Mishchenko A., Kim Y.-J., Gorbachev R.V., Georgiou T., Morozov S.V., Grigorenko A.N., Geim A.K., Casiraghi C., Castro Neto A.H., Novoselov K.S. (2013) Strong light–matter interactions in heterostructures of atomically thin films. *Science.* 340: 1311–1314.
21. Liu H., Du Y., Deng Y., Ye P.D. (2015) Semiconducting black phosphorus: synthesis, transport properties and electronic applications. *Chem. Soc. Rev.* 44: 2732–2743.
22. Liu H., Neal A.T., Zhu Z., Luo Z., Xu X., Tomanek D., Ye P.D. (2014) Phosphorene: An unexplored 2D semiconductor with a high hole mobility. *ACS Nano* 8: 4033–4041.
23. Moreno-Moreno M., Lopez-Polin G., Castellanos-Gomez A., Gomez-Navarro C., Gomez-Herrero J. (2016) Environmental effects in mechanical properties of few-layer black phosphorus. *2D Mater.* 3: 031007 (1–6).
24. Chen Y., Chen C., Kealhofer R., Liu H., Yuan Z., Jiang L., Suh J., Park J., Ko C., Choe H.S., Avila J., Zhong M., Wei Z., Li J., Li S., Gao H., Liu Y., Analytis J., Xia Q., Asensio M.C., Wu J. (2018) Black arsenic: A layered semiconductor with extreme in-plane anisotropy. *Adv. Mater.* 30: 1800754 (1–6).
25. Zhang S., Xie M., Li F., Yan Z., Li Y., Kan E., Liu W., Chen Z., Zeng H. (2016) Semiconducting group 15 monolayers: A broad range of band gaps and high carrier mobilities. *Angew. Chem.* 55: 1666–1669.
26. Ha S.T., Liu X., Zhang Q., Giovanni D., Sum T.C., Xiong Q. (2014) Synthesis of organic–inorganic lead halide perovskite nano platelets: Towards high-performance perovskite solar cells and optoelectronic devices. *Adv. Opt. Mater.* 2: 838–844.
27. Song J., Xu L., Li J., Xue J., Dong Y., Li X., Zeng H. (2016) Monolayer and few-layer all-inorganic perovskites as a new family of two-dimensional semiconductors for printable optoelectronic devices. *Adv. Mater.* 28: 4861–4869.
28. Zheng K., Zhu Q., Abdellah M., Messing M.E., Zhang W., Generalov A., Niu Y., Ribaud L., Canton S.E., Pullerits T. (2015) Exciton binding energy and the nature of emissive states in organo metal halide perovskites. *J. Phys. Chem. Lett.* 6: 2969–2975.
29. Parveen S., Paul K.K., Giri P.K. (2020) Precise tuning of the thickness and optical properties of highly stable 2D organo-metal halide perovskite nanosheets through a solvo-thermal process and its applications as a white LED and a fast photo detector. *Appl. Mater. Interfaces.* 12: 6283–6297.
30. Vogt P., Padova P.D., Quaresima C., Avila J., Frantzeskakis E., Asensio M.C., Resta A., Ealet B., Lay G.L. (2012) Silicene: compelling experimental evidence for graphenelike two-dimensional silicon. *Phys. Rev. Lett.* 108: 155501 (1–5).
31. Anasori B., Lukatskaya M.R., Gogotsi Y. (2017) 2D metal carbides and nitrides (MXenes) for energy storage. *Nat. Rev. Mater.* 2: 16098 (1–17).
32. Jiang X., Kuklin A.V., Baev A., Ge Y., Ågren H., Zhang H., Prasad PN (2020) Two-dimensional MXenes: From morphological to optical, electric and magnetic properties and applications. *Phys. Rep.* 848: 1–58.
33. Zimmerman J.D., Diev V.V., Hanson K., Lunt R.R., Yu E.K., Thompson M.E., Forrest S.R. (2010) Porphyrin-tape/C_{60} organic photodetectors with 6.5% external quantum efficiency in the near infrared. *Adv. Mater.* 22: 2780–2783.
34. Huo L., Hou J., Zhang S., Chen H.Y., Yang Y. (2010) A polybenzo [1,2-b:4,5-b′] dithiophene derivative with deep HOMO level and its application in high-performance polymer solar cells. *Angew. Chem.* 49: 1500–1503.
35. Jiang W., Li Y., Wang Z. (2014) Tailor-made Rylene arrays for high performance n-channel semiconductors. *Acc. Chem. Res.* 47: 3135–3147.
36. Dong H., Zhu H., Meng Q., Gong X., Hu W. (2012) Organic photo response materials and devices. *Chem. Soc. Rev.* 41: 1754–1808.
37. Mishra A., Bäuerle P.(2012) Small molecule organic semiconductors on the move: promises for future solar energy technology. *Angew. Chem.* 51: 2020–2067.

38. Bao W., Cai X., Kim D., Sridhara K., Fuhrer M.S. (2013) High mobility ambipolar MoS_2 field-effect transistors: Substrate and dielectric effects. *Appl. Phys. Lett.* 102: 042104 (1–5).
39. Halls J.J.M., Walsh C.A., Greenham N.C., Marseglia E.A., Friend R.H., Moratti S.C., Holmes A.B. (1995) Efficient photodiodes from interpenetrating polymer networks. *Nature.* 376: 498–500.
40. Qin L., Wu L., Kattel B., Li C., Zhang Y., Hou Y., Wu J., Chan W.L. (2017) Using bulk heterojunctions and selective electron trapping to enhance the responsivity of perovskite–graphene photodetectors. *Adv. Funct. Mater.* 27:1704173 (1–8).
41. Lemme M.C., Koppens F.H.L., Falk A.L., Rudner M.S., Park H., Levitov L.S., Marcus C.M. (2011) Gate-activated photo response in a graphene p-n junction. *Nano Lett.* 11: 4134–4137.
42. Zhang J., Li Y., Zhang X., Yang B. (2010) Colloidal self-assembly meets nano fabrication: From two-dimensional colloidal crystals to nanostructure arrays. *Adv. Mater.* 22: 4249–4269.
43. Aruna P., Joseph C.M. (2021) Optical and photo sensing properties of gold nano particles doped poly (3-hexylthiophene-2, 5-diyl) thin films. *Mater. Lett.* 293: 129726.
44. Peng Z.Y., Xu J.L., Zhang J.Y., Gao X., Wang S.D. (2018) Photodetectors: Solution-processed high performance hybrid photodetectors enhanced by perovskite/MoS_2 bulk heterojunction. *Adv. Mater. Interfaces* 5: 1870089 (1–7).
45. Ye Y., Ye Z., Gharghi M., Zhu H., Zhao M., Wang Y., Yin X., Zhang X. (2014) Exciton-dominant electroluminescence from a diode of monolayer MoS_2. *Appl. Phys. Lett.* 104: 193508 (1–4).
46. Yao Y., Liang Y., Shrotriya V., Xiao S., Yu L., Yang Y. (2007) Plastic near-infrared photodetectors utilizing low band gap polymer. *Adv. Mater.* 19: 3979–3983.
47. Aruna P., Joseph C.M. (2018) Spectral sensitivity of fullerene doped P3HT thin films for color sensing applications. *Mater. Today Proc.* 5: 2412–2418.
48. Hansen M.P., Malchow D.S. (2008) Overview of SWIR detectors, cameras, and applications. *SPIE Proc.* 6939: 693901 (1–11).
49. Soderblom L.A., Britt D.T., Brown R.H., Buratti B.J., Kirk R.L., Owen T.C., Yelle R.V. (2004) Short-wavelength infrared (1.3 –2.6 μm) observations of the nucleus of comet 19P/Borrelly. *Icarus.* 167: 100–112.
50. Mueller T., Xia F., Avouris P. (2010) Graphene photodetectors for high-speed optical communications, *Nat. Photonics.* 4: 297–301.
51. Yin Z., Li H., Li H., Jiang L., Shi Y., Sun Y., Lu G., Zhang Q., Chen X., Zhang H. (2012) Single-layer MoS_2 phototransistors. *ACS. Nano.* 6: 74–80.
52. Huang M., Wang M., Chen C., Ma Z., Li X., Han J., Wu Y. (2016) Broad band black-phosphorus photodetectors with high responsivity. *Adv. Mater.* 28: 3481–3485.
53. Tan W.C., Huang L., Ng R.J., Wang L., Hasan D.M.N., Duffin T.J., Kumar K.S., Nijhuis C.A., Lee C., Ang K.W. (2018) A black phosphorus carbide infrared phototransistor. *Adv. Mater.* 30: 1705039 (1–8).
54. Chen Y., Lin H., Hu J., Li M. (2014) Heterogeneously integrated silicon photonics for the mid-infrared and spectroscopic sensing. *ACS. Nano.* 8: 6955–6961.
55. Rodrigo D., Limaj O., Janner D., Etezadi D., de Abajo F.J.G., Pruneri V., Altug H. (2015) Mid-infrared plasmonic bio sensing with graphene. *Science* 349: 165–168.
56. Liu C.H., Chang Y.C., Norris T.B., Zhong Z. (2014) Graphene photodetectors with ultra-broad band and high responsivity at room temperature. *Nat. Nanotechnol.* 9: 273–278.
57. Guo Q., Pospischil A., Bhuiyan M., Jiang H., Tian H., Farmer D., Deng B., Li C., Han S.J., Wang H., Xia Q., Ma T.P., Mueller T., Xia F. (2016) Black phosphorus mid-infrared photodetectors with high gain. *Nano Lett.* 16: 4648–4655.

58. Youngblood N., Chen C., Koester S.J., Li M. (2015) Waveguide-integrated black phosphorus photo detector with high responsivity and low dark current. *Nat. Photonics* 9: 247–252.
59. Amani M., Regan E., Bullock J., Ahn G.H., Javey A. (2017) Mid-wave infrared photoconductors based on black phosphorus-arsenic alloys. *ACS. Nano.* 11: 11724–11731.
60. Zak A., Andersson M.A., Bauer M., Matukas J., Lisauskas A., Roskos H.G., Stake J. (2014) Antenna-integrated 0.6 THz FET direct detectors based on CVD Graphene. *Nano Lett.* 14: 5834–5838.

18 2D Nanomaterials for Electrocatalytic Hydrogen Production

Arun Prasad Murthy and Nihila Rahamathulla
Vellore Institute of Technology

CONTENTS

18.1 INTRODUCTION

Nanotechnology has made revolutionary developments in the advancement of industrial processes, materials, and applications by completely exploiting the unique surface phenomena that matter exhibits when reduced to their nanoscale. Surprisingly, nanomaterials exhibit entirely different properties from their bulk materials. Nanomaterials are classified based on their nanoscopic dimensions as zero-dimensional (0D) nanomaterials

DOI: 10.1201/9781003178453-18

(nanoparticles), one-dimensional nanomaterials (1D) (nanotubes, nanorods, and nanowires), two-dimensional nanomaterials (2D), and so on.

2D nanomaterials are ultrathin (less than 5 nm) nanosheets consisting of one or few atomic layers. In 2004, 2D single-atom-thick graphene crystallites, the first 2D nanomaterial was isolated from bulk graphite by Prof Andre Geim and Prof Kostya Novoselov [1]. Since then, 2D nanomaterials such as transition metal dichalcogenides (TMDCs), transition metal oxides, carbides, nitrides, carbonitrides (MXenes), 2D metal-organic frameworks (2D-MOF), and 2D polymers have been extensively explored by the researchers. Different structures of 2D nanomaterials are shown in Figure 18.1. The 2D nanomaterials have numerous applications in the fields of nanoelectronics, catalysis, biomedical, bioimaging, photothermal therapy, separation, transistors, sensors, and energy storage due to their atomic thickness, electron confinement in two dimensions, sheet-like structure, flexibility as well as high anisotropic, optical, mechanical, thermal, and physicochemical properties. 2D nanomaterials exist in many crystalline forms and therefore they exhibit electric conductivity properties ranging from insulators to super metallic conductors. The structural and conductive properties of 2D nanomaterials have drawn significant research attention to catalytic applications such as electrocatalysis. The presence of dopants or defects in 2D nanomaterials can tune the energy bandgap to obtain desired electric conductivity [2]. In addition, the high surface-to-volume ratio with abundant catalytically active sites exposed on the surface renders 2D metal nanomaterials to be employed as electrocatalysts in water splitting, fuel cells, metal–air batteries, CO_2 reduction, etc., and have proved to be of having great potential in replacing the commercial electrocatalysts. This chapter discusses in detail the importance of 2D nanomaterials employed as electrocatalysts in water splitting, specifically in hydrogen evolution reaction (HER).

The global energy crisis is expected to increase with the burgeoning global population in addition to climate change due to increased CO_2 emissions from fossil

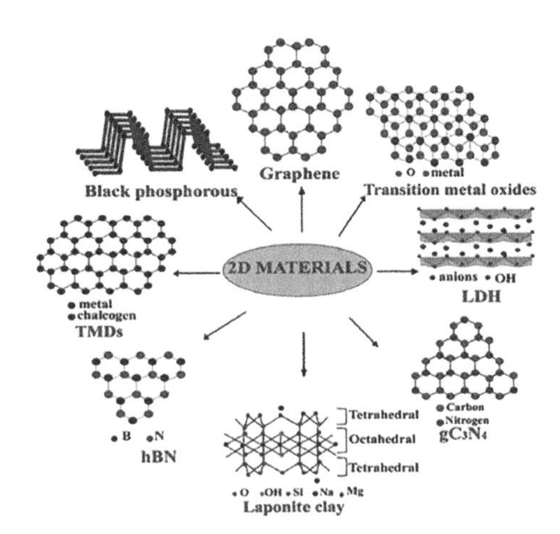

FIGURE 18.1 Structures of different 2D nanomaterials such as graphene, transition metal oxides, layered double hydroxide (LDH), gC_3N_4, hBN, TMDs, and black phosphorous. Adapted with permission from ref [2]. Copyright (2020) Elsevier B.V.

fuel combustion [3–5]. This has led to ever-increasing demands for sustainable and carbon-free clean energy sources. However, renewable energy sources necessitate highly efficient technologies to capture, convert, and store as they are intermittent in nature. Solar energy is the supreme source of renewable energy source though intermittent in nature. Hydrogen is considered to be the future energy carrier. Solar-driven electrocatalytic water splitting is a promising method for the conversion of solar energy into chemical energy and its further storage, for example, as fuel hydrogen. At present, however, hydrogen is produced mainly from steam reforming of coal, petroleum, and natural gas, which still leads to the global CO_2 emission [6,7]. Another propitious method for clean hydrogen production is electrocatalytic water splitting. Water electrolysis consists of two half reactions: hydrogen and oxygen evolution reactions that require a thermodynamic potential of 1.23 V at 25°C to produce hydrogen at the cathode and oxygen at the anode [3]. The oxygen produced can be utilized for industrial or medical use and the significance of the later use can be understood in this pandemic period while the hydrogen can be stored as a fuel. This high pure hydrogen when utilized as fuel to generate energy, it has the potential for near-zero greenhouse gas emission as the by product is only water vapor and warm air. Hydrogen possesses high energy density per unit mass, low toxicity, and the ability to transport it safely qualifies hydrogen to be the near future's sustainable energy source [8–10].

Mechanism of HER consists of combination of elementary steps such as Volmer step and either Heyrovsky step or the Tafel recombination step as given below [12,13].

a. Volmer step: Initial discharge of proton or

$$\text{In acidic media: } H_3O^+ + e^- \rightarrow H_{ads} + H_2O \tag{18.1}$$

$$\text{In basic media: } H_2O + e^- \leftrightarrows H_{ads} + OH^- \tag{18.2}$$

b. Heyrovsky or Tafel step: Hydrogen formation

$$\text{In acidic media: } H_{ads} + H_3O^+ + e^- \rightarrow H_2 + H_2O\left(\text{Heyrovsky step or desorption step}\right) \tag{18.3}$$

$$\text{In basic media: } H_{ads} + H_2O + e^- \leftrightarrows H_2 + OH^- \tag{18.4}$$

Or

$$H_{ads} + H_{ads} \rightarrow H_2\left(\text{Tafel step or recombination step}\right) \tag{18.5}$$

Therefore, two possible mechanisms are Volmer–Tafel and Volmer–Heyrovsky based on the elementary steps. Different pathways of HER reaction in acidic and basic medium are illustrated in Figure 18.2. All these steps require an efficient electrocatalyst with highly active adsorption sites. An important contrivance in determining the mechanism and rate-determining step (RDS) is the Tafel equation $\eta = b \log(j) + a$ where η is the overpotential, b is the Tafel slope, and a is the Tafel constant [14]. When the Tafel slope reaches 120 mV dec^{-1}, the RDS becomes a Volmer reaction, a desorption reaction at 40 mV dec^{-1}, and a Tafel reaction at 30 mV dec^{-1} [3]. A lower Tafel slope is always

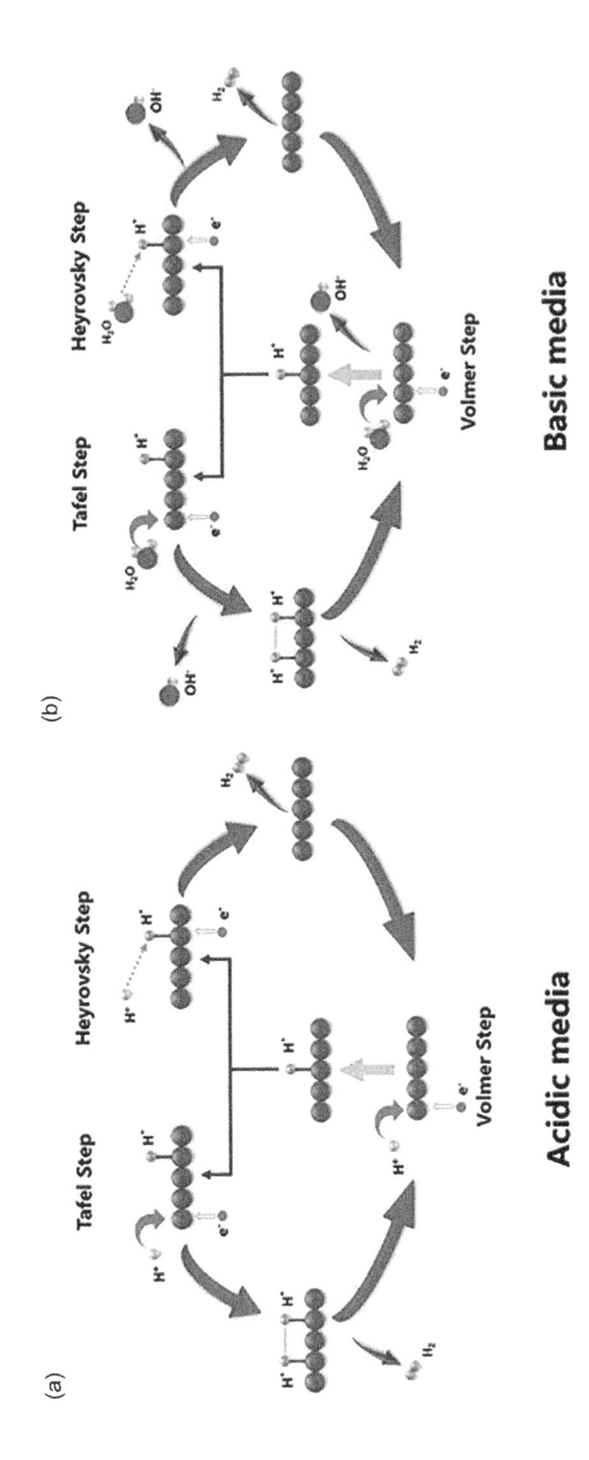

FIGURE 18.2 Different pathways for HER: (a) in acidic media and (b) in basic media. Adapted with permission from ref [11]. Copyright (2020) WILEY-VCH Verlag GmbH & Co. KGaA, Weinheim.

preferred for an electrocatalyst since greater catalytic current can be attained at moderately lower overpotentials. The kinetic barrier for water electrolysis, however, necessitates a greater potential overpotential (η) than the thermodynamic potential (1.23 V) to overcome the kinetic barrier. Overpotential is another important parameter to evaluate the HER activity of the electrocatalysts. Generally, the overpotential required to attain a current density of 10 mA cm^{-2} (η_{10}) is used to compare the HER activities among different electrocatalysts. Under equilibrium conditions, the exchange current density (j_0) is an account of the intrinsic charge transfer [15]. A higher exchange current density indicates lower charge transfer resistance (R_{ct}) and therefore, a lower reaction barrier. In brief, a lower Tafel slope, overpotential, and R_{ct} result in faster reaction kinetics and enhanced HER performance. Noble metals like Pt and Pt-based materials are the best-known electrocatalysts for the HER with negligible overpotential even at high reaction rates. But, its utilization for large-scale use is limited due to its high cost and limited abundance. Non-noble electrocatalysts with low cost, abundance, high activity, lower overpotential, and long-term stability have been under investigation in recent past years. In this context, due to the distinctive physical and chemical properties derived from their structural and electronic features of 2D ultrathin materials with a single- or few-layered thickness, they have been explored as HER electrocatalysts [16]. Especially, when the bulk materials are reduced to 2D nanosheets, their reduced thickness results in enhanced HER activity. The superiority of ultrathin 2D nanomaterials as electrocatalysts arises as they have more exposure to active sites by decreasing the energy barrier, which is achieved by introducing defects in the basal plane or by doping with conductive heteroatoms. Thus, 2D electrocatalysts with their unique anisotropic properties open a new path for enhanced HER performance.

18.2 2D NANOSTRUCTURES AS HER ELECTROCATALYSTS

18.2.1 HOLEY PT NANOSHEETS ON NIFE-HYDROXIDE LAMINATES

The most active and benchmark electrocatalyst for HER is well-known Pt. However, many researchers have been ardently investigating the reaction mechanisms and innovating the designs for efficient Pt-based electrocatalysts due to their limited performance in alkaline solutions. Jang et al. [17] fabricated a 2D–2D interfacially integrated nanoplatform of nanoporous 2D-Pt on NiFe-LDH (layered double hydroxide) nanosheets and employed it as an effective electrocatalyst for HER. The electrocatalyst synergistically hastened the water dissociation with maximum Pt-atomic utilization efficiency, furnishing 6.1 times higher Pt mass activity than the commercial 20 wt.% Pt/C and exhibited HER operational stability for about 50 hours. A high HER activity was exhibited by the electrocatalyst with a Tafel slope of 32.3 mV dec^{-1}. It exhibited η_{100} of 61 mV, which was remarkably lower than the commercial 20 wt.% Pt/C catalyst (η_{100} = 342 mV). The in situ-generated hybrids of the electrocatalyst showed 2.57 times higher electrochemically active surface area (ECSA) than that of the physically mixed electrocatalyst (LDH + 2D-Pt), suggesting that the synthesized hybrids possessed a higher number of active sites when compared to (LDH + 2D-Pt). The authors attributed the high HER activity of the electrocatalyst to the structural lamellar design with 2D–2D interfacial 2D-Pt layers clamped on NiFe-LDH sheets.

18.2.2 Two-Dimensional Transition Metal Dichalcogenides

2D transition metal dichalcogenides (2D TMDC) have been a promising electrocatalyst for HER. The 2D layered structure of TMDC nanosheets offers a number of reaction sites and has less in-plane resistivity through the basal plane, facilitating easier electron transport. For enhanced HER performance, Chen et al. [18] employed a monolayer of $MoTe_2$ as a representative for activation of the basal plane of 2D TMDCs. The structural stabilities and energetics of possible 2H/1T′ phase boundary configurations were examined by comprehensive first-principle calculations. Te, Mo, and hollow sites were identified as energetically stable phase boundaries and potential catalytic centers for HER by employing basal plane activation, resulting in increased HER activity over the pristine basal lattice. In specific, the hollow sites, a new group of sites generated by phase boundaries exhibited a Gibbs free energy (ΔG_H) near the thermoneutral value, similar to that of Pt for hydrogen adsorption. The mechanisms of hydrogen adsorption at phase boundaries were analyzed, which were attributed to their distinct electronic structure and local hydrogen adsorption geometries.

18.2.3 MoS_2 Quantum Dots on Graphitic Carbon Nitride

Graphitic carbon nitride (g-C_3N_4) is an important metal-free-carbon-based semiconductor electrocatalyst for HER due to its electronic structure, nitrogen content, and stability. However, to enhance the catalytic activity of pristine g-C_3N_4, Liu et al. [19] decorated it with MoS_2 quantum dots. MoS_2 having a similar structure to that of g-C_3N_4 promoted the formation of the 2D hybrid nanosheets with a p-n heterojunction interface. Unsaturated Mo and S atoms at the edges also served a greater number of catalytic active sites. Electrochemical studies were conducted by a standard three-electrode system in an acidic medium. Clearly, with the introduction of MoS_2 quantum dots on g-C_3N_4, lowering of activation barriers, decrease in onset potential and enhanced HER activity were observed.

18.2.4 2D Metal-Organic Frameworks

2D-MOF nanosheets, integrating the advantages of 2D material and MOF have been spurring research attention in catalysis application on account of its ultrathin thickness, high surface-to-volume atom ratios, tunability, and large surface area. MOF contains metal ions and organic ligands via coordination bonds. MOF-derived materials have been used in electrocatalytic HER in alkaline media on account of their high conductivity and better electrochemical properties.

18.2.4.1 2D Porous NiCoSe Nanosheet Arrays on Ni Foam

Zhou et al. [20] developed in situ growth of porous NiCoSe nanosheet arrays based on 2D MOFs on Ni foam (NiCoSe-NF) and employed them as an effective HER electrocatalyst. The characteristic porous 2D framework of the electrocatalyst provided high surface area for HER, facile electrolyte penetration, high conductivity, and enhanced release of gaseous products. The 2D porous NiCoSe nanosheet arrays on Ni foam were synthesized using ZIF-67 as a precursor, which was further subjected to react with x mmol of Se powder (x=2, 4, 6, 8,) to fabricate different compositions of NiCoSe-NF. The HER activity was evaluated using a three-electrode system in an alkaline medium. For comparison, the commercial Pt/C loaded on the Ni foam, bare Ni foam, NiCo LDH-NF were also analyzed under the same conditions. NiCoSe (8 mmol) had a Tafel slope of 82.3 mV dec^{-1}, which was found to be less than those of NF (153.4 mV dec^{-1}), NiCo LDH (128.3 mV dec^{-1}), NiCoSe (2 mmol) (130.9 mV dec^{-1}), NiCoSe (4 mmol) (130.3 mV dec^{-1}), and NiCoSe (6 mmol) (107.4 mV dec^{-1}). The η_{10} of NiCoSe (8 mmol) was 170 mV, which was lower than those of other compositions as well as NiCo LDH (229 mV). R_{ct} was evaluated using electrochemical impedance spectroscopy (EIS) analysis. The NiCoSe (8 mmol) nanosheets also exhibited the smallest R_{ct} of 2.614 Ω, facilitating the highest electron transferability among the investigated electrocatalysts. Superior oxygen evolution reaction (OER) performance was also exhibited by NiCoSe (2 mmol) with lower Tafel slope (92 mV dec^{-1}) and R_{ct} values (2.325 Ω). A system of two-electrode cell was constructed by the authors to study the water electrolysis in 1 M KOH solution in which NiCoSe samples served as both cathode and anode. The electrocatalyst with high electrochemical surface area, mesoporous structure, and synergic effect between Ni foam and NiCoSe nanosheet arrays served not only as a HER electrocatalyst but also as an unprecedented bifunctional electrocatalyst in overall water splitting.

18.2.4.2 2D CoNi Bimetallic MOF

2D CoNi-MOF nanoplate array grown on the Cu foil having direct exposure of metal nodes present in the porous structure was investigated as a promising HER electrocatalyst of MOF group. Liu et al. [21] synthesized the bimetallic CoNi-MOF (1:1,Co:Ni) and also annealing 1:1 CoNi-MOF through annealing with NH_3 forming Ni nitride composites on amorphous carbon (CoNiN-C). Electrochemical studies were conducted for both OER and HER activities in 1 M KOH. The CoNiN-C required η_{10} of 120 mV, which was lower than that of 1:1 CoNi-MOF (376 mV). The lower higher charge transfer resistance of CoNiN-C (4.62 Ω) than 1:1 CoNi-MOF (102.4 Ω) suggests that CoNiN-C has better HER activity than 1:1 CoNi-MOF. Stability tests also demonstrated the better durability of CoNiN-C as a remarkable water-splitting electrocatalyst. On account of better OER and HER activities, CoNiN-C was employed as cathode CoNiN-C and CoNi(1:1)-MOF as anode for overall water splitting. Faradaic efficiency of 99.0% was achieved for H_2 production on CoNiN-C similarly 98.9% was obtained for O_2 production on CoNi(1:1)-MOF [21].

18.2.4.3 NiFe(dobpdc)-MOF with an Extended Organic Linker

Various analogues of M_2(dobpdc) (where M-metal and dobpdc^{4-} -4,4′-dioxidobiphenyl-3,3′ - dicarboxylate) have been widely used for HER electrocatalysis. Qi et al. [22] fabricated NiFe(dobpdc) as an electrocatalyst for overall water splitting and carried out various electrochemical studies. NiFe(dobpdc) has a yarn-ball-like structure and the extended organic linker (dobpdc) can increase the Brunauer, Emmett and Teller (BET) surface area, which in turn can enhance the interaction between NiFe(dobpdc) electrocatalyst and water. The HER activity was evaluated using a three-electrode system in an alkaline medium. NiFe(dobpdc) required η_{10} of 233 mV and η_{100} of 170 mV. The Tafel slopes of NiFe(dobpdc) was 69 mV dec^{-1}, which was smaller than those of Fe_2(dobpdc) and Ni_2(dobpdc). Notable OER activity was also shown by NiFe(dobpdc) encouraging the setup of a two-electrode system in 1.0 M KOH for overall water splitting.

18.2.4.4 2D Layered CuS-C

Rong et al. [23] synthesized 2D layered CuS-C as a Cu-based MOF (Cu-BDC nanosheets) via sulfidation of Cu-BDC nanosheets. HER electrocatalytic investigation of layered CuS-C, bulk CuS-C, and Cu-C was examined in acidic medium by linear sweep voltammetry (LSV). The Tafel slope of layered CuS-C was found to be 44 mV dec^{-1}, which was lower than those of bulk CuS (91 mV dec^{-1}) and Cu-C (102 mV dec^{-1}). In contrast, layered Cu-C exhibited a much smaller onset potential of 128 mV to attain η_{10} than bulk CuS (370 mV). A charge transfer resistance (R_{ct}) of 55 Ω was exhibited by CuS-C, which was lower than those of bulk CuS-C (147 Ω) and Cu-C (272 Ω). Layered CuS-C gave excellent stability and exhibited negligible decay even after 1,000 potential cycles and also HER current was retained to 96.5% of the initial value which marks its excellent stability.

18.2.5 Wrinkled Rh_2P Nanosheets

Metal phosphides and sulfides supported on carbon are Pt-free electrocatalysts that have drawn attention for HER electrocatalytic activity. Wang et al. [24] fabricated a new category of wrinkled Rh_2P ultrathin nanosheets as an universal-pH-efficient HER electrocatalyst. Wrinkled Rh_2P and Rh nanosheets were supported on carbon, which was denoted as w-Rh_2P/C and Rh/C, respectively. Electrochemical studies of these electrocatalysts were conducted in alkaline (0.1 M KOH), acidic (0.1 M $HClO_4$), and neutral (0.5 M phosphate-buffered saline (PBS) solution) media. In alkaline medium, w-Rh_2P/C exhibited η_{10} of 18.3 mV, which is lower than 68.9 mV of Rh/C. The Tafel slopes obtained for w-Rh_2P/C and Rh/C were 61.5 and 113.0 mV dec^{-1}, respectively. Under similar conditions, commercially available Pt/C exhibited η_{10} of 58.4 mV and the Tafel slope of 87.4 mV dec^{-1}. The HER electrocatalytic properties exhibited by w-Rh_2P/C were found to be superior to commercially available (20%) Pt/C. In acidic medium, w-Rh_2P/C required overpotential of 15.8 mV (η_{10}) lower than those of Rh/C (41.6 mV) and commercial Pt/C (22.1 mV). The Tafel slope of w-Rh_2P/C was 29.9 mV dec^{-1}, which was similar to Pt/C 29.6 mV dec^{-1} and lower than that of Rh/C (37.4 mV dec^{-1}). Further, HER activity was evaluated in PBS solution (pH 7), which resulted in η_{10} of 21.9 mV and a Tafel slope of 78.4 mV dec^{-1}. The HER performance of w-Rh_2P/C was better than those of Rh/C and commercial Pt/C in neutral medium. Wrinkled Rh_2P exhibited η_{10} of 18.3 mV and Tafel slope of 61.5 mV dec^{-1}, which was better than commercial Pt/C used widely as a benchmark electrocatalyst for HER in alkaline medium.

18.2.6 Graphene Hybrid Systems

Graphene is a monolayered carbon with honeycomb structure that has been widely employed as an electrocatalyst or as support for electrocatalytic water splitting. Researches on graphene suggest that graphene hybrid by doping it with various heteroatoms and compounds enhance the electrochemical properties of graphene. Chemical doping on graphene significantly increases the electronic structural properties of active sites leading to enhanced electric conductivity and stability under drastic reaction conditions. Li et al. [25] investigated HER and OER properties of certain graphene hybrids to fabricate highly active electrocatalysts for overall water splitting. Nitrogen, Phosphorous, or Sulfur (N, P, or S)-doped graphene potentially improved HER performance compared to intrinsic graphene. N/P binary-doped graphene exhibited ΔG_{H*} of 0.53 eV, which was smaller than that of N-doped graphene (ΔG_{H*}=0.81 eV) and the smallest ΔG_{H*} of 0.23 eV was obtained for N/S binary-doped graphene.

18.2.6.1 Metal-Doped Graphene

Li et al. [25] investigated atomic Co on N-graphene (Co/N-graphene) for the HER activity of metal-doped graphene. Co/N-graphene exhibited an onset potential and η_{10} of 30 and 147 mV, respectively. Tafel slope obtained was 82 mV dec^{-1}. Further, the electrocatalyst showed excellent stability for about 10 hours in 0.5 M H_2SO_4 in a stability test. Ni-doped graphene also manifested excellent electrocatalytic activity for HER with a Tafel slope of 45 mV dec^{-1} and stability of 5 hours in 0.5 M H_2SO_4.

18.2.6.2 Metal Sulfide or Metal Selenide–Graphene Hybrids

Li et al. [25] synthesized MoS_2 on reduced graphene (MoS_2/rGO) hybrids by solvothermal method and its electrochemical measurements were conducted in an acidic medium. MoS_2/rGO decreased the aggregation of MoS_2 during HER and provided abundant active sites at edges for HER. MoS_2/rGO showed a Tafel slope of 41 mV dec^{-1} and maintained the catalytic activity for about 1,000 potential cycles in a stability test. $MoSe_2$ on layered graphene also manifested high HER activity than intrinsic graphene. Presence of synergetic effect on $MoSe_2$/graphene hybrids bestowed an onset potential of 125 mV and a Tafel slope of 67 mV dec^{-1}. Excellent stability was also maintained, which was evaluated by a measuring current density over 6,000 seconds in 0.5 M H_2SO_4.

18.2.6.3 Metal Phosphides or Metal Carbides–Graphene Hybrids

Transition metal phosphides (TMPs) and carbides on graphene have made extensive advances in HER studies. Li et al. [25] synthesized porous CoP/graphene hybrids and subjected to electrochemical analysis in alkaline media (1.0 M KOH). The electrocatalyst exhibited a smaller Tafel slope of 38 mV dec^{-1} than that of bare CoP (60 mV dec^{-1}) and a stable catalytic current density for nearly 20 hours. Iron phosphides on graphene exhibited a remarkable HER activity with an onset potential of 30 mV, Tafel slope of 50 mV dec^{-1}, and η_{10} of 123 mV. The synergistic effects between metal phosphides and graphene resulted in improved electrical conductivity with and abundant exposure of active sites.

18.2.6.4 Bimetallic Phosphides on Reduced Graphene

Li et al. [25] synthesized cobalt-doped nickel phosphides on reduced graphene oxide (NiCoP/rGO) and employed as a bifunctional hybrid catalyst for the overall water splitting in all pH ranges. Tafel slopes of NiCoP/rGO hybrids were 91.0 mV dec^{-1}

in 1.0 M PBS buffer (neutral medium) and 124.1 mV dec^{-1} in 1.0 M KOH. However, the electrocatalyst showed excellent HER activity in 0.5 M H_2SO_4 with Tafel slope of 45.2 mV dec^{-1} and η_{20} of 55 mV. High catalytic stability over 18 hours was also exhibited by the electrocatalyst. On account of the bifunctional catalytic properties exhibited by NiCoP/rGO hybrids for both OER and HER, a two-electrode system of NiCoP/rGO||NiCoP/rGO in 1.0 M KOH was constructed, which showed η_{10} of 1.59 V for more than 75 hours with ~100% Faradic efficiency. Even though graphene-based electrocatalysts are widely studied for improving overall water splitting, and it is still challenging for large-scale practical applications.

18.2.7 CoP Nanosheet Aerogel

TMPs are one of the highly active classes of HER electrocatalysts. CoP nanostructures, as an example of TMPs were subjected to distinct morphologies to obtain better HER catalytic activity. Li et al. [26] synthesized 2D ultrathin CoP Nanosheets on aerogel by ice-templating method as illustrated in Figure 18.3 and demonstrated enhanced HER performance at all pH ranges. The porous structure of aerogel and lower thickness of CoP (<1.5 nm) augmented higher mass transfer and also prevented agglomeration of CoP nanosheets. The HER performance was evaluated using a three-electrode system in all pH ranges. The Tafel slope and η_{10} values of CoP aerogels were 67 mV dec^{-1} and 113, respectively, in 0.5 M H_2SO_4. The HER properties of CoP aerogels were exceptional when compared to previously reported Co-based electrocatalysts. The authors investigated HER activities of the electrocatalyst in 1 M KOH and 1 M PBS. The electrocatalyst exhibited Tafel slopes of 72 and 81 mV dec^{-1} in the 1 M KOH and PBS media, respectively. Compared to 0.5 M H_2SO_4 electrolyte, CoP aerogels exhibited higher η_{10} in 1 M KOH (154 mV) and 1 M PBS (161 mV). ECSA studies were carried out through double-layer capacitance (C_{dl}) measurements. CoP aerogels showed C_{dl} values of 19.16 mF cm^{-2} in 0.5 M H_2SO_4 and 7.66 mF cm^{-2} in 1 M KOH. The electrocatalyst also showed remarkable stability in all pH range with negligible deactivation during the catalytic cycles.

18.2.8 Ruthenium-Doped Bimetallic Phosphide on Ni Foam

Heteroatom-doped electrocatalysts have proved to lower the activation energy resulting in accelerated reaction kinetics enhancing HER performance. Metal-hydrogen bond strength and water-dissociation properties of Ru are similar to those of Pt. Lin et al. [27] fabricated HER electrocatalyst by doping trace amount (0.6 wt.%) of Ru-bimetallic phosphide on Ni foam (Ru-NiFeP/NF). It was derived from 2D MIL-53(NiFe) MOF nanosheets and synthesized via hydrothermal process followed by phosphorization. The schematic representation of synthesis of Ru-NiFeP/NF is illustrated in Figure 18.4. HER and OER electrochemical studies of Ru-doped NiFeP/NF were conducted in acidic (0.5 M H_2SO_4), alkaline (1 M KOH), and neutral (PBS) media. For comparison, the electrochemical studies of NiFeP/NF and Pt/C/NF electrocatalysts were conducted under similar conditions. Ru-NiFeP/NF demonstrated enhanced HER activity in acidic medium with η_{10} of 29.3 mV while that of NiFeP/NF was 146.6 mV. It may be noted that η_{100} of

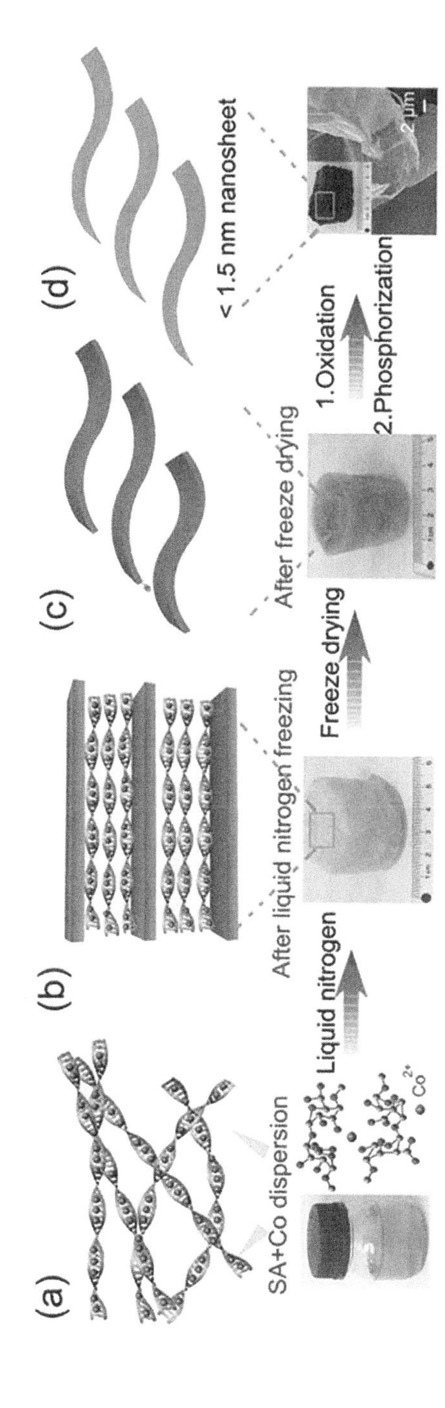

FIGURE 18.3 Schematic representation of preparation of CoP aerogel nanosheets. Adapted with permission from ref [25]. (a) SA+Co dispersion. (b) After liquied freezing. (c) After freeze drying. (d) Nanosheets and microstructure. Copyright (2018) WILEY-VCH Verlag GmbH & Co. KGaA, Weinheim.

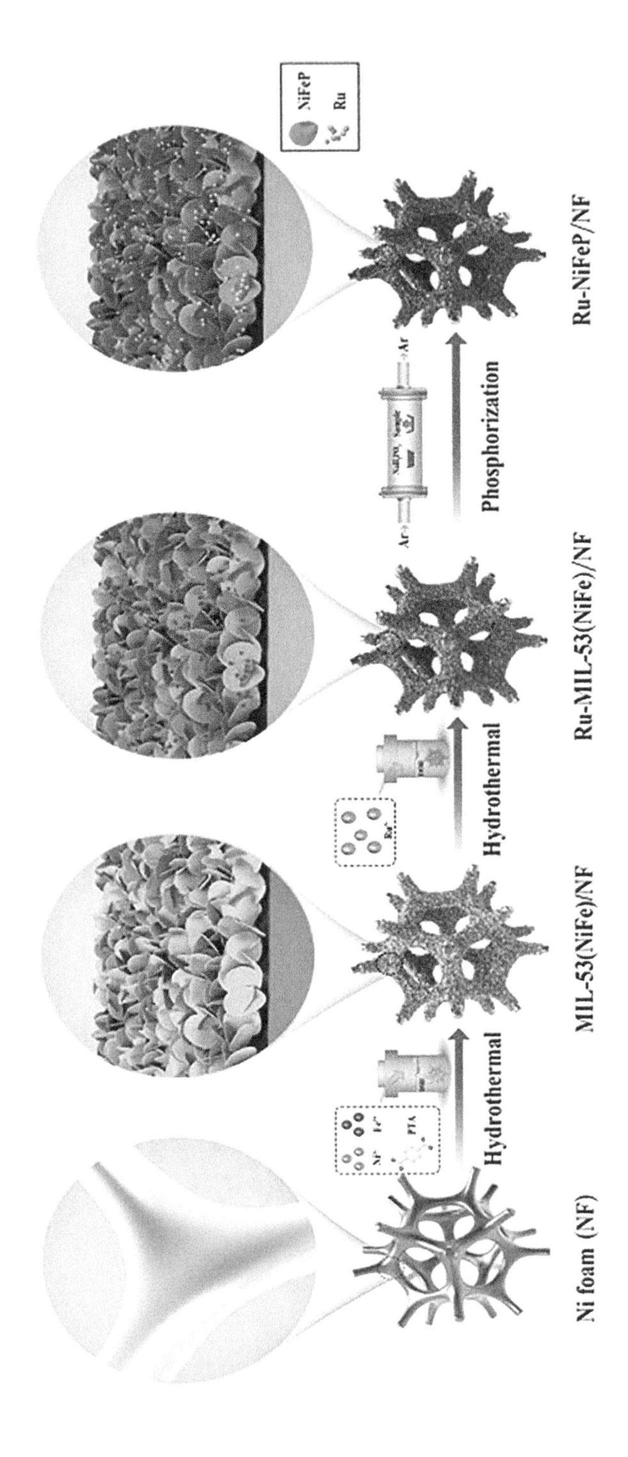

FIGURE 18.4 Schematic representation of synthesis of Ru-NiFeP/NF nanosheets. Adapted with permission from ref [26]. Copyright (2020) Elsevier B.V.

Ru-NiFeP/NF (88.2 mV) was lower than that of Pt/C/NF (136.2 mV). The Tafel slope of Ru-NiFeP/NF was found to be 55.7 mV dec^{-1}. When compared with other electrocatalysts studied by the authors, Ru-NiFeP/NF possessed smaller R_{ct} indicating faster electron transfer. In neutral medium, η_{10} of Ru-NiFeP/NF was 105.1 mV, which was lower than that of NiFeP/NF (218.1 mV). The Tafel slope of Ru-NiFeP/NF in the same medium was 82.7 mV dec^{-1}, but was greater than that of Pt/C/NF (53.7 mV dec^{-1}). Ru-NiFeP/NF exhibited η_{100} of 139.5 mV, which was lower than that of Pt/C/NF (153.4 mV) in alkaline medium. Ru-NiFeP/NF showed excellent long-term stability with no apparent in its initial activity. The above studies indicated that Ru-NiFeP/NF was superior to previously reported HER electrocatalysts such as Ru-MoO_2, Ru-MoS_2/CC, and $NiCo_2Px$ nanowires in a wide pH range.

18.2.9 Porous W-Doped CoP Nanoflake Arrays

TMPs have been used extensively as efficient HER electrocatalysts due to their remarkable intrinsic properties and prolonged stability. Doping TMPs with heteroatom can modulate their electronic structures and facilitate their electronic conductivity, which can significantly enhance their HER performance in a wide pH range. TMPs-based electrocatalysts exist in powder form having complex preparation methods involving polymeric binders. These polymeric binders can block the catalytic active sites and result in poor catalytic activity. In this context, an efficient synthesis route with hierarchical nanoarrays grown on conducting substrate was designed by Wang et al. [28]. The authors synthesized W-doped CoP nanoflake arrays on carbon cloth (W-CoP/CC) and employed it as a superior HER electrocatalyst in a wide pH range. Schematic representation of synthesis of W-CoP NAs/CC is illustrated in Figure 18.5. W-CoP/CC HER performance was evaluated in acidic, alkaline, and neutral media. For

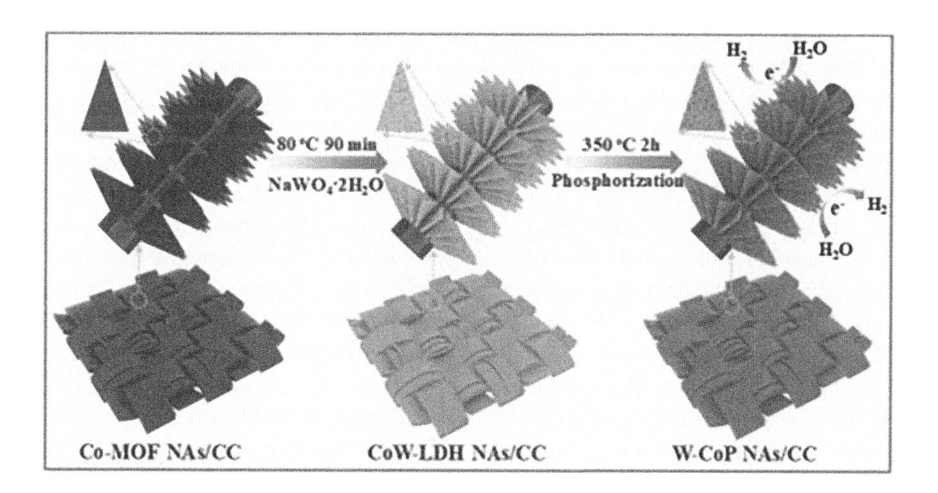

FIGURE 18.5 Schematic representation of synthesis of W-CoP NAs/CC. Adapted with permission from ref [27]. Copyright (2019) WILEY-VCH Verlag GmbH & Co. KGaA, Weinheim.

comparison, the HER performance of Pt/C/CC, CoP/CC, and Co_3O_4/CC electrocatalysts was also studied under similar conditions. In 0.5 M H_2SO_4, W-CoP/CC exhibited an onset potential of 31 mV and the η_{10} was 89 mV, which was lower than that of CoP/CC (123 mV), but greater than that of Pt/C/CC (35 mV). The Tafel slopes of W-CoP/CC, Pt/C/CC, and CoP /CC were 58, 32, and 97 mV dec^{-1}, respectively. In 1.0 M KOH, η_{10} of W-CoP/CC was 94 mV, which was lower than those of Co/CC (129 mV) and Co_3O_4/CC (305 mV). The Tafel slope of W-CoP/CC (63 mV dec^{-1}) was also lower than those of CoP/CC (74 mV dec^{-1}) and Co_3O_4/CC (128 mV dec^{-1}). Among the above three electrocatalysts, W-CoP/CC exhibited the best catalytic activity in alkaline medium. In 1.0 M PBS, the onset potentials of W-CoP/CC, CoP NAs/CC, and Co_3O_4/CC were 51, 79, and 313 mV and the η_{10} values were 102, 132, and 397, respectively. Also, the Tafel slope of W-CoP/CC (87 mV dec^{-1}) was lower than those of CoP/CC (98 mV dec^{-1}) and Co_3O_4 /CC (117 mV dec^{-1}). W-CoP/CC exhibited about 100% Faradaic efficiency and also high stability with no appreciable degradation in activity for about 36 hours in all three media. The excellent electrocatalytic performance of W-CoP/CC could be attributed to the hierarchically porous nanoflake array configuration and heteroatom doping with W, resulting in highly exposed active sites facilitating the diffusion of H_2 produced and efficiently enhancing the HER performance.

18.2.10 Single Atom on the 2D Matrix

It is well-known that noble metals-based electrocatalysts are the most efficient electrocatalysts for HER but their scarcity and high cost limit their utilization for large-scale hydrogen production. An alternative way is to reduce the size of noble metals-based electrocatalysts to single-atom catalysts (SACs) [29]. SACs act as a link between heterogenous and homogenous catalysis. Examples of SAC present in nature are Mg in chlorophyll entailed in photosynthesis and other metals present in enzymes that carry out biocatalysis. Reducing the size of catalysts to SACs results in an enhanced electronic and atomic structure with a greater number of exposed surface atoms thereby enhancing catalytic performance. But, reducing the size may lead to aggregation and the particle set off instability. Therefore, stable SACs need to be synthesized efficiently using solid support with a larger surface area, for example, 2D matrix of N-doped graphene, LDH, graphdiyne, 2D MOFs, and transition metal dichalcogenides (TMDs). These SAC hybrids are highly stabilized by the covalent interaction present between 2D-matrix and single atoms. These interactions will construct a new electronic state, which leads to a new set of active sites and maximize atom utilization efficiency thereby enhancing catalytic performance to a large extent.

Yu et al. [30] supported zero-valent palladium atoms on graphdiyne nanosheets (GDY-Pd^0) at a mass loading of 0.2%. GDY-Pd^0 HER studies were conducted in 0.5 M H_2SO_4. For comparison, pristine GDY, Pd nanoparticles decorated on GDY (Pd NP/GDY), and commercial Pt/C (20 wt.%) were also studied by the authors under similar conditions. GDY-Pd^0 exhibited excellent HER activity with η_{10} of 55 mV, which was lower than those of commercial Pt/C (62 mV), Pd NP/GDY (115 mV), and pure GDY (481 mV). The mass activity of GDY-Pd^0 was 26.9 times higher than that of Pt/C. The Tafel slope of GDY-Pd^0 was (47 mV dec^{-1}) lower than those of Pd NP/GDY (139 mV dec^{-1}) and GDY (212 mV dec^{-1}). R_{ct} of GDY-Pd^0 3.37 Ω was very much lower than that of the pristine GDY (30.97 Ω) indicating faster electronic transfer. Thus, zero-valent palladium atoms on 2D graphdiyne nanosheets fabricated by the authors delivered exceptional HER performance compared to Pd NP/GDY and commercial Pt/C (20 wt.%).

18.3 CONCLUDING REMARKS

In water electrolysis, HER has both of practical and fundamental importance and has gained captivating utilization as hydrogen fuel in the future hub of renewable energy. HER is acknowledged as one of the paramount technologies for hydrogen production as a greener method of production of hydrogen. Materials based on noble metal like Pt are the best-known electrocatalysts for HER. Due to the scarcity and high cost of precious metal compounds, the large-scale HER production still remains a challenge. In this chapter, we have briefly discussed the recent significant role of 2D nanomaterials that are employed as an alternative to Pt-based electrocatalysts for HER facilitating the electrocatalytic reduction of water to molecular hydrogen. The electrochemical properties of various 2D nanomaterials have been demonstrated and proved to be of great potent for HER performance in a wide pH range. Due to their unique physicochemical properties derived from atomically thin dimensions, 2D nanomaterials have been at the forefront of research in recent years. 2D electrocatalysts comprise of 2D MOFs, phosphorene, TMDs, transition metal chalcogenides (TMCs), graphene, graphydine, hybrids of 2D SACs, and so on, which have shown remarkable HER performance, wherein some of them have proved to be even better than commercial Pt/C. Extensive studies on altering compositions, morphological structure, and other properties of electrocatalysts are still in progress to maximize the HER performance. In recent years, a wide-ranging family of 2D nanomaterials have been fabricated in the field of electrochemistry. This has prompted intense and diverse research projects comprising both fundamental studies and technology applications [3]. In this context, the successful application of 2D nanomaterial electrocatalyst for HER can be attributed to its exceptional functional capabilities. 2D hybrid electrocatalysts are well-known for exhibiting enhanced HER activity. For the next-generation renewable energy systems, large-scale hydrogen production and development are of great significance.

Electrocatalysts with 2D layered structures with the benefit of release of stress can buffer the shock of electrolyte convection, hydrogen bubble rupture, and evolution, assuring prolonged stability [31]. 2D electrocatalysts with rich porosity enhance the infiltration of electrolyte and facile evolution of hydrogen at the high catalytic current. Graphene, Phosphorene, etc. have been subjected to doping with heteroatoms, which has led to the modulation of the electronic structure induced by interfacial charge and thereby they have exhibited enhanced HER performance. Electrocatalysts with 2D nanosheet structure have unique characteristics like high morphological anisotropy, larger electrochemical surface area with a large number of surface-active sites, mesoporous structure, open ion diffusion channels, synergic effect, along with distinctive electronic structures. While, 2D-MOF nanosheets have integrated the advantages of both 2D material and MOF, which are coordination of metal ions and organic ligands possess several advantages on account of their ultrathin thickness, high surface-to-volume atom ratios, and large surface area [32]. Another class of HER electrocatalysts is TMDC nanosheets, which have less in-plane resistivity throughout the basal plane [11] and relatively high mobility of the charge carriers, which in turn facilitate easier electron transport. MXenes have emerged recently as a new class of 2D nanomaterials with conductivity similar to that of metals or semiconductors and also possess similar physical properties like that of graphene. However, the 2D surface morphology with distinct electronic properties needs to be fully investigated and comprehended for progressive HER performance. Further exploration of 2D nanomaterials as electrocatalysts requires rational design and synthetic methods for optimal HER performance. It is highly significant to further understand their structural and electrochemical properties. In addition, creating effective hybrids of 2D nanomaterial-based structures will furnish more avenues for broader range of electrolytic water splitting. Hence, the significance of in-depth studies on electrocatalytic properties of 2D nanomaterials, especially for HER, can be understood.

REFERENCES

1. Novoselov, K. S. Electric field effect in atomically thin carbon films. *Science* **2004**, *306* (5696), 666–669. https://doi.org/10.1126/science.1102896.
2. Singh, N. B.; Shukla, S. K. Properties of two-dimensional nanomaterials. In *Two-Dimensional Nanostructures for Biomedical Technology*; Elsevier, 2020; pp. 73–100. https://doi.org/10.1016/B978-0-12-817650-4.00003-6.
3. Murthy, A. P.; Madhavan, J.; Murugan, K. Recent advances in hydrogen evolution reaction catalysts on carbon/carbon-based supports in acid media. *Journal of Power Sources* **2018**, *398*, 9–26. https://doi.org/10.1016/j.jpowsour.2018.07.040.
4. Theerthagiri, J.; Lee, S. J.; Murthy, A. P.; Madhavan, J.; Choi, M. Y. Fundamental aspects and recent advances in transition metal nitrides as electrocatalysts for hydrogen evolution reaction: A review. *Current Opinion in Solid State and Materials Science* **2020**, *24* (1), 100805. https://doi.org/10.1016/j.cossms.2020.100805.
5. Theerthagiri, J.; Murthy, A. P.; Lee, S. J.; Karuppasamy, K.; Arumugam, S. R.; Yu, Y.; Hanafiah, M. M.; Kim, H.-S.; Mittal, V.; Choi, M. Y. Recent progress on synthetic strategies and applications of transition metal phosphides in energy storage and conversion. *Ceramics International* **2021**, *47* (4), 4404–4425. https://doi.org/10.1016/j.ceramint.2020.10.098.
6. Murthy, A. P.; Govindarajan, D.; Theerthagiri, J.; Madhavan, J.; Parasuraman, K. Metal-doped molybdenum nitride films for enhanced hydrogen evolution in near-neutral strongly buffered aerobic media. *Electrochimica Acta* **2018**, *283*, 1525–1533. https://doi.org/10.1016/j.electacta.2018.07.094.
7. Murthy, A. P.; Theerthagiri, J.; Madhavan, J. Highly water dispersible polymer acid-doped polyanilines as low-cost, nafion-free ionomers for hydrogen evolution reaction. *ACS Applied Energy Materials* **2018**, *1* (4), 1512–1521. https://doi.org/10.1021/acsaem.7b00315.
8. Murthy, A. P.; Theerthagiri, J.; Premnath, K.; Madhavan, J.; Murugan, K. Single-step electrodeposited molybdenum incorporated nickel sulfide thin films from low-cost precursors as highly efficient hydrogen evolution electrocatalysts in acid medium. *The Journal of Physical Chemistry C* **2017**, *121* (21), 11108–11116. https://doi.org/10.1021/acs.jpcc.7b02088.
9. Murthy, A. P.; Theerthagiri, J.; Madhavan, J.; Murugan, K. Enhancement of hydrogen evolution activities of low-cost transition metal electrocatalysts in near-neutral strongly buffered aerobic media. *Electrochemistry Communications* **2017**, *83*, 6–10. https://doi.org/10.1016/j.elecom.2017.08.011.

10. Murthy, A. P.; Theerthagiri, J.; Madhavan, J.; Murugan, K. Electrodeposited carbon-supported nickel sulfide thin films with enhanced stability in acid medium as hydrogen evolution reaction electrocatalyst. *Journal of Solid State Electrochemistry* **2018**, *22* (2), 365–374. https://doi.org/10.1007/s10008-017-3763-4.
11. Fu, Q.; Han, J.; Wang, X.; Xu, P.; Yao, T.; Zhong, J.; Zhong, W.; Liu, S.; Gao, T.; Zhang, Z.; Xu, L.; Song, B. 2D transition metal dichalcogenides: Design, modulation, and challenges in electrocatalysis. *Advanced Materials* **2021**, *33* (6), 1907818. https://doi.org/10.1002/adma.201907818.
12. Murthy, A. P.; Theerthagiri, J.; Madhavan, J.; Murugan, K. Highly active MoS_2/carbon electrocatalysts for the hydrogen evolution reaction – insight into the effect of the internal resistance and roughness factor on the tafel slope. *Physical Chemistry Chemical Physics* **2017**, *19* (3), 1988–1998. https://doi.org/10.1039/C6CP07416B.
13. Shanmugam, P.; Murthy, A. P.; Theerthagiri, J.; Wei, W.; Madhavan, J.; Kim, H.-S.; Maiyalagan, T.; Xie, J. Robust bifunctional catalytic activities of n-doped carbon aerogel-nickel composites for electrocatalytic hydrogen evolution and hydrogenation of nitrocompounds. *International Journal of Hydrogen Energy* **2019**, *44* (26), 13334–13344. https://doi.org/10.1016/j.ijhydene.2019.03.225.
14. Murthy, A. P.; Theerthagiri, J.; Madhavan, J. Insights on tafel constant in the analysis of hydrogen evolution reaction. *The Journal of Physical Chemistry C* **2018**, *122* (42), 23943–23949. https://doi.org/10.1021/acs.jpcc.8b07763.
15. Wang, S.; Lu, A.; Zhong, C.-J. Hydrogen production from water electrolysis: Role of catalysts. *Nano Convergence* **2021**, *8* (1), 4. https://doi.org/10.1186/s40580-021-00254-x.
16. Cai, Z.; Yao, Q.; Chen, X.; Wang, X. Nanomaterials with different dimensions for electrocatalysis. In *Novel Nanomaterials for Biomedical, Environmental and Energy Applications*; Elsevier, 2019; pp. 435–464. https://doi.org/10.1016/B978-0-12-814497-8.00014-X.
17. Jang, S. W.; Dutta, S.; Kumar, A.; Hong, Y.-R.; Kang, H.; Lee, S.; Ryu, S.; Choi, W.; Lee, I. S. Holey Pt nanosheets on nife-hydroxide laminates: Synergistically enhanced electrocatalytic 2D interface toward hydrogen evolution reaction. *ACS Nano* **2020**, *14* (8), 10578–10588. https://doi.org/10.1021/acsnano.0c04628.
18. Chen, Y.; Ou, P.; Bie, X.; Song, J. Basal plane activation in monolayer $MoTe_2$ for the hydrogen evolution reaction *via* phase boundaries. *Journal of Materials Chemistry A* **2020**, *8* (37), 19522–19532. https://doi.org/10.1039/D0TA06165D.
19. Liu, Y.; Zhang, H.; Ke, J.; Zhang, J.; Tian, W.; Xu, X.; Duan, X.; Sun, H.; O Tade, M.; Wang, S. 0D (MoS_2)/2D (g-C_3N_4) heterojunctions in Z-scheme for enhanced photocatalytic and electrochemical hydrogen evolution. *Applied Catalysis B: Environmental* **2018**, *228*, 64–74. https://doi.org/10.1016/j.apcatb.2018.01.067.
20. Zhou, Y.; Chen, Y.; Wei, M.; Fan, H.; Liu, X.; Liu, Q.; Liu, Y.; Cao, J.; Yang, L. 2D MOF-derived porous NiCoSe nanosheet arrays on Ni foam for overall water splitting. *CrystEngComm* 2021, *23* (1), 69–81. https://doi.org/10.1039/D0CE01527J.
21. Liu, M.; Zheng, W.; Ran, S.; Boles, S. T.; Lee, L. Y. S. Overall water-splitting electrocatalysts based on 2D CoNi-metal-organic frameworks and its derivative. *Advanced Materials Interfaces* **2018**, *5* (21), 1800849. https://doi.org/10.1002/admi.201800849.
22. Qi, L.; Su, Y.-Q.; Xu, Z.; Zhang, G.; Liu, K.; Liu, M.; Hensen, E. J. M.; Lin, R. Y.-Y. Hierarchical 2D yarn-ball like metal–organic framework NiFe(Dobpdc) as bifunctional electrocatalyst for efficient overall electrocatalytic water splitting. *Journal of Materials Chemistry A* **2020**, *8* (43), 22974–22982. https://doi.org/10.1039/D0TA08094B.
23. Rong, J.; Xu, J.; Qiu, F.; Fang, Y.; Zhang, T.; Zhu, Y. 2D metal-organic frameworks-derived preparation of layered CuS@C as an efficient and stable electrocatalyst for hydrogen evolution reaction. *Electrochimica Acta* **2019**, *323*, 134856. https://doi.org/10.1016/j.electacta.2019.134856.

24. Wang, K.; Huang, B.; Lin, F.; Lv, F.; Luo, M.; Zhou, P.; Liu, Q.; Zhang, W.; Yang, C.; Tang, Y.; Yang, Y.; Wang, W.; Wang, H.; Guo, S. Wrinkled Rh_2 P nanosheets as superior PH-universal electrocatalysts for hydrogen evolution catalysis. *Advanced Energy Materials.* **2018**, *8* (27), 1801891. https://doi.org/10.1002/aenm.201801891.
25. Li, J.; Zhao, Z.; Ma, Y.; Qu, Y. Graphene and their hybrid electrocatalysts for water splitting. *ChemCatChem* **2017**, *9* (9), 1554–1568. https://doi.org/10.1002/cctc.201700175.
26. Li, H.; Zhao, X.; Liu, H.; Chen, S.; Yang, X.; Lv, C.; Zhang, H.; She, X.; Yang, D. Sub-1.5 Nm Ultrathin CoP nanosheet aerogel: Efficient electrocatalyst for hydrogen evolution reaction at all PH values. *Small* **2018**, *14* (41), 1802824. https://doi.org/10.1002/smll.201802824.
27. Lin, Y., Wang, L., Cao, D., Gong, Y.; Zhang, M.; Zhao, L. Ru doped bimetallic phosphide derived from 2D metal organic framework as active and robust electrocatalyst for water splitting. *Applied Surface Science*, **2020**, *536*, 147952. https://www.sciencedirect.com/science/journal/01694332/536/supp/C. https://doi.org/10.1016/j.apsusc.2020.147952.
28. Wang, X.; Chen, Y.; Yu, B.; Wang, Z.; Wang, H.; Sun, B.; Li, W.; Yang, D.; Zhang, W. Hierarchically porous W-doped CoP nanoflake arrays as highly efficient and stable electrocatalyst for PH-universal hydrogen evolution. *Small* **2019**, *15* (37), 1902613. https://doi.org/10.1002/smll.201902613.
29. Sun, J.-F.; Xu, Q.-Q.; Qi, J.-L.; Zhou, D.; Zhu, H.-Y.; Yin, J.-Z. Isolated single atoms anchored on N-doped carbon materials as a highly efficient catalyst for electrochemical and organic reactions. *ACS Sustainable Chemistry & Engineering* **2020**, *8* (39), 14630–14656. https://doi.org/10.1021/acssuschemeng.0c04324.
30. Yu, H.; Xue, Y.; Huang, B.; Hui, L.; Zhang, C.; Fang, Y.; Liu, Y.; Zhao, Y.; Li, Y.; Liu, H.; Li, Y. Ultrathin nanosheet of graphdiyne-supported palladium atom catalyst for efficient hydrogen production. *iScience* **2019**, *11*, 31–41. https://doi.org/10.1016/j.isci.2018.12.006.
31. Zhang, S.; Wang, W.; Hu, F.; Mi, Y.; Wang, S.; Liu, Y.; Ai, X.; Fang, J.; Li, H.; Zhai, T. 2D CoOOH sheet-encapsulated Ni2P into tubular arrays realizing 1000 MA Cm^{-2}-level-current-density hydrogen evolution over 100h in neutral water. *Nano-Micro Letters* **2020**, *12* (1), 140. https://doi.org/10.1007/s40820-020-00476-4.
32. Zheng, W.; Tsang, C.-S.; Lee, L. Y. S.; Wong, K.-Y. Two-dimensional metal-organic framework and covalent-organic framework: Synthesis and their energy-related applications. *Materials Today Chemistry* **2019**, *12*, 34–60. https://doi.org/10.1016/j.mtchem.2018.12.002.

19 Application of Graphene Family Materials for High-Performance Batteries and Fuel Cells

Chen Shen and S. Olutunde Oyadiji
The University of Manchester

CONTENTS

19.1 INTRODUCTION

Graphene is an allotrope of carbon, whose micro-structure is a repeating hexagonal pattern based on benzene ring and only has one layer of carbon atoms. Due to this special micro-structure, it is treated as one of the representative 2D materials. Once the word 'graphene' captures people's imagination, the discussion of this material never fades away within both professionals and ordinary people. In the public's view, graphene is an epic material. After all, ordinary people's understanding of

DOI: 10.1201/9781003178453-19

graphene mainly comes from the Nobel Physics Prize awarded to the co-discoverers of graphene in 2010, and the news about graphene industrial parks been built around the world. Besides, the public's perception is also influenced by advertisements on TV and websites, which claim their advanced products apply the latest graphene technology.

On the contrary, in the researchers' opinions, graphene is not as 'wonderful' as claimed in the current consumer product advertisements. Indeed, graphene has outstanding physical properties and stable chemical properties to some extent. However, these properties are only theoretical values, which are realizable only at a small-scale micro-level, and that are not likely to be directly applied in real life. Also, it is not simple to fabricate a flawless and high-quality graphene piece, which further increases the difficulties of its application. In the researcher's view, the majority of commercial graphene products in the market are just a 'stunt'. Researchers tend to believe that these products are just conventional products marked with higher retail prices, which only use this fancy word to attract customers' attention, especially for a consumer who does not have a strong background in science.

Among all commercial graphene products, graphene-based electronics, including power suppliers, capacitors, and electrodes, are, perhaps, the products with the most voluminous commercial applications, with batteries attracting the most attention. In the smartphone industry, manufactures are expecting a type of battery with higher capacity, faster-charging velocity, and thinner thickness to make their product outstanding. In the electric vehicle industry, a type of battery with a high capacity, long lifetime, and the reliable output voltage is needed to dominate the power supply field in this new field.

In this chapter, the application of graphene and graphene-based material in the battery and fuel cell industries is introduced. Section 19.2 covers the basic characteristics of graphene. How graphene will benefit the performance of batteries and fuel cells are introduced in Section 19.3, especially lithium-ion batteries (LIBs) and fuel cells.

19.2 GRAPHENE AND GRAPHENE-BASED MATERIALS

19.2.1 Classification and Fabrication

Graphene consists of carbon, in its thickness direction it only contains one layer of atoms. Therefore, graphene can be treated as a two-dimensional material. The research on this material can be traced back to the 1960s. In 2004, Andre Geim and Konstantin Novoselov from the University of Manchester, using ductile tape, successfully peeled graphene flakes from highly oriented pyrolytic graphite and were awarded the Nobel Physics Prize in 2010. Graphene's essential microstructure is a sp^2 C-C bond under a hexagonal benzene-ring shape (honeycomb structure, shown in Figure 19.1). Theoretically, graphene should be single-layered. Nevertheless, because of the fabrication difficulty, a local multi-layer or acceptable range of doped elements (commonly hydrogen or oxygen) is allowed. In practice, the word 'graphene' is often wrongly used to refer to different materials that can be classified as belonging to the graphene-based material family. It is correctly used to refer to pristine graphene (pure graphene, it has the best quality). However, it is wrongly used to refer to graphene

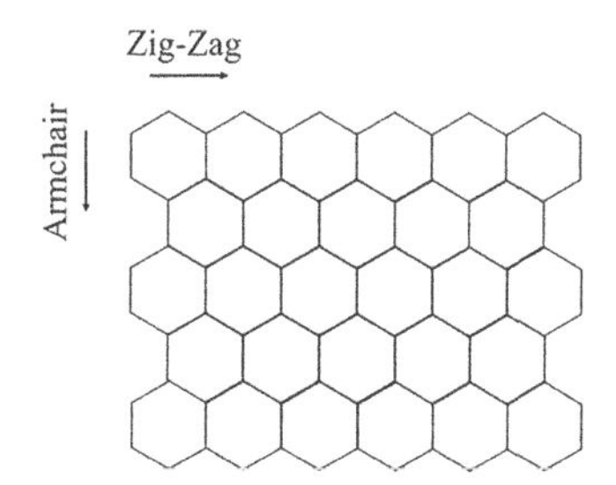

FIGURE 19.1 Microstructure of graphene.

oxide (the oxidation of graphene, it contains a large amount of oxygen atoms) or reduced graphene oxide (rGO; the product obtained after removing the majority of oxygen atoms from graphene oxide, it contains a small amount of oxygen).

The methods of fabricating graphene can be classified as direct methods and indirect methods. Direct methods include physical approaches (e.g., mechanical peeling and ultrasonic peeling from graphite) and chemical approaches (e.g., chemical vapor deposition and SiC epitaxial growth), which are used to fabricate continuous pristine graphene sheets directly and typically require a relatively complicated operation. The product is expected to be of high quality. Indirect methods produce rGO from graphene oxide, which can be fabricated following the Hummer's method or can be directly purchased from the market. By removing the majority of the oxygen atoms from graphene oxide, rGO can be obtained. Commonly used indirect methods include photo induction (UV, laser, etc.), chemical induction (hydrazine, sodium ascorbate, etc.), and thermal induction (annealing, temperature bath, etc.). The fabrication process of rGO is relatively simpler to be performed though the quality of the product is not easy to be fully controlled. The best quality rGO has properties that are very similar to those of pristine graphene. Indirect methods are popular within research establishments as they can be used as a replacement for expensive pristine graphene. The majority of 'graphene' produced by researchers themselves and reported in journal articles refers to 'reduced graphene oxide'.

Besides, based on the thickness, graphene can be classified as one-layer graphene, bilayer graphene, and few-layer (3–9 layer) graphene. Graphene with more than 10 layers of carbon atoms is classified as graphite. A more general name of graphene with multi-layer is graphene nano-platelet (GNP), which refers to the mixture of graphene whose thickness varies from 3 to 100 nm. GNP is a popular material to be used in industrial products, especially for products that do not require the continuity of graphene. High-quality graphene with only one layer of atoms is also defined as graphene nano-sheet (GN).

Apart from the thickness and quality, for GN, especially for the sheets smaller than 50 nm, based on its boundary morphology, it can be further classified as zig-zag direction and armchair direction. As shown in Figure 19.1, a zig-zag edge is similar to the tetracene structure while the armchair edge is close to chrysene. It should be noted that the differences in boundary directions will not have significant influences on the properties of graphene; it will only slightly affect the electronic properties at the edge [1].

19.2.2 Properties

Table 19.1 lists the properties of graphene. As a form of carbon, graphene is hydrophobic. Thus, its dissolvability is low but can form stable dispersion in polarized liquids. Typically, graphene is black, no matter whether it is in solid form or dispersion form. For relatively low concentration dispersion, sometimes it can be light brown. Its chemical properties are stable so it cannot simply react with other materials under normal conditions. Graphene has a low electric resistance rate (200 $\Omega\ m^{-1}$) and high intrinsic mobility (200,000 $cm^2V^{-1}{\cdot}s$), which makes it ideal for battery and supercapacitors. The high melting temperature (4,510 K) and high thermal conductivity (5,000 W mK^{-1}) also make it suitable to be applied in the thermal field. Overall, graphene has outstanding physical properties.

The mechanical properties of graphene have made graphene to be well known. The widely accepted values are 1 TPa for its Young's modulus and 130 GPa for ultimate tensile stress. These two values are very high. Comparatively, Young's modulus of graphene can reach around five times and the ultimate tensile stress is more than three times the corresponding properties of steel. However, it should be noticed that these values are just theoretical values at the micro-level. The samples used for mechanical tests are flawless micro-flakes. They were tested via the nano-indentation method under an atomic force microscope (AFM) by Lee et al. [2]. The methods used to derive these values are not the same as those used at the macro-level, and the macro-level derived values are not the same as those derived by other methods that we are familiar with. Researchers also performed macro-level mechanical tests on graphene sheets though the results are relatively low in value. Apart from the methodologies to obtain the values, another reason for this difference is the difficulty in controlling the quality of the macro-sized graphene. It usually contains a large quantity of micro-cracks and local overlap, which will significantly influence the

TABLE 19.1
Properties of Graphene [1]

Properties	Value
Color	Black
Specific surface area	2,630 m^2g^{-1}
Intrinsic mobility	200,000 $cm^2V^{-1}\cdot s$
Theoretical young's modulus	1 TPa
Theoretical tensile strength	130 GPa
Thermal conductivity	Around 5,000 W mK^{-1}
Electric resistance rate	200 $\Omega\, m^{-1}$
Optical transmittance	97.7%
Melting temperature	4,510 K
Dissolvability	Hydrophobic
Chemical stability	Prominent

mechanical properties obtained via a traditional tensile test. There is still an argument about the effective values of the mechanical properties of graphene.

19.2.3 Characterization Techniques

As introduced in Section 2.1, the majority of 'graphene' mentioned in journal articles, as a matter of fact, is 'reduced graphene oxide'. The morphology of these two materials is similar under macro views. Besides, the quality of rGO, including the quantity of micro-cracks, oxygen and hydrogen contents, the local overlapping (thickness) as well as the reduction level cannot be determined accurately using bare eyes. Therefore, special characterization equipment and techniques are required. Based on the expected outcomes, the characterization methods to qualify graphene can be divided into two groups, for morphology and contents. Typically, morphology observation equipment that is used includes electronic microscope, transmission electron microscope, scanning electron microscope, scanning tunneling microscope (STM), and AFM. The equipment can magnify the observation area 10^3–10^9 times to observe the morphology. Specifically, AFM and STM can be used to output detailed 3D images and thickness profiles to further characterize the morphology of graphene. To verify and analyze the content of the sample, X-ray diffraction (XRD), Fourier transforms infrared spectrum (FT-IR), and Raman test are typically used, with Raman test and XRD being used to verify the material type. The working principle of the Raman test is based on observation of the location and intensity of the characteristic band on the reflection spectrum. As for XRD, it relies on the location and intensity of the Eigen peak on the X-ray spectra. FT-IR is used to find out the content of the material. The functional groups within the sample can be verified through the reflected infrared light spectrum. The techniques mentioned above are the most commonly used approaches to characterize graphene/rGO. The quality of graphene/rGO should be verified before application.

19.2.4 General Applications of Graphene

With various outstanding properties, graphene can be applied in various areas. With good electric conductivity, graphene can be used to synthesize electric conductive composite and low resistant lead wire. Its high thermal conductivity had already been applied by smartphone factories as a rapid cooling coating. Besides, the high Young's modulus and ultimate tensile stress of G/rGO enables it to be used as micro fillers to enhance the mechanical properties of composites. Apart from these, graphene has been successfully used to fabricate a seawater filter, which can directly transfer seawater into drinking water and provides a relatively cheap method for water desalination. However, limited by the cost of high-quality samples, graphene is more likely to be applied in military, medical and advanced material fields where performance is the first consideration rather than price. Currently, it is not realistic to use this material in our daily life [1]. The application of graphene in batteries and fuel cells is introduced in detail in the next section.

19.3 APPLICATION OF GRAPHENE AND GRAPHENE-BASED MATERIALS IN BATTERIES AND FUEL CELLS

19.3.1 Introduction

In this section, apart from the use of graphene-based material in batteries and fuel cells, details about their history, mechanism, and key parameters are also included. To start with, the definitions of cell, battery, and fuel cell are provided. Cells and batteries are both used to supply power to electrical appliances, from small electric watches to large electric automobiles. A cell is the basic electrochemical/galvanic unit to store and transform the chemical power stored in the electrode or fuel, while a battery refers to the device, which consists of piles of cell units for longer output. The storage battery is most commonly seen in our daily life. A storage battery stores electrochemical energy in its electrodes, which will be transformed into electrical energy when external connections are made to the battery. The modern battery[1] was developed over the past two centuries. It was first invented by the Italian physicist Alessandro Volta in 1800 with Cu and Zn units, and it can output stable electricity for an acceptable length of time. In the following century, batteries including Zn-Cu, Zn-acid, iron-carbon, lead-acid, Zn-Mn, Zn-carbon, and Ni-Cd batteries were gradually invented.

On the other hand, a fuel cell is an electrochemical cell that converts the chemical energy of a fuel (hydrogen, diesel, ethanol, etc.) and an oxidizing agent (often oxygen) into electricity through a pair of redox reactions. The concept of the fuel cell was first proposed in the 1830s and first developed in 1932 by the English engineer Francis Thomas Bacon. Due to the 'green' product (water), a H-O fuel cell is expected to play a significant role in the next energy revolution. Fuel cells are similar to batteries but require a continuous source of fuel. They will continue to produce electricity as long as fuel is available [3].

19.3.2 Structures, Basic Mechanisms, and Key Parameters of Cells/Batteries

No matter the type, mechanism, and appearance of the cell/battery, it contains three essential components: anode, cathode, and electrolyte. For a battery, the anode (marked as '−') is the electrode where electrons are lost (shown conventionally as the electrode into which current flows at the macro level). This occurs whilst the battery is in its discharging mode. Typically, the anode material is more active and is oxidized whilst the battery is functioning. On the contrary, reduction happens on the cathode (marked as '+'), where the cathode material receives the electrons from the anode. It is generally shown as the end where current flows out. The electrolyte creates an atmosphere to assist the movement of ions. The output voltage of the entire cell equals the differences of the half-cell reduction potential between the cathode and anode. Technically, the more amount of anode material, the longer a battery can operate.

[1] To avoid confusion, the 'cell' and 'battery' in the usage of the terms 'fuel cell' and 'storage battery' (hereinafter abbreviated as battery unless otherwise specified) in the following sections are independent of the 'quantity of basic elements or galvanic cells'.

To describe the properties of a cell/battery, three of the most commonly used parameters are: capacity, energy, and voltage. Capacity denotes the electric energy stored within the device. The value is described in Ampere-hour (A·h), where 1 A·h equals to the quantity of electric charge in a 1 Ampere steady current flow for 1 hour. Energy is described in W·h and denotes the value of power, which can be output under specific environmental conditions. Voltage represents the value of output voltage while the device is functioning. Theoretically, energy should equal to the voltage multiplied by capacity. The information about other parameters, which can be used to assess the quality of a battery is listed in Table 19.2.

Among storage batteries, LIB, which is widely applied in electric devices, needs to be highlighted. It has a long lifetime, thin dimensions, high capacity, and can output stable power. The LIB should be distinguished from lithium battery, whose electrodes are lithium and are highly active. It is risky while functioning and in storage, therefore it is not widely applied in our daily life. The electrode material for LIB is a compound of lithium, which is relatively stable and safe to be used. The LIB reaction is reversible and achieved by the repeatable intercalate/deintercalate process of lithium-ion. While discharging, the lithium-ion will be oxidized at the anode and will be reduced while charging. The co-inventors of LIB were awarded the Nobel Prize in Chemistry in 2019. Nowadays, different kinds of cells/batteries can be found everywhere.

TABLE 19.2
Parameters of Battery

Parameter	Unit	Meaning
Rated capacity	A·h	Value of energy battery can store under test condition
Rated energy	W·h	Value of energy battery can output under test condition, equals to the rated voltage × rated capacity
Rated voltage	V	Magnitude of voltage while battery is functioning
Energy density	W·h/L	Value of energy per unit volume contained
Power density	W/L	Value of maximum power per unit volume contained
Specific power	W/kg	Value of maximum power per unit mass contained
Specific capacity	A·h/g (mAh/g)	Value of energy unit electrode material contains
Specific energy	W·h/g (mWh/g)	Value of energy unit electrode material can output, equals to the output voltage × specific capacity
C-rate/E-rate	%	Discharging rate of current/power against the capacity of battery
Lifetime	-	Quantity of cycle a battery can reach before its capacity is reduced to a given value
Internal resistance	Ω	Value of resistance of the battery
Self-discharge	%	Loss of capacity during long-term storage
State of charge (SOC)	%	Current battery capacity as a percentage of maximum capacity
Depth of discharge (DOD)	%	Ratio of output capacity to rated capacity while working
State of charge (SOD)	%	Ratio of remaining energy to rated capacity while functioning

The convenience and flexibility they provide have significant benefits to people's life and the development of science. Currently, limited by its stability and output efficiency, the commercialization of H-O fuel cells is not yet 100% mature. LIB has a very high market share. Battery factories are investigating heavily in this field and expecting the next generation of batteries to appear soon. The effort scientists are making to optimize these two types of energy suppliers through graphene, which will be introduced in the next section.

19.3.3 The Application of Graphene/Graphene Family Materials on Batteries

19.3.3.1 Graphene/Graphene Family Materials in Lithium-ion Batteries

19.3.3.1.1 Working Principle of Lithium-ion Batteries

As briefly introduced in the previous sections, LIB is the most commonly used battery to power mobile devices and small electric devices in daily applications. Early forms of LIB' cathode materials include Li-TiS_2, Li-MoS_2, and Li_xMnO_2. Currently, the most popular LIB electrode materials are $LiCoO_2$-graphite. The working schematics of LIB are shown in Figure 19.2. The left-to-right process is the discharging process while the right-to-left direction is the charging process. In the discharging process, Li-ions leave the anode and move to the cathode where they are reduced. This translates generally to an external electron flow from the graphite anode to the $LiCoO_2$ cathode whereas the current flows from the cathode to the anode. While

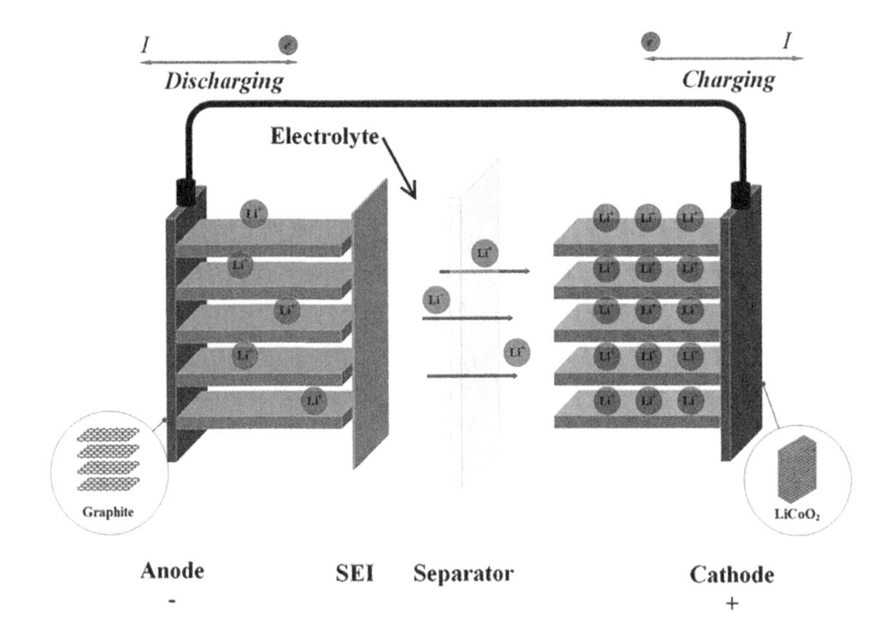

FIGURE 19.2 The charging/discharging process of Li-ion battery.

discharging, the reduction potential for the cathode is 1 V. The cathode half-reaction can be written as:

$$CoO_2 + Li^+ + e^- = LiCoO_2$$

As for the anode, the half-reaction is:

$$LiC_6 = Li^+ + C_6 + e^-$$

The reduction potential is around –3V. Therefore, the theoretical cell voltage for this type of LIB is around 4 V, which is suitable to be used for charging. More statistics about the reduction potential can be found in references [4–6]. The full reaction at the electrodes can be written as:

$$CoO_2 + LiC_6 \leftrightharpoons LiCoO_2 + C_6$$

Apart from the theoretical output voltage, the theoretical value of battery capacity is an important parameter as well. It can be calculated using:

$$Q_{\text{Theoretical}} = n \cdot \frac{F}{M}$$

In which, n is the quantity of electron transferred within the reaction. F is the Faraday constant, represents the value of charges per mole of electron carries. This value equals to the product of the Avogadro constant N_A (6.02×10^{23}/mol) and the elementary charge (1.6×10^{-19}C), which is 96,500 C mol^{-1}. M is the molecular weight of the active material, typically described under g mol^{-1}. The value of capacity is described in C mol^{-1}. To transfer the value into commonly used mAh g^{-1}, the value needs to be multiplied by 1/3.6 (1,000 mAh/3,600 s).

As an example, the theoretical value for $LiCoO_2$-graphite battery's capacity is given by:

$$Q_{\text{Theoretical}} = n \cdot \frac{F}{M} = 1 \times \frac{96,500}{72} \times \frac{1}{3.6} \approx 372 \text{ mAh g}^{-1}$$

in which, $n = 1$ due to one electron transferred during the reaction, $M = 72$ is the molecular weight of C_6.

The theoretical capacity value can only be obtained when the deintercalate coefficient of Li-ion equals to 1. In real life, the real deintercalate coefficient of the battery needs to be multiplied with the theoretical value to obtain the practical capacity. If other parameters such as the environmental factors and the loss of cathode material are taken into account, the more accurate method to acquire the practical capacity of the battery is to integrate the discharging current curve over time, written as:

$$Q = \int_0^t I(\tau)\, d\tau$$

in which, t: Total discharging time, $I(\tau)$: The value of electric current, τ: Time.

Moreover, as shown in Figure 19.2, a passivation film or solid electrolyte interface (SEI) will be slowly generated at the interface between the liquid electrolyte and solid phase. It covers the anode during the first charging process of the battery. This layer is permselectable for Li-ion and can avoid the unwanted reaction between intermediate products. The SEI film can act as a protective film to extend the lifetime of the anode, whose thickness is typically 100–150 nm. As for the cathode, a passivation layer can be observed at the interface between the electrode and the electrolyte as well, though it is significantly thinner than the SEI at the anode (usually 1–2 nm), and it does not function the same way as the SEI. Besides, a physical separator is required between the anode and the cathode to prevent unwanted shortcuts caused by the contact of the two electrodes. As the separator is immersed into the electrolyte, it should be made from a material that has good corrosion resistance under an organic environment. Thus, the most commonly used materials to make a separator are polymer film/polymer composite films (polypropylene (PP), tetrafluoroethylene (TFE), polyvinyl chloride (PVC), polyethylene (PE), etc.). In recent years, advanced materials including nylon, glass, ceramic, non-woven fibers, and polyester are gradually being applied. This is because uniformly distributed microporosity is also required on the separator to ensure low internal resistance and good permeability of electrons and Li-ions. Moreover, the separator is supposed to have reliable mechanical properties, thermal stability including low heat shrink rate, good wettability by the electrolyte, and smooth and flat surface.

Furthermore, for the electrode reaction under real circumstances, due to the uncontrollable factors such as reaction level, environmental condition, overcharging, over-discharging, defect on the cell, unexpected impurity, the output voltage/current, and lifetime will be influenced. For instance, if over-charged/discharged, the reaction products on the electrodes will be peroxided and irreversible, which will influence the lifetime of the electrodes. Over-charging/discharging will also potentially cause damage to the SEI. The electrolyte will then directly contact with electrodes and result in the swelling of the battery. If Li-ion functions under low temperature continuously, the viscosity of the electrolyte will be significantly increased, which will directly influence the activity of the Li-ion at the cathode and its ability to pass through the SEI. This will result generally in a short working time of LIB in cold weather/climate (winter). Nowadays, commercial LIB usually comes with a protective circuit to prevent negative influences from unexpected factors.

19.3.3.1.2 Approaches to Improve the Performance of Lithium-ion Batteries

Capacity is one of the most important parameters to describe the performance of a battery. However, for commercial batteries in real applications, especially for mobile phones, considering the limitation of the dimensions, the output stability, rechargeable stability, and energy density are the priorities. After all, for public customers, a cheap and reliable battery is always a better option than a heavy and giant single-use battery. To optimize the performance of LIB, the commonly used methods are listed below. The methods can be easily operated though they are complicated to apply.

1. **Structure:** Optimize the dimension of the battery to increase the ratio of the active material. It can be achieved by modifying the outer dimensions, or by optimizing the barrier and interface design within the battery to save internal space. Besides, ion transferability can also be improved by drilling micro-holes on the barrier and interface.
2. **Electrolyte:** Theoretically, the electrolyte is inconsumable during the battery reaction. It only acts as an agent to assist the transportation of Li^+. Therefore, by improving the conductivity of the electrolyte, the quantity of electrolyte can be reduced to save more space for the active materials. It can be achieved by changing the electrolyte materials, or by adding conductive materials to improve the existing electrolyte's electrical conductivity. Moreover, the performance of batteries can also be improved by the optimization of SEI, which can directly be achieved by adding conductive materials within the electrolyte. This approach will increase the electron transfer rate and result in the improvement of the battery properties.
3. **Electrode:** The chemical reactions in a battery are determined by the materials of the electrode. Therefore, using advanced electrode materials can optimize the reaction mechanism to ensure more electron transportation and a stable reaction process. So far, for LIB, it is unreal to find a new method and entirely replace the existing mechanism. It is preferred to optimize the electrodes by doping them with active addictive materials. For the anode, the doping material should ensure that the electrode has a greater ability to accumulate Li^+ (it can also be achieved by increasing the interlayer spacing of graphite). As for the cathode, the level of reaction can be improved by doping the electrode to increase the level of reactivity, which, in turn, will lead to higher output of electricity. Apart from doping, the same effect can also be achieved by surface coating or surface cladding of the electrodes using carbon or metallic materials. The coating should simplify the transportation process of Li^+ and increase the quantity of exchanged ions in unit time. The properties of the electrodes can also be optimized by physical treatment. By refining the particle of the electrodes or by nanocrystallization, the path for the transportation of Li^+ ions will be optimized via the reduction of the diffusion distance. Increasing the micro-porosity of the electrodes will also enhance the performance of the batteries. However, instead of using advanced electrode materials with higher porosity, the common electrode materials can be used by subjecting them to surface treatment including fluorination and oxidization.

19.3.3.1.3 Approaches to Improve the Performance of Lithium-ion Batteries using Graphene Family Materials

Based on the methods introduced in the previous section, the following approach can be applied to improve the key properties of LIB by using graphene.

Using Graphene on Anode: In a LIB, graphite electrodes typically act as the anode, which collects the Li^+ ions from the cathode while it is charging. Compared with a graphite electrode, the use of graphene as anode will change the reaction products at the anode, which will result in a different value of the output voltage and potentially higher values of theoretical capacity. This is due to the ability of a multi-layer graphene sheet to accommodate more Li^+. It is pointed out by Kaskhedikar and Maier in 2008, that the interlayer spacing of graphite anode is 3.35 Å while it can be observed for a multi-layer graphene anode that the layer-by-layer spacing is close to 4 Å, which provides more space for the Li^+ to store. Besides, compared with graphite, Li^+ can be adsorbed on both sides of graphene, which further increases the Li^+ content. Ideally, a product such as LiC_3 will be generated instead of LiC_6, which will double the capacity to around 740 mAh g^{-1} [7]. Moreover, in real life, graphene anode is mainly fabricated by rGO, whose microstructure typically contains lots of defects and impurities. The defects within the microstructure will result in a reduction of electrical conductivity. Nevertheless, it will benefit the accommodation of Li^+ to some extent and further improve the capacity.

According to statistics compiled by Al Hassan et al. (2019), with defects within the microstructure, the capacity of the battery can be increased from 800 to 2,000 mAh g^{-1}. On the other hand, it should be noted that the defect within the anode is uncontrollable. Furthermore, although the capacity of the first charging/discharging cycle is being significantly increased, the capacity drops clearly after 50–100 cycles. Besides, compared with a graphite anode, the cost of graphene (rGO) anode is significantly higher. Therefore, graphene anode is not yet being widely used in commercial battery products [8]. Apart from directly using graphene as an anode, using graphene composites/alloys/nanohybrids as the anode material is another method to take both the cost and the performance of the battery into consideration. As a form of carbon, graphene can be used to easily form a good quality compound. With the addition of other elements, the cost can be decreased. Due to the interface properties, the capacity, cyclic stability, and capacity retention can be improved. Successfully applied composites materials include Si, SiC, SiO, Mo, MoS_2, Ge, and Sn; metal oxides such as Co_3O_4, Fe_3O_4, Fe_2O_3, SnO_2, TiO_2, SnO_2, CuO, and MoO_3; alloys like silicon/graphene, tin/graphene, germanium/graphene; nanohybrids such as Mn_3O_4, SnO_2, SnO_2 and phosphorus, Sb_2O_3, Ag. However, despite the improvements in the performance of batteries, there is a long waiting period before graphene-based anode can go into commercial application.

Use of Graphene on Electrolyte: The traditional electrolyte for commercial LIB is typically liquid phase (some shown as crystals under room temperature) solvent plus lithium salt ($LiPF_6$ is the most common option). Apart from electrodes, the electrolyte will influence the internal resistance and the capacity of batteries. Over the past two decades, with the development of electrolyte materials, the capacity of LIB has improved significantly. The graphene (rGO or GO) additives in the electrolyte can increase the conductivity of the electrolyte. It can also benefit the wettability of the electrolyte on the electrodes. The problem which limits the large-scale application of graphene electrolyte is the same as graphene anode: price and stability.

Still, considering the commercial profit and scientific progress, industries and researchers invest more time and resources in the research of solid-phase electrolytes, which is expected to be the breakthrough of the next-generation LIB.

19.3.3.2 Graphene/Graphene Family Materials on Fuel Cells

19.3.3.2.1 Working Principle of Fuel Cells

Apart from rechargeable LIB, a fuel cell is another type of electrochemical power supplier, which is also a major focus of research investigations. Compared with a rechargeable LIB, fuel cells are typical of large volume and non-rechargeable. Unlike rechargeable batteries, the materials used for the electrode reaction are required to be fed continuously into the system. Therefore, the working life of a fuel cell is significantly longer than rechargeable batteries. The working principle for a fuel cell is shown in Figure 19.3. The schematic shows a proton exchange membrane (PEM) fuel cell, whose most suitable working temperature is around 80 (determined by different materials of the membrane). The oxidation reaction of the fuel (typically hydrogen) will take place around the anode, where hydrogen loses ions and is oxidized to H^+. The generated H^+ will transfer to the cathode through the polymer electrolyte membrane and assist the reduction of the oxidizing agent (oxygen is the most commonly used material). Unused fuel and oxygen as well as the reaction product (H_2O) will then leave the cell and can be recycled afterward.

To ensure a safe and compact structure, the electrodes are typically designed as bipolar plates, with flow channels uniformly distributed on them. Figure 19.4 shows various designs of the flow channels including (a) straight parallel, (b) interdigitated, (c) pin-type, (d) spiral, (e) single-channel serpentine, and (f) multiple-channel serpentine flow channel configurations [9]. The flow channels assist the

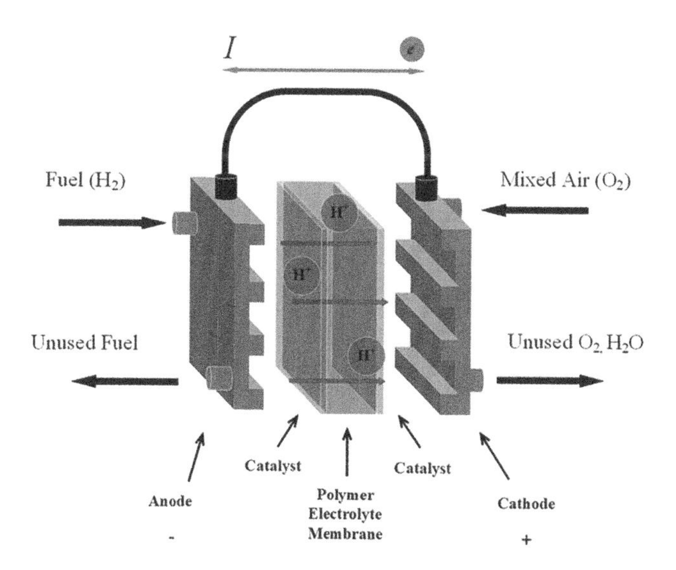

FIGURE 19.3 Schematic of hydrogen–oxygen proton-exchange membrane fuel cell.

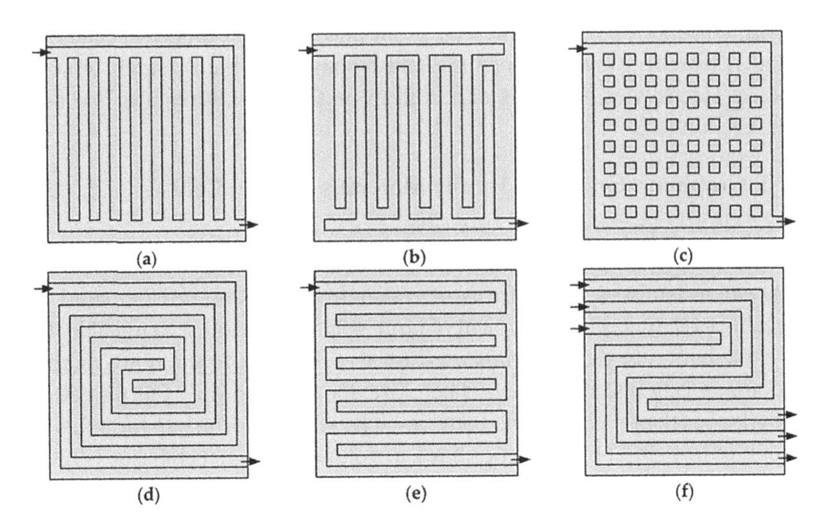

FIGURE 19.4 Scheme of typical flow field configurations (a) straight parallel; (b) interdigitated; (c) pin-type; (d) spiral; (e) single-channel serpentine; (f) multiple-channel (triple) serpentine [9]. Adapted with permission from [9]. Copyright (2019) Copyright The Authors, some rights reserved; exclusive licensee MDPI.

flow of gas and prevent direct contact between oxygen and hydrogen. A thin layer of catalyst is also required, which is typically made of platinum. The material used for the polymer electrolyte membrane is required to only allow hydrogen ions to pass through from the anode to the cathode while preventing the oxygen from passing through. It also should have reasonable strength and temperature resistance to fit multiple working conditions. For the PEM fuel cell, the half-reaction on the anode is written as:

$$H_2 = 2H^+ + 2e^-$$

As for the cathode reaction, it is written as:

$$2H^+ + \frac{1}{2}O_2 + 2e^- = H_2O$$

The overall electrode reaction is written as:

$$H_2 + \frac{1}{2}O_2 \leftrightharpoons H_2O$$

Apart from H-O PEM fuel cells, there are many other fuel cells with different fuel and electrode reaction mechanisms. Commonly applied fuel cells include solid oxide fuel cells, alkaline fuel cells, molten carbonate fuel cells, and direct methanol fuel cells. Different types of fuel cells have different efficiencies, output power, and working temperatures while the reaction product from the electrode reaction is clean material like water, which makes them environment-friendly. Theoretically, as long as the fuel and oxygen are fed ceaselessly, the cell can function continuously. Therefore, in applications, fuel cells are usually designed to be large to store more fuel. Another advantage of a fuel cell is the fuel unit that can be replaced or refilled simply and swiftly. Unlike compact LIBs, fuel cells are ideal for power systems of automobiles, ships, aircraft, and submarines. However, to achieve maximum usage of fuel cells, there is the need to establish a fuel (H_2) refilling system (similar to the current gas station network). Some countries (such as Japan) have decided to develop hydrogen supply systems and networks to replace the existing fossil fuel supply system and network, and they expect to achieve an eco-friendly society in the next few decades.

19.3.3.2.2 Approaches to Improve the Performance of Fuel Cells

Unlike LIB whose key parameters are the capacity, output voltage, and lifetime, with the large dimension and simplicity of refilling the fuel, the operation time and capacity are not that significant for a fuel cell. Instead, the low-temperature startup/functionality, efficiency as well as output rate to reach the maximum capacity are its key parameters. The performance of a fuel cell can be optimized using the following approaches:

1. **Fuel**: As the most essential constituent of a fuel cell, a fuel that contains a large amount of chemical energy needs to be used. However, the development time, which is typically 20,000h+ per prototype, and the development cost that can be £10M+ (for fuel only) make the development of the fuel for fuel cells a rather expensive undertaking. Therefore, the researches on the type of fuel to use in fuel cells researches are focused on hydrogen. Developing a new type of fuel is not realistic currently.
2. **Structure**: Optimizing the structure of a fuel cell to optimize the gas flow is the most effective method to enhance a fuel cell's performance. The internal gas flow can be guided by multiple channel configurations on the bipolar plates, as shown in Figure 19.4, to increase the contact area of gas as well as controlling the internal heat radiation to maintain a suitable working temperature. Besides, the pressure, gas composition, reactant utilization, internal resistance, and the current density within the cell will also be influenced by the structure design, which will further result in the enhancement of the fuel cell's performance.
3. **Catalyst**: A catalyst layer is typically located between the electrodes and the PEM, to boost the velocity of reaction and optimize the reaction utility of raw materials. The main catalyst material is required to have a large active surface area and activity. Platinum and platinum alloy are the most commonly used catalyst materials. The catalytic properties of the catalyst layer degrade over time due to corrosion/oxidation of platinum or aggravation of carbon impurity. This results in the reduction of catalyst activity and slows down the reaction velocity, which influences the output of a fuel cell. Therefore, the optimization of the material of the catalyst layer is an approach to enhance a fuel cell's properties.
4. **PEM**: The PEM between the electrodes is one of the core components of a fuel cell. It assists the transfer of H^+ (protons) from the anode to the cathode

and resists the direct passage of electrons across it. The PEM acts similarly to the electrolyte in LIBs. Besides, the PEM should have low gas permeation rate to prevent direct contact between oxygen and hydrogen to prevent the explosion. Good mechanical properties and good thermal/chemical stabilities are also required to ensure its integrity while working. The chemical or mechanical corrosion of PEM will increase its micro or macroporosity, which will result in low efficiency and increase gas permeation rate, which increases the potential explosion risk of the cell. Thus, selecting a suitable membrane material based on its working condition is a significant factor related to a fuel cell's performance and safety. For instance, a solid oxide fuel cell uses a ceramic material to allow it to operate at high temperatures, while for normal temperature conditions, a long-chain polymer is used to make the cell compact.

19.3.3.2.3 Approaches to Improve the Performance of Fuel Cells through Graphene Family Materials

Based on the principles outlined, the following methods are the most commonly applied approaches to improve the performances of fuel cells using graphene.

Use of Graphene in Catalyst/Electrodes: The output requirement for a fuel cell needs it to have a high velocity of the reaction. It is mainly determined by the reaction velocity on the cathode. Thus, it needs a catalyst layer to ensure the reaction speed. Rare metals have been proved to have a good catalytic effect on fuel cell reactions. In this regard, platinum and its alloys are the most common catalyst materials. Due to the high price of both the platinum and its alloys, an approach to produce a not-to-highly price catalytic layer is to dope the microstructure of a reduced quantity of titanium and its alloys with a small quantity of graphene. This will enhance the velocity of oxygen reduction reaction induced by the doped catalyst (by improving the carrier transportation process of the H^+ across the PEM and, thereby, boost the oxidation–reduction reaction speed) and reduce its cost at the same time. Furthermore, graphene doping in the catalyst layer on the anode will also optimize the oxygen reduction reaction process, though not as effective as the improved performance achievable by the graphene doping in the catalyst layer on the cathode [10].

Use of Graphene on Bipolar Plate: The bipolar plate in a fuel cell has multifunction. Apart from controlling the flow of the fuel gas and air, it also functions as a medium to conduct electric currents. The most commonly applied bipolar plate materials are copper and nickel, whose properties have a high probability of degrading during while the operating process of the cell unit. Although it does not significantly influence the general performance of the cell, stability and durability still need to be ensured while the fuel cell is functioning. Adding some quantity of graphene to the bipolar plate is a practical method to reduce the electrical resistance of the bipolar plate to a relatively low value, and to slow down the corrosion of the bipolar plate during the oxidation process. Moreover, using graphene-based polymer composites as the plate material is a good method to optimize the performance of the cell unit, although the cost of the plate might be higher.

Use of Graphene on PEM/Electrolyte: As introduced in the previous section, the PEM needs to have good mechanical properties and selective permitted ability. The most commonly used PEM material includes poly(vinyl alcohol)-based composites, Nafion copolymer, and sulfonated-polyetheretherketones (SPEEK) [11]. The properties of these plastic-based materials can be modified easily by graphene micro-filler through the wet mixing method. The graphene-modified composites have better ionic conductivity and enhanced mechanical properties, which results in a higher power density and longer lifetime of the cell. Besides, graphene can also enhance PEM with optimized thermal stability and water absorption capacity, allowing the cell unit to function under various environmental conditions. Alternatively, instead of using graphene as micro-fillers to optimize the solid-state electrolytes, it can be directly used as a bulk or solid piece electrolyte. Nevertheless, the bulk/solid piece electrolyte technologies are not as mature as modifying PEM using micro-fillers.

19.3.4 Discussion

In this section, the application of graphene in the battery and fuel cell fields has been introduced. Although graphene has good electrical conductivity and a large surface area, which is ideal for battery and fuel cell performance enhancement, due to limitations of current technology and price, graphene has not been widely applied in commercial battery or fuel cell products yet. Offsetting aside the issue of current technology, the future applications of graphene in batteries or fuel cells depend on the categories of these electrochemical units. Table 19.3 shows the categorization of electrochemical cells based on their application scenario, accessibility, and price.

TABLE 19.3
Categories of Electrochemical Power Suppliers

Category	Properties	Price
Commercial	Usually cheap and disposable (some are rechargeable), low power density. Can be directly used in daily life on low-power electric devices such as controller, mouse, and toy.	Less than £5
Moderate	The majority can be represented by a lithium-ion battery, widely applied on mobile devices, tablets, and wearable devices. Rechargeable, able to constantly function for more than 10 hours. Usually sealed and requires assembly by an experienced technician.	£10–100, depending on properties
Advanced	Usually has special requirements such as high working temperature, high output power (e.g., fuel cell), and long working time (more than 24 hours). It is usually large and is unlikely to be sold separately to private customers as it is usually in-built as part of a system such as a vehicle.	More than £1,000

For the commercial category, the cost of graphene enhancement of their properties far outweighs the performance enhancement. From the mechanism or technology point of view, graphene should be able to improve its performance. However, from cost consideration, with one gram of graphene worth more than 50 pieces of batteries, the cost–benefit ratio is very small. Besides, such batteries can only be used once, which is a waste of precious graphene material. Therefore, graphene is not suitable to be used to enhance their performance. As for advanced high power density electrochemical units like H-O fuel cells, because of their large dimension, large quantities of graphene will be required to improve their performance. This will further increase the cost of the cell unit, not to mention their original high price. Apart from this, for a fuel cell, the simpler and more cost-effective approach is to optimize its structure.

Graphene is more likely to be applied in the optimization of moderate category batteries like LIB as long as the cost of graphene can be further reduced and its enhancement effect can be further improved. Technically, increasing the interlayer distances in a graphite electrode can improve the performance of a LIB, which can be achieved by using advanced manufacture technologies to produce graphene electrodes in which the layer-by-layer distances are increased. Commercially, there are three main reasons for graphene to be applied in a moderate category battery like LIB. First, the quantity of graphene used to optimize a LIB is relatively small, which makes the increment of cost also be relatively small. Second, the recycling industry of LIB is already mature. With the development of technology, the expensive graphene content is possible to be re-used. Third, as the LIB is widely applied in mobile phones, product customers are willing to pay more money, to experience the latest technology, which makes it have bright market potential. Researchers and factories are currently trying hard to bring the graphene-enhanced LIB into daily use.

19.4 SUMMARY AND PROSPECTIVE

In this chapter, the information on graphene family material, basic knowledge of battery and fuel cell, and the applications of graphene in the battery and fuel cell field have been briefly introduced. Nowadays, with investment from national governments around the world, more and more graphene industrial parks are being built globally. Researchers are trying their best to overcome the technical difficulties and to seek opportunities for the application of graphene. As a star material over the past decade, graphene has already proven its bright application potential in various fields including material, energy, and medical fields. However, due to its high price, mature graphene-based battery products can be hardly found in the market, even if the market is looking for a next-generation mobile device battery. The authors predict that with the development of technology to solve the technical challenge of mass production of high-quality graphene, graphene-based LIBs can be expected to become ubiquitous within the next decade.

REFERENCES

1. C. Shen and S. O. Oyadiji, "The processing and analysis of graphene and the strength enhancement effect of graphene-based filler materials: A review," *Mater. Today Phys.*, vol. 15, p. 100257, 2020. doi: 10.1016/j.mtphys.2020.100257.
2. C. Lee, X. Wei, J. W. Kysar, and J. Hone, "Measurement of the elastic properties and intrinsic strength of monolayer graphene," *Science*, vol. 321, no. 5887, p. 385, 2008. doi: 10.1126/science.1157996.
3. K. Saikia, B. K. Kakati, B. Boro, and A. Verma, "Current advances and applications of fuel cell technologies BT - recent advancements in biofuels and bioenergy utilization," P. K. Sarangi, S. Nanda, and P. Mohanty, Eds. *Recent Advancements in Biofuels and Bioenergy Utilization.* Singapore: Springer, 2018, pp. 303–337.
4. A. J. Bard, R. Parsons, and J. Jordan, *Standard Potentials in Aqueous Solution.* Marcel Dekker, Inc., New York, 1985.
5. G. Milazzo, S. Caroli, and R. D. Braun, "Tables of standard electrode potentials," *J. Electrochem. Soc.*, vol. 125, no. 6, pp. 261C–261C, 1978. doi: 10.1149/1.2131790.
6. E. H. Swift and E. A. Butler, *Quantitative Measurements and Chemical Equilibria.* San Francisco, CA: W. H. Freeman, 1972.
7. N. A. Kaskhedikar and J. Maier, "Lithium storage in carbon nanostructures," *Adv. Mater.*, vol. 21, no. 25–26, pp. 2664–2680, 2009. doi: 10.1002/adma.200901079.
8. M. R. Al Hassan, A. Sen, T. Zaman, and M. S. Mostari, "Emergence of graphene as a promising anode material for rechargeable batteries: S review," *Mater. Today Chem.*, vol. 11, pp. 225–243, 2019. doi: 10.1016/j.mtchem.2018.11.006.
9. D. M. Neto, M. C. Oliveira, J. L. Alves, and L. F. Menezes, "Numerical study on the formability of metallic bipolar plates for proton exchange membrane (PEM) fuel cells," *Metals,* vol. 9, no. 7. 2019. doi: 10.3390/met9070810.
10. H. Su and Y. H. Hu, "Recent advances in graphene-based materials for fuel cell applications," *Energy Sci. Eng.*, vol. 9, no. 7, pp. 958–983, Jul. 2021. doi: 10.1002/ese3.833.
11. Y. Wang, D. F. Ruiz Diaz, K. S. Chen, Z. Wang, and X. C. Adroher, "Materials, technological status, and fundamentals of PEM fuel cells – A review," *Mater. Today*, vol. 32, pp. 178–203, 2020. doi: 10.1016/j.mattod.2019.06.005.

20 2D Transition Metal Dichalcogenides (TMD)-Based Nanomaterials for Lithium/Sodium-ion Batteries

Tian Wang, Ashok Kumar Kakarla, and Jae Su Yu
Kyung Hee University

CONTENTS

20.1 INTRODUCTION

Fast-growing consumption of energy has drawn great attention to exploring sustainable, clean, and low-cost energy storages devices. The traditional fossil fuels-based energy supply systems are expected to be replaced by renewable energy, such as solar energy, wind energy, and so on [1,2]. However, the intermittent issue of renewable energy hinders its practical application. As an alternative solution, electrochemical energy storage (EES) devices, such as rechargeable batteries, become increasingly important for research and development. The application areas range from portable electronics to the electrification of transportation and coupling with renewable energy sources for powering the electrical grid [3,4]. Nowadays, rechargeable battery devices are playing an important role in smart electronics, portable electronic devices, and electric vehicles, by which our lifestyle is transformed [5].

DOI: 10.1201/9781003178453-20

Meanwhile, the substantial progress in battery technology will also build a stable foundation for the successful realization of a carbon-neutral society in the energy transition [6].

Lithium-ion batteries (LIBs) have been more interested in energy storage systems as the power source for portable electronic devices since their first commercialization by Sony in 1991. LIBs been also considered as the most promising energy storage system for large-scale applications. The energy density of LIBs is also steadily increasing at a rate of 7–8 Wh kg^{-1} $year^{-1}$. Up to now, the state-of-the-art LIBs have reached an energy density of 250 Wh kg^{-1} at the battery level (for 18,650 type cells). Meanwhile, the energy densities of 235 Wh kg^{-1} and 500 Wh L^{-1} at the battery pack level demanded by the market are also increasing [7,8]. Sodium-ion batteries (SIBs) have captured widespread attention because of their abundant resources, low cost, and similar electrochemistry to LIBs [9,10]. Figure 20.1a displays the components of a rechargeable LIB/SIB. During the discharge process, Li/Na ions de-intercalate from the anode after crossing the separator and intercalate into the cathode material. The charge process is reversed. There are many achievements for LIBs that can be easily applied for SIBs, which makes the rapid development of SIBs within only a few years. However, the low energy density and limited cycle life of electrode materials are of great challenge for the commercialization of SIBs [11]. Therefore, researchers are looking for more suitable electrochemically active materials to develop higher energy density and power density batteries.

Since the successful preparation of graphene in 2004, two-dimensional (2D) nanomaterials attracted worldwide attention due to their unique properties. Layered transition metal dichalcogenides (TMDs), as typical graphene-like 2D nanomaterials, have gained great interest in energy storage due to their X-M-X structure and unique electronic and mechanical properties. The unique 2D architecture can enhance the contact area between the electrode and electrolyte, offer more active sites for foreign ions, and increase the ion migration kinetics. Meanwhile, compared with the graphite anode, the higher voltage platform (usually above 1.0 V) of TMDs can effectively avoid the formation of dendrite and ensure battery safety. These advantages suggest that the TMDs are the ideal candidates for next-generation anode materials. In this chapter, we present the recent progress in the synthesis and characteristics of 2D TMD materials as an anode in Li/Na-ion batteries (Figure 20.1b). Meanwhile, the

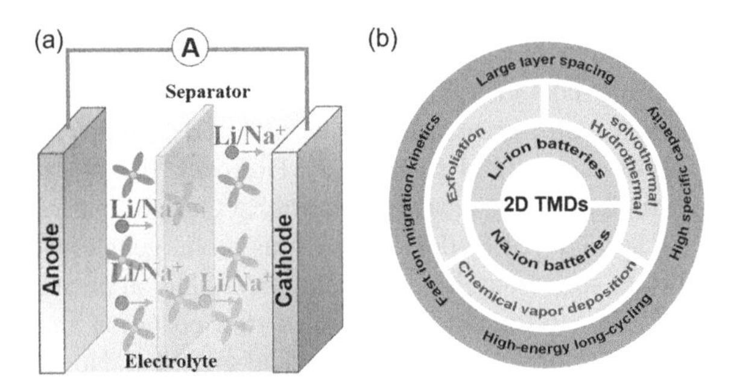

FIGURE 20.1 (a) Components of a rechargeable LIB/SIB. (b) Synthesis and characteristics of 2D TMD materials as an anode in Li/Na-ion batteries.

exciting progress of these 2D TMD materials used as an anode in Li/Na-ion batteries is displayed. Finally, we conclude the improvement and prospects in the Li/Na-ion storage performance of 2D TMD materials.

20.2 STRUCTURE AND PROPERTIES OF 2D TMDs

The layered TMDs usually possess two polymorphs, i.e., 1T and 2H which stand for trigonal and hexagonal phases, respectively. The layered structures are similar to graphite and the thickness of each monolayer is around 0.6–0.7 nm. Meanwhile, they also have a common formula of MX_2 in which M is a transition metal, from groups 4 to 10 of the periodic table, and X is the chalcogens, respectively. The X-M-X sheets are bonded by covalent force, while individual layers are held together by Van der Waals force [12]. A variety of polytypes of the bulk crystals are formed by the variations in stacking order and metal atom coordination. Therefore, a monolayer MX_2 displays different phases (trigonal prismatic or octahedral) according to the coordination environment of M with X. For the H phase, M is in trigonal prismatic coordination with X. For the T phase, M is in octahedral coordination with X. The distorted T phase is named the T′ phase, which has the same coordination environment as the T phase [13].

Different polymorphs endow layered TMDs with different properties, leading to the change in performance when being used in EES devices. For example, the 2H and 1T phase MoS_2 structures are widely used in energy storage and conversion devices due to their characteristic structure and rich physical and chemical properties [14]. However, the structure of the 2H MoS_2 suffers from the large volume change during the charging/discharging process and has a poor electronic conductivity, which hinders the development of 2H MoS_2 in Li/Na-ion batteries. Compared with 2H phase MoS_2, the 1T phase MoS_2 presents a metallic transport behavior, which delivers fast electron conduction. Besides, 1T MoS_2 has a bigger interlayer spacing than 2H MoS_2, thereby enhancing the insertion/extraction of the foreign ion [14,15]. Therefore, 1T phase 2D TMDs have been studied extensively in recent years.

20.3 PREPARATION PROCESSES

20.3.1 Hydrothermal/Solvothermal Method

The hydrothermal/solvothermal method is a low-cost and high-efficiency synthetic method that can synthesize most TMD nanomaterials, which has been intensively investigated. The materials synthesized via hydrothermal/solvothermal method have advantages of high phase purity, uniform size distributions, and controllable morphology. Compared with conventional synthetic methods, hydrothermal synthesis is more excellent in several aspects. For example, the compounds with elements in oxidation states are difficult to attain, especially important for transition metal compounds. However, the hydrothermal method is easy to achieve in closed systems [16]. Over the past decade, many 2D TMD materials with different morphologies have been reported by the hydrothermal/solvothermal method. The typical representatives are MoS_2 and WS_2, and their corresponding carbon nanocomposite materials, etc.

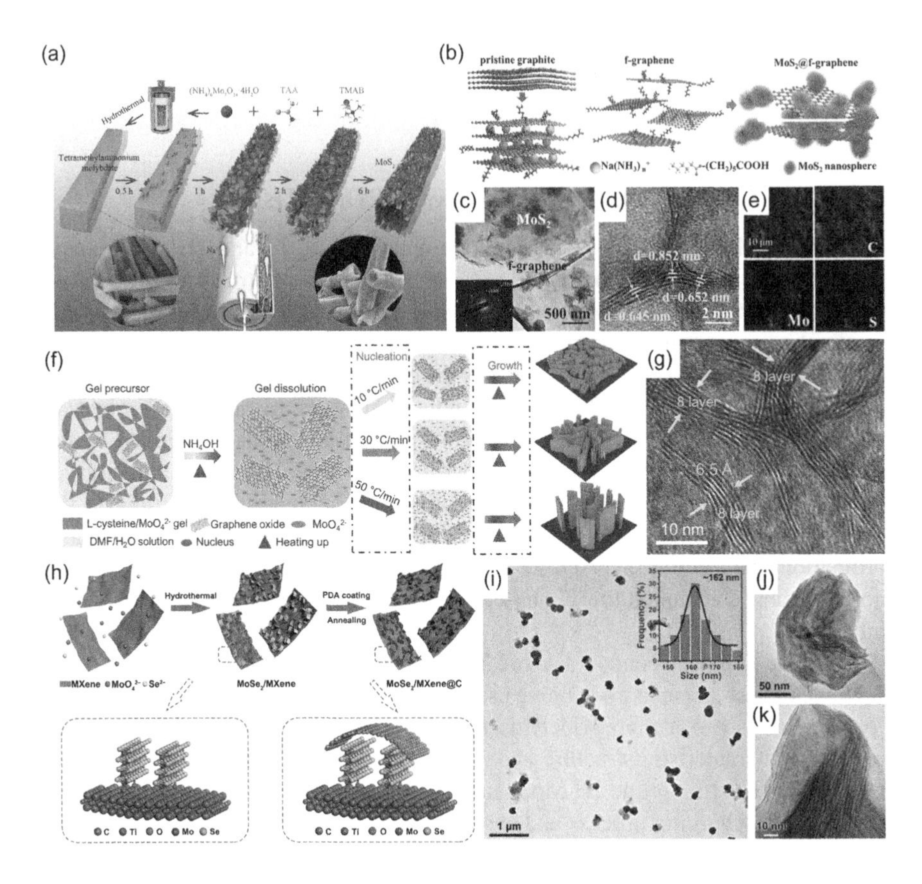

FIGURE 20.2 (a) Schematic illustration of the tubular MoS_2 hierarchical structures. (Adapted with permission from [17]. Copyright (2020) American Chemical Society.) (b) Formation process of MoS_2/graphene heterostructure and (c, d) TEM and HRTEM images and (e) the corresponding elemental mapping images. (Adapted with permission from [18]. Copyright (2017) American Chemical Society.) (f) Schematic illustration and (g) TEM images of the VO-MoS_2/N-RGO. Adapted with permission from [19]. Copyright (2018) Wiley-VCH. (h) Schematic illustration for the preparation of $MoSe_2$/MXene@C. Adapted with permission from [20]. Copyright (2019) American Chemical Society. (i) TEM and (j, k) HRTEM images of the WS_2 nanochains. Adapted with permission from [21]. Copyright (2018) Elsevier.

As shown in Figure 20.2a, the tubular MoS_2 hierarchical structure was prepared by using ($(NH_4)_6Mo_7O_{24}·4H_2O$) and thioacetamide (TAA) as a precursor and tetramethylammonium bromide (TMAB) as an additive. In this process, the addition of TMAB helped to form a layered solid stick-like tetramethylammonium molybdate, which acts as a template and intermediate during the formation of tubular MoS_2 hierarchical structure [17]. Huang et al. reported a facile method to prepare the MoS_2/graphene heterostructure with MoS_2 nanospheres grown on the graphene. At first, the graphite was exfoliated by ions and modified with functional groups to form a functionalized graphene. In the following hydrothermal reaction, the functional groups on the graphene led to the MoS_2 nucleation and growth (Figure 20.2b).

In this architecture, the MoS_2 nanospheres anchored on the functionalized graphene not only decreased the aggregate of MoS_2 but also prevented graphene layers from restacking. The transmission electron microscope (TEM) (Figure 20.2c) and high-resolution TEM (HRTEM) images (Figure 20.2d) displayed the hexagonal structure of the MoS_2/graphene heterostructure, which indicated a large layer spacing (larger than 0.64 nm). The elemental mappings also suggested a uniform distribution of MoS_2 on graphene (Figure 20.2e) [18].

High-efficiency ion-based secondary battery electrode materials need to meet rapid ion mobility during the design process. In this regard, designing electrode materials with a vertical ion diffusion path is an ideal choice. Therefore, Park et al. introduced a vertically oriented MoS_2 with spatially controlled geometry on nitrogenous graphene sheets (VO-MoS_2/N-RGO) by the solvothermal method. In their research, the VO-MoS_2/N-RGO was synthesized by an L-cysteine/MoO_4^{2-} gel precursor in a mixed *N, N*-dimethylformamide and water solvent. In the case of selecting a suitable precursor, by controlling the heating rate of the nucleation process, the vertical arrangement structure was obtained (Figure 20.2f). Figure 20.2g displayed the TEM image of the VO-MoS_2/N-RGO. The uniform layer number distribution and large layer spacing are availed for the insertion/extraction of the foreign ion. As a consequence, with a longer nucleation time, a denser MoS_2 on the basal surface was formed [19].

The 2D TMDs have high surface energy usually, which causes the nanosheets to agglomerate, which results in the rapid capacity fading when used as electrode materials. Zhang et al. presented a rational design and fabrication of hierarchical carbon-coated $MoSe_2$/MXene hybrid nanosheets ($MoSe_2$/MXene@C) as an anode material to prevent the 2D $MoSe_2$ sheets from restacking (Figure 20.2h). In this architecture, the highly conductive MXene substrate could effectively improve electronic conductivity. Meanwhile, the carbon layer further enhances the stability of the composite structure [20]. Du et al. introduced a metallic 1T′ phase dominated WS_2 (1T′-D WS_2) nanostructure by a one-pot colloidal synthesis approach. In this study, oleic acid and oleylamine were used as the reaction solvent and 1T′-D WS_2 nanostructure was successfully prepared at 280°C [21]. Figure 20.2i showed the TEM image of the 1T′-D WS_2 with a dominant size of ~162 nm. Meanwhile, the HRTEM images revealed that the entire WS_2 was composed of closely stacked thin nanosheets (Figure 20.2j and k).

Recently developed ternary sulfides, such as Ni-Co-S, have attracted wide attention in EES due to their high stability, rich active sites, and excellent electronic conductivity. In our previous work, we reported nanosliver decorated $Ni_{0.67}Co_{0.33}S$ forest-like nanostructures on Ni foam (nano-Ag@NCS FNs/Ni foam) via a facile wet-chemical method, followed by a light-induced growth of nano-Ag onto NCS [22]. In desirable growth, the NCS FNs were obtained on Ni foam by rapid nucleation and precipitation of nickel (Ni^{2+}), cobalt (Co^{2+}), and thiosulfate ($S_2O_3^{2-}$) ions (Figure 20.3a and b). Then, the nano-Ag was anchored on NCS FNs/Ni foam by a facile photoreduction method at room temperature (Figure 20.3c). Hetero-network-based MoS_2@Cu_2MoS_4_210 (MS@CMS obtained at 210°C) 3D nanoflowers (NFs) and Co-nanoparticles-containing MS@CMS-210 (Co-MS@CMS-210) 3D NFs were successfully prepared by a hydrothermal synthesis method [23]. As shown in Figure 20.3d, copper oxide (Cu_2O) nanospheres (NSs) were prepared by the co-precipitation

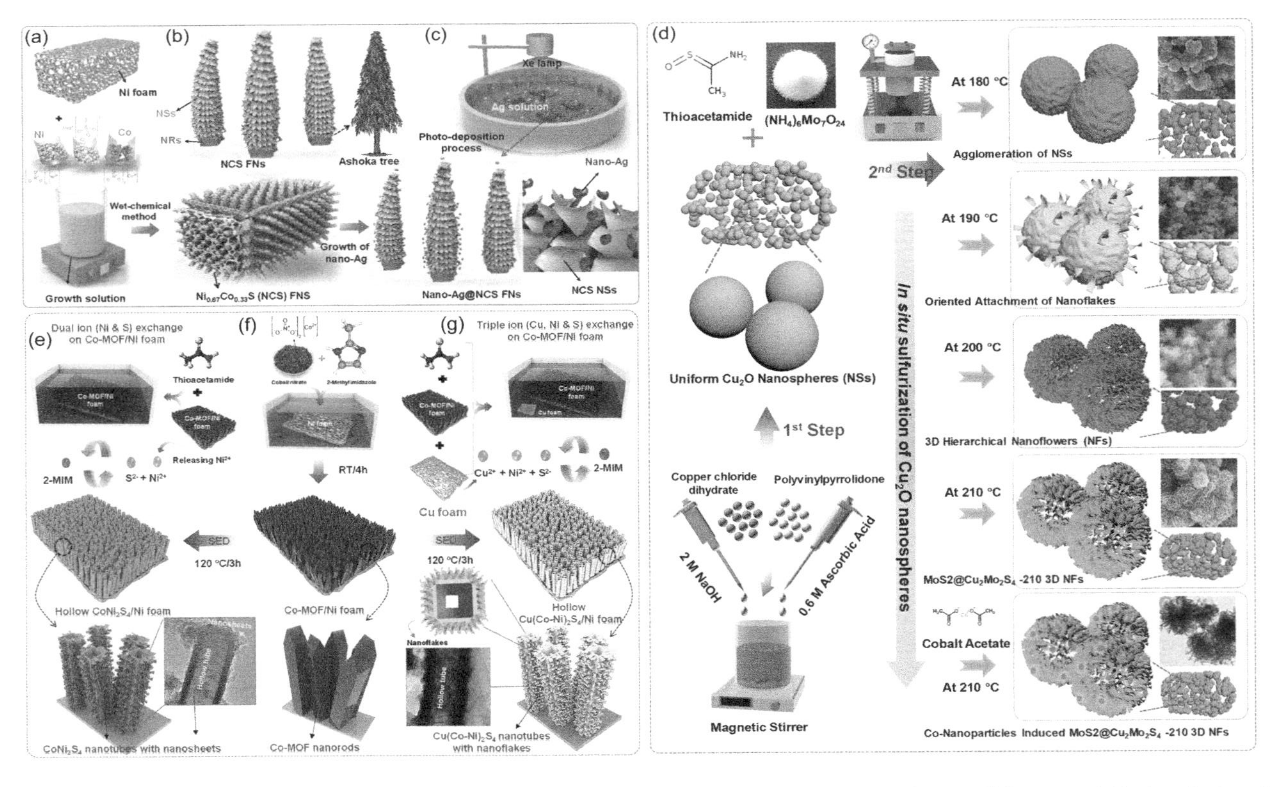

FIGURE 20.3 (a) Preparation of growth environment, (b) schematic illustration for the growth of NCS NFs/Ni foam, and (c) photoinduced growth of Ag nanoparticles on the NCS NFs/Ni foam. Adapted with permission from [22]. Copyright (2019) Wiley-VCH. (d) Schematic illustration for the preparation of the Cu_2O NSs, MS@CMS, and Co-MS@CMS-210 hybrid products at different temperatures. (Adapted with permission from [23]. Copyright (2021) Elsevier.) Schematic illustration for the preparation process of (e) Co-MOF/Ni foam, (f) $CoNi_2S_4$ NTs/Ni foam, and (g) $Cu(Co-Ni)_2S_4$ NTs/Ni foam. Adapted with permission from [24]. Copyright (2021) Elsevier.

method. Then, the formed uniform Cu_2O NSs were employed as a template to produce a series MoS_2@Cu_2MoS_4 under different temperatures. Finally, the Co-MS@CMS-210 was obtained by using the Co-nanoparticles induced MS@CMS-210 3D NF. Metal-organic framework (MOF), as a new class of organic–inorganic hybrid materials, has shown great potential as an attractive template for the field of energy storage devices. Yu et al. introduced Cu(Co-Ni)$_2$S$_4$ nanotubes on Ni foam (Cu(Co-Ni)$_2$S$_4$ NTs/Ni foam) using synchronous etching and multi-ion doping enabled MOF precursor [24]. Initially, the MOF precursor was synthesized on Ni foam with a facile aqueous solution process, which provided Co^{2+} and Ni^{2+} ions during the solvothermal process (Figure 20.3e). Figure 20.3f was the preparation of the $CoNi_2S_4$ NTs/Ni foam by the multi-ion-exchange process. Then, the MOF/Ni foam and Cu foam were added into the sulfur-contained solution to form hollow structured Cu(Co-Ni)$_2$S$_4$ NTs/Ni foam (Figure 20.3g). In this architecture, the in situ doping Cu enhanced the electrochemical conductivity. The hollow structure increased the electroactive surface area. Therefore, this ternary architecture system indicates significant potential for Li/Na-ion battery electrode materials.

20.3.2 Chemical Vapor Deposition Method

Chemical vapor deposition (CVD), which is a bottom-up method for preparing 2D TMD materials can be used to prepare high-quality 2D materials on different types of substrates. Meanwhile, the CVD method can also be used to grow layered TMD nanomaterials on other kinds of electrochemically active materials, resulting in hybrid electrode materials. In this method, the precursors are exposed to the substrate under conditions of high temperature. When the kinetic and thermodynamic conditions required for the chemical reaction are reached, the 2D ultrathin products are deposited on the substrate. As shown in Figure 20.4a, Wang et al. reported a large-area multilayer MoS_2/WS_2 heterostructures on the SiO_2/Si substrate via the CVD method. The reaction process was mainly based on the instability of MoO_3 and WoO_3 at 650°C, which could be easily volatilized into the gaseous state. Then, a thin layer of MoS_2/WS_2 film was deposited on the substrate after the reaction of gaseous MoO_3 and WoO_3 with sulfur vapor [25]. Wang et al. discovered that strong light absorption and fast intralayer mobility could not be well developed in the usual reported monolayer/few-layer MoS_2 structures. Therefore, large-area, high-quality, and vertically oriented few-layer MoS_2 (V-MoS_2) nanosheets were prepared by CVD method [26]. Figure 20.4b displayed the morphology of the prepared MoS_2 nanosheets. The cross-sectional image of the V-MoS_2 (Figure 20.4c) indicated that the height was about 2 μm with a bright and nearly transparent vertically oriented structure [26]. The V-MoS_2/Si heterojunction displayed an excellent lateral photovoltaic performance in the wide range of visible-near-infrared light and prominent linearity without applying an external bias voltage.

Recently, Ly et al. reported the CVD growth of 2D MoS_2 into a deep kinetic regime via KCl as a catalyst and the plasma pretreatment on growth substrates (Figure 20.4d). By this method, they achieved an unprecedented nonequilibrium high-index faceting and unusual high-symmetry shapes in 2D materials [27]. During the preparation of the 2D TMD materials via the CVD method, the substrates played a significant

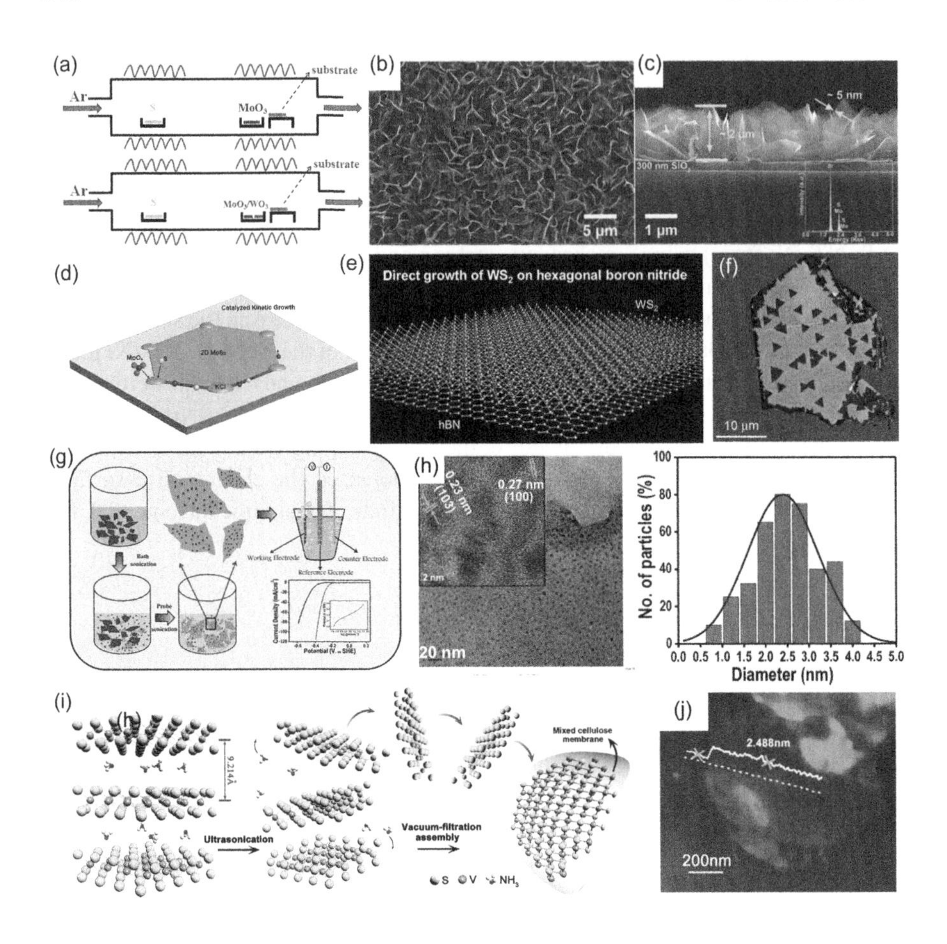

FIGURE 20.4 (a) Schematic diagram of the CVD grown MoS_2 and MoS_2/WS_2 films. (Adapted with permission from [25]. Copyright (2018) Elsevier.) (b) Top-view and (c) cross-sectional SEM images of V-MoS_2. Adapted with permission from [26]. Copyright (2018) Wiley-VCH. (d) Schematic diagram of the catalyzed kinetic growth. Adapted with permission from [27]. Copyright (2020) American Chemical Society. (e) Schematic diagram and (f) SEM image of triangular-shaped WS_2 crystals grown onto a h-BN flake. Adapted with permission from [28]. Copyright (2014) American Chemical Society. (g) Synthesis progress and (h) of the MoS_2 quantum dots interspersed in MoS_2 nanosheets. Adapted with permission from [29]. Copyright (2014) American Chemical Society. (i) VS_2 NH_3 precursor with NH_3 molecules intercalated into the S-V-S layers and (j) AFM image of the exfoliated VS_2 nanosheets. Adapted with permission from [30]. Copyright (2011) American Chemical Society.

role in limiting the physical properties of atomic-layer materials. Therefore, a suitable substrate directly determines the morphology and properties of the 2D TMDs. Kitaura et al. reported direct CVD growth of WS_2 onto high-quality hBN by a 3-furnace CVD setup [28]. Figure 20.4e showed the atomic structure diagram of the WS_2/

hBN heterojunction. As shown in Figure 20.4f, there were some dark triangles with about 3 μm on hBN, which indicates the formation of WS_2 crystals with a triangular shape.

20.3.3 Exfoliation Method

Exfoliation method is a top-down method to prepare TMD nanosheets. TMD nanosheets are easily obtained by exfoliation when the Van der Waals force between layers weakens. Among all the exfoliation methods, solvent-assisted and ion (Li^+) intercalation-assisted exfoliation are two common methods. Compared with the ion intercalation-assisted method, the solvent-assisted exfoliation method is simple to implement. Figure 20.4g displayed the synthesis process of MoS_2 quantum dots interspersed in few-layered MoS_2 sheets. The MoS_2 quantum dots were obtained by a liquid exfoliation technique in organic solvents. At first, the MoS_2 powder was added in 1-methyl-2-pyrrolidone and sonicated in an ultrasonic bath continuously for 3.5 h. Then, the dispersion was sonicated with a sonic tip for another 3.5 h. After standing 12 hours, the liquid exfoliation of MoS_2 was obtained by centrifugal treatment [29]. Figure 20.4h showed the TEM image of MoS_2 quantum dots (~2 nm) interspersed in MoS_2 nanosheets.

In addition to the mechanical stirring technology to prepare single/few-layer 2D materials, ion intercalation technology is also used to exfoliate layered 2D materials. However, the traditional ion exfoliation technology suffered from the residual of the nonvolatile ions, which hindered the performance of the prepared materials. NH_3 is a micromolecule with very positive physical activity and chemical reactivity. Meanwhile, the excellent volatilization property makes no residue after the evaporation. Zhang et al. introduced a unique NH_3-assisted strategy to exfoliate VS_2 flakes into ultrathin VS_2 nanosheets stacked with less than five S-V-S single layers (Figure 20.4i) [30]. Atomic force microscope (AFM) image showed that the *c* parameter of VS_2 was 5.73 Å and the thickness was 2.488 nm (Figure 20.4j), which denoted that the product consisted of 4–5 single layers of S-V-S.

20.4 APPLICATIONS

20.4.1 Application in Lithium-ion Batteries

LIBs have been applied as a major power source for portable electronic devices in the past decades. Graphite is employed as a predominant anode material for commercial LIBs. However, the low theoretical capacity of 372 mAh g^{-1} hinders its large-scale application. It is a considerable research direction to develop alternative anode materials with improved Li-ion storage performance [31,32].

In recent years, the 2D TMDs have attracted considerable attention in LIBs due to their unique properties [33]. Onion-like crystalline WS_2 nanoparticles anchored on graphene sheets as high-performance anode materials for LIBs were reported by Inha Kim et al. [34]. The onion-like crystalline WS_2 nanoparticles evenly coated on graphene sheets (WS_2@Gs) were synthesized via ball milling technique applying WO_3 nanoparticles (~15 nm) and following sulfidation. The ball milling method enabled uniform covering of WO_3 nanoparticles on graphene sheets without aggregation and the sulfidation induced phase transformation of the WO_3 to WS_2 nanoparticles (Figure 20.5a). The unique structured WS_2@Gs nanocomposites exhibited excellent Li-ion storage capability of 587.1 mA h g^{-1} at 200 mA g^{-1}. Meanwhile, it delivered an excellent reversible capacity of 371.9 mA h g^{-1} at 1,000 mA g^{-1} and good stability with a capacity retention of 62% after 500 cycles.

As shown in Figure 20.5b, 1T-$MoSe_2$ with expanded interlayer spacing of 10.0 Å in situ grown on single-wall carbon nanotube (SWCNT) film was prepared through a solvothermal technique [35]. Related with X-ray absorption near-edge structures, 1T-$MoSe_2$/SWCNTs amalgamated structures could provide robust chemical and electrical inclusion between SWCNT film and 1T-$MoSe_2$ nanosheets in a form of

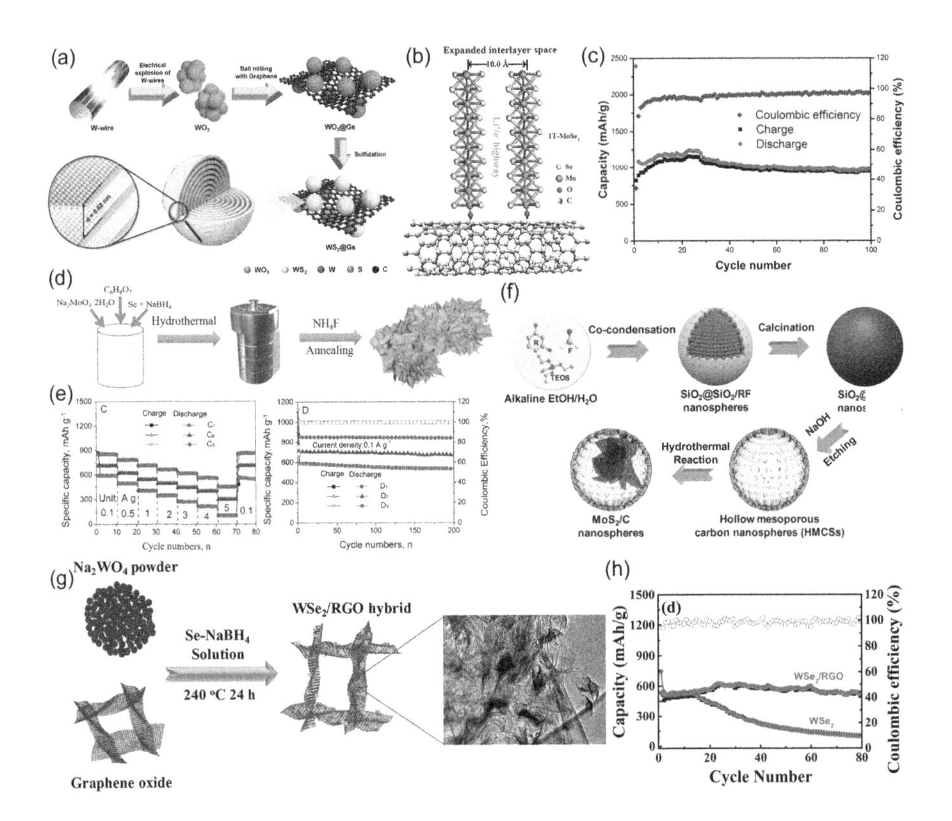

FIGURE 20.5 (a) Schematic illustration of the synthesis process and crystalline features of WS_2@Gs. Adapted with permission from [34]. Copyright (2019) Elsevier. (b) Schematic illustration of the electrochemical process and (c) cycle performance of the 1T-$MoSe_2$/SWCNTs. Adapted with permission from [35]. Copyright (2017) American Chemical Society. (d) Synthesis process and (e) rate and cycle performances of the $MoSe_2$/NFC nanomaterials. Adapted with permission from [36]. Copyright (2020) American Chemical Society. (f) Formation mechanism of the MoS_2@C nanospheres. Adapted with permission from [37]. Copyright (2017) American Chemical Society. (g) Schematic illustration and SEM image of the WSe_2/RGO. (h) Cycle performance of the WSe_2/RGO. Adapted with permission from [38] Copyright (2018) Elsevier.

C-O-Mo bonding, which gains good efficiency electron/ion transport pathway and structural stability, thus directly enabling high-performance Li-storage properties. The cycling performance of 1T-$MoSe_2$/SWCNTs delivered a capacity of 971 mAh g^{-1} at 300 mA g^{-1} after 100 cycles with a capacity retention of 89% (Figure 20.5c), owing to the strong C-O-Mo bonding and 1T-$MoSe_2$/SWCNTs accommodates volume variation during the continuous lithiation/de-lithiation process. The poor cycle performance induced by the vast capacity alteration on the charge/discharge method severely limited its practical application in LIBs. Recently, Liang et al. reported 3D hierarchical $MoSe_2$/N, F co-doped carbon ($MoSe_2$/NFC) heterostructure assembled by ultrathin nanosheets for LIBs [36]. Here, 3D hierarchical $MoSe_2$/NFC heterostructure was obtained by one-pot hydrothermal process, followed by N, F co-doping method that utilized NH_4F as fluoride and nitrogen sources (Figure 20.5d). N, F co-doped carbon was executed to enhance the conductivity and the constrain volume expansion of $MoSe_2$ during the charge/discharge process. Figure 20.5e depicted the rate capability of all the electrodes at different current densities. As compared to the $MoSe_2$ (C1) and $MoSe_2$/C (C2), the $MoSe_2$/NFC (C3) displayed the capacities of 853.9, 782.1, 713.2, 663.2, 621.2, 563, and 452.9 mAh g^{-1} at 0.1, 0.5, 1, 2, 3, 4, and 5 A g^{-1}, respectively. Such improved rate capability was assigned to the 3D hierarchical structure assisted Li-ion diffusion and the integrated F, N co-doped carbon increased the electronic conductivity. The cycling performance for all the electrodes was obtained at a rate of 0.1 A g^{-1}. $MoSe_2$ (D1) and $MoSe_2$/C (D2) showed lower capacities of 535.3 and 669.9 mAh g^{-1} and the $MoSe_2$/NFC (D3) provided excellent capacity of 838.9 mAh g^{-1} after 200 cycles. From the above results, the $MoSe_2$/NFC again confirms that the 3D hierarchical structures can deliver more active sites for Li-ion storage.

Zhang et al. prepared the petal-like MoS_2 nanosheets by the combination of hydrothermal, co-condensation, and annealing methods [37]. As shown in Figure 20.5f, a modern methodology for cost-effective preparation of petal-like MoS_2 nanosheets internal hollow mesoporous carbon spheres (HMCSs) and yolk–shell architecture MoS_2@C was developed. The HMCSs could effectively control the expansion of MoS_2 nanosheets and enhanced the electronic conductivity and structural solidity of the hybrid material. The yolk–shell MoS_2@C delivered an excellent discharge capacity of 993 mAh g^{-1} at 1 A g^{-1} and also provided a good high-rate performance at 10 A g^{-1} of 595 mAh g^{-1}. The yolk–shell structure ensured good cycling capability and excellent rate performance, which are desirable for LIB applications. In another investigation, the WSe_2/RGO hybrid nanostructures were constructed by the WSe_2 nanosheet template in situ grown on RGO nanosheets via a simple solvothermal route (Figure 20.5g) [38]. Compared with WSe_2 nanosheets, the capacity of WSe_2/RGO reached about 528 mAh g^{-1} at 0.2 C after 80 cycles (Figure 20.5h). The key role of graphene nanosheets effectively prevented the WSe_2 nanosheets from aggregation.

Yang et al. reported the $MoSe_2$ nanosheet arrays with layered MoS_2 heterostructures for superior hydrogen evolution and Li-storage performance. The $MoSe_2$ nanosheet arrays with layered MoS_2 heterostructures were fabricated by the exfoliation of MoS_2, followed by solvothermal method. The MoS_2 nanosheets were grown during the exfoliation process. Subsequently, a solvothermal growth was applied to prepare $MoSe_2$ (syn-$MoSe_2$) [39]. Schematic illustration of the synthesis route was

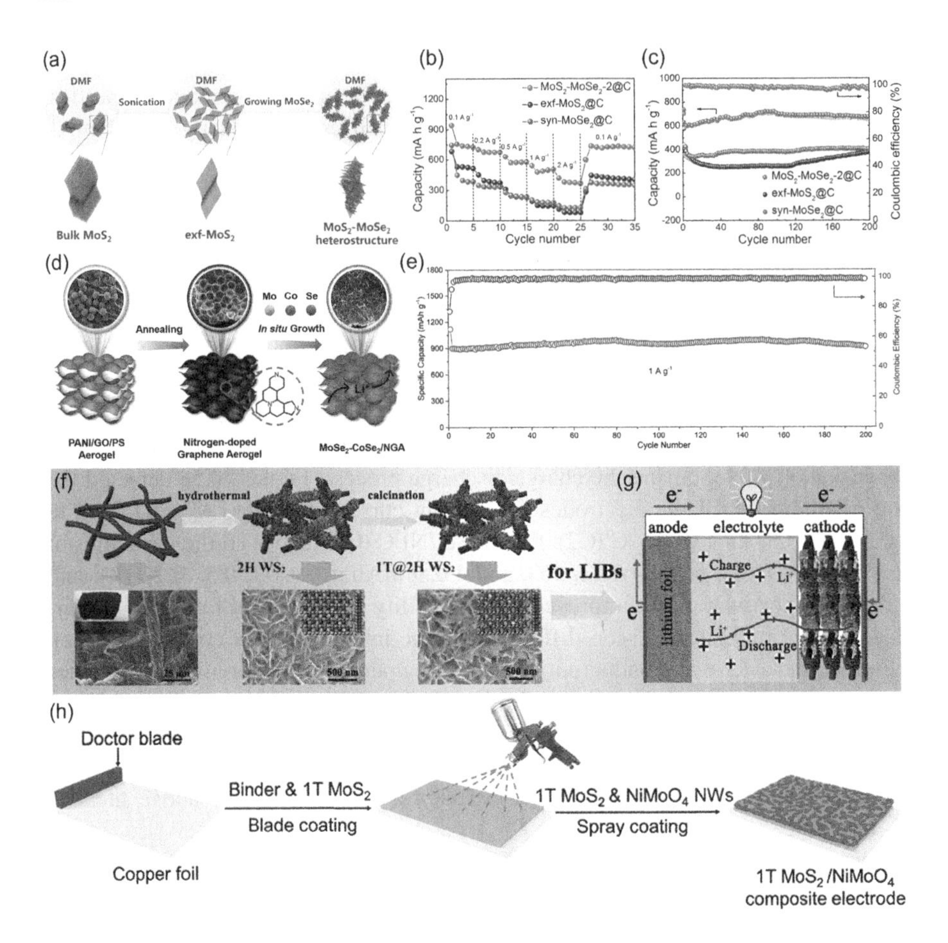

FIGURE 20.6 (a) Synthesis route and (b) rate and (c) cycle performances of the MoS_2-$MoSe_2$ heterojunction. Adapted with permission from [39]. Copyright (2017) American Chemical Society. (d) Schematic illustration of the preparation route and (e) cycle performance of $MoSe_2$-$CoSe_2$/NGA. Adapted with permission from [40]. Copyright (2019) Elsevier. (f) Fabrication process of 1T@2H WS_2@CFC and (g) schematic illustration of the composite for LIBs. Adapted with permission from [41]. Copyright (2018) Elsevier. (h) Preparation of the 1T MoS_2/$NiMoO_4$ composite electrode. Adapted with permission from [42]. Copyright (2019) American Chemical Society.

displayed in Figure 20.6a. The MoS_2-$MoSe_2$ heterostructures showed attractive electrochemical performance as well as good cycle life and rate capability. With different current rates increasing from 0.1 to 1 A g^{-1}, the MoS_2-$MoSe_2$@C delivered good performance. Meanwhile, a lithiation capacity of 736 mA g^{-1} was obtained when the current density returned to the initial current density, which implied the outstanding rate capacity of MoS_2-$MoSe_2$@C electrode (Figure 20.6b). Figure 20.6c displayed the cycling performances at 0.2 A g^{-1} for all the electrodes. The MoS_2-$MoSe_2$@C electrodes delivered a capacity of 680 mAh g^{-1} and maintained a stable capacity over 200 cycles with an excellent coulombic efficiency of 98%. These properties of MoS_2-$MoSe_2$ heterostructures make it a promising low-cost applicant for the anode material of LIBs. Zhang et al. reported the $MoSe_2$-$CoSe_2$/N-doped graphene aerogel nanocomposites by a single-step hydrothermal method [40]. Figure 20.6d showed the preparation route of the $MoSe_2$-$CoSe_2$ nanocomposites. The porous (N-doped graphene aerogel) NGA with lofty specific surface area delivered a stable and conductive network, which could accelerate electron transmission and shorten the diffusion distance of Li ions. Figure 20.6e delivered the long-term cycling stability of the $MoSe_2$-$CoSe_2$/NGA nanocomposites with a high current density of 1 A g^{-1}. It performed outstanding specific capacity of 914 mAh g^{-1} after 200 cycles.

Wang et al. reported 1T@2H WS_2 nanosheet arrays anchored on carbon fibers as a flexible electrode for LIBs. The 1T phase WS_2 unveiled inimitable performances and represented the microstructure compared with 2H phase WS_2 [41]. The unveiling of N caused the phase change from 2H to 1T and the robust N-W bonds enhanced the solidity of 1T WS_2. Figure 20.6f and g represented the synthesis mechanism of 1T@2H WS_2@carbon fiber cloth (CFC) and composite for LIBs. Here, a carbonization technique was used to make flexible (CFC), and via hydrothermal method, the 2H WS_2@CFC was fabricated. The evenly distributed carbon fibers are closely interwoven, and it was not only provided as the carbon matrix but also increased the rapid transfer of Li ions during the lithiation/delithiation process. Li et al. developed a conductive additive-free composite electrode using a composition of blade and spray coating activities to create a two-layer composite electrode. Figure 20.6h displayed the schematic procedure for forming the 1T MoS_2/$NiMoO_4$ composite electrode. A mixture of binder and 1T MoS_2 was coated on the copper foil and dried in a vacuum oven. Then, the 1T MoS_2/$NiMoO_4$ composite electrode was obtained by spraying a 1T MoS_2/$NiMoO_4$ solution on the electrode [42]. When the 1T MoS_2/$NiMoO_4$ composite was employed as an anode in LIBs, it delivered an excellent charge capacity of 940.1 mAh g^{-1}. The excellent performance is attributed to both the dexterously designed electrode structure and the intrinsic electrochemical properties. This strategy provides an effective method for improving the poor conductivity of active materials as an electrode in lithium metal batteries (LMBs).

20.4.2 Application in Sodium-ion Batteries

Rechargeable SIBs have shown great potential as a promising substitute for LIBs due to their abundance and low cost. However, the large ionic radius of Na ion (0.106 nm) compared with that of Li ion (0.076 nm) leads to the slow kinetic characteristics of Na-related electrode reactions, resulting in poor rate capability and low capacity. Recently, 2D TMDs (such as MoS_2, SnS_2, WS_2, and $MoSe_2$) have been widely employed for promising SIB electrodes due to their suitable interlayer spacing [43]. Xu et al. reported a controllable design of MoS_2 nanosheets via a solvothermal method [44]. Figure 20.7a showed the fabrication process and morphologies of MoS_2 nanosheets. Here, solvothermal method was helpful to construct a sequence of MoS_2 nanosheets (dozens of layered MoS_2: DL-MoS_2, few-layered MoS_2: FL-MoS_2, and ultrasmall MoS_2: US-MoS_2) anchored on a nitrogen-doped graphene (NG) to form hybrid electrodes. The prepared MoS_2 nanosheets showed tunable size and number of layers with interplanar spacing, owing to the switching of solvents and raw materials. The excellent electrochemical behavior resulted in the controllable design of

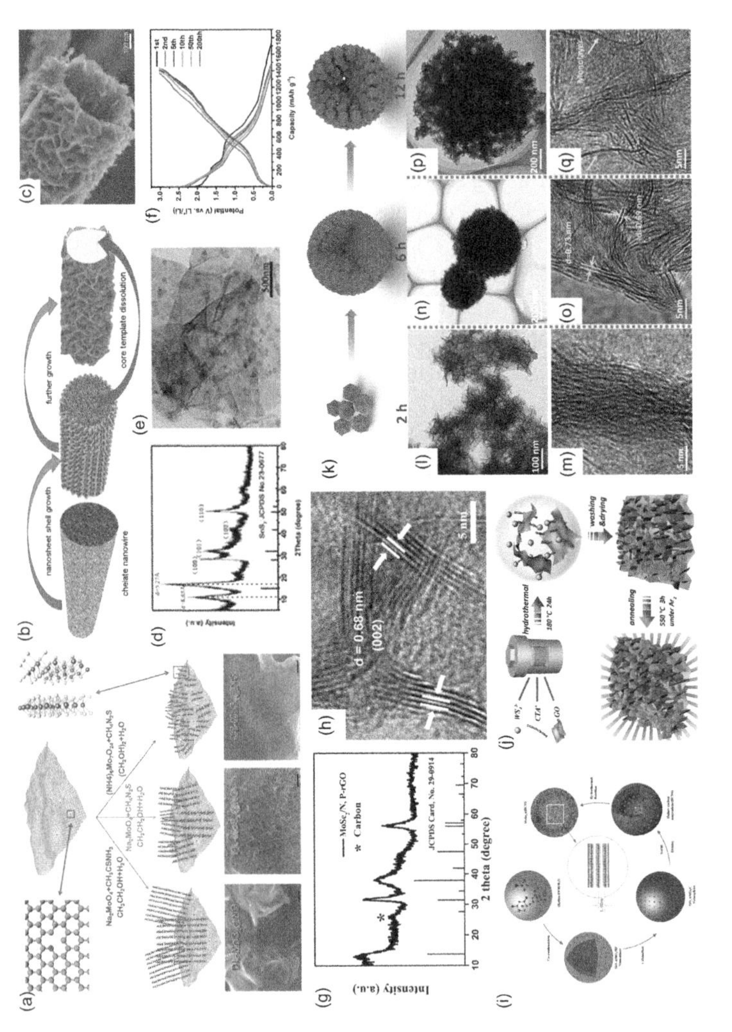

FIGURE 20.7 (a) Synthesis procedure and morphologies of the DL-MoS_2@NG, FL-MoS_2@NG, and US-MoS_2@NG (Scale bars, 100 nm). Adapted with permission from [44]. Copyright (2018) Wiley-VCH. (b) Schematic illustration of the fabrication and (c) SEM image of the SnS_2 nanosheets assemble hierarchical tubular structure. Adapted with permission from [45]. Copyright (2018) Elsevier. (d) XRD pattern and (e) TEM image of the SnS_2/rGO/SnS_2 composite. (f) Discharge/charge curves of the SnS_2/rGO/SnS_2 electrode. (Adapted with permission from [46]. Copyright (2019) American Chemical Society.) (g) XRD pattern and (h) TEM image of the $MoSe_2$/N, P-rGO. (Adapted with permission from [47]. Copyright (2017) Wiley-VCH.) (i) Synthesis route of the nanostructured $MoSe_2$@HCNS. Adapted with permission from [48]. Copyright (2018) Wiley-VCH. (j) Synthesis process of WS_2/rGO nanomaterials. Adapted with permission from [49]. Copyright (2019) Elsevier. (k) Schematic illustration for the formation of the H-WS_2@NC. (l–q) TEM images and HRTEM images of the H-WS_2@NC. Adapted with permission from [50]. Copyright (2020) Elsevier.

MoS_2 nanosheets with more active S edges on the surface and increased interplanar spacing for Na-ion storage.

In another study, the SnS_2 nanosheets assembled hierarchical tubular structures using metal chelate nanowires were fabricated via solvothermal synthesis by Zhao et al. [45]. Figure 20.7b represented the schematic illustration for the preparation of the SnS_2 nanosheets-assembled hierarchical tubular structures ($SnS_2NS\subset HTSs$). During synthesis, the SnS_2 nanosheets were grown on cobalt-nitrilotriacetic acid chelate nanowires (Co-NTA CNWs), and the Co-NTA CNWs were completely dissolved, extending to the formation of SnS_2 $NS\subset HTSs$ without further calcination and/or chemical etching process. Figure 20.7c showed the SEM image of the SnS_2 $NS\subset HTSs$. The image exhibited well structural solidity without obvious distortion in presence with sustaining the high-temperature annealing. Therefore, this electrode material delivered an excellent electrochemical performance.

Owing to the layered structure with high capacity, the SnS_2 materials have attracted much attention in the EES. Recently, Jiang et al. reported a sandwich-like SnS_2/graphene/SnS_2 with expanded interlayer distance via hydrothermal synthesis [46]. Figure 20.7d showed the XRD pattern of the as-obtained sandwich-like SnS_2/rGO/SnS_2 composite. The enlarged interlayer spacing of SnS_2/rGO/SnS_2 composite showed 5.27 and 8.03 Å by the insertion/extraction of Li/Na ions with fast transport dynamics. Figure 20.7e suggested the sandwich-like structures, which was formed by SnS_2 nanosheets evenly deposited on both the sides of rGO surface. The SnS_2/rGO/SnS_2 nanocomposites exhibited a high revocable capacity of 1,860 mAh g^{-1} at 0.1 A g^{-1}. After 100 cycles, they even provided a revocable capacity of 1,133 mAh g^{-1} (Figure 20.7f). Such a high reversible capacity and good cycling performance indicated that the SnS_2/rGO/SnS_2 nanocomposites have a huge promise in rechargeable SIBs.

Owing to the large spacing (0.64 nm), the $MoSe_2$ electrodes showed improved features for SIBs. Niu et al. reported $MoSe_2$ treated N, P co-doped carbon via the combination of hydrothermal technique and calcination process [47]. Figure 20.7g displayed the XRD pattern of the $MoSe_2$/N, P-rGO. All the remaining peaks were well matched with the hexagonal phase $MoSe_2$. Figure 20.7h was the high-magnification SEM image of $MoSe_2$/N, P/rGO sample. The 3D network would help the electrolyte penetration into the core and extended the space to assist the volume variation upon cycling. The lattice fringe interlayer spacing was calculated to be 0.68 nm, matched well with the (002) plane of $MoSe_2$. In another investigation, Liu et al. reported few-layer $MoSe_2$ nanosheets with extended (002) planes confined in hollow carbon nanospheres (HCNS) (Figure 20.7i) [48]. In this architecture, the HCNS could prohibit the restacking of $MoSe_2$, which allowed the development of evenly enclosed few-layer $MoSe_2$ nanosheets with an extended interlayer spacing formation. The interlayer spacing was expanded to 1.02 nm, which was substantially larger than its intrinsic (002) plane of 0.64 nm. The large interlayer spacing reduced the barrier of Na ions and improved the migration kinetics of Na ions, thus exhibiting excellent Na-storage performance.

WS_2 has a large interlayer spacing (0.62 nm), which is conducive to the storage of Na ions. However, WS_2 exposes weak conductivity with decreased oxidoreduction kinetics and huge volume change during cycling, so it is difficult to achieve as

TABLE 20.1
Summary of the Electrochemical Performance in the Reported Literature

Materials	Method	Current Density (mA g^{-1})	Cycle Number	Capacity (mAh g^{-1})	CE/CR (%)	Ref.
2D TMD nanomaterials for LIBs						
$C@WS_2@Gs$	Electrical explosion and Ball milling	1,000	500	496.1	61.9 (CR)	[34]
1T-$MoSe_2$/SWCNTs	Solvothermal	300	100	971	~98.0	[35]
$MoSe_2$/NFC	Hydrothermal	100	200	838.9	~99.0	[36]
$MoS_2@C$	Hydrothermal	1,000	200	993	~100	[37]
WSe_2/RGO	Solvothermal	62.8 (0.2 C)	80	528	~99.0	[38]
MoS_2-$MoSe_2$	Liquid phase ultrasonic-assisted exfoliation	200	200	~680	~98.0	[39]
$MoSe_2$-$CoSe_2$/NGA	Solvothermal	1,000	200	914	~99.0	[40]
1T@2H WS_2@CFC	Hydrothermal	100	200	1130	~100	[41]
2D TMD nanomaterials for SIBs						
MoS_2@NG	Solvothermal	1,000	1,000	198	~100	[44]
SnS_2	Solvothermal	50	50	414	58.5	[45]
SnS_2/rGO/SnS_2	Hydrothermal	100	200	1357	96.6	[46]
$MoSe_2$/N,P-rGO	Solvothermal	500	1,000	378	87	[47]
$MoSe_2$@HCNS	Solvothermal	1,000	1,000	502	~98.3	[48]
WS_2/rGO	Hydrothermal	100	70	522.3	99	[49]
H-WS_2@NC	Hydrothermal	100	200	473	99.1	[50]

an anode material for SIBs. In order to solve these issues, Song et al. reported self-assembled nanohoneycomb WS_2 modified with graphene [49]. Figure 20.7j illustrated the production method of WS_2/rGO by a self-assembly development under hydrothermal activity. The synthesized cetyltrimethylammonium bromide (CTAB) acted as a surfactant and helped to form WS_2 nanofibers, and these nanofibers were grown on the rGO to form WS_2/rGO composites. The unique structure not only enhanced oxidation–reduction kinetics but also buffered the volume extension, thereby achieving an outstanding electrochemical performance.

Hu et al. reported the self-assembled conductive interlunar-extended WS_2 nanosheets anchored on nitrogen-doped carbon matrix (H-WS_2@NC) for rapid and steady sodium storage [50]. Figure 20.7k showed the schematic illustration for the formation of the H-WS_2@NC microflower buds and the time-dependent tests were conducted out to examine the structural growth. The ultrathin WS_2 nanosheets were randomly constructed for 2 hours. When the reaction time was increased to 6 h, the randomly constructed WS_2 nanosheets were self-assembled to form 3D hierarchical solid microflowers (S-WS_2@NC). As the reaction proceeded to 12 hours, the WS_2 microflower buds and a few of monolayer WS_2 nanosheets were obtained (Figure 20.7l–q). When the reaction time was increased to 24 hours, the well-defined 3D hollow hierarchical microflower buds were formed (H-WS_2@NC). The nano-microstructure collectively incorporated the performances of few layers and increased interplanar space of 2D WS_2 nanosheets as well as N-doped carbon amalgamation 3D hierarchical hollow porous structure. The H-WS_2@NC showed excellent structural stability with high revocable performance, prominent cycling stability, and rate performance as an anode for SIBs.

20.5 CONCLUSION AND OUTLOOKS

This chapter reviewed recent progress in the preparation of 2D TMD-based nanomaterials and their applications in Li/Na-ion batteries. A series of TMD and TMD-based nanomaterials prepared by hydrothermal/solvothermal, CVD, and exfoliation techniques were summarized. Meanwhile, we summarized the current studies of the 2D TMDs for Li/Na-ion anode. Some related electrochemical performances were summarized in Table 20.1. Although 2D TMD materials are considered to be an ideal choice for the anode of Li/Na-ion batteries, some issues need to be improved. (a) The large-scale preparation method to obtain 2D TMD materials is still a challenge. (b) Even though the large specific surface area of 2D TMDs is conducive to the pseudocapacitive behavior, thereby providing battery capacity, a series of side reactions with electrolytes cause the irreversible consumption of electrolytes, which results in low efficiency. (c) Although the introduction of carbon materials can improve the stability and conductivity of the 2D TMD materials, the tap density of the electrode material will be reduced, resulting in a low volumetric energy density. Therefore, it is necessary to find a simple, low-cost, and high-efficiency synthetic method for large-scale production of 2D TMDs. Meanwhile, rational design of 2D TMDs-based architectures will also be a promising mothed to improve the electrochemical performance.

REFERENCES

1. Yun, Q.; Li, L.; Hu, Z.; Lu, Q.; Chen, B.; Zhang, H., Layered transition metal dichalcogenide-based nanomaterials for electrochemical energy storage. *Adv. Mater.* **2020**, *32*, 1903826.
2. Nagaraju, G.; Sekhar, S. C.; Yu, J. S., Utilizing waste cable wires for high-performance fiber-based hybrid supercapacitors: an effective approach to electronic-waste management. *Adv. Energy Mater.* **2018**, *8*, 1702201.
3. Randau, S.; Weber, D. A.; Kötz, O.; Koerver, R.; Braun, P.; Weber, A.; Ivers-Tiffée, E.; Adermann, T.; Kulisch, J.; Zeier, W. G.; Richter, F. H.; Janek, J., Benchmarking the performance of all-solid-state lithium batteries. *Nat. Energy* **2020**, *5*, 259–270.
4. Kim, H. S.; Cook, J. B.; Lin, H.; Ko, J. S.; Tolbert, S. H.; Ozolins, V.; Dunn, B., Oxygen vacancies enhance pseudocapacitive charge storage properties of MoO_{3-x}. *Nat. Mater.* **2017**, *16*, 454–460.
5. Pomerantseva, E.; Gogotsi, Y., Two-dimensional heterostructures for energy storage. *Nat. Energy* **2017**, *2*. 1–6.
6. Grey, C. P.; Tarascon, J. M., Sustainability and in situ monitoring in battery development. *Nat. Mater.* **2016**, *16*, 45–56.
7. Fang, S.; Bresser, D.; Passerini, S., Transition metal oxide anodes for electrochemical energy storage in lithium-and sodium-ion batteries. *Adv. Energy Mater.* **2019**, *10*, 1902485.
8. Schmuch, R.; Wagner, R.; Hörpel, G.; Placke, T.; Winter, M., Performance and cost of materials for lithium-based rechargeable automotive batteries. *Nat. Energy* **2018**, *3*, 267–278.
9. Wang, T.; Shen, X.; Huang, J.; Xi, Q.; Zhao, Y.; Guo, Q.; Wang, X.; Xu, Z., Tulip-like MoS_2 with a single sheet tapered structure anchored on N-doped graphene substrates via C-O-Mo bonds for superior sodium storage. *J. Mater. Chem. A* **2018**, *6*, 24433–24440.
10. Zhao, C.; Wang, Q.; Yao, Z.; Wang, J.; Sánchez-Lengeling, B.; Ding, F.; Qi, X.; Lu, Y.; Bai, X.; Li, B.; Li, H.; Aspuru-Guzik, A.; Huang, X.; Delmas, C.; Wagemaker, M.; Chen, L.; Hu, Y. S., Rational design of layered oxide materials for sodium-ion batteries. *Science* **2020**, *370*, 708–711.
11. Liu, Q.; Hu, Z.; Chen, M.; Zou, C.; Jin, H.; Wang, S.; Chou, S. L.; Liu, Y.; Dou, S. X., The cathode choice for commercialization of sodium-ion batteries: layered transition metal oxides versus prussian blue analogs. *Adv. Funct. Mater.* **2020**, *30*, 1909530.
12. Wang, Q. H.; Kalantar-Zadeh, K.; Kis, A.; Coleman, J. N.; Strano, M. S., Electronics and optoelectronics of two-dimensional transition metal dichalcogenides. *Nat. Nanotechnol.* **2012**, *7*, 699–712.
13. Yang, E.; Ji, H.; Jung, Y., Two-dimensional transition metal dichalcogenide monolayers as promising sodium ion battery anodes. *J. Phys. Chem. C* 2015, 119, 26374–26380.
14. Bai, J.; Zhao, B.; Zhou, J.; Si, J.; Fang, Z.; Li, K.; Ma, H.; Dai, J.; Zhu, X.; Sun, Y., Glucose-induced synthesis of 1T-MoS_2/C hybrid for high-rate lithium-ion batteries. *Small* **2019**, *15*, 1805420.
15. Xiang, T.; Fang, Q.; Xie, H.; Wu, C.; Wang, C.; Zhou, Y.; Liu, D.; Chen, S.; Khalil, A.; Tao, S.; Liu, Q.; Song, L., Vertical 1T-MoS_2 nanosheets with expanded interlayer spacing edged on a graphene frame for high rate lithium-ion batteries. *Nanoscale* **2017**, *9*, 6975–6983.
16. Rabenau, A., The role of hydrothermal synthesis in preparative chemistry. *Angew. Chem. Int. Ed. Engl.* **1985**, *24*, 1026–1040.
17. Wang, P.; Sun, S.; Jiang, Y.; Cai, Q.; Zhang, Y. H.; Zhou, L.; Fang, S.; Liu, J.; Yu, Y., Hierarchical microtubes constructed by MoS_2 nanosheets with enhanced sodium storage performance. *ACS Nano* **2020**, *14*, 15577–15586.

18. Wang, B.; Zhang, Y.; Zhang, J.; Xia, R.; Chu, Y.; Zhou, J.; Yang, X.; Huang, J., Facile synthesis of a MoS_2 and functionalized graphene heterostructure for enhanced lithium-storage performance. *ACS Appl. Mater. Interfaces* **2017**, *9*, 12907–12913.
19. Li, P.; Jeong, J. Y.; Jin, B.; Zhang, K.; Park, J. H., Vertically oriented MoS_2 with spatially controlled geometry on nitrogenous graphene sheets for high-performance sodium-ion batteries. *Adv. Energy Mater.* **2018**, *8*, 1703300.
20. Huang, H.; Cui, J.; Liu, G.; Bi, R.; Zhang, L., Carbon-coated $MoSe_2$/MXene hybrid nanosheets for superior potassium storage. *ACS Nano* **2019**, *13*, 3448–3456.
21. Liu, Z.; Li, N.; Su, C.; Zhao, H.; Xu, L.; Yin, Z.; Li, J.; Du, Y., Colloidal synthesis of 1T' phase dominated WS_2 towards endurable electrocatalysis. *Nano Energy* **2018**, *50*, 176–181.
22. Nagaraju, G.; Sekhar, S. C.; Ramulu, B.; Yu, J. S., An integrated approach toward renewable energy storage using rechargeable $Ag@Ni_{0.67}Co_{0.33}S$-based hybrid supercapacitors. *Small* **2019**, *15*, 1805418.
23. Khaja Hussain, S.; Vamsi Krishna, B. N.; Nagaraju, G.; Chandra Sekhar, S.; Narsimulu, D.; Yu, J. S., Porous $Co-MoS_2@Cu_2MoS_4$ three-dimensional nanoflowers via in situ sulfurization of Cu_2O nanospheres for electrochemical hybrid capacitors. *Chem. Eng. J.* **2021**, *403*, 126319.
24. Nagaraju, G.; Sekhar, S. C.; Ramulu, B.; Yu, J. S., High-performance hybrid supercapacitors based on MOF-derived hollow ternary chalcogenides. *Energy Stor. Mater.* **2021**, *35*, 750–760.
25. Shan, J.; Li, J.; Chu, X.; Xu, M.; Jin, F.; Fang, X.; Wei, Z.; Wang, X., Enhanced photoresponse characteristics of transistors using CVD-grown MoS_2/WS_2 heterostructures. *Appl. Surf. Sci.* **2018**, *443*, 31–38.
26. Cong, R.; Qiao, S.; Liu, J.; Mi, J.; Yu, W.; Liang, B.; Fu, G.; Pan, C.; Wang, S., Ultrahigh, ultrafast, and self-powered visible-near-infrared optical position-sensitive detector based on a CVD-prepared vertically standing few-layer MoS_2/Si heterojunction. *Adv. Sci.* **2018**, *5*, 1700502.
27. Huang, L.; Thi, Q. H.; Zheng, F.; Chen, X.; Chu, Y. W.; Lee, C. S.; Zhao, J.; Ly, T. H., Catalyzed kinetic growth in two-dimensional MoS_2. *J. Am. Chem. Soc.* **2020**, *142*, 13130–13135.
28. Okada, M., Direct chemical vapor deposition growth of WS_2 atomic layers on hexagonal boron nitride. *ACS Nano* **2014**, *8*, 8273–8277.
29. Gopalakrishnan, D.; Damien, D.; Shaijumon, M.M., MoS_2 quantum dot-interspersed exfoliated MoS_2 nanosheets. *ACS Nano* **2014**, *8*, 5297–5303.
30. Feng, J.; Sun, X.; Wu, C.; Peng, L.; Lin, C.; Hu, S.; Yang, J.; Xie, Y., Metallic few-layered VS_2 ultrathin nanosheets: High two-dimensional conductivity for in-plane supercapacitors. *J. Am. Chem. Soc.* **2011**, *133*, 17832–17838.
31. Sekhar, S. C.; Ramulu, B.; Narsimulu, D.; Arbaz, S. J.; Yu, J. S., Metal-organic framework-derived $Co_3V_2O_8@CuV_2O_6$ hybrid architecture as a multifunctional binder-free electrode for Li-ion batteries and hybrid supercapacitors. *Small* **2020**, *16*, 2003983.
32. Goodenough, J. B., How we made the Li-ion rechargeable battery. *Nat. Electron.* **2018**, *1*, 204–204.
33. Sekhar, S. C.; Ramulu, B.; Arbaz, S. J.; Hussain, S. K.; Yu, J. S., One-pot hydrothermal-derived $NiS_2-CoMo_2S_4$ with vertically aligned nanorods as a binder-free electrode for coin-cell-type hybrid supercapacitor. *Small Methods* **2021**, *5*, 2100335.
34. Kim, I.; Park, S.-W.; Kim, D.-W., Onion-like crystalline WS_2 nanoparticles anchored on graphene sheets as high-performance anode materials for lithium-ion batteries. *Chem. Eng. J.* **2019**, *375*, 122033.
35. Xiang, T.; Tao, S.; Xu, W.; Fang, Q.; Wu, C.; Liu, D.; Zhou, Y.; Khalil, A.; Muhammad, Z.; Chu, W.; Wang, Z.; Xiang, H.; Liu, Q.; Song, L., Stable 1T-$MoSe_2$ and carbon nanotube hybridized flexible film: binder-free and high-performance Li-ion anode. *ACS Nano* **2017**, *11*, 6483–6491.

36. Liang, Q.; Zhang, L.; Zhang, M.; Pan, Q.; Li, Y.; Tan, C.; Zheng, F.; Huang, Y.; Wang, H.; Li, Q., Three-dimensional hierarchical $MoSe_2$/N, F co-doped carbon heterostructure assembled by ultrathin nanosheets for advanced lithium-ion batteries. *ACS Sustain. Chem. Eng.* **2020**, *8*, 14127–14136.
37. Zhang, X.; Zhao, R.; Wu, Q.; Li, W.; Shen, C.; Ni, L.; Yan, H.; Diao, G.; Chen, M., Petal-like MoS_2 nanosheets space-confined in hollow mesoporous carbon spheres for enhanced lithium storage performance. *ACS Nano* **2017**, *11*, 8429–8436.
38. Wang, X.; He, J.; Zheng, B.; Zhang, W.; Chen, Y., Few-layered WSe_2 in-situ grown on graphene nanosheets as efficient anode for lithium-ion batteries. *Electrochim. Acta* **2018**, *283*, 1660–1667.
39. Yang, J.; Zhu, J.; Xu, J.; Zhang, C.; Liu, T., $MoSe_2$ Nanosheet array with layered MoS_2 heterostructures for superior hydrogen evolution and lithium storage performance. *ACS Appl. Mater. Interfaces* **2017**, *9*, 44550–44559.
40. Zhang, X.; Zhou, J.; Zheng, Y.; Chen, D., $MoSe_2$-$CoSe_2$/N-doped graphene aerogel nanocomposites with high capacity and excellent stability for lithium-ion batteries. *J. Power Sources* **2019**, *439*, 227112.
41. Wang, T.; Sun, C.; Yang, M.; Zhang, L.; Shao, Y.; Wu, Y.; Hao, X., Enhanced reversible lithium ion storage in stable 1T@2H WS_2 nanosheet arrays anchored on carbon fiber. *Electrochim. Acta* **2018**, *259*, 1–8.
42. Li, Z.; Zhan, X.; Zhu, W.; Qi, S.; Braun, P. V., Carbon-free, high-capacity and long cycle life 1D-2D $NiMoO_4$ nanowires/metallic 1T MoS_2 composite lithium-ion battery anodes. *ACS Appl. Mater. Interfaces* **2019**, *11*, 44593–44600.
43. Yao, K.; Xu, Z.; Huang, J.; Ma, M.; Fu, L.; Shen, X.; Li, J.; Fu, M., Bundled defect-rich MoS_2 for a high-rate and long-life sodium-ion battery: Achieving 3D diffusion of sodium ion by vacancies to improve kinetics. *Small* **2019**, *15*, 1805405.
44. Xu, X.; Zhao, R.; Ai, W.; Chen, B.; Du, H.; Wu, L.; Zhang, H.; Huang, W.; Yu, T., Controllable Design of MoS_2 nanosheets anchored on nitrogen-doped graphene: Toward fast sodium storage by tunable pseudocapacitance. *Adv. Mater.* **2018**, *30*, 1800658.
45. Zhao, J.; Yu, X.; Gao, Z.; Zhao, W.; Xu, R.; Liu, Y.; Shen, H., One step synthesis of SnS_2 nanosheets assembled hierarchical tubular structures using metal chelate nanowires as a soluble template for improved Na-ion storage. *Chem. Eng. J.* **2018**, *332*, 548–555.
46. Jiang, Y.; Song, D.; Wu, J.; Wang, Z.; Huang, S.; Xu, Y.; Chen, Z.; Zhao, B.; Zhang, J., Sandwich-like SnS_2/graphene/SnS_2 with expanded interlayer distance as high-rate lithium/sodium-ion battery anode materials. *ACS Nano* **2019**, *13*, 9100–9111.
47. Niu, F.; Yang, J.; Wang, N.; Zhang, D.; Fan, W.; Yang, J.; Qian, Y., $MoSe_2$-Covered N, P-Doped carbon nanosheets as a long-life and high-rate anode material for sodium-ion batteries. *Adv. Funct. Mater.* **2017**, *27*, 1700522.
48. Liu, H.; Guo, H.; Liu, B.; Liang, M.; Lv, Z.; Adair, K. R.; Sun, X., Few-layer $MoSe_2$ nanosheets with expanded (002) planes confined in hollow carbon nanospheres for ultrahigh-performance Na-ion batteries. *Adv. Funct. Mater.* **2018**, *28*, 1707480.
49. Song, Y.; Liao, J.; Chen, C.; Yang, J.; Chen, J.; Gong, F.; Wang, S.; Xu, Z.; Wu, M., Controllable morphologies and electrochemical performances of self-assembled nano-honeycomb WS_2 anodes modified by graphene doping for lithium and sodium ion batteries. *Carbon* **2019**, *142*, 697–706.
50. Hu, X.; Liu, Y.; Li, J.; Wang, G.; Chen, J.; Zhong, G.; Zhan, H.; Wen, Z., Self-assembling of conductive interlayer-expanded WS_2 nanosheets into 3D hollow hierarchical microflower bud hybrids for fast and stable sodium storage. *Adv. Funct. Mater.* **2019**, *30*, 1907677.

Index

For Product Safety Concerns and Information please contact our EU representative GPSR@taylorandfrancis.com Taylor & Francis Verlag GmbH, Kaufingerstraße 24, 80331 München, Germany

Printed and bound by CPI Group (UK) Ltd, Croydon, CR0 4YY
01/05/2025
01858556-0003